BARYONS AS SKYRME SOLITONS

INTERNATIONAL WORKSHOP ON

BARYONS AS SKYRME SOLITONS

Siegen, Germany 28 – 30 September 1992

Editor

G Holzwarth

Universität Gesamthochschule Siegen

World Scientific

Singapore • New Jersey • London • Hong Kong

Published by

World Scientific Publishing Co Pte Ltd
5 Toh Tuck Link, Singapore 596224
USA office: 27 Warren Street, Suite 401-402, Hackensack, NJ 07601
UK office: 57 Shelton Street, Covent Garden, London WC2H 9HE

British Library Cataloguing-in-Publication Data
A catalogue record for this book is available from the British Library.

INTERNATIONAL WORKSHOP BARYONS AS SKYRME SOLITONS

ISBN-13 978-981-02-1408-1
ISBN-10 981-02-1408-1

PREFACE

Almost ten years have passed since the reanimation of the Skyrme soliton model for baryons by Witten in 1983 initiated a surge of papers on this alternative aspect of low-energy hadron physics. Meanwhile the initial rush has faded away; nevertheless there has been steady progress in this field over the past years.

Through the remarkable achievements of Chiral Perturbation Theory and extensive work on bosonization of specific fermionic models the conceptual foundations for effective mesonic actions have been profoundly strengthened, and the interplay between quark degrees of freedom and the concept of topological solitons carrying the baryon number has become quite transparent.

For a wide range of realistic applications the soliton model has provided impressive results, and tedious numerical analyses extended our knowledge about the rich variety of classical configurations favored by the model. The endeavours to overcome evidently unsatisfactory features have induced numerous extensions of the Skyrme model which have pushed it far beyond its original simplicity. Inclusion of higher-order gradient terms in the lagrangian, extension to higher flavor-symmetry groups and incorporation of corresponding symmetry breakers, explicit coupling to additional scalar and vector mesons and to electro-weak fields have greatly increased the complexity of the soliton model for baryons.

In the course of this process many of the early difficulties encountered by the Skyrme model, like formfactors being too soft, rising meson-baryon phase shifts, or missing intermediate-range attraction in the nucleon-nucleon force, have found satisfactory solutions. In most of these instances, it turned out that the idea of baryons as solitons in an effective meson theory provides correct physical information about baryon dynamics if one uses an effective lagrangian which correctly descibes the mesonic sector, and if one can find trustworthy approximations to extract this information. Unfortunately, this may quickly become forbiddingly difficult; e.g. multiple meson production in baryon-baryon collisions seems naturally contained in the dynamics but almost impossible to calculate reliably. At the same time, this increased complexity also has generated more and more ambiguity in the effective action which may have brought the model to the limits where an effective theory looses its cogency.

Some stubborn problems have remained over the years, like the mass problem, the value of g_A, the isoscalar magnetic moment, the low-energy πN P-waves. But there are clear indications that these problems are closely related to each other and require careful treatment of the next order terms in the $1/N_c$ expansion. Comparatively little work has been done in that respect and it is on this direction that future research should focus.

So this seems a good point in time to critically assess what has been achieved so far, to work out the hard core of the various model extensions, and to delineate the necessary developments for going beyond the semiclassical schemes mostly used at present.

For a justification of the soliton concept from the underlying fundamental theory of strong interactions it is very welcome that the Nambu-Jona-Lasinio model which is believed to reflect important features of QCD, favors a solitonic baryon directly from the fermion determinant without reference to a gradient expansion. Interestingly, this only happens if vector mesons are included. If corroborated, these recent developments could really serve as profound support of Skyrme's original idea, and they provided an excellent starting point for the workshop.

Skyrme's concept has not been confined to the baryon it was invented for. Two other prominent examples of fruitful application in related fields were included here: the skyrmion in the Weinberg-Salam theory, and Karliner's presentation of quark and antiquark solitons in multi-flavor QCD.

The body of the workshop was devoted to four major areas of investigations:

The first reviewed electromagnetic and parity-violating properties of baryons, with critical emphasis on some unsatisfactory features as presently obtained from soliton models. Structural difficulties, as particularly stressed by Kaisér's comparison with constraints from Chiral Perturbation Theory clearly point to the necessity of going beyond semiclassical methods.

The second part specifically dealt with the various extensions of the Skyrme model to $n > 2$ flavors, and the different methods to deal with the strong symmetry breaking. Apparently the soliton model can provide the essential flavor structure of nucleons and hyperons, tells about the resulting observables, in relation to the symmetry breaking on the quark level. Strikingly, as in the NJL Model, vector mesons play a crucial role for these considerations, and, if eliminated, provide the guidelines for extending the fourth-order Gasser-Leutwyler parametrization to higher orders in powers of gradients of the pseudoscalar Goldstone fields. Even the recent discussion of heavy-quark symmetries fits naturally into the skyrmion picture, as demonstrated in Mannque Rho's exciting contribution.

The third part covered recent developments in the description of the πN system. In this field where several years ago the soliton model had brought a sweep of amazing results the intricacies of dealing with constrained variables have still prevented fully consistent solutions. But it has become clear from the analyses of low-energy πN scattering that proper inclusion of $1/N_c$ corrections is mandatory for further progress and will in fact lead to satisfactory results. It also appears that our better understanding of the pionic fluctuations around the soliton has brought the long-standing problem with the nucleon mass very close to its solution.

In the fourth part systems with baryon number $B \neq 1$ were discussed. According to Walhout's review the long story of the nucleon-nucleon interaction has finally reached a good end. For $B = 2$ the fancy (and may be slightly embarrassing) torus configuration has not gone away; on the contrary, its remnants seem important even in the deuteron, and Snyderman's movies on baryon-baryon scattering brought it to real life. At present one can only speculate what surprises may still be hidden in the rich functional space of dibaryons in $SU(3)$. This experience is a bit discourageing for attempts to look into the proton-antiproton system.

The contributions to this workshop clearly demonstrate that the soliton model for baryons has passed the scrutinity of ten years' research very much alive, and emerges as solid and powerful alternative approach to low-energy hadron physics. It has become quite clear what should be done next, and that the outlook for the task lying ahead is promising. This lively status of the field was reflected in the atmosphere of the meeting and it is documented in this collection of contributions. We hope that this volume will prove very useful as a detailed survey of what has been achieved so far and as guideline for future work. So we are especially grateful to all the speakers for the careful preparation of their manuscripts and we are happy to include here also the contributions of J.Kunz and N.Scoccola, who unfortunately could not participate.

The meeting was generously supported by the Symposienprogramm der Stiftung Volkswagenwerk, and by the Institut für Geistes- und Sozialwissenschaften at Siegen University. The smooth coordination of events and administrative tasks was efficiently and reliably handled by our Conference Secretaries, Ms. Steinleitner and Ms. Krebs.

November 1992 G. Holzwarth, Siegen University

CONTENTS

III. Hyperons

IV. Meson-Baryon System

V. Systems with $B \neq 1$

I. Quarks and Skyrmions

EFFECTIVE CHIRAL MESON THEORY OF QCD

H. REINHARDT *

Institut für Theoretische Physik, Universität Tübingen

Auf der Morgenstelle 14, D-7400 Tübingen, Germany

ABSTRACT

Attempts to derive the effective meson theory of QCD are briefly reviewed. It is argued that its low-energy behaviour is entirely determined by chiral symmetry. With this in mind the effective meson theory is obtained from the bosonization of the NJL model. The physical content of the effective meson theory is discussed with emphasis on the soliton sector. The emergence of baryons as chiral solitons of the meson fields is explained as a consequence of the polarization of the Dirac sea by topologically non-trivial chiral fields.

1. Introduction

The aim of my talk is to give a brief introduction to the subject of effective meson theories, which are the basis for the soliton description of baryons. I will briefly review some attempts to derive the effective meson theory of QCD and discuss some general features of this theory: I will put an emphasis on those aspects which are relevant for the chiral solitons.

Effective meson theories are based on 'tHooft's pioneering work on large-N QCD [1]. He discovered that when QCD is generalized from the gauge group SU(3) to an arbitrary number of colours N the theory contains a hidden expansion parameter, namely $\frac{1}{N}$. Furthermore in the limit where the number of colours goes to infinity QCD reduces to an effective theory of weakly interacting mesons and glueballs. Later, Witten conjectured that in this effective meson theory the baryons appear as topological solitons of the meson fields [2].

'tHooft's results on large-N QCD rely on combinatorial arguments which are based on the fact that for a large number of colours the gauge group SU(N) can be replaced by the group U(N). In the U(N) gauge theory there are N quarks and N^2 gluons. Therefore, concerning the colour structure, the gluons can be thought of as being built up from a quark and an antiquark. Replacing in a Feynman diagram gluon lines by pairs of quark and antiquark lines and taking into account that each closed quark group yields a factor N, one easily shows that quark loops are suppressed compared to gluon loops by $\frac{1}{N}$. In the same way one shows that nonplanar diagrams are suppressed by $\frac{1}{N^2}$, compaired to planar diagrams.

In 1+1 dimensions QCD can be exactly bosonized and the corresponding effective "meson" theory is explicitly known [3]. In our physical world of 3+1 dimension an exact bosonization of QCD is impossible due to the selfinteraction of gluons. At low

*supported by Deutsche Forschungsgemeinschaft under contract 856/2-1

energies only the mesons with the largest Compton wave length are relevant. These are the pseudoscalar mesons, the would-be Goldstone bosons of spontaneous breaking of chiral symmetry, which in a non-linear representation are given by the chiral field. Guided by symmetry arguments and meson phenomenology Balachandran et al. [4] and Witten [5] have argued that the low-energy approximation to the effective meson theory in 3+1 dimension should be given by the non-linear σ model supplemented by the Wess-Zumino term. We will later see that this model is in fact the leading (in the number of derivatives of meson fields) order term of the effective meson theory. But since the non-linear model does not allow for stable solitons one has to include higher order terms and Witten proposed to use the Skyrme model [6] as the minimal approximation to the effective meson theory of QCD.

There have been many attempts to derive the effective meson theory in $D = 4$ directly from QCD [7-16]. In refs. 7-9 it was tried to construct the effective meson action from the chiral anomaly. The essential idea is to make a chiral rotation of the quarks

$$q \rightarrow q^{\Omega} = (P_R\Omega + P_L\Omega^{\dagger})q \tag{1}$$

$$\Omega^2 = U = \exp i\pi(x) \tag{2}$$

and interpret to the increase in the action as the action of the chiral field:

$$\exp\Gamma[U = \Omega^2] = \frac{\int DADqD\bar{q}\,\exp\int L_{QCD}(A, q^{\Omega}, \bar{q}^{\Omega})}{\int DADqD\bar{q}\,\exp\int L_{QCD}(A, q, \bar{q})} \tag{3}$$

In fact, making a shift of integration variables from the q to the chirally rotated field q^{Ω} and taking into account that according to Fujikawa [17] the integration measure is not invariant but gives rise to a Jacobian $J(\Omega)$, one obtains that the above defined effective meson action is nothing else but the integrated chiral anomaly [18]:

$$\exp\Gamma[U = \Omega^2] = J(\Omega) \tag{4}$$

The evaluation of the integrated chiral anomaly proceeds as follows: one first considers infinitesmaly small chiral rotations for which the Jacobian can be expressed directly in terms of the (differential) chiral anomaly of Adler, Bell and Jackiw [18]

$$J(\Omega) = \exp \mathrm{Tr}(\pi\mathcal{A}) \tag{5}$$

When the differential chiral anomaly is integrated to find the anomaly for finite chiral rotations, so-called non-minimal terms arise which are artifacts of the regularization. These terms have the normal parity behaviour and give rise to both the non-linear σ-model term and higher order Skyrme terms. Let me emphasize, however, that usually such terms are eliminated from the anomaly since they are artifacts of the regularization and in fact, such terms appear from the non-anomalous part of the effective action, as we will see later. Furthermore let me stress that these approaches assume that spontaneous breaking of chiral symmetry takes place but do not explain it.

Fancier approaches and in my opinion also more sophisticated approaches to find the effective meson action of QCD are the approach of [14] and a similar approach, which in spirit is the same as [14], is presented in [15]. These approaches try to extract the structure of the effective meson theory without really calculating it from QCD. In fact, these approaches exploit only the chiral symmetry of QCD and would in principle give the same result for any other chiral invariant theory with fermions, which has the same group structure as QCD. They do not explain spontaneous break-down of chiral symmetry but assume it.

There is yet another approach to obtain the effective meson theory which uses the Nambu-Jona-Lasinio model [10,11] or extensions of it [12,13] in an intermediate step of the derivation. In these approaches one assumes that the flavour dynamics of QCD can be approximated at low energies by a Nambu-Jona-Lasinio type of model, i.e. by a chiral invariant fourfermion interaction. Refs. [10,11] use directly the Nambu-Jona-Lasinio model as starting point while [12,13] use a current expansion of QCD to motivate an effective 4-fermion interaction, which is essentially the Nambu-Jona-Lasinio model in [12] while in [13] the quark interaction is mediated by an effective gluon propagator, which is phenomenologically chosen. The approaches of [10 - 13, 16] have the advantage that they in fact explain the spontaneous breaking of chiral symmetry and not assume it.

What is common to all approaches is that they exploit global chiral invariance of QCD and they result in the same effective meson theory for the chiral field, which is given by

$$S_q = \operatorname{tr}\log(i\not{D} - m(U)^{\gamma_5}) \tag{6}$$

where m is the constituent quark mass arising from spontaneous breaking of chiral symmetry. We can conclude that the effective meson theory of QCD is almost entirely determined by the global chiral invariance of QCD.In fact, this is the basic assumption underlying chiral perturbation theory [20]. I will therefore use the Nambu-Jona-Lasinio model to mimic the low-energy flavour dynamics of QCD and demonstrate in the context of this model the derivation of the effective meson theory.

The Nambu-Jona-Lasinio model is also supported by the strong coupling lattice expansion [21]. The NJL model was motivated from the instanton liquid picture [19] of the QCD vacuum [16]. Furthermore, recent work in the context of the so-called field strength approach to Yang-Mills theories [22] also indicates that at low energies QCD becomes essentially an effective quark theory with a quark interaction of the Nambu-Jona-Lasinio type.

2. Path Integral Bosonization

The Nambu-Jona-Lasinio model is defined by the Lagrangian:

$$\mathcal{L} = \bar{q}(i\not{\partial} - m_0)q + \frac{g}{2}(\bar{q}\Gamma_i q)^2 \tag{7}$$

where g is a dimensionful coupling constant and Γ_i contains the generators of the flavour group and also the Lorentz tensors.The quantum field theory of this model is then defined by the functional integral

$$Z = \int DqD\bar{q} \exp \int \mathcal{L} d^4x. \tag{8}$$

In order to bosonize the theory we have to integrate out the quark fields. Due to the quark interaction this cannot be done in a straightforward manner. For this reason we first linearize the quark interaction by means of an auxiliary field ϕ

$$\exp \frac{g}{2}(\bar{q}\Gamma_i q)^2 = \int D\phi \exp\left(-\frac{1}{2g}\phi_i^2 - \bar{q}\phi_i\Gamma_i q\right) \tag{9}$$

Inserting this identity into equation (8) the quark fields can be integrated out. Defining

$$\phi = \phi_i \Gamma_i \tag{10}$$

one obtains the following effective meson theory

$$Z = \int D\phi \exp\left(-\frac{1}{2g}\mathrm{tr}\phi^2 + \mathrm{tr}\log(i\partial\!\!\!/ \; - m_0\phi)\right) \tag{11}$$

Finally, making a shift in the integration variable

$$\phi + m_0 \to \phi \tag{12}$$

we find that the effective meson theory, which is an exact mapping of the Nambu-Jona-Lasinio model, is given by

$$Z = \int d\phi \exp S[\phi] \tag{13}$$

where

$$S[\phi] = -\frac{1}{2g}\mathrm{tr}(\phi - m_0)^2 + \mathrm{Tr}\log iD\!\!\!/, \quad iD\!\!\!/ = i\partial\!\!\!/ - \phi \tag{14}$$

is the effective meson action. The quadratic part of the action will later give rise to the meson mass terms. The difficult part to evaluate is the quark loop

$$S_q[\phi] = \mathrm{Tr}\log iD\!\!\!/ \tag{15}$$

This term arises from integrating out the quark fields and represents the quantum transition amplitude of a Fermi system moving in the "external" meson field ϕ

$$S_q = log \int DqD\bar{q} \exp \int \bar{q} i D\!\!\!/ \, q \tag{16}$$

$$= \log\langle\psi|\bar{T}\exp[-\int_{-T/2}^{T/2} d\tau\, h(\tau)]|\psi\rangle \tag{17}$$

In this transition amplitude T is the Euclidean time interval and $\bar{T}$ denotes time-ordering. Furthermore h is the corresponding Dirac hamiltonian defined by

$$i\not{D} = \beta(-\partial_\tau - h) \tag{18}$$

$$h = \alpha\nabla + \beta\phi \tag{19}$$

The reference state Ψ is hidden in the boundary condition to the fermionic functional integral. Let me emphasize that the quantum transition amplitude and therefore also the quark loop (16) is defined only after specifying this reference state. In general, the reference state will contain not only the negative energy Dirac sea, but in addition also quarks or anti-quarks in valence orbits. Therefore the quark loop can be split in a vacuum part and a valence quark or anti-quark part

$$S_q[\phi] = S_q^0[\phi] + S_q^{val}[\phi]. \tag{20}$$

This splitting holds for any quark determinant and follows from the single particle nature of the Dirac operator (18), see [23]. By definition the vacuum is the state with all negative energy states occupied and all positive energy states empty. In the presence of an external field like a chiral meson field, states which have positive energy in the absence of the external field may be lowered so much that their energy becomes negative and they join the Dirac sea. Although of valence quark origin, these states must then be occupied in the new vacuum. Accordingly this vacuum becomes charged and carries quantum numbers of valence quarks like spin, isospin or baryon charge.

In the following I will concentrate on the vacuum parts since the mesons represent collective (quark-antiquark) excitations of the vacuum.

3. The Effective Meson Action

The evaluation of the vacuum part is most easily accomplished by letting the Euclidean time interval T go to infinity. This projects the quantum transition amplitude onto the vacuum transition amplitude, and the specific boundary conditions in the function integral become irrelevant. In other words large Euclidean times automatically pick up the vacuum as the reference state

$$|\psi\rangle = |0\rangle. \tag{21}$$

In Euclidean space the action of the quark loop (16), (23) is in general a complex quantity

$$S_q^0 = S_R^0 + iS_I^0. \tag{22}$$

Real and imaginary part are conveniently written as

$$S_R = \frac{1}{2}\mathrm{Tr}\,\log \not{D}^\dagger \not{D} \tag{23}$$

$$iS_I = \frac{1}{2}\mathrm{Tr}\log(\not{D})^2/\not{D}^\dagger\not{D} \tag{24}$$

while the imaginary part is a finite quantity, the real part is ultraviolet divergent and needs regularization. A convenient regularization scheme is the proper time regularization [24]. This regularization exploits the representation of the logarithm by the incomplete Γ-function

$$\Gamma(0,x) = \int_x^\infty \frac{d\tau}{\tau}e^{-\tau} \approx \mathrm{const} - \log x, \quad x \to 0. \tag{25}$$

In this regularization we find for the real part of the quark loop

$$S_R = -\frac{1}{2}\int_1^\infty \frac{d\tau}{\tau}\mathrm{Tr}\exp\left(-\tau \not{D}^\dagger \not{D}/\Lambda^2\right) \tag{26}$$

where Λ is an ultraviolet cut-off. Note that this regularization procedure preserves gauge and Lorentz invariance.

3.1. The gauged linear σ-model: mesons

In the meson sector one does not need an exact evaluation of the quark determinant, a low-energy approximation is usually sufficient. At low energies the quark loop can be evaluated in the gradient expansion, which is an expansion in the momenta of the "external" meson fields. This expansion is used also in chiral perturbation theory. A more elegant low-energy evaluation is provided by the heat-kernel method [25]. The heat kernel expansion is the relativistically covariant analog of the Wigner-Kirkwood expansion used in many-body physics. Like the latter the heat-kernel expansion is a semiclassical type of expansion, which can be characterized as a modified gradient expansion. It agrees with the gradient expansion only in leading order, but reshuffles terms of the gradient expansion in higher order.

In leading order of the gradient expansion the real part of the effective meson action defined by (14) and (20) is given by the gauged linear σ-model

$$\begin{aligned} S_R^0[\Phi] &= \int \mathcal{L}_R\, d^4x \\ \mathcal{L}_R &= \frac{1}{4}F_\pi^2(\Phi)\,\mathrm{tr}(\nabla_\mu\Phi U)^\dagger(\nabla_\mu\Phi U) - W(\Phi,U) \\ &\quad -\frac{1}{2}g_V^2(\Phi)tr((F_{\mu\nu}^V)^2 - (F_{\mu\nu}^A)^2) - \frac{1}{2}M_\nu^2(V_\mu)^2 - \frac{1}{2}M_A^2(A_\mu)^2 \end{aligned} \tag{27}$$

Here

$$\nabla_\mu M = \partial_\mu M + [V_\mu, M] - \gamma_5\{A_\mu, M\} \tag{28}$$

denotes the covariant derivative of the scalar and pseudo-scalar meson field

$$\phi = S + i\gamma_5 P i\not{V} - i\not{A}\gamma_5, \quad M = S + iP = \Phi U, \tag{29}$$

and

$$\begin{aligned} F^{v}_{\mu\nu} &= \partial_\mu V_\nu - \partial_\mu V_\nu + [V_\mu, V_\nu] + [A_\mu, A_\nu] \\ F^{A}_{\mu\nu} &= \partial_\mu A_\nu - \partial_\nu A_\mu + [A_\mu, V_\nu] + [V_\mu, A_\nu] \end{aligned} \tag{30}$$

are the vector and axial-vector field strength tensors. Furthermore M_V and M_A are the masses of the vector- and axial-vector mesons, which are inversely proportional to the coupling constant of the model (8). The potential for the scalar and pseudoscalar mesons $W(\Phi, U)$ is given by

$$W(\Phi, U) = +\frac{1}{4g} tr(M - m_0)^\dagger (M - m_0) + \frac{1}{4} F_\pi^2(\Phi) tr(\Phi^4 - 2\phi^2\Lambda^2 e^{-(\Phi/\Lambda)^2}) \tag{31}$$

and has a typical Mexican hat shape which describes the spontaneous break-down of chiral symmetry. The radial degree of freedom is given by the field Φ, which is the corresponding order parameter for chiral symmetry breaking while the pseudoscalar meson fields parametrize the phase of M. Finally, for large cut-offs Λ the quantities $F_\pi(\Phi)$ and $g_V(\Phi)$ depend logarithmically on Φ^2/Λ^2. On the chiral circle, i.e. at the minimum of the potential i.e. at the minimum of the potential $\Phi = m$ (m denotes the constituent quark mass) these quantities become the pion decay constant and the universal a vector coupling constant.

The gauged linear σ model describes quite satisfactorily the meson masses and decay constants of the pseudoscalar and vector mesons. Furthermore this model respects all the low-energy theorems known from current algebra like the Goldberger-Trieman relation,

$$m = F_\pi g_{qq\pi}, \tag{32}$$

Weinberg's relation,

$$M_{A_1} = \sqrt{2} M_\rho, \tag{33}$$

and the KSFR relation ,

$$M_\rho^2 = 2 g_V^2 F_\pi^2. \tag{34}$$

This is not surprising since the starting Nambu-Jona-Lasinio model from which the effective meson theory was derived obeys chiral symmetry, which is the basis for current algebra. Furthermore, when the electromagnetic field is included the model shows vector dominance.

Table 1 shows the numerical results for meson masses and decay constants obtained in leading order gradient expansion of a bosonized Nambu-Jona-Lasinio model where also the 'tHooft determinant was included in order to simulate the effect of the U(1) anomaly [26]: For more details see [26].

3.2. Skyrme terms and the Skyrme model.

So far we have restricted ourselves to the leading terms (quadratic in the meson fields and their derivatives) of the effective meson lagrangian. Let us now consider the next-to- leading order derivative terms confining ourselves to the chiral field which dominates at low energies due to the smallness of the masses of the pseudoscalar mesons. We will set the higher- lying scalar, vector and axialvector meson fields equal to their vacuum value, i.e.

$$\Phi = m, \qquad V_\mu = A_\mu = 0 \tag{35}$$

The next-to-leading order terms in the gradient expansion of the effective meson lagrangian read [11]

$$\mathcal{L}_4 = \frac{1}{32e^2}\mathrm{tr}\{[L_\mu, L_\nu]^2 + 2(L_\mu L_\mu)^2 - 4(\partial_\mu L_\mu)^2\}, \quad L_\mu = U^\dagger \partial_\mu U \tag{36}$$

where the strength constant e is finite for $\Lambda \to \infty$ and approaches 2π in this limit. The first term on the right-hand side is the celebrated Skyrme term. In fact the predicted strength parameter $e = 2\pi$ is not terribly far off the value $e = 5.45$ obtained by Adkins et al. [27] by fitting the nucleon–Δ mass splitting within the Skyrme model. If we had not obtained the two additional terms the next to leading order gradient expansion of the chiral quark loop (13) would have provided us with the Skyrme model, a very welcome result. However, the second and third term undo the job of the Skryme term and destabilize the chiral soliton. Furthermore the third term leads even to a vacuum instability [28, 29]. Nevertheless the next to leading order terms including the second and third ones with the undesired consequences in the soliton sector are physically relevant and can in fact be measured in low energy $\pi - \pi$ and $\pi - K$ scattering. The numerical coefficients of these terms predicted by the gradient expansion of the quark loop in units of 10^{-3} are 0.79, 1.58 and -3.71 respectively. They compare rather well with the values obtained by a phenomenological analysis of the meson data in chiral perturbation theory [20]: 0.9 ± 0.3, 1.7 ± 0.7 and -4.4 ± 2.5, respectively . Hence we may conclude that the gradient expansion of the effective meson action (13) is in satisfactory agreement with low-energy meson data, but neither in leading nor next to leading order it provides stable soliton solutions of the chiral field.

The Skyrme model can, however, be obtained in leading order of the gradient expansion of the effective meson action if one includes not only the pseudoscalar field, but also the vector and axial-vector meson fields. For this purpose one has to eliminate the latter ones in the static limit treating the kinetic energy of the vector and axial-vector mesons as perturbation. From (22) one then obtains precisely the Skyrme model with explicit breaking of chiral symmetry

$$\mathcal{L} = \frac{1}{4}F_\pi^2\,\mathrm{tr}L_\mu L_\mu + \frac{1}{32e^2}\,\mathrm{tr}[L_\mu, L_\nu]^2 + \frac{1}{2}M_\pi^2 F_\pi^2\,\mathrm{tr}(U-1) \tag{37}$$

where the strength of the Skyrme model is given by the universal vector coupling constant

Table 1: *Meson masses and decay constants calculated in the gradient expansion of the effective meson lagrangian following from the bosonization of a Nambu-Jona-Lasinio model supplemented by the 'tHooft determinant* [26]. Φ *denotes the* $\eta - \eta'$ *mixing angle.*

	calc.	exp.
m_u/MeV	313	$\approx$ 315
m_d/MeV	315	$\approx$ 315
m_s/MeV	414	$\approx$ 500
$-\langle\bar{u}u\rangle^{1/3}$/MeV	366.9	200 - 250
$-\langle\bar{d}d\rangle^{1/3}$/MeV	367.3	200 - 250
$-\langle\bar{s}s\rangle^{1/3}$/MeV	378.5	100 - 200
f_π/MeV	93*	93.3
f_K/MeV	97	110 - 115
f_η/MeV	98	$\approx$ 112
$f_{\eta'}$/MeV	95	$\approx$ 100
$m_{\pi\pm}$/MeV	140*	140
$m_K \pm$ /MeV	493*	493
m_K^0/MeV	497*	497
m_π^o/MeV	140	135
m_η/MeV	498	549
$m_{\eta'}$/MeV	957 *	957
m_δ/MeV	782	983
m_κ/MeV	919	1350
m_ϵ/MeV	1012	1300 or 1587
m_S/MeV	621	975
m_ρ/MeV	781	770
m_ω/MeV	783 *	783
m_{κ^*}/MeV	840	892
m_ϕ/MeV	884	1020
m_{A_1}/MeV	1096	1275
m_D/MeV	1100	1283
m_Q/MeV	1218	1270 or 1400
m_E/MeV	1346	1422
Θ	-22 ^{0}	$\approx 23^0$

$$e \sim g_V \tag{38}$$

For more details see ref. 30.

If one uses instead of the gradient expansion the heat-kernel expansion, one finds in next to leading order besides the fourth-order derivative terms given in (26) also a term with six derivatives

$$\mathcal{L}_6 = -\frac{1}{32e^2}\frac{2}{5m^2}\,\mathrm{tr}(\Box\partial_\mu U^\dagger)(\Box\partial_\mu U) \tag{39}$$

It can be shown [31] that this term is sufficient to stabilize the chiral soliton. This shows that the heat-kernel expansion, which resums the gradient expansion, might be sufficient in the soliton sector of the effective meson theory.

So far we have restricted our attention to the real part of the fermion determinant (25), (26). The evaluation of the imaginary part is a bit more involved and is most conveniently done using powerful differential geometric methods. From a physical point of view the imaginary part represents the field theoretic analog of the non-adiabatic extension [32] of the Berry phase [33]. In leading order of the gradient expansion one finds for the imaginary part the action, the so-called Wess-Zumino term [34]

$$S_I^0 = S_{WZ}(U) + \ldots. \tag{40}$$

This term vanishes for trivial chiral fields $U = const$. In the bosonized theory all observables become, like the action, functionals of the meson fields. In particular this is true for the baryon current. The bosonized form of the baryon current can be easily obtained from the effective action by taking the variation with respect to the vector meson field. Since the action consists of a vacuum part and valence quark part the same must be true for observables like the vector current

$$j_B^\mu(x) = (j_B^\mu)^0 + (j_B^\mu)^v. \tag{41}$$

In leading order of the gradient expansion the vacuum part of the baryon current receives a contribution only from the Wess-Zumino term, which is precisely given by the topological current of the Skyrme model

$$(j_B^\mu(x))^0 = B^\mu(U) + \cdots, \quad B^\mu(x) = \frac{1}{24\pi^2}\epsilon^{\mu\nu\kappa\lambda}\,\mathrm{tr}(L_\nu L_\mu L_x). \tag{42}$$

4. Baryons as chiral solitons

The above considerations show that the topological current of the Skyrme model is only a low-energy approximation to the vacuum part of the baryon current and can hence not describe the baryon current of valence quarks. On the other hand in all purely mesonic soliton models (e.g. the Skyrme model) one identifies the topological current with the full baryon current. How can these models describe baryons, which consist of three valence quarks if the topological current does not include the valence quark contribution? To answer this question we consider the spectrum of quarks in a topologically non-trivial external chiral field. Let us assume that the chiral field has winding number $n = 1$ so that the baryon charge N_B following from the topological current is $N_B = 1$. Fig. 1 shows the spectrum of the quarks in this field as function of the strength of the chiral field, which is given by the product of the constituent quark mass m times the spatial extension of the chiral field. (Note that the constituent quark mass is the coupling constant between the quarks and the chiral field.) For weak chiral fields the spectrum of the quarks is only slightly modified. However, as

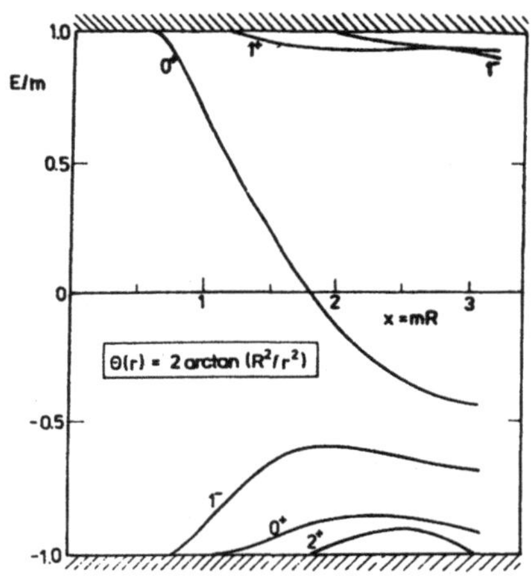

Figure 1: Spectrum of the quarks in a chiral nedghog field with winding number one. The states are classified according to the grand spin $\vec{K} = \vec{j} + \vec{t}$ and the parity.

the strength of the chiral field increases the lowest-lying valence quark state becomes strongly bound and eventually joins the Dirac sea. As soon as its energy is negative, this state becomes part of the Dirac sea and is occupied in the physical vacuum. The vacuum is then charged and carries non-zero baryon number. Since our observations do not resolve the Dirac sea we attribute this charge of the Dirac sea to the meson field which causes the polarization of the Dirac sea. But strictly speaking it is not the meson field which carries the baryon charge but the polarized Dirac sea.

The above considerations show: For weak chiral fields the valence quark energy is positive and the vacuum is uncharged. A baryon number $N_B = 1$ system is obtained here by adding three valence quarks on top of the vacuum. On the other hand for large chiral fields with topological winding number $n = 1$ the stable physical vacuum is strongly polarized and contains already three valence quarks. It is this domain of the coupling constant where the topological current can be identified with the baryon current of the nucleon, while for small coupling constants the topological current is only an insufficient approximation to the vacuum part of the baryon current. For small coupling constants the vacuum carries no baryon charge and three valence quarks have explicitly to be added on top of the vacuum to obtain a $N_B = 1$ state.

The validity of the soliton picture of the nucleon can be tested within the Nambu-Jona-Lasinio model by doing a fully selfconsistent calculation for the chiral soliton. This was first done in [35] and subsequently in [36] and [37]. In the preliminary calculations only the chiral field was included. It was found that the valence quarks in the self-consistent chiral soliton have a positive energy. This implied that the nucleon is not a Skyrmion but rather dominated by three valence quarks. The picture of the nucleon changes, however, when one includes vector [38] and axial-vector mesons [39, 40]. One then finds that the valence quarks have indeed a negative energy. This result obtained within the Nambu-Jona-Lasinio model strongly supports Witten's conjecture that the nucleon is a Skyrmion.

5. Concluding remarks

In my talk I have discussed how the effective meson theory of QCD can be extracted from the microscopic quark degrees of freedom. I have pointed out that the effective meson theory seems to be entirely determined by chiral invariance. I have also explained how a soliton picture of the nucleon emerges in the effective mesonic theories from the underlying microscopic quark dynamics. The microscopic calculations within the Nambu-Jona-Lasinio model show that the soliton picture of the nucleon is obtained only when one includes vector and axial-vector mesons. Furthermore the Skyrme model arises from the low-energy effective Lagrangian only when one includes vector and axial-vector mesons and eliminates the latter in the infinite mass limit. This shows that probably vector and axial-vector mesons are necessary for a realization of the soliton picture. Furthermore the Skyrme model obviously represents only a crude approximation to the true chiral soliton, since it relies on the infinite vector and axial-vector meson mass limit. This comes out also in explicit calculations done within the Skyrme model and the gauged non-linear σ- model [41], [42]. The inclusion of vector mesons, sticking to the leading order gradient expansion, gives an essential improvement compared to the pure Skyrme model for most of the physical observables of the baryons calculated so far within the soliton picture.

The soliton picture is most successful in describing low-energy properties of the baryons which are sensitive only to the structure of the outer part of the baryons. The drawback of the soliton picture is that the classical soliton does not have the correct quantum numbers of baryons. Before the chiral soliton can be confronted with experimental data of baryons, the soliton has to be quantized. This has been accomplished so far only in a semiclassical fashion. Recent calculations performed within the Skyrme model show that the semiclassical quantization is insufficient and one loop corrections are relevant [43]. These one loop corrections also eliminate deficiencies of the soliton description of baryons, in particular, the fact that the baryon masses come out much too large when the parameters of the effective meson theory are adjusted in the meson sector from the physical data [44].

6. Acknowledgements

The author is indebted to Dr. Ch. Weiss for a careful reading of the manuscript and critical remarks. He would like to thank the organizers of this meeting, in particular Professor G. Holzwarth, for the warm hospitality and the stimulating atmosphere at this workshop.

References

1. G. 'tHooft, *Nucl. Phys.* **B72** (1974) 461
2. E. Witten, *Nucl. Phys.* **B160** (1979) 56
3. E. Witten, *Commun. Math. Phys.* **92** (1984) 455;
   D. Gepner, *Nucl. Phys.* **B252** (1985) 481
4. A.P. Balachandran, V.P. Nair, S.G. Rajeev and A. Stern, *Phys. Rev. Lett.* **49** (1982) 1124
5. E. Witten, *Nucl. Phys.* **B223** (1983) 422, 433
6. T.H.R. Skyrme, *Proc. Roy. Soc.* **A260** (1961) 127; *Nucl. Phys.* **31** (1962) 556
7. D.I. Dyakonov and M.I. Eides, *Prisma Zh. Eksp. Teor. Fiz.* **38** (1983) 358
8. J. Balog, *Phys. Lett.* **B149** (1984) 197
9. A. Adriana and L. Bonora, *Nucl. Phys.* **B233** (1984) 232
10. A. Dhar, R. Shankar and S.R. Wadia *Phys. Rev.* **D31** (1985) 3256
11. D. Ebert and H. Reinhardt, *Nucl. Phys.* **B271** (1986) 188; *Phys. Lett.* **B173** (1986) 453
12. R.D. Ball, Workshop on Skyrmions and Anomalies, Mogilany, Poland (1987)
13. R.T. Cahill and C.D. Roberts, *Phys. Rev.* **D36** (1985) 2419
14. P. Simic, *Phys. Rev. Lett.* **55** (1985) 40; *Phys. Rev.* **D34** (1986) 1903
15. N. Karchev and A. Slavnov, *Teor. Mat. Fiz. (USSR)* **65** (1985) 192
16. D. Dyakonov and Y. V. Petrov, *Nucl. Phys.* **B272** (1986) 457
17. K. Fujikawa, *Phys. Rev.* **D21** (1980) 2848; **D23** (1981) 2262; **D29** (1984) 285; A.P. Balachandran *et al.*, *Phys. Rev.* **D25** (1982) 2713
18. S. Adler, *Phys. Rev.* **177** (1969) 2426; J.S. Bell and R. Jackiw, *N. Cimento* **60A** (1969) 47
19. E. Shuryak, *Phys. Rep.* **61** (1980) 72
20. J. Gasser and H. Leytwyler, *Ann. Phys.* **158** (1984) 142
21. H. Kluberg-Stern, A. Morel and B. Petersson, *Nucl. Phys.* **B215** (1983) 527; N. Kawamoto and J. Smit, *Nucl. Phys.* **B192** (1981) 100
22. M. Schaden, H. Reinhardt, P. Amundsen and M. Lavelle, *Nucl. Phys.* **B339** (1990) 595; H. Reinhardt, *Phys. Lett.* **B248** (1990) 365
23. H. Reinhardt, *Nucl. Phys.* **A503** (1989) 825
24. J.S. Schwinger, *Phys. Rev.* **82** (1951) 664
25. see Appendix A of ref. 11
26. H. Reinhardt and R. Alkofer, *Phys. Lett.* **B207** (1988) 482; R. Alkofer and H. Reinhardt, *Z. Phys.* **C45** (1989) 275

27. H. Adkins, C. Nappi and E. Witten, *Nucl. Phys.* **B223** (1983) 552
28. H. Reinhardt and D. Ebert, *Phys. Lett.* **B173** (1986) 459; I. Aitchinson, C. Frazer, E. Tador and J. Zuk, *Phys. Lett.* **B165** (1985) 163; H. Reinhardt, *Phys. Lett.* **B188** (1987) 482
29. H. Reinhardt and A. Wirzba, Contribution to the PANIC Conference, Tokyo, 1987, Proceedings, p. 64
30. H. Reinhardt and B.V. Dang, *Nucl. Phys.* **A500** (1989) 563
31. B. Kämpfer and H. Reinhardt, *Ann. Phys.* **1** (1992) 296
32. Aharanov and Anandan, *Phys. Rev. Lett.* **58** (1987) 1593; H. Reinhardt, *Phys. Rev. Lett.* **59** (1987) 2823
33. M. Berry, *Proc. Roy. Soc. Ser.* **A392** (1984) 45
34. J. Wess and B. Zumino, **Phys. Lett. B37** (1971) 95
35. H. Reinhardt and R. Wünsch, *Phys. Lett.* **B215** (1988) 577, *Phys. Lett.* **B230** (1989) 93
36. Th. Meissner, F. Grümmer and K. Goeke, *Phys. Lett.* **B227** (1989) 296
37. R. Alkofer, *Phys. Lett.* **B236** (1990) 310
38. R. Alkofer and H. Reinhardt, *Phys. Lett.* **B244** (1990) 461
39. R. Alkofer, H. Reinhardt, H. Weigel and U. Zückert, *Phys. Rev. Lett.* **69** (1992) 1874, see also R. Alkofer, this meeting
40. K. Goeke *et al.*, *Z. Phys. A* in press
41. U.-G. Meissner, N. Kaiser and W. Weise, *Nucl. Phys.* **A466** (1987) 685
42. H. Weigel, this meeting; N.W. Park and H. Weigel, *Nucl. Phys.* **A541** (1992) 453; B. Schwesinger, H. Weigel, G. Holzwarth and A. Hayashi, *Phys. Rep.* **173** (1989) 173
43. B. Moussalam, this meeting
44. G. Holzwarth, this meeting

The Chiral Soliton of the Nambu–Jona-Lasinio Model

REINHARD ALKOFER
Institute for Theoretical Physics, Tübingen University,
Auf der Morgenstelle 14, D-W7400 Tübingen, Germany

ABSTRACT

The baryon number one soliton of the NJL model is investigated. The Euclidian effective action is regularized using Schwinger's proper time method. Numerical results for the static soliton energy are presented for the cases without and with vector and axialvector fields. It is found that the baryon number is carried by the polarized Dirac sea of the quarks if the a_1 meson is included in contrast to the case without a_1 where the baryon number resides on valence quarks. Thus the model strongly supports the picture of the baryon as a topological soliton of the meson fields. Including the ω vector meson leads to a non–vanishing imaginary part of the Euclidian effective action even for static solitons. Leaving the chiral circle, i.e. including the isoscalar scalar meson, the soliton first seems unstable. However, we demonstrate that a ϕ^4 term stabilizes the soliton. Finally, the self–consistent two–flavor chiral soliton is embedded in the SU(3) flavor group and collectively quantized. The predicted mass splittings for the low–lying baryons with different hypercharge are in excellent agreement with the experimental data if the physical value for the kaon decay constant is substituted.

1. Introduction

As we have learned during this workshop there were many attempts to "derive" the effective meson Lagrangian from QCD in some low–energy approximation from the dynamics of the quarks.[1] Unfortunately this effective meson theory is not strictly known. However, phenomenological observations tell us that this effective theory should embody the very successful concept of spontaneously broken chiral symmetry which is manifest in the low energy hadron spectra. In fact, at low energies the effective meson Lagrangian is almost entirely determined by chiral symmetry. Any chirally invariant quark Lagrangian, which breaks chiral symmetry spontaneously in the vacuum, leads essentially to the same effective meson theory. This suggests to replace (for the purpose of obtaining the effective meson theory) QCD by a simpler chiral invariant model of chiral flavor dynamics. In this respect the Nambu–Jona-Lasinio (NJL) model[2] has been very successful. Bosonization of this model[3] leads to an effective meson Lagrangian which is in satisfactory overall agreement with low–energy meson data. Furthermore, there are investigations which show that the NJL model can be understood as low–energy approximation to QCD.[4] Given its success in the meson sector[5,6,7] it is natural to also investigate its soliton solutions in order to describe baryons. Unfortunately, the gradient expansion up to forth order in the derivatives of the quark determinant fails to provide stable soliton solutions.[8,9] Although this expansion yields in the next to leading order the desired Skyrme term two more terms appear which undo the job of the Skyrme term and destabilize the chiral soliton. On the other hand, once the gradient expansion is abandoned and the fermion determinant is treated explicitly, stable soliton solutions are obtained.

Taking into account pseudoscalar fields only self-consistent soliton solutions of the NJL model with full inclusion of the Dirac sea have been found.[10,11,12] Recently, these calculations have been extended to include vector and axialvector mesons.[13,14,15,16] Especially, the axialvector meson a_1 proves to be important. Its incorporation leads to the "diving" of the valence quark into the Dirac sea, and the baryons look indeed like Skyrmions. As we will see below the scalar meson field plays a very special role concerning the stability of the NJL soliton.[17,18,19] In fact, the NJL model has to be suitably generalized in order to provide stable solitons.[18,19]

As is already known from the Skyrmion the static soliton does not display the correct quantum numbers. The usual cranking procedure projects on the nucleon and the Δ.[20,21,22] Using the Yabu and Ando (YA) approach[23] to treat symmetry breaking strange baryons as chiral solitons of the NJL model have been studied recently.[24,25,26]

This talk is organized as follows: First, the static energy functional for the baryon number one soliton of the NJL model is discussed. Then the question whether the scalar meson field renders the soliton unstable is considered. Furthermore, numerical results for the soliton of the $\pi - \rho - a_1$ and the $\pi - \omega$ systems are presented and discussed. In the next section the semiclassical quantization within SU(3) (and pseudoscalar fields only) is shown. Finally, the main conclusions are summarized.

2. The Static Energy Functional

In this section the basic features of the underlying NJL model are discussed. Special emphasis is hereby put on the question of regularization which proves to be far from trivial as the action in terms of meson fields is in general complex, *i.e.* it contains also an imaginary part. The main result of this section will be the energy functional (35) for static meson field configurations. Extremizing this functional under the constraint of having baryon number one will yield the static soliton.

2.1. *The Action*

The starting point for the following considerations is the chirally invariant NJL model[2,3,5,6,7]

$$\begin{aligned}\mathcal{L} = \bar{q}(i\partial\!\!\!/ - m^0)q \quad &+ \quad 2g_1 \sum_{i=0}^{N_f^2-1} \left((\bar{q}\frac{\lambda^i}{2}q)^2 + (\bar{q}\frac{\lambda^i}{2}i\gamma_5 q)^2 \right) \\ &- \quad 2g_2 \sum_{i=0}^{N_f^2-1} \left((\bar{q}\frac{\lambda^i}{2}\gamma_\mu q)^2 + (\bar{q}\frac{\lambda^i}{2}i\gamma_5\gamma_\mu q)^2 \right), \qquad (1)\end{aligned}$$

wherein q denotes the quark spinors and m^0 the current quark mass matrix. Here we will work in the isospin limit, *i.e.* $m_u^0 = m_d^0 = m^0$. The matrices $\lambda^i/2$ are the generators of the flavor group ($\lambda^0 = \sqrt{2/N_f \mathbf{1}}$). For the case of the static soliton we will restrict ourselves to two flavors using $\lambda^i = \tau^i$, $i = 0,\ldots 3$. The coupling constants g_1 and g_2 will be determined from mesonic properties.

Applying standard functional integral bosonization techniques the model (1) can be rewritten in terms of composite meson fields[3]

$$\begin{aligned} \mathcal{A} &= \mathcal{A}_F + \mathcal{A}_m, \\ \mathcal{A}_F &= \mathrm{Tr}\log(i\not{D}) = \mathrm{Tr}\log\left(i(\not{\partial} + \not{V} + \gamma_5\not{A}) - (P_R M + P_L M^\dagger)\right), \\ \mathcal{A}_m &= \int d^4x \left(-\frac{1}{4g_1}\mathrm{tr}(M^\dagger M - m_0(M + M^\dagger) + m_0^2) - \frac{1}{4g_2}\mathrm{tr}(V_\mu V^\mu + A_\mu A^\mu)\right). \end{aligned} \tag{2}$$

Here $P_{R,L} = (1 \pm \gamma_5)/2$ are the projectors on right– and left–handed quark fields, respectively. $V_\mu = -i\sum_{a=0}^{3} V_\mu^a \tau^a/2$ and $A_\mu = -i\sum_{a=0}^{3} A_\mu^a \tau^a/2$ denote the vector and axial vector meson fields. The complex field M describes the scalar and pseudoscalar meson fields, $S_{ij} = S^a \tau_{ij}^a/2$ and $P_{ij} = P^a \tau_{ij}^a/2$:

$$M = S + iP = \xi_L^\dagger\ \Phi\ \xi_R, \tag{3}$$

wherein we already introduced the angular decomposition of the complex field M into a Hermitean field Φ and unitary fields ξ_L and ξ_R which are related to the chiral field by $U = \xi_L^\dagger \xi_R$. The latter is conveniently expressed in terms of a chiral angle Θ

$$U(x) = \exp\left(i\Theta(x)\right). \tag{4}$$

The quark determinant $\mathcal{A}_F$ diverges and therefore needs to be regularized. As in a study of the mesonic sector of the NJL model [3] as well as in previous studies of the soliton sector [10,11,12] we will use Schwinger's proper time regularization [27] which introduces an $O(4)$-invariant cut-off Λ after continuation to Euclidean space. For the regularization procedure it is necessary to consider the real and imaginary part of $\mathcal{A}_F$ independently

$$\begin{aligned} \mathcal{A}_F &= \mathcal{A}_R + \mathcal{A}_I \\ \mathcal{A}_R &= \frac{1}{2}\mathrm{Tr}\log(\not{D}_E^\dagger \not{D}_E) \\ \mathcal{A}_I &= \frac{1}{2}\mathrm{Tr}\log((\not{D}_E^\dagger)^{-1}\not{D}_E). \end{aligned} \tag{5}$$

The real part $\mathcal{A}_R$ diverges like $\log p^2$ for large momenta p whereas the imaginary part $\mathcal{A}_I$ does not contain any divergencies, *i.e.* it is finite without regularization. However, we believe that it has to be regularized also in order to have a consistent model. After all, the occurrence of the cutoff is a very crude way of mimicing the asymptotic freedom of QCD.

For the real part the proper time regularization consists in replacing the logarithm by a parameter integral

$$\mathcal{A}_R = -\frac{1}{2}\int_{1/\Lambda^2}^{\infty} \frac{ds}{s}\mathrm{Tr}\exp\left(-s\not{D}_E^\dagger \not{D}_E\right), \tag{6}$$

which for $\Lambda \to \infty$ reproduces the logarithm up to an irrelevant constant. Since the operator $\not{D}_E^\dagger \not{D}_E$ is Hermitean and positive definite this integral is well defined. For the imaginary part the regularization procedure is equivalent,

$$\mathcal{A}_I = -\frac{1}{2}\int_{1/\Lambda^2}^{\infty} \frac{ds}{s} \mathrm{Tr}\exp\left(-s(\not{D}_E^\dagger)^{-1}\not{D}_E\right), \tag{7}$$

however, in this case one has to be careful concerning the convergence of the integral. We will see later that for the cases of interest there is no problem.

2.2. *Determination of Model Parameters*

In order to determine the coupling constants g_1 and g_2 from the meson sector it is sufficient to only inspect $\mathcal{A}_R$. Varying the regularized effective action with respect to the scalar and pseudoscalar fields yields the Dyson–Schwinger or gap equations

$$\begin{aligned} \langle S_{ij} \rangle &= \delta_{ij} M_i \\ M_i &= m_i^0 - 2g_1 \langle \bar{q}q \rangle_i \\ \langle \bar{q}q \rangle_i &= -M_i^3 \frac{N_c}{4\pi^2} \Gamma(-1, M_i^2/\Lambda^2). \end{aligned} \tag{8}$$

The quantities M_i, $i = u, d, s$ are the dynamically generated constituent quark masses and the $\langle \bar{q}q \rangle_i$ the quark condensates reflecting spontaneous breaking of chiral symmetry.

Employing the unitary gauge $\xi_L^\dagger = \xi_R = \exp(i\varphi)$ and expanding the effective action up to second order in the fluctuations $\varphi = \varphi^a(\lambda^a/2)$ allows to extract the inverse propagator for the pseudoscalar mesons.[26] The Bethe–Salpeter equation which determines the physical meson masses m_φ is equivalent to the condition that the meson propagator has a pole:

$$D^{-1}(q^2 = m_\varphi^2) = 0. \tag{9}$$

The residue allows also to extract the corresponding meson decay constant

$$f_{ij}^2(q^2) = \frac{1}{4}(M_i + M_j)^2 \frac{N_c}{4\pi^2} \int_0^1 dx\ \Gamma\Big(0, [(1-x)M_i^2 + xM_j^2 - x(1-x)m_\varphi^2]/\Lambda^2\Big). \tag{10}$$

We want to emphasize here that our Bethe–Salpeter equation is the one in ladder approximation and that no other approximations as *e.g.* gradient or heat kernel expansions have been made. This also implies that the decay constants are evaluated on the corresponding meson mass shell. In the isospin limit ($m_u^0 = m_d^0$ and therefore $M_u = M_d$) the pion decay constant is then given by

$$f_\pi^2 = M_u^2 \frac{N_c}{4\pi^2} \int_0^1 dx \Gamma\Big(0, [M_u^2 - x(1-x)m_\pi^2]/\Lambda^2\Big). \tag{11}$$

One clearly sees that in the chiral limit ($m_\pi = 0$) the expression

$$f_\pi^2 = M_u^2 \frac{N_c}{4\pi^2} \Gamma(0, (M_u/\Lambda)^2) \tag{12}$$

calculated by means of a derivative expansion becomes exact.

For the case without vector mesons we determine the parameters of the model, the coupling constant g_1, the cutoff Λ and the two current masses m_u^0 and m_s^0 by fitting pseudoscalar meson properties. First, we want to reproduce the experimental value of the pion decay constant $f_\pi = 93$ MeV. This fixes the ratio Λ/M_u. The current masses are determined by the pion and kaon masses, $m_\pi = 135$ MeV and $m_K = 495$ MeV. This leaves one adjustable parameter, *e.g.* the coupling constant g_1. However, as g_1 is not very transparent we will use the gap equations to reexpress it in terms of the up constituent mass M_u which we will then choose in the range between 350 and 500 MeV. In table 1 the corresponding values are displayed. Note that the current masses are somewhat larger than in other regularization schemes. The ratio m_s^0/m_u^0 is, as usually, varying from 23 to 25. Note also that the kaon decay constant f_K comes out about 15 % smaller than its experimental value $f_K = 1.24 f_\pi \approx 114$ MeV.

Table 1. The up and strange constituent and current masses, the cutoff and the kaon decay constant for the parameters used later. All numbers are in MeV.

M_u	M_s	m_u^0	m_s^0	Λ	f_K
350	590	14.7	366	641	104.1
400	626	16.2	391	631	99.9
450	663	17.1	409	633	97.0
500	700	17.6	421	642	94.9

The inclusion of vector and axialvector mesons leads to pseudoscalar–axialvector meson mixing, especially $\pi - a_1$–mixing. This changes the expression for the pion decay constant. In the chiral limit ($m_u^0 = m_d^0 = 0$) it is given by

$$f_\pi^2 = \frac{6M_u^2}{g_V^2} \frac{1}{1 + 6M_u^2/m_\rho^2} \tag{13}$$

where

$$m_\rho^2 = \frac{g_V^2}{4g_2} \quad \text{and} \quad g_V = \left(\frac{1}{8\pi^2} \Gamma(0, \frac{M_u^2}{\Lambda^2}) \right)^{-1/2} \tag{14}$$

is the universal vector coupling constant and $M_u = M_d$ was used. Using this procedure the cutoff Λ is now much larger than in the case without $\pi - a_1$–mixing, *e.g.* for $M_u = 350$MeV we have then $\Lambda = 1274$MeV instead of $\Lambda = 649$MeV if we fix Λ from the pion decay constant f_π. g_2 is determined from the ρ meson mass m_ρ leaving g_1 (or, equivalently, the constituent mass M_u) as a free parameter.

2.3. *The Energy Functional for Static Meson Fields*

Next we will consider the energy functional of the static soliton. After Wick rotation the Euclidian Dirac operator corresponding to Eq. 2 is given by

$$\begin{aligned} i\beta \not{D}_E &= -\partial_\tau - h, \\ h &= \boldsymbol{\alpha}\cdot\boldsymbol{p} + iV_4 + i\gamma_5 A_4 + \boldsymbol{\alpha}\cdot\boldsymbol{V} + \gamma_5\boldsymbol{\alpha}\cdot\boldsymbol{A} + \beta(P_R M + P_L M^\dagger) \end{aligned} \tag{15}$$

wherein τ denotes the Euclidian time. In Euclidian space τ, V_4 and A_4 have to be considered as real quantities. This leads to a non–Hermitean Hamiltonian h even for static configurations (*i.e.* $[\partial_\tau, h] = 0$) if non–vanishing time components of vector or axialvector meson fields are included.

For the chiral field we impose the hedgehog ansatz:

$$U(\boldsymbol{x}) = \exp\big(i\boldsymbol{\tau}\cdot\hat{\boldsymbol{r}}\Theta(r)\big). \tag{16}$$

This configuration has vanishing 'grand spin' $\boldsymbol{G} = \boldsymbol{l} + \boldsymbol{\sigma}/2 + \boldsymbol{\tau}/2$, *i.e.* $[\boldsymbol{G}, U] = 0$. For the scalar field only the isoscalar part may deviate from its vacuum expectation value M_u. Requiring vanishing grand spin it depends only on the radius r,

$$\Phi(\boldsymbol{x}) = \phi(r)\mathbf{1}. \tag{17}$$

The only possible ansatz for the ω field with grand spin zero has vanishing spatial components($\omega_i = 0$):

$$\omega_\mu = \omega(r)\delta_{\mu 4}. \tag{18}$$

For the ρ and a_1 meson we use the spherically symmetric *ansätze*

$$\begin{aligned} V_\mu &= -i\rho^a_\mu \tau^a, \quad \rho^a_0 = 0, \quad \rho^a_i = -\epsilon^{aik}\hat{r}^k G(r), \\ A_\mu &= -i a^a_\mu \tau^a, \quad a^a_0 = 0, \quad a^a_i = \hat{r}^i\hat{r}^a F(r) + \delta^{ia} H(r) \end{aligned} \tag{19}$$

where the index a runs from 1 to 3.

Thus the Euclidian Dirac Hamiltonian reads

$$\begin{aligned} h &= \boldsymbol{\alpha}\cdot\boldsymbol{p} + i\omega(r) + \phi(r)\beta(\cos\Theta(r) + i\gamma_5\boldsymbol{\tau}\cdot\hat{\boldsymbol{r}}\sin\Theta) \\ &+ \frac{1}{2}(\boldsymbol{\alpha}\times\hat{r})\boldsymbol{\tau}G(r) + \frac{1}{2}(\boldsymbol{\sigma}\cdot\hat{r})(\boldsymbol{\tau}\cdot\hat{r})F(r) + \frac{1}{2}(\boldsymbol{\sigma}\cdot\boldsymbol{\tau})H(r). \end{aligned} \tag{20}$$

which obviously is not Hermitean since $\omega(r)$ is real giving rise to complex eigenvalues of h. For static configurations the eigenvalues of ∂_τ, $i\Omega_n = i(2n+1)\pi/T$, $(n = 0, \pm1, \pm2, ..)$ may be separated.* Thus the eigenvalues $\lambda_{n,\nu}$ of the operator $\partial_\tau + h$ read:

$$\lambda_{n,\nu} = -i\Omega_n + \epsilon_\nu = -i\Omega_n + \epsilon^R_\nu + i\epsilon^I_\nu. \tag{21}$$

---

*The eigenfunctions of ∂_τ assume anti-periodic boundary conditions in the Euclidian time interval T. The Ω_n are the analogues of the Matsubara frequencies with T figuring as inverse temperature.

The fermion determinant is expressed in terms of the eigenvalues $\lambda_{n,\nu}$:

$$\mathcal{A}_R = \frac{1}{2}\sum_{\nu,n}\log(\lambda_{n,\nu}\lambda^*_{n,\nu}) \qquad \text{and} \qquad \mathcal{A}_I = \frac{1}{2}\sum_{\nu,n}\log(\frac{\lambda_{n,\nu}}{\lambda^*_{n,\nu}}). \tag{22}$$

Using (21) the real part reads:

$$\begin{aligned}\mathcal{A}_R &= \frac{1}{2}\sum_{\nu,n}\log((\Omega_n-\epsilon^I_\nu)^2+(\epsilon^R_\nu)^2)\\ &\to -\frac{1}{2}\sum_{\nu,n}\int_1^\infty \frac{d\tau}{\tau}\exp\{-\frac{\tau}{\Lambda^2}((\Omega_n-\epsilon^I_\nu)^2+(\epsilon^R_\nu)^2)\}\end{aligned} \tag{23}$$

according to the proper time regularization scheme (6). For large Euclidian time intervals ($T\to\infty$) the temporal part of the trace may be performed

$$\mathcal{A}_R = -\frac{T}{2}\sum_\nu\int_{-\infty}^\infty\frac{dz}{2\pi}\int_1^\infty\frac{d\tau}{\tau}\exp\{-\frac{\tau}{\Lambda^2}(z^2+(\epsilon^R_\nu)^2)\} \tag{24}$$

where we have shifted the integration variable $z-\epsilon^I_\nu\to z$. For $T\to\infty$ we may read off the Dirac sea contribution to the real part of the energy functional from $\mathcal{A}_R\to -TE^R_{\text{vac}}$:

$$E^R_{\text{vac}} = \frac{N_C}{4\sqrt{\pi}}\sum_\nu|\epsilon^R_\nu|\Gamma(-\frac{1}{2},(\epsilon^R_\nu/\Lambda)^2). \tag{25}$$

For soliton configurations with vanishing ω (*i.e.* $\epsilon^R_\nu=\epsilon_\nu$) there is no contribution from the imaginary part and eq. (25) is the expression for the energy of the Dirac sea.

For the imaginary part we obtain

$$\mathcal{A}_I = \frac{1}{2}(\sum_\nu\sum_{n=-\infty}^\infty\log(\lambda_{\nu,n})-\sum_\nu\sum_{n=-\infty}^\infty\log(\lambda^*_{\nu,n})) = \frac{1}{2}\sum_\nu\sum_{n=-\infty}^\infty\log\frac{i\Omega_n-\epsilon_\nu}{i\Omega_n-\epsilon^*_\nu} \tag{26}$$

where we have reversed the sign in the first sum over the integer variable n. Next we express $\mathcal{A}_I$ in terms of a parameter integral:

$$\mathcal{A}_I = \frac{1}{2}\sum_\nu\sum_{n=-\infty}^\infty\int_{-1}^1 d\lambda\frac{-i\epsilon^I_\nu}{i\Omega_n-\epsilon^R_\nu-i\lambda\epsilon^I_\nu}. \tag{27}$$

In analogy to (24) we may carry out the temporal trace in the limit $T\to\infty$:

$$\mathcal{A}_I = \frac{-i}{2}\sum_\nu\int_{-1}^1 d\lambda\, T\int_{-\infty}^\infty\frac{dz}{2\pi}\epsilon^I_\nu[i(z-\lambda\epsilon^I_\nu)-\epsilon^R_\nu]^{-1}. \tag{28}$$

Shifting the integration variable $z-\lambda\epsilon^I_\nu\to z$ the integral over λ may be done

$$\mathcal{A}_I = \frac{-i}{2}\sum_\nu\epsilon^I_\nu\int_{-\infty}^\infty\frac{dz}{2\pi}\frac{-2\epsilon^R_\nu}{z^2+(\epsilon^R_\nu)^2}. \tag{29}$$

$\mathcal{A}_I$ is regularized in proper time by expressing the integrand as a parameter integral:

$$\frac{-1}{z^2+(\epsilon_\nu^R)^2} \to \int_{1/\Lambda^2}^\infty d\tau \exp\{-\tau(z^2+(\epsilon_\nu^R)^2\} \tag{30}$$

which obviously is finite for $\Lambda \to \infty$. Continuing the evaluation of $\mathcal{A}_I$ in analogy to eqs. (23-25) we find for the contribution of the Dirac sea to the imaginary part of the Euclidian energy E^I_{vac}

$$E^I_{\text{vac}} = \frac{-N_C}{2}\sum_\nu \epsilon^I_\nu \text{sign}(\epsilon^R_\nu)\begin{cases} 1, & \mathcal{A}_I \quad \text{not} \quad \text{regularized} \\ \mathcal{N}_\nu, & \mathcal{A}_I \quad \text{regularized}\end{cases} \tag{31}$$

where $\mathcal{N}_\nu = \frac{1}{\sqrt{\pi}}\Gamma(\frac{1}{2},(\epsilon^R_\nu/\Lambda)^2)$ are the vacuum "occupation numbers" in the proper time regularization scheme. The upper case, of course, corresponds to the limit $\Lambda \to \infty$. Obviously only the real part of the one-particle energy eigenvalue is relevant for the regularization of $\mathcal{A}_I$. Eq. (31) reveals that we have succeeded in finding a regularization scheme for $\mathcal{A}_I$ that only involves quantities which are strictly positive definite. This is not evident from the definition of $\mathcal{A}_I$ (5).

The total energy functional contains besides E^R_{vac} and E^I_{vac} also the valence quark energy

$$E^R_{\text{val}} = N_C\sum_\nu \eta_\nu|\epsilon^R_\nu| \qquad E^I_{\text{val}} = N_C\sum_\nu \eta_\nu \text{sign}(\epsilon^R_\nu)\epsilon^I_\nu \tag{32}$$

with $\eta_\mu = 0,1$ being the occupation numbers of the valence quark and anti-quark states. Furthermore the meson energy is obtained by substituting the ansätze (16-19) into (2):

$$\begin{aligned} E_m &= 4\pi\int drr^2\Big(\frac{\langle\bar{u}u\rangle}{M_u-m_u^0}(\phi^2(r)-M_u^2)+m_\pi^2 f_\pi^2(1-\frac{\phi(r)}{M_u}\cos\Theta(r)) \\ &\quad +af_\pi^2(G^2(r)+\frac{1}{2}F^2(r)+F(r)H(r)+\frac{3}{2}H^2(r)-2\omega^2(r))\Big) \end{aligned} \tag{33}$$

where $a = 1+m_\rho^2/6M_u^2$ if $\pi - a_1$–mixing is included and $a = m_\rho^2/6M_u^2$ if it is not. Note that we are working in the isospin limit which implies $m_\omega = m_\rho$. Continuing back to Minkowski space we find for the total energy functional:

$$E[\phi,\Theta,\omega,G,F,H] = E^R_{\text{val}} + E^I_{\text{val}} + E^R_{\text{vac}} + E^I_{\text{vac}} + E_m. \tag{34}$$

The equations of motion are now obtained by extremizing the static Minkowski energy. They are solved by iteration. We start off with test meson profiles to calculate the matrix elements of the Hamiltonian. The resulting eigenvalues and eigenvectors are substituted into the equations of motion yielding new profile functions. This process is repeated until convergence is reached. If h is not Hermitean (in Euclidian space) which is the case if the ω meson is included one has to distinguish between left and right eigenvectors of h. For more details see ref. [16]. Otherwise h can be written in a real and symmetric form using eigenfunctions of the Hamiltonian with vanishing meson fields and of the grand spin as basis. The momenta are discretized using a finite spherical box. An important point are the boundary conditions used. A discussion of the advantages of two types of boundary conditions may be found in ref. [26].

3. Instability due to scalar mesons?

The first calculations have been done with restriction to the chiral field only [10,11,12]. Then one finds stable solitons for constituent masses $M_u > 325$MeV where $f_\pi = 93$MeV is used to fix the cutoff. The static soliton mass is approximately 1200 MeV almost independent of M_u. For physical values of M_u, $M_u \approx 350$MeV, the valence quark energy is positive and half of the soliton energy is carried by the valence quark. Even more, quantities like the moments of inertia are dominated by the valence quark contribution.[26]

Restricting oneself to the pseudoscalar field only the stability of the soliton can be easily understood. The value of the chiral angle at the origin is fixed, $\Theta(r=0) = \pi$. Assuming Θ to be monotonically decreasing it can be characterized by its spatial extension. The valence quark favors a large spatial extension of the soliton because it then becomes stronger bound, *i.e.* its energy is decreased. On the other hand, the Dirac sea is much less distorted for small soliton sizes and therefore its contribution to the energy behaves oppositely to the one of the valence quark. Given both features one obtains a minimum in the energy.

Recently it has been shown that including the scalar isoscalar field renders the soliton unstable.[17] Allowing this field to develop a narrow and high peak near the origin leads to the collapse of the NJL soliton. Note that this kind of instability is qualitatively different to the "breathing mode instability" which could have occurred for the case of the chiral field only (and which is the one discussed for Skyrme type models). Especially, the narrow and high peak which develops in the scalar field leads to the fact that very high momentum components of this field are involved.

For a number of reasons, the scalar–isoscalar sector of the NJL model is the least well understood of the effective dynamics. The absence of clearly interpretable resonances in this channel makes it difficult to determine parameters of effective models in the meson sector.† Moreover, it is likely that $\bar{q}q$–scalars strongly mix with scalar glueballs, whose presence in the low–energy dynamics is required by the scale anomaly of QCD. In an effective Lagrangian language these are described by a scalar field which is subject to an effective potential leading to a non–vanishing vacuum value. The feature which interests us here most is the fact that the mixing of the scalar glueball with the scalar meson Φ leads to a Φ^4 term in the action. As will be demonstrated in the following such a term prevents the collapse of the soliton.[18,19]

Employing the special parametrization $\Phi U = \sigma + \boldsymbol{\tau}\cdot\boldsymbol{\pi}$ we use the *ansatz*

$$\sigma(r) = M_u[1 + Wf(r/R)\cos\Theta(r/R)], \quad \boldsymbol{\pi}(\mathbf{r}) = \hat{\mathbf{r}} M_u W f(r/R)\sin\Theta(r/R) \tag{35}$$

† Additionally, the physical scalar mesons mix with states composed of two pseudoscalar mesons making the extraction of the properties of "bare" scalars extremely complicated.

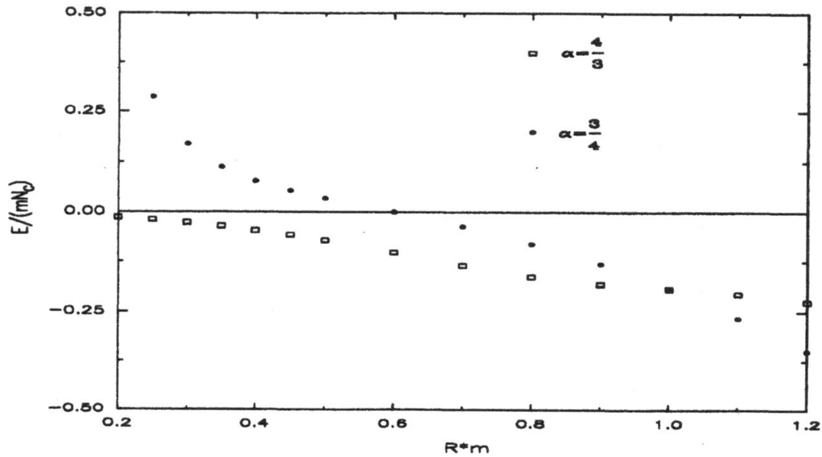

Fig. 1. The energy of the fermion determinant for two different meson field configurations. Note that the mesonic part of the energy vanishes for $R \to 0$ in both cases.

with $W(RM_u)^\alpha = const.$ R denotes a characteristic radius which parametrizes the size of the field configuration. In ref.[17] a Wood-Saxon type shape was chosen for $f(r/R)$ while a straight line profile was assumed for $\Theta(r/R)$. The constant α is determined such that the mesonic part of the energy vanishes in the limit $R \to 0$

$$\alpha < \begin{cases} \frac{3}{2} & \text{for a } \phi^2 \text{ potential} \\ \frac{3}{4} & \text{for a } \phi^4 \text{ potential} \end{cases} \tag{36}$$

In figure 1 we display the contribution of the fermion determinant to the static energy. Obviously the meson profile collapses in case there is only a ϕ^2 potential, however, in the presence of a ϕ^4 term in the mesonic energy functional the fermion determinant is strongly repulsive and prevents the soliton from collapsing.

To be more explicit we calculated the self–consistent soliton including a contribution to the mesonic energy of the form

$$a\phi^4(r) + bM_u^2\phi^2(r) \tag{37}$$

where the ratio between the parameters a and b (the unconventional coefficient bM_u^2 was chosen in order to have dimensionless parameters) has been determined by fixing simultaneously the constituent mass and the quark condensate. The cutoff Λ was,

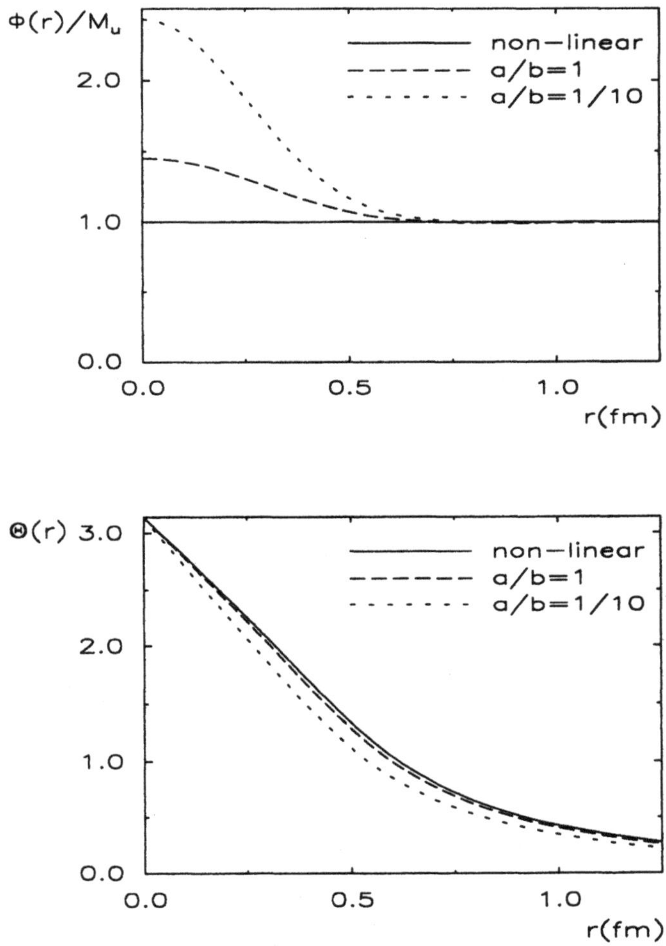

Fig. 2. The scalar meson profile $\phi(r)$ and the chiral angle $\theta(r)$ for the non–linear case (full line), for $a/b = 1$ (dashed line) and $a/b = 1/10$ (dotted line).

as usual, fixed using the pion decay constant as input. The profile function $\phi(r)$ as well as the chiral angle Θ are compared for two different values of a/b to some older calculation using pseudoscalar fields only are shown in figure 2. One clearly sees that $\Theta(r)$ does not change much even if a/b is as small as 1/10. Note that $a \to \infty$ corresponds to the non–linear case, *i.e.* to the restriction to the chiral circle. The soliton energy decreases from 1214 MeV in the non–linear case to 1179 MeV for the case $a/b = 1/10$, *i.e.* we have only a three percent decrease even for very small a/b. Therefore one may conclude that the NJL soliton based on a generalized model with Φ^4 term is very similar to the one where the scalar field is constrained to its vacuum value.

4. The soliton of the $\pi - \rho - a_1$–system

In the following we will use a fixed $\phi(r) = M_u$. However, from eq. (3) it is obvious that also vector and axialvector meson fields are present in the soliton for which we will use the profile functions of eqs. (18) and (19). Postponing the difficulties which arise due to the imaginary part of the action generated by the ω meson to the next section we restrict ourselves for the moment to the $\pi - \rho - a_1$–system.

Additionally, we consider only the chiral limit $m_u^0 = m_d^0 = 0$. The cutoff Λ and the coupling constant g_2 are determined as described in section 2.2. As input we use $f_\pi = 93\text{MeV}$ and $m_\rho = 770\text{MeV}$, see eqs. (13) and (14). In the numerical method we used two different boundary conditions (for a detailed description see appendix B of ref.[26]) in order to be sure that spurious contributions are negligible. For the cases $M_u = 300$, 350 and 400 MeV the total soliton energy E as well as its different contributions are shown in table 2. The energy eigenvalue of the valence quark is negative in all three cases. Its contribution is therefore already fully included in the vacuum part of the soliton energy. Note that there is also no (direct) contribution of the valence quark to other physical variables, especially the baryon current: **The baryon number is carried by the vacuum.** As the vacuum is unobservable this means that the baryon number "resides" in the meson profiles, *i.e.* these results support Witten's conjecture that baryons may be described as solitons within purely mesonic models.

Table 2. The soliton energy E as well as its Dirac sea and mesonic contributions E_{sea} and E_{mes} for different values of the constituent quark mass M. Shown is also the energy of the 'dived' level.

M (MeV)	300	350	400
E (MeV)	1104	1010	938
E_{sea} (MeV)	736	615	544
E_{mes} (MeV)	368	395	394
e_{val}/M	-0.04	-0.38	-0.54

In table 3 we compare the contributions to the soliton energy for three types of calculations: In the first only the pseudoscalar field is taken into account, in the second one includes the ρ meson and in the third one the $\pi - \rho - a_1$-system is considered.

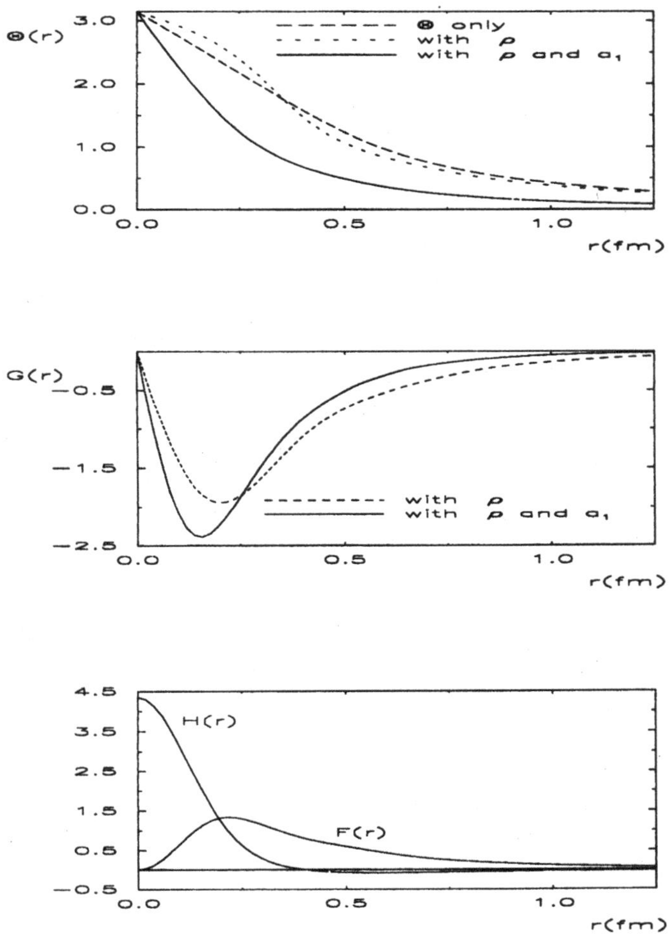

Fig. 3. The chiral angle Θ, the ρ meson profile $G(r)$ and the a_1 meson profiles $F(r)$ and $H(r)$ for the three cases of table 3.

Table 3. The contributions to the soliton energy for the case where one considers the chiral angle Θ only, with Θ and ρ and with Θ,ρ and a_1 for a constituent mass $M = 350$MeV. For completeness the values of the cutoff Λ are shown.

	only Θ	Θ and ρ	Θ,ρ and a_1
E (MeV)	1214	957	1010
E_{val} (MeV)	653	147	0
E_{sea} (MeV)	561	655	615
E_{mes} (MeV)	0	155	395
Λ (MeV)	649	649	1274

One sees clearly that the presence of the a_1 field leads to the "diving" of the valence quark level. This can be easily understood from the a_1 meson profiles, see figure 3. As the valence quark level is a s-wave with respect to grand spin it will be mostly influenced by meson profiles which are non-vanishing at the origin as $H(r)$ does in contrast to $G(r)$ and $F(r)$. Furthermore, as can be seen from figure 3 the chiral angle is "squeezed" if the a_1 meson is included.

We want to emphasize here that the NJL soliton including the ρ and a_1-mesons in addition to the chiral field supports Witten's conjecture that baryons may be described as solitons within purely mesonic models. Earlier calculations without the a_1-meson seemed to indicate that baryonic properties are dominated by valence quarks in contradiction to the assumption of a purely mesonic topological soliton description of baryons. This situation is in agreement with the experience one has with the pure Skyrme model where only the chiral field is included and the (axial-) vector mesons are left out. One knows that the Skyrmion based on pseudoscalars only has serious deficiencies, as *e.g.* wrong electromagnetic properties due to missing vector dominance [28] or wrong "high energy behavior" of $\pi - N$ phase shifts due to the higher order stabilization terms.[29]

5. The soliton of the $\pi - \omega$–system

We have calculated the chiral soliton with the ω vector meson for a wide range of values for the constituent mass. In contrast to the results of ref.[15] no instability is obtained, especially we find stable soliton solutions for the parameters which are extracted from mesonic data (see eqs. 13,14). This difference is most likely due to the fact that in ref.[15] eigenvalues of Minkowski space Hamilton operators have been used in order to regularize. However, regularization can be defined only in Euclidean space, see eqs. 6 and 7. Therefore the treatment of ref.[15] is inconsistent. A more detailed discussion can be found in ref.[16].

From figure 4 we see that the soliton energy (34) varies only mildly as a function of the constituent mass. The decrease of the real part is compensated by the increase of the imaginary part. The mesonic part is almost independent of the constituent mass. Noting that the energy of the NJL soliton with the chiral field only typically is of the order 1.2GeV [10,11,12] we recognize that the inclusion of the ω meson increases

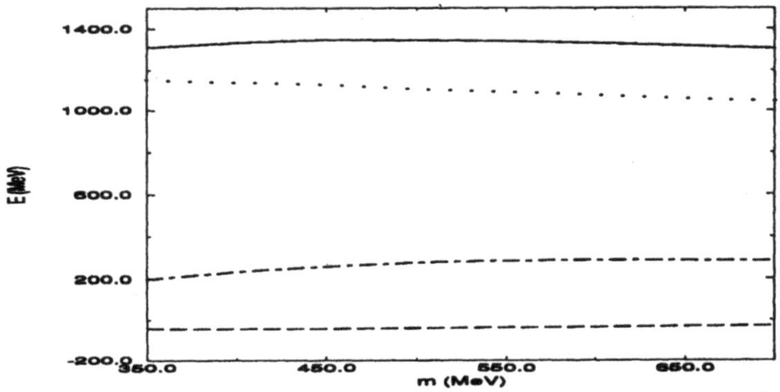

Fig. 4. The total static energy of the soliton as a function of the constituent mass. The various contributions are: real part (dotted), imaginary part (dashed-dotted) and the mesonic part (dashed).

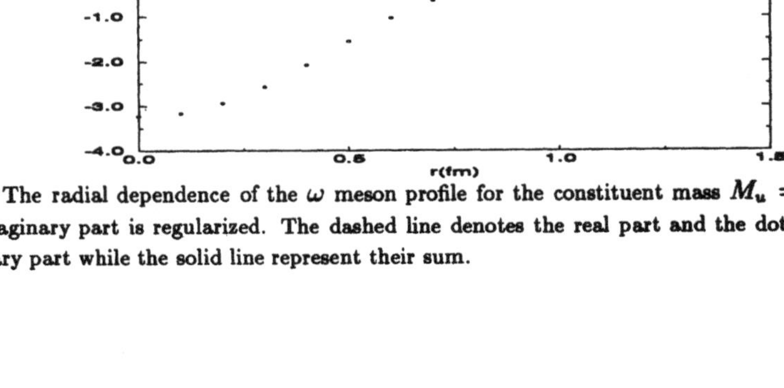

Fig. 5. The radial dependence of the ω meson profile for the constituent mass M_u = 600MeV. The imaginary part is regularized. The dashed line denotes the real part and the dotted line the imaginary part while the solid line represent their sum.

the total energy. Actually we find that the real part is lowered by the inclusion of ω, however, the additional imaginary part pushes the result for the total energy above the value obtained in the soliton model with only Θ. This indicates that equations of motion for time components of vector fields prove to be constraints which is realized by the complex nature of the energy in Euclidian space.

Furthermore, the results demonstrate that the non-regularization of $E^I_{\rm vac}$ does not strongly effect the result for the total energy. Merely the increase of the imaginary part with M_u is somewhat more pronounced since the exponential damping in (31) drops. However, the (non-) regularization of $E^I_{\rm vac}$ has a significant effect on the ω-meson profile; especially we find that for $M_u \geq 600\text{MeV}$ $\omega(r)$ does not assume its maximal value at the origin when $E^I_{\rm vac}$ is not regularized. We furthermore recognize that there exists a large cancellation between the real part of $\omega(r)$ and its imaginary part, see figure 5. In contrast to the naïve expectation we find for small M_u that the real part of the baryon density dominates $\omega(r)$ in the vicinity of the origin.

Of course, the final goal is to incorporate all vector mesons in the calculation of the soliton of the NJL model and work in this direction is in process. This implies a fusion of the work presented in this and the preceding section. Actually we do not expect the conclusion that the whole baryon charge is carried by the asymmetry of the polarized Dirac sea to change since including the ω meson we find that the critical constituent mass at which the real part of the valence quarks' energy changes sign $M_{\rm crit.} = 545\text{MeV}$ is already about 180MeV lower than in the NJL model including the chiral field only.[10,11,12]

6. Semiclassical Quantization in SU(3)

The chiral hedgehog soliton has neither the good spin nor flavor quantum numbers of the physical baryons. To project on baryon quantum numbers we employ the cranking procedure and impose the following ansatz:

$$\xi(\boldsymbol{x},t) = R(t)\xi(\boldsymbol{x})R^\dagger(t) \tag{38}$$

where $R(t)$ describes the (adiabatic) rotation in $SU(3)$ flavor space. Of course, we only have zero modes in the subspace $SU(2)_I \times U(1)_Y$ of $SU(3)$, nevertheless (38) is reasonable since we consider $SU(3)$ as an approximate symmetry. Elevating $R(t)$ to be $SU(3)$ valued furthermore allows us to easily make contact with the phenomenology of baryon representations. This approach has received intensed recognition after Yabu and Ando (YA)[23] demonstrated that the resulting collective Hamiltonian may be diagonalized exactly. More recently the YA approach has been extented considerably to more complicated symmetry breaking schemes.[30,31,32]

The main purpose of the investigations in this section is to explain how semi-classical quantization can be applied to the NJL soliton and furthermore provides reasonable results for baryon mass differences. Therefore we shall restrict ourselves to the approach which contains pseudoscalar fields only.

6.1. *Description of the approach*

In the following it will be convenient to transform the fermion determinant to the flavor rotation frame $q(\mathbf{x},t) = R(t)q'(\mathbf{x},t)$. Obviously this transformation cancels the rotation matrices in our *ansatz* (38) at the expense of induced terms in the quark Hamiltonian due to the time dependence of the rotations and symmetry breaking:

$$h = h_0 + h_{rot} + h_{SB} \tag{39}$$

wherein h_0 is the static $SU(2)$ quark Hamiltonian introduced in section 2. The induced Hamiltonian h_{rot} originates from the time dependence of the collective rotation

$$h_{rot} = -iR^\dagger(t)\dot{R}(t) = \frac{1}{2}\sum_{a=1}^{8}\lambda_a\Omega_a = \begin{pmatrix} \Omega_\pi + \Omega_\eta & \Omega_K \\ \Omega_K^\dagger & -2\Omega_\eta \end{pmatrix}. \tag{40}$$

The contribution to h due to $SU(3)$ symmetry breaking is more difficult to handle:

$$\begin{aligned} h_{SB} &= \mathcal{T}\beta[R^\dagger(t)\langle S\rangle R(t) - \langle S\rangle]\mathcal{T}^\dagger \\ &= \frac{\Delta M}{\sqrt{3}}\mathcal{T}[\beta\sum_{i=1}^{3}(D_{8i}\lambda_i + \sum_{\alpha=4}^{7}D_{8\alpha}\lambda_\alpha + (D_{88}-1)\lambda_8)]\mathcal{T}^\dagger \end{aligned} \tag{41}$$

with $\mathcal{T} = \xi(\mathbf{x})P_L + \xi^\dagger(\mathbf{x})P_R$ and $\Delta M = M_u - M_s$ denotes the difference of the up and strange constituent masses. The adjoint representation $D_{ij} = \frac{1}{2}\mathrm{tr}(\lambda_i R\lambda_j R^\dagger)$ of the rotation matrices clearly exhibits the transformation properties of the symmetry breaking part of the Hamiltonian.

Our goal is to expand the energy in both ΔM and the angular velocities Ω_a. The latter actually corresponds to an expansion in $1/N_C$. In this expansion we keep terms quadratic in the perturbation $h_P = h_{rot} + h_{SB}$. Then, due to isospin invariance, the most general form of the collective Lagrangian reads:

$$\begin{aligned} L = & - E + \frac{1}{2}\alpha^2\sum_{i=1}^{3}\Omega_i^2 + \frac{1}{2}\beta^2\sum_{\alpha=4}^{7}\Omega_\alpha^2 - \frac{\sqrt{3}}{2}B\Omega_8 + \alpha_1\sum_{i=1}^{3}D_{8i}\Omega_i + \beta_1\sum_{\alpha=4}^{7}D_{8\alpha}\Omega_\alpha \\ & - \frac{1}{2}\gamma(1-D_{88}) - \frac{1}{2}\gamma_1(1-D_{88}^2) - \frac{1}{2}\gamma_2\sum_{i=1}^{3}D_{8i}D_{8i} - \frac{1}{2}\gamma_3\sum_{\alpha=4}^{7}D_{8\alpha}D_{8\alpha}. \end{aligned} \tag{42}$$

E hereby is the classical soliton energy. In the following we will describe the origin of all terms in the collective Lagrangian.

The valence quarks are treated by standard perturbation techniques, however, they are only explicitly added as long as the corresponding energy eigenvalue is positive. Due to the embedding of the static soliton solution in the $SU(2)$ subgroup only isosinglet operators contribute non-trivially in first order perturbation theory. Since λ_8 acts like $\frac{1}{\sqrt{3}}\times$ unit − matrix in the isospin subgroup we obtain:

$$B^{\mathrm{val}} = \eta^{\mathrm{val}} \tag{43}$$

which is the valence quark contribution to the baryon number. By the same argument

$$\gamma^{\text{val}} = -\frac{2}{3}N_C\Delta M\eta^{\text{val}}\langle\text{val}|\mathcal{T}\beta\mathcal{T}^\dagger|\text{val}\rangle. \tag{44}$$

The valence quark contribution to the expressions quadratic in angular velocities, $\frac{1}{2}\Theta_{ab}\Omega_a\Omega_b$ is the well known cranking result:

$$\Theta_{ab}^{\text{val}} = \frac{N_C}{2}\eta^{\text{val}}\sum_{\mu\neq\text{val}}\frac{\langle\text{val}|\lambda_a|\mu\rangle\langle\mu|\lambda_b|\text{val}\rangle}{e_\mu - e_{\text{val}}}. \tag{45}$$

Isospin invariance demands that the matrix for the moment of inertia, Θ_{ab}^{val} is diagonal and only two independent components, Θ_{33}^{val} and Θ_{44}^{val} corresponding to α^2 and β^2 respectively, exist. The same argument, of course, applies to the terms quadratic in ΔM contributing to γ_2 and γ_3 as well:

$$\Gamma_{ab}^{\text{val}} = \frac{2}{3}N_C(\Delta M)^2\eta^{\text{val}}\sum_{\mu\neq\text{val}}\frac{\langle\text{val}|\mathcal{T}\beta\lambda_a\mathcal{T}^\dagger|\mu\rangle\langle\mu|\mathcal{T}\beta\lambda_b\mathcal{T}^\dagger|\text{val}\rangle}{e_\mu - e_{\text{val}}} \tag{46}$$

and also to those linear in both Ω_a and ΔM:

$$\Delta_{ab}^{\text{val}} = \frac{N_C}{\sqrt{3}}\Delta M\eta^{\text{val}}\sum_{\mu\neq\text{val}}\frac{\langle\text{val}|\lambda_a|\mu\rangle\langle\mu|\mathcal{T}\beta\lambda_b\mathcal{T}^\dagger|\text{val}\rangle}{e_\mu - e_{\text{val}}}. \tag{47}$$

Δ_{ab}^{val} contributes to α_1 and β_1. In all three cases the 88-component vanishes since the classical hedgehog commutes with λ_8.

Next we turn to the evaluation of the vacuum contribution to the fermion determinant in presence of the perturbation h_P. As for the calculation of the static soliton solution we transform to Euclidean space and take the vacuum ($\xi = 1$) as reference:

$$\mathcal{A}_F = \text{Tr}\log\left(i\not{D}_E(\xi(\boldsymbol{x},t))\right) - \text{Tr}\log\left(i\not{D}_E(\xi = 1)\right). \tag{48}$$

In the flavor rotating frame we have

$$\bar{q}\not{D}_E q = \bar{q}'\not{D}'_E q', \qquad \text{with} \qquad i\not{D}'_E = \beta(\partial_\tau - (h_0 + h_{rot} + h_{SB})). \tag{49}$$

Attention has to be paid to the fact that in Euclidean space the angular velocities Ω_a are to be considered anti-Hermitean quantities.

Again it proves most convenient to treat real and imaginary parts separately. First we consider the real part in the flavor rotation frame:

$$\mathcal{A}_R = -\frac{1}{2}\int_{1/\Lambda^2}^{\infty}\frac{ds}{s}\text{Tr}\exp(-s\not{D}'^\dagger_E\not{D}'_E). \tag{50}$$

This quantity is expanded up to second order in h_P. Here we will confine ourselves to the presentation of the resulting coefficients in the collective Lagrangian. Their derivation is presented in appendix C of ref.[26].

The underlying assumption of all these calculations is that of a strictly adiabatic rotation. Then we recover the old result for the moments of inertia:

$$\Theta_{ab}^{vac} = \frac{N_C}{4}\sum_{\mu\nu} f_\Theta(e_\mu, e_\nu; \Lambda)\langle\mu|\lambda_a|\nu\rangle\langle\nu|\lambda_b|\mu\rangle \tag{51}$$

where the cut-off function $f_\Theta(e_\mu, e_\nu; \Lambda)$ is given by

$$f_\Theta(e_\mu, e_\nu; \Lambda) = \frac{1}{\Lambda\sqrt{\pi}}\frac{e^{-E_\mu^2} - e^{-E_\nu^2}}{E_\nu^2 - E_\mu^2} - \frac{2}{\Lambda}\frac{1}{E_\mu - E_\nu}[\text{sign}(E_\nu)\mathcal{N}_\mu - \text{sign}(E_\mu)\mathcal{N}_\nu] \tag{52}$$

with $E_\mu = e_\mu/\Lambda$. Due to different up and strange constituent masses the strange moment of inertia Θ_{44}^{vac} does not vanish for $\xi = 1$ and the corresponding subtraction (48) has to be performed.

The expansion in ΔM yields a linear contribution

$$\gamma^{vac} = -\frac{2N_C\Delta M}{\sqrt{3}}\sum_\mu \text{sign}(e_\mu)\mathcal{N}_\mu\langle\mu|\mathcal{T}\beta\lambda_8\mathcal{T}^\dagger|\mu\rangle \tag{53}$$

wherein after subtracting the $\xi = 1$ result only the sum over non-strange states remains. The term quadratic in ΔM reads

$$\Gamma_{ab}^{vac} = \frac{N_C}{3}(\Delta M)^2\sum_{\mu\nu} f_\Gamma(e_\mu, e_\nu; \Lambda)\langle\mu|\mathcal{T}\beta\lambda_a\mathcal{T}^\dagger|\nu\rangle\langle\nu|\mathcal{T}\beta\lambda_b\mathcal{T}^\dagger|\mu\rangle \tag{54}$$

with the cut-off function

$$f_\Gamma(e_\mu, e_\nu; \Lambda) = \frac{\text{sign}(e_\mu)\mathcal{N}_\mu - \text{sign}(e_\nu)\mathcal{N}_\nu}{e_\mu - e_\nu}. \tag{55}$$

The expansion of the real part of the action does not provide any term linear in the angular velocities. This may be understood by noting that an expansion of $\mathcal{A}_R$ only yields terms even in time derivatives. However, these linear terms are provided by the imaginary part since it is known from the derivative expansion of $\mathcal{A}_I$ that only odd powers of the time derivative operator appear.

In contrast to the real part of the Euclidean action the imaginary part, which in the flavor rotating frame reads:

$$\mathcal{A}_I = \frac{1}{2}\text{Tr}\log\left((\not{D}_E'^\dagger)^{-1}\not{D}_E'\right) \tag{56}$$

vanishes for $\xi = 1$ and therefore no vacuum subtraction is needed. We get contributions from $\mathcal{A}_I$ to the baryon number

$$B^{vac} = \sum_\mu \text{sign}(e_\mu)\mathcal{N}_\mu \tag{57}$$

and to the symmetry breaker linear in ΔM:

$$\Delta_{ab}^{\rm vac} = \frac{-N_C}{2\sqrt{3}}\Delta M \sum_{\mu\nu} f_\Gamma(e_\mu, e_\nu; \Lambda)\langle\mu|\lambda_a|\nu\rangle\langle\nu|\mathcal{T}\beta\lambda_b\mathcal{T}^\dagger|\mu\rangle. \tag{58}$$

Let us perform a small detour to Skyrme type, *i.e.* purely mesonic, models. There it is well known that kaon fields are induced by the collective rotation into strange direction. Parametrizing[33]

$$\xi(\boldsymbol{x}, t) = R(t)\xi_k\xi(\boldsymbol{x})\xi_k R^\dagger(t); \ \xi_k = \exp(iZ) \tag{59}$$

the induced kaon fields are contained in Z:

$$Z = \begin{pmatrix} 0 & K \\ K^\dagger & 0 \end{pmatrix} \quad , \qquad K = W(r)\sum_{i=1}^{3}(\hat{x}_i\lambda_i)\Omega_K. \tag{60}$$

The strange moment of inertia, β^2, turns out to be a functional of the radial function $W(r)$. The inclusion of the Wess-Zumino term Γ_{WZ} is important in this context since it provides the source terms for $W(r)$. In the Skyrme model it is found[33] that the induced components contribute about 50% to the strange moment of inertia which demonstrates that these fields may play a crucial role. Of course, these excitations are also present in the NJL model. A complete treatment would provide an integral equation for $W(r)$ as the stationary condition. This, however, is a very tedious task since it involves complicated matrix elements of ξ_k. We therefore approximate the influence of the induced fields by employing the derivative expansion. That is, we include all terms up to second order in the derivatives, especially also the terms which splits pion and kaon decay constants. The *ansatz* (59,60) is then substituted yielding the strange moment of inertia a functional of $W(r)$: $\beta_I^2[W]$. This represents a space integral over linear and quadratic expressions in $W(r)$. The corresponding coefficient functions involve the chiral angle for which the self-consistent static soliton is substituted.

Finally we substitute our ansatz for the rotating meson fields (38) into the expression for the mesonic part of the action, $\mathcal{A}_m$ (2) and subtract the $\xi = 1$ contribution. This may be done easily and gives additional contributions to the symmetry breaking parameters $\gamma, \gamma_1, \gamma_2$ and γ_3.

As in Skyrme type models Noether charges corresponding to right $SU(3)$ transformations may be constructed leading to the quantization prescription for the right $SU(3)$ generators, R_a

$$R_a = -\frac{\partial L}{\partial \Omega_a} = \begin{cases} -(\alpha^2\Omega_a + \alpha_1 D_{8a}) = -J_a, & \text{a=1,2,3} \\ -(\beta^2\Omega_a + \beta_1 D_{8a}), & \text{a=4,..,7} \\ \frac{\sqrt{3}}{2}B, & \text{a=8} \end{cases} \tag{61}$$

wherein J_i $(i = 1, 2, 3)$ denote the spin operators.

The Hamiltonian operator:

$$H = -\sum_{a=1}^{8} R_a \Omega_a - L \tag{62}$$

may be diagonalized exactly which is done along the lines of the original Yabu-Ando approach yielding the energy expression for baryon B:

$$E_B = E + \frac{1}{2}\left(\frac{1}{\alpha^2} - \frac{1}{\beta^2}\right) J(J+1) - \frac{3}{8\beta^2} + \frac{1}{2\beta^2}\epsilon_{SB}, \tag{63}$$

wherein ϵ_{SB} is the eigenvalue of

$$\begin{aligned} C_2 &+ \beta^2\gamma(1 - D_{88}) \\ &+ \beta^2(\alpha_1/\alpha^2)\sum_{i=1}^{3} D_{8i}(2R_i + \alpha_1 D_{8i}) + \beta_1 \sum_{\alpha=4}^{7} D_{8\alpha}(2R_\alpha + \beta_1 D_{8\alpha}) \\ &+ \beta^2\gamma_1(1 - D_{88}^2) + \beta^2\gamma_2 \sum_{i=1}^{3} D_{8i}D_{8i} + \beta^2\gamma_3 \sum_{\alpha=4}^{7} D_{8\alpha}D_{8\alpha}. \end{aligned} \tag{64}$$

$C_2 = \sum_{a=1}^{8} R_a^2$ denotes the quadratic Casimir operator of $SU(3)$. The eigenvalues ϵ_{SB} are determined using a generalized YA approach[‡](*c.f.* ref.[32]).

The constraint for the right hypercharge $Y_R = \frac{2}{\sqrt{3}}R_8 = B$ confines the possible eigenstates to those which carry half integer spin for $B = 1$, *i.e.* fermions.

6.2. *Numerical results*

We find that the valence quark dominates the vacuum contribution to both the non-strange moment of inertia α^2 as well as the strange moment of inertia β^2. For the latter we also obtain a significant contribution due to the induced kaon fields. This result compares with the findings in the Skyrme model.[33] We would like to stress that for $M_s \neq M_u$ the vacuum contribution to the strange moment of inertia Θ_{44}^{vac} becomes a non-diverging quantity only after subtracting the reference value $\Theta_{44}^{vac}(\xi = 1)$. Actually we find that the subtracted value decreases as the difference $M_s - M_u$ increases.

For the coefficients of the expressions linear in the angular velocities Δ_{ab} the dominance of the valence quarks is even more drastically pronounced and the contribution due to the polarized Dirac sea is almost negligible. For the relevant range of constituent masses we only find a moderate dependence of Δ_{ab}^{vac} on the cut-off Λ. This is not surprising since the imaginary part of the Euclidean action stays finite as $\Lambda \rightarrow \infty$. The linear terms are found to have a significant influence on the energy eigenvalues of the baryons. Since both, α_1 as well as β_1 turn out be negative the linear terms increase the $SU(3)$ symmetry breaking. In ref.[25] these linear terms have been incorporated via an expansion in the angular velocity as well as the strange current mass m_s^0 without

‡Yabu and Ando only considered $C_2 + \beta^2\gamma(1 - D_{88})$.

Table 4. The mass differences for the low-lying $\frac{1}{2}^+$ and $\frac{3}{2}^+$ baryon states as functions of the up-constituent mass M_u. All numbers are in MeV.

M_u	350.	400.	450.	500.	Expt.
M_N	1684.	1725.	1726.	1717.	938.
$M_\Lambda - M_N$	149.	116.	94.	78.	177.
$M_\Sigma - M_N$	182.	157.	134.	116.	254.
$M_\Xi - M_N$	320.	256.	210	177.	379.
$M_\Delta - M_N$	209.	296.	350.	392.	293.
$M_{\Sigma^*} - M_N$	350.	401.	433.	460.	446.
$M_{\Xi^*} - M_N$	493.	506.	515.	526.	591.
$M_\Omega - M_N$	637.	611.	596.	592.	733.

performing the corresponding shift in the scalar field (*c.f.* section 2). Of course, in an evaluation to all orders this approach should be compatible to ours. However since (at least in the proper time regularization scheme) we find $M_s - M_u \ll m_s^0$ we expect the expansion in $M_s - M_u$ to be more reasonable.

In the context of symmetry breaking on the effective Lagrangian level γ - the coefficient of $(1 - D_{88})$ - turns out to be most important. The deviation of M_s/M_u from unity increases γ^m by about 25%[26] compared to our earlier results[24] where the approximation of identical constituent masses was made. This increase is compensated by the decrease of the multiplying integral due to the introduction of the finite pion mass in the equation of motion for the chiral soliton. While the contribution of the valence quarks to γ is large and positive the polarized Dirac tends to cancel this effect.

Finally we would like to note that the symmetry breakers involving the non-diagonal matrix elements of the $SU(3)$ rotations ($\sum_{i=1}^{3} D_{8i}D_{8i}$ and $\sum_{\alpha=4}^{7} D_{8\alpha}D_{8\alpha}$) only play a minor role for the baryon mass differences but nevertheless they will be included for completeness and consistency.

Now we are ready to discuss our predictions for the baryon mass spectrum which are displayed in table 4. As in almost all types of soliton models the result for the absolute value of the masses turns out to be too large when the physical value for f_π is used. However, our result is already several hundred MeV lower than that in the Skyrme model.[33,34] In the two flavor Skyrme model it has been demonstrated[35,36,37] that quantum fluctuations of the classical soliton significantly lower the absolute energy. It is important to mention that the dominant subtractions are of $O(N_C^0)$ and therefore identical for nucleon and Δ. Thus it seems appropriate to concentrate on mass differences only.

As may been seen from table 4 the mass splittings between states of different spins decrease with increasing constituent mass. This is mainly linked to the decrease of α^2 with M_u. On the contrary the mass differences between members of the same spin multiplet get smaller as the constituent mass gets larger since also γ decreases with M_u. As mentioned above, allowing for $m_\pi \neq 0$ in the evaluation of the classical

Table 5. The mass differences for the estimate described in the text. The particles' names refer to the corresponding mass difference with respect to the nucleon. All numbers are in MeV.

$M_u = 390MeV$	Λ	Σ	Ξ	Δ	Σ^*	Ξ^*	Ω
calculated	121.	162.	267.	283.	394.	504.	615.
$\gamma_{\rm rescaled}$	175.	248.	396.	291.	449.	608.	765.
Expt.	177.	254.	379.	293.	446.	591.	733.

soliton lowers α^2 and therefore increases the mass splitting of the $s = \frac{1}{2}^+$ and $s = \frac{3}{2}^+$ multiplets. Note that consistency requires to use a non-vanishing current mass, *i.e.* a non-vanishing pion mass, in the meson as well as in the baryon sector.

We furthermore recognize form table 4 that the mass splittings between baryons of the same spin are predicted too low. This shortcoming is not completely unexpected since already in the meson sector of the model we have found that *e.g.* for $M_u = 400MeV$ f_K is about 15% lower than the experimental value (cf. table 1). That this deficiency transfers to the baryon sector may be seen from the following estimate. We choose $M_u = 390MeV$ to reproduce the experimental nucleon-Δ mass difference. Then we scale the dominant symmetry breaking parameter $\gamma^m + \gamma^{vac}$ by $(\frac{f_K^{expt.}}{f_K^{pred.}})^2 = (\frac{114}{100})^2$ to calculate the baryon mass differences. The valence quark contribution to γ should be described only by the constituent masses but not by the predicted value for f_K. The results stemming from this estimate are shown in table 5 and are found to be in excellent agreement with experimental data.

Thus we may conclude that the too small predictions for the baryon mass splittings are directly connected to the too small result for f_K. We furthermore expect that a more complete model (*i.e.* (axial-) vector mesons included) which reproduces the experimental value for f_K will in fact yield agreement for the baryon mass differences.

7. Conclusions

The main conclusions of this talk may be summarized as follows:

- The exact evaluation of the quark determinant for the baryon number one soliton is possible. No derivative or gradient expansion is needed in order to investigate stability/instability of the soliton.

- Restriction of the scalar meson field to its vacuum expectation value yields stable solitons; without this restriction the soliton is unstable if the original NJL model is used. If this model is supplemented with a Φ^4 term the soliton becomes stable again. Even more, for a reasonable size of this term it is only slightly different from the soliton with constraint on the chiral circle. There are several possible origins for such a term. The physically best motivated one is the generalization of the NJL model such that the scale anomaly of QCD is described on an effective level.

- The inclusion of the vector meson ρ and the axialvector meson a_1 leads to a valence quark level which has joined the Dirac sea. Thus **the NJL soliton supports Witten's conjecture, *i.e.* the Skyrmion picture of the baryon.**

- The isoscalar vector meson ω can only be included if the imaginary part of the Euclidian action is taken into account. Despite the fact that the original problem is formulated in Minkowski space regularization enforces the continuation back and forth to Euclidian space. The results obtained so far do not indicate that the above conclusion concerning the Skyrmion picture of the baryon is in jeopardy. A full calculation including π, ω, ρ and a_1 is in progress.

- The collective approach to the SU(3) chiral soliton gives reasonable results for the mass splittings of the low–lying $\frac{1}{2}^+$ and $\frac{3}{2}^+$ baryons. The agreement is even excellent after the kaon decay constant has been rescaled to its physical value.

Acknowledgements

I would like to thank Prof. Holzwarth for his warm hospitality and the other participants of the workshop for the stimulating atmosphere and a lot of helpful discussions.

Furthermore, I thank my colleagues H. Reinhardt, H. Weigel, C. Weiss and U. Zückert who participated in parts of the work underlying this talk. Helpful contributions to this manuscript by H. Weigel and U. Zückert are gratefully acknowledged.

This work is supported by the Deutsche Forschungsgemeinschaft (DFG) under contract Re 856/2-1.

References

1. H. Reinhardt, these proceedings.

2. Y. Nambu and G. Jona-Lasinio, *Phys. Rev.* **122** (1961) 345; **124** (1961) 246.

3. D. Ebert and H. Reinhardt, *Nucl. Phys.* **B271** (1986) 188.

4. M. Schaden, H. Reinhardt, P. A. Amundsen and M. J. Lavelle, *Nucl. Phys.* **B339** (1990) 595;

   R. Alkofer and H. Reinhardt, *Z. Phys.* **A343** (1992) 79.

5. H. Reinhardt and R. Alkofer, *Phys. Lett.* **207B** (1988) 482;

   R. Alkofer and H. Reinhardt, *Z. Phys.* **C45** (1989) 245.

6. V. Bernard, R. L. Jaffe and Ulf-G. Meißner, *Nucl. Phys.* **B308** (1988) 753.

   V. Bernard, Ulf-G. Meißner, A. H. Blin and B. Hiller, *Phys. Lett.* **B253** (1991) 443.

7. S. Klimt, M. Lutz, U. Vogl and W. Weise, *Nucl. Phys.* **A516** (1990) 429.

8. I. Aitchinson, C. Frazer, E. Tador and J. Zuk, *Phys. Lett.* **B165** (1985) 163.

9. H. Reinhardt and D. Ebert, *Phys. Lett.* **B215** (1986) 459.

10. H. Reinhardt and R. Wünsch, *Phys. Lett.* **B215** (1988) 577.

11. T. Meissner, F. Grümmer and K. Goeke, *Phys. Lett.* **B227** (1989) 296.

12. R. Alkofer, *Phys. Lett.* **B236** (1990) 310.

13. R. Alkofer and H. Reinhardt, *Phys. Lett.* **B244** (1991) 461.

14. R. Alkofer, H. Reinhardt, H. Weigel and U. Zückert, *Phys. Rev. Lett.* **69** (1992) 1874.

15. C. Schüren, E. Ruiz–Arriola and K. Goeke, *Phys. Lett.* **B287** (1992) 283.

16. R. Alkofer, H. Reinhardt, H. Weigel and U. Zückert, "The isoscalar vector meson ω in the Nambu–Jona-Lasinio soliton", Tübingen preprint August 1992; submitted to Phys. Lett. **B**.

17. P. Sieber, T. Meißner, F. Grümmer and K. Goeke, *Nucl. Phys.* **A547** (1992) 459.

18. T. Meißner et al., "Scale Invariance and the Stability of a Hedgehog Soliton", Bochum preprint RUB-TPII-30/92, Sept. 1992.

19. C. Weiss, H. Weigel and R. Alkofer, "The Nambu–Jona-Lasinio soliton with generalized scalar interactions", Tübingen preprint, October 1992.

20. H. Reinhardt, *Nucl. Phys.* **A503** (1989) 825.

21. K. Goeke et al., *Phys. Lett.* **B256** (1991) 321.

22. M. Wakamatsu and H. Yoshiki, *Nucl. Phys.* **A524** (1991) 561.

23. H. Yabu and K. Ando, *Nucl. Phys.* **B301** (1988) 601.

24. H. Weigel, R. Alkofer and H. Reinhardt, *Phys. Lett.* **B284** (1992) 296.

25. A. Blotz et al., *Phys. Lett.* **B287** (1992) 29.

26. H. Weigel, R. Alkofer and H. Reinhardt, "Strange Baryons as Chiral Solitons of the Nambu–Jona-Lasinio model", *Nucl. Phys.* **B**, in press.

27. J. Schwinger, *Phys. Rev.* **82** (1951) 664.

28. U.-G. Meißner, N. Kaiser and W. Weise, *Nucl. Phys.* **A466** (1987) 685.

29. G. Eckart, A. Hayashi and G. Holzwarth, *Nucl. Phys.* **A448** (1986) 732;

B. Schwesinger, H. Weigel, G. Holzwarth and A. Hayashi, *Phys. Rep.* **173** (1989) 173.

30. J. Schechter and H. Weigel, *Phys. Lett.* **B261** (1991) 235; *Phys. Rev.* **D44** (1991) 2916.

31. B. Schwesinger, H. Weigel, *Phys. Lett.* **B267** (1991) 438; *Nucl. Phys.* **A540** (1992) 461.

32. N. W. Park and H. Weigel, *Phys. Lett.* **B268** (1991) 155; *Nucl. Phys.* **A541** (1992) 453.

33. H. Weigel, J. Schechter N. W. Park and Ulf-G. Meißner, *Phys. Rev.* **D43** (1991) 869.

34. G. Pari, B. Schwesinger and H. Walliser, *Phys. Lett.* **B255** (1991) 1.

35. I. Zahed, A. Wirzba and Ulf-G. Meißner, *Phys. Rev.* **D33** (1986) 830;

A. Dobado and J. Terron, *Phys. Lett.* **B247** (1990) 581.

36. B. Moussallam and D. Kalafatis, *Phys. Lett.* **B272** (1991) 196;

B. Moussallam, these proceedings.

37. G. Holzwarth, these proceedings.

STRANGENESS IN SKYRME AND IN NAMBU–JONA-LASINIO MODELS

M. PRASZAŁOWICZ*
Institute for Theor. Physcs II, Ruhr-University
4630 Bochum, Germany

ABSTRACT
We discuss the phenomenology of SU(3) Skyrme and Nambu–Jona-Lasinio models. We show that while the Skyrme model, in its simplest version, is unable do reproduce the hyperon spectrum, the NJL model describes both hyperon and isospin splittings with satisfactory accuracy. The difference between the two models lies in new *anomalous* moments of inertia, which vanish in the Skyrme model and get non-zero contribution from the *valence* quarks in the NJL model

1. Introduction

Although QCD is now commonly accepted as the ultimate theory of strong interactions, the low energy properties of the hadronic states are usually calculated in terms of various *effective* models, which are believed to follow from QCD. Since there is however no rigorous direct derivation (see *e.g.*Ref.[1]), one has some freedom in choosing ones preferable model and preferable lagrangian.

In this talk we will discuss the phenomenology of the SU(3) Skyrme model[1–5] and the Nambu–Jona-Lasinio (NJL) model.[6–9] In order to make the discussion transparent we will stick to the simplest versions of the models; the more refined versions are discussed here by other speakers. Our attention will be focused on the splittings both in SU(3) multiplets and in isospin multiplets. We will not consider absolute masses which come out always too high.

We will see that in the simplest version of the Skyrme model the pattern of splittings is generically wrong: $\Lambda - N = \Sigma - \Lambda = 2\,(\Xi - \Sigma)$ in contradiction with experimental situation where $\Lambda - N \approx \Xi - \Sigma > \Sigma - \Lambda$. To cure this disease one usually escapes to more complicated lagrangians and, at the same time, one applies diagonalization procedure of Yabu and Ando.[10] Although, as we will see on the example of the explicit perturbative calculation, this method shifts the spectrum in the right direction, one has to express a criticism against it, as it sums up an arbitrary subseries in the strange quark masses, neglecting other terms of the same order.

In the semibosonised NJL model (see *e.g.* Ref.[11,12]), in which quarks interact *via* a self-consistent meson field, the spectrum is satisfactorily reproduced[6,7] in the linear order in m_s. In the NJL model the energy gets contribution from the *valence* and *sea* quarks. The *classical* part, *i.e.* the energy of the soliton, is exactly the same as in the two flavor case. The *quantum* corrections are calculated by adiabatical rotation of the soliton resulting in a hamiltonian analogous to the one of the Skyrmion. A novelty is

*Alexander von Humboldt Fellow, on leave of absence from the Institute of Physics, Jagellonian University, ul. Reymonta 4, 30-059 Kraków. Poland

due to the mixed terms linear in the current quark mass and in the rotational velocity. These terms vanish in an ordinary Skyrme model; in the present model they get main contribution from the *valence* part. The resulting spectrum fits the data with a 10 % accuracy. At the same time the isospin splittings due to the $m_d - m_u$ mass difference are reproduced within experimental errors.[9] Keeping in mind the simplicity of the starting lagrangian these results are surprisingly accurate.

2. Gell-Mann–Okubo Mass Formulae

Any dynamical model of light baryons has to reproduce and also explain the Gell-Mann–Okubo[13,14] mass formulae which are derived assuming that the SU(3) breaking mass operator ΔM transforms like a $Y = 0$, $I = 0$ and $I_3 = 0$ component of the octet tensor operator. Then, due to the Wigner–Eckhart theorem, matrix elements of this operator are given by:

$$\Delta M_B^{(8)} = F \begin{pmatrix} 8 & 8 & 8_- \\ 000 & B & B \end{pmatrix} + D \begin{pmatrix} 8 & 8 & 8_+ \\ 000 & B & B \end{pmatrix} \tag{1}$$

for the octet, and

$$\Delta M_B^{(10)} = C \begin{pmatrix} 8 & 10 & 10 \\ 000 & B & B \end{pmatrix} \tag{2}$$

for the decuplet. The reduced matrix elements F, D and C are free constants. $B = Y, I, I_3$ for the baryon in question. The SU(3) Clebsch–Gordan coefficients can be written in terms of the diagonal SU(3) operators:[15]

$$\Delta M_B^{(8)} = -\frac{F}{2}\mathbf{Y} - \frac{D}{\sqrt{5}}(1 - \mathbf{I}^2 + \frac{1}{4}\mathbf{Y}^2), \qquad \Delta M_B^{(10)} = -\frac{C}{2\sqrt{2}}\mathbf{Y}. \tag{3}$$

The predictive power of Eq.(3) consists in the fact, that the number of free parameters, which could in principle parametrize the mass splittings, is reduced from 3 to 2 for the octet and to 1 for the decuplet.

Equations (3) yield the relations:

$$F = M_\Xi - M_N, \tag{4}$$

$$\frac{1}{\sqrt{5}}D = \frac{1}{2}(M_\Sigma - M_\Lambda) = \frac{1}{3}(2M_\Sigma - M_\Xi - M_N) = M_\Xi + M_N - 2M_\Lambda,$$

for the octet, and equal level spacing for the decuplet:

$$\frac{1}{2\sqrt{2}}C = M_{\Sigma^*} - M_\Delta = M_{\Xi^*} - M_{\Sigma^*} = M_\Omega - M_{\Xi^*}. \tag{5}$$

From these relations we can estimate the values of parameters F, D and C:

$$F = 379, \quad D = 79 \pm 17 \text{ and } C = 415 \pm 15 \quad \text{MeV}. \tag{6}$$

The resulting spectrum is presented in the first column of Fig.1. A small admixture of other operators like $\mathbf{Y}^2$ shifts the spectrum to the experimental position.

The mass splittings of the baryons belonging to the same isospin multiplet consist of two parts: hadronic and electromagnetic,[16] namely:

$$\Delta m_B = (\Delta m_B)_{\rm h} + (\Delta m_B)_{\rm e}. \tag{7}$$

If the isospin breaking is assumed to be driven by an octet isovector tensor operator corresponding to $I_3 = 0$ then, in analogy to the Gell-Mann–Okubo mass formulae (1,2), one gets:

$$(\Delta m_B^{(8)})_{\rm h} = \frac{1}{\sqrt{3}} f \begin{pmatrix} 8 & 8 & 8_- \\ 010 & B & B \end{pmatrix} + \sqrt{\frac{5}{3}} d \begin{pmatrix} 8 & 8 & 8_+ \\ 010 & B & B \end{pmatrix} \tag{8}$$

for the octet, and

$$(\Delta m_B^{(10)})_{\rm h} = \sqrt{\frac{2}{3}} c \begin{pmatrix} 8 & 10 & 10 \\ 010 & B & B \end{pmatrix} \tag{9}$$

for the decuplet (normalization factors in front of the f, d and c are chosen for future convenience). Evaluating the SU(3) Clebsch-Gordan coefficients gives:[15]

$$(\Delta m_B^{(8)})_{\rm h} = -\frac{1}{3} f\, \mathbf{I}_3 + d\, \mathbf{Y}\, \mathbf{I}_3, \qquad (\Delta m_B^{(10)})_{\rm h} = -\frac{1}{3} c\, \mathbf{I}_3. \tag{10}$$

Electromagnetic part of the isospin splittings was estimated by Gasser and Leutwyler[16] for the octet. Their estimate confirms a reasonable assumption that $\Sigma^- - \Sigma^+$ mass difference has no electromagnetic contribution. This assumption allows us to determine coefficient f, and also c for the decuplet, where no estimate of the electromagnetic part of $\Sigma^{*-} - \Sigma^{*+}$ exists. Coefficient d can be determined from the hadronic part of the n–p mass difference.[9] Altogether we get:

$$f = 12.11 \pm 1.14\ , \quad d = 1.73 \pm 0.38 \quad \text{and} \quad c = 6.6 \pm 1.0 \quad \text{MeV}. \tag{11}$$

Symmetry considerations alone are not able to provide us with any relations between the reduced matrix elements. Dynamical models, like Skyrme model or NJL model, make specific predictions for these constants. In the next sections we will calculate coefficients F, D, C, f, d, and c within the framework of the simplest version of the SU(3) Skyrme model and subsequently in the simplest version of the semibosonised SU(3) NJL model.

3. SU(3) Skyrme Model

Let us start by specifying the effective lagrangian proposed by Skyrme[17,18] and later generalized by Witten:[19,20]

$$\begin{aligned} \int dt\, L_{\rm Sk} \;=\;& \frac{F_\pi^2}{16} \int dt\, d^3r\, {\rm Tr}(\partial_\mu U^\dagger \partial^\mu U) + \frac{1}{32e^2} \int dt\, d^3r\, {\rm Tr}([\partial_\mu U\, U^\dagger, \partial_\nu U\, U^\dagger]^2) \\ &+ \int dt\, d^3r\, {\rm Tr}\left\{a\, \{U + U^\dagger - 2\} + b\, \{(U + U^\dagger)\lambda_8\}\right\} + N_c \Gamma_{\rm WZ}. \end{aligned} \tag{12}$$

Here N_c is a number of colors, parameters F_π and e, if taken from meson physics, are equal to $F_\pi \sim 186$ MeV and $e \sim 5.5$ respectively. Γ_{WZ} denotes the Wess-Zumino term[21,22] and

$$a = \frac{F_\pi^2}{32}(m_\pi^2 + m_\eta^2), \quad b = \frac{\sqrt{3}F_\pi^2}{24}(m_\pi^2 - m_K^2). \tag{13}$$

A *time-independent* Anzatz for the SU(3) matrix U_0 takes the form of a *hedgehog*:

$$U_0 = \begin{bmatrix} \cos P(r) + i\vec{n}\vec{\tau}\sin P(r) & & \begin{matrix}0\\ \\ 0\end{matrix} \\ 0 \quad & 0 & 1 \end{bmatrix}, \tag{14}$$

where $P(r)$ is a profile function. For the purpose of this talk we will take

$$P(r) = 2\arctan\left(\frac{r}{r_0}\right)^2 \tag{15}$$

and calculate analytically all relevant quantities as functions of the soliton size r_0. The energy of the solution (14) is then given by:[23]

$$M_{\rm cl} = \frac{F_\pi}{e}\pi^2\frac{3\sqrt{2}}{16}(4x_0 + \frac{15}{x_0}) + \frac{\mu^2}{e^3F_\pi}\pi^2\frac{\sqrt{2}}{2}x_0^3. \tag{16}$$

with

$$\mu^2 = \frac{m_\pi^2 + 2m_K^2}{3} \approx (412 \ \ \text{MeV})^2, \tag{17}$$

where $x_0 = eF_\pi r_0$.

Minimizing Eq.(16) with respect to x_0 we find in the chiral limit ($\mu^2 = 0$) $x_0 = \sqrt{15/4}$ and $M_{\rm cl} = 40.54\ F_\pi/e$ *i.e.* approximately 1370 MeV. This certainly unsatisfactory result (going off chiral limit shifts $M_{\rm cl}$ further up by about 280 MeV) is common for all chiral models which tend to overestimate the classical soliton mass. There is a hope that various other effects like gluonic corrections[24] or Casimir effect[25,26] may bring this value down to less than 1 GeV. In what follows we will abandon the idea to fit the absolute masses, instead we will concentrate on the mass splittings.

The splittings are calculated by rotating the static solution

$$U_0 \to A(t)\, U_0\, A^\dagger(t), \tag{18}$$

where $A \in SU(3)/U(1)$, since $[\lambda_8, U_0] = 0$. Therefore matrix A is defined up to a *local* U(1) factor $h = \exp(i\lambda_8\phi)$,*i.e.* A and Ah are equivalent. This leads to a constraint which has to be imposed on the physical spectrum. Introducing 8 collective coordinates:[2–4]

$$A^\dagger(t)\,\frac{d}{dt}A(t) = \frac{i}{2}\sum_{\alpha=1}^{8}\lambda_\alpha\Omega_\alpha; \tag{19}$$

one gets the following quantum mechanical lagrangian:[2-4]

$$L = -M_{\rm cl}[P] + \frac{I_A[P]}{2}\sum_{i=1}^{3}\Omega_i^2 + \frac{I_B[P]}{2}\sum_{k=4}^{7}\Omega_k^2 + \frac{N_c}{2\sqrt{3}}\Omega_8 + \Delta m[A,P], \tag{20}$$

where $\Delta m[A,P]$ corresponds to the symmetry breaking piece. The moments of inertia can be again calculated analytically:[23]

$$\begin{aligned} I_A &= \frac{1}{e^3F_\pi}\pi^2\frac{\sqrt{2}}{12}\left(6x_0^3+25x_0\right) &\approx \frac{107}{e^3F_\pi}, \\ I_B &= \frac{1}{e^3F_\pi}\pi^2\frac{\sqrt{2}}{16}\left(4x_0^3+9x_0\right) &\approx \frac{40.5}{e^3F_\pi}. \end{aligned} \tag{21}$$

The standard quantization procedure[2-4] leads to the following hamiltonian:

$$M_{\rm B} = M_{\rm cl} + H_{\rm SU(2)} + H_{\rm SU(3)} + H_{\rm br}, \tag{22}$$

with

$$H_{\rm SU(2)} = \frac{C_2({\rm SU(2)_R})}{2I_A}, \qquad H_{\rm SU(3)} = \frac{C_2({\rm SU(3)_L}) - C_2({\rm SU(2)_R}) - \frac{N_c^2}{12}}{2I_B}, \tag{23}$$

where C_2 denotes the quadratic Casimir operator of the *right* SU(2) symmetry corresponding to spin and of the *left* SU(3) corresponding to flavor.

The wave function of the baryon state can be written as an SU(3) rotation matrix:

$$\psi_{\rm B}(A) = \sqrt{\dim(p,q)}\, D^{(p,q)}_{B\,S}(A) = \sqrt{\dim(p,q)}\left\langle Y, I, I_3 \mid D^{(p,q)}(A) \mid Y_{\rm R}, S, -S_3\right\rangle, \tag{24}$$

where quantum numbers B and S denote now hypercharge, isospin and its third component, and right hypercharge, spin and its third component (with minus sign), respectively. SU(3) representations are labeled by (p,q), however not all p and q are allowed. The system is constrained since lagrangian (20) does not contain terms quadratic in the 8-th velocity. The constraint $Y_{\rm R} = N_c/3$ selects the representations of triality zero:

$$8,\ 10,\ \overline{10},\ 27,\ 35,\ \overline{35},\ 64,\ \ldots \tag{25}$$

for $N_c = 3$. The success of the model is the prediction that the lowest baryonic states belong to the octet and decuplet representations of SU(3).

The symmetry breaking Hamiltonian:

$$H_{\rm br} = \alpha\, D^{(8)}_{8,8}(A) \tag{26}$$

splits the hyperon spectrum (here representation (1,1) is denoted as (8) and index 8 corresponds to $Y=0$ and $I=0$). Constant α is given as:[23]

$$\alpha = -\frac{\Delta\mu^2}{e^3F_\pi}\pi^2\frac{\sqrt{2}}{2}x_0^3 \approx -\frac{7.7\ {\rm GeV}^2}{e^3F_\pi}, \tag{27}$$

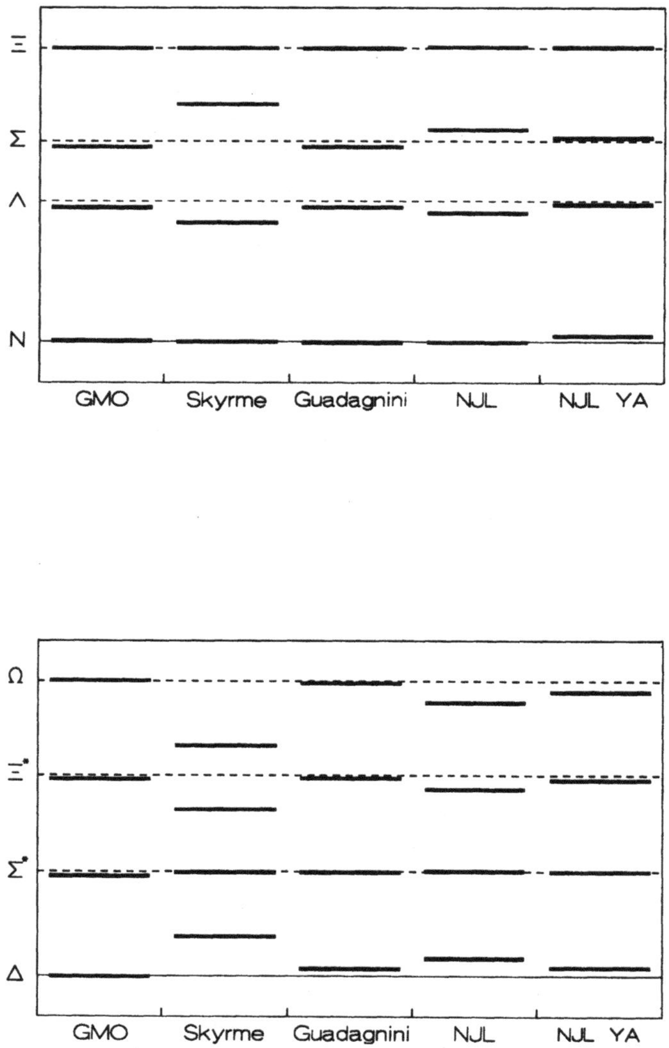

Fig. 1. Spectrum of light baryons. Dashed lines represent experimental data, solid lines correspond to theoretical predictions described in the text.

where we have used

$$\Delta\mu^2 = \frac{2}{3}(m_K^2 - m_\pi^2) \approx (388 \ \ \mathrm{MeV})^2. \tag{28}$$

The first and probably the only good news about the breaking hamiltonian (26) is that it automatically fulfills the Gell-Mann–Okubo mass relation, since the *left* SU(3) Clebsch-Gordan coefficients coincide with the ones of Eqs(1, 2). The *right* C-G coefficients give specific predictions for the reduced matrix elements:

$$F = -\frac{1}{2}\,\alpha, \qquad D = -\frac{1}{2\sqrt{5}}\,\alpha, \qquad C = -\frac{1}{2\sqrt{2}}\,\alpha. \tag{29}$$

The spectrum however looks rather odd since the two ratios:

$$\frac{F}{D} = \sqrt{5} = 2.24 \ \ (\mathrm{exp.}\ \ 4.40), \qquad \frac{F}{C} = \sqrt{2} = 1.41 \ \ (\mathrm{exp.}\ \ 0.91) \tag{30}$$

are far from their experimental values. If we choose $\alpha = -758$ MeV in such a way that F is reproduced then we get spectrum shown in the second column of Fig.1. However α is not a free parameter. The 10–8 splitting (230 MeV) requires $e^3F_\pi = 16.41$ GeV. Through Eqs(21, 27) we get that $1/I_B = 405$ MeV and $\alpha \approx -470$ MeV, approximately 40 % smaller than the value required to fit F. One could in principle enlarge α by tuning the kaon mass, but the problem of Eq.(30) remains.

Guadagnini[2] in the first paper on the SU(3) quantization of the Skyrme model proposed to add *by hand* a term proportional to the hyphercharge:

$$H_{\mathrm{G}} = \alpha\ D^{(8)}_{8,8}(A) + \beta\,\mathbf{Y}. \tag{31}$$

β is a free parameter and the reduced matrix elements read:

$$F = -\frac{1}{2}\,\alpha - 2\,\beta, \quad D = -\frac{1}{2\sqrt{5}}\,\alpha, \quad C = -\frac{1}{2\sqrt{2}}(\,\alpha + 8\,\beta). \tag{32}$$

An excellent fit to the data (column 3 in Fig.1) is obtained with $\alpha = -382$ MeV and $\beta = -94$ MeV. This time α is smaller than the value required by 10–8 splitting, but only by about 20 %.

One way to generate *effective* β is to calculate the $O(\alpha^2)$ correction to baryon energy (this approach for not too large α is in fact equivalent to the diagonalization procedure of Yabu and Ando[10] which sums up the perturbative series in α):

$$\begin{aligned}
\Delta M_B^{(8)} &= \left(\frac{1}{4}\alpha - \frac{1}{60}\alpha^2 I_B\right)\mathbf{Y} + \left(\frac{1}{10}\alpha + \frac{2}{75}\alpha^2 I_B\right)\left(1 - \mathbf{I}^2 + \frac{1}{4}\mathbf{Y}^2\right) \\
&- \frac{1}{750}\,\alpha^2 I_B\,\mathbf{Y}^2 - \frac{47}{750}\,\alpha^2 I_B, \\
\Delta M_B^{(10)} &= \left(\frac{1}{8}\alpha - \frac{29}{672}\alpha^2 I_B\right)\mathbf{Y} - \frac{1}{168}\,\alpha^2 I_B\,\mathbf{Y}^2 - \frac{13}{168}\,\alpha^2 I_B.
\end{aligned} \tag{33}$$

Constants F, D and C can be immediately read off from Eqs(33), moreover new terms not present in the original Gell-Mann–Okubo mass formula (3) are generated. Numerically they are small and we will neglect them in the present discussion.

In order to fit F and D one needs $\alpha \approx -677$ MeV and $1/I_B = 378$ MeV; with these values one gets a bit too small $C = 327$ MeV. These values of α and I_B are again larger than the ones required by 10–8 splitting.

This was the status of the SU(3) Skyrme model in the mid eighties. A philosophical question whether one should trust perturbative expansion in the strange quark mass arose. Some people decided to work within an approach which breaks the SU(3) flavor symmetry already at the level of the Anzatz for the U_0 field,[27,28] others tried to push up α and I_B either by enlarging the original Skyrme lagrangian or by employing another model (like NJL model discussed in the next section).

4. Semibosonised SU(3) Nambu–Jona-Lasinio Model

It was already shown in the talk of Alkofer[12] that the solitonic solutions of the semibosonised NJL model are studied in terms of an effective action:

$$S_{\rm eff} = -{\rm Sp}\log(i\partial\!\!\!/ \; - \; m \; - \; M\cdot U^{\gamma_5}). \tag{34}$$

This time, however, U is an SU(3) matrix of Eq.(14). M is the constituent quark mass, which is in fact the only free parameter of the model. The bare quark mass matrix can be written in a form:

$$m = \mu_0\, \lambda_0 - \mu_8\, \lambda_8 - \mu_3\, \lambda_3, \tag{35}$$

where λ_i are Gell-Mann SU(3) matrices ($\lambda_0 = \sqrt{2/3}\; \mathbf{1}$) and

$$\mu_0 = \frac{1}{\sqrt{6}}(m_{\rm u} + m_{\rm d} + m_{\rm s}), \quad \mu_8 = \frac{1}{\sqrt{12}}(2\,m_{\rm s} - m_{\rm u} - m_{\rm d}), \quad \mu_3 = \frac{1}{2}(m_{\rm d} - m_{\rm u}). \tag{36}$$

The energy of the soliton consists of two parts: the energy of the *continuum*, *i.e.* the energy corresponding to the effective action (34), and the energy of the *valence* level. In what follows, for simplicity, we confine our discussion to the *continuum* part only, but one always has to remember that the pertinent *valence* contribution has to be added.

The effective action (34) can be rewritten in terms of the Euclidean spectral representation:[6,7]

$$S_{\rm eff} = -N_c T \int \frac{d\omega}{2\pi} {\rm Tr}\log\left(i\omega + H\right)\left[1 + \frac{1}{i\omega + H}(-i\gamma_4 A^\dagger m\, A + A^\dagger \dot{A})\right], \tag{37}$$

where H is the hermitean static hamiltonian: $H = \gamma_4(\gamma_i\partial_i + MU_0)$, and appropriate regularization is understood. The static soliton solution for H reduces to the one found in the SU(2) case. Formula (37) is already written in a form ready to be expanded in a power series in m and in generalized velocities Ω. Let us for the moment forget about the mass matrix m and expand (37) in powers of Ω. We get (back in Minkowski metric) familiar lagrangian (20) with:

$$I_{ab} = \frac{N_c}{4}\int\frac{d\omega}{2\pi}\mathrm{Tr}\left[\frac{1}{i\omega+H}\lambda_a\frac{1}{i\omega+H}\lambda_b\right] = \begin{cases} I_A\delta_{ab} & \text{for } a,b=1...3 \\ I_B\delta_{ab} & \text{for } a,b=4...7\,. \\ 0 & \text{for } a,b=8 \end{cases} \tag{38}$$

The above expression is assumed to be properly regularized and the full moments of inertia have also a *valence* part.

The quantization proceeds exactly as in the case of the Skyrme model,[2–4] and as a result one arrives at the hamiltonian (22).

The novelty comes from the expansion in powers of the rotated matrix m:

$$L_m = -\sigma[\sqrt{6}\mu_0 - \sqrt{3}(\mu_8 D^{(8)}_{88} + \mu_3 D^{(8)}_{38})] - 2(\mu_8 D^{(8)}_{8a} + \mu_3 D^{(8)}_{3a})K_{ab}\Omega_b, \tag{39}$$

where constant σ

$$i\frac{N_c}{4}\int\frac{d\omega}{2\pi}\mathrm{Tr}\left[\frac{1}{i\omega+H}\gamma_4\lambda_a\right] = \begin{cases} \sqrt{6}\,\sigma & \text{for } a=0 \\ \sqrt{3}\,\sigma & \text{for } a=8 \\ 0 & \text{for } a=1...7 \end{cases} \tag{40}$$

is related to the pion-nucleon sigma term $\Sigma = 3/2(m_u + m_d)\sigma$ and *anomalous* tensor K_{ab} is defined as:

$$K_{ab} = i\frac{N_c}{4}\int\frac{d\omega}{2\pi}\mathrm{Tr}\left[\frac{1}{i\omega+H}\gamma_4\lambda_a\frac{1}{i\omega+H}\lambda_b\right] = \begin{cases} K_A\delta_{ab} & \text{for } a,b=1...3 \\ K_B\delta_{ab} & \text{for } a,b=4...7\,. \\ 0 & \text{for } a,b=8 \end{cases} \tag{41}$$

We call K_{ab} *anomalous* since it comes from the imaginary part of the effective action, which is related to anomaly, and as such does not require regularization. In fact K_{ab} gets contribution almost entirely from the *valence* level.

The quantized hamiltonian H_{br} is not as simple as in the Skyrme model and reads:

$$\begin{aligned} H_{br} &= \alpha_8\, D^{(8)}_{88}(A) + \beta_8\,\mathbf{Y} + \gamma_8 \sum_{a=1}^{3} D^{(8)}_{8a}(A)\mathbf{S}_a \\ &+ \alpha_3\, D^{(8)}_{38}(A) + \beta_3\,\frac{2}{\sqrt{3}}\mathbf{I}_3 + \gamma_3 \sum_{a=1}^{3} D^{(8)}_{3a}(A)\mathbf{S}_a, \end{aligned} \tag{42}$$

where

$$\alpha_i = \sqrt{3}\left(-\sigma + \frac{K_B}{I_B}\right)\mu_i, \quad \beta_i = -\sqrt{3}\frac{K_B}{I_B}\,\mu_i, \quad \gamma_i = 2\left(\frac{K_A}{I_A} - \frac{K_B}{I_B}\right)\mu_i. \tag{43}$$

Index $i = 8$ corresponds to $Y = 0$, $I = 0$ and index $i = 3$ to $Y = 0$, $I = 1$, $I_3 = 0$.

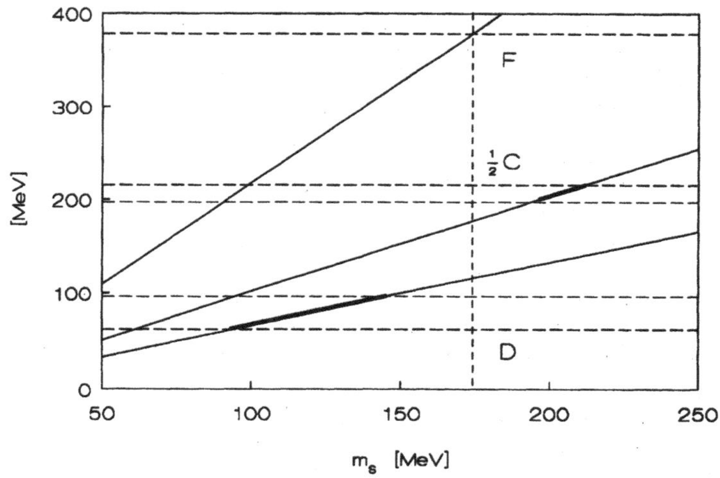

Fig. 2. Constants F, D and $1/2\,C$. Solid lines represent model predictions as functions of m_s. Dashed line correspond to the error bars of Eq.(6).

We see that (42) contains a term proportional to $\mathbf{Y}$ which naturally arises in this model, moreover there is another term proportional to the product of a D function and the spin operator.

We adopt the following numerical procedure: first we find the solitonic solution for a range of constituent masses M, then we find the optimal value of M which reproduces the 10–8 splitting due to the rotational hamiltonian $H_{\mathrm{SU}(2)}$ (23). It turns out[6,7] that $M = 390$ MeV and the corresponding moments of inertia take the following values: $1/I_A = 157$, $1/I_B = 234$, $1/K_A = 469$ and $1/K_B = 704$ MeV, and $\sigma = 3.07$.[29] We will see that not only hyperon splittings but also isospin splittings are well reproduced. To this end let us let us define the following quantities:

$$\varphi = \sigma + 2\frac{I_B}{K_B} + \frac{I_A}{K_A}, \qquad \gamma = \sigma + 2\frac{I_B}{K_B} + 5\frac{I_A}{K_A}, \qquad \delta = \sigma + 2\frac{I_B}{K_B} - 3\frac{I_A}{K_A}. \tag{44}$$

Then we get:

$$F = \frac{1}{2}\varphi\, m_s, \qquad D = \frac{1}{2\sqrt{5}}\,\delta\, m_s, \qquad C = \frac{1}{2\sqrt{2}}\gamma\, m_s, \tag{45}$$

$$f = \frac{3}{4}\varphi\,(m_{\mathrm{d}} - m_{\mathrm{u}}), \qquad d = \frac{3}{20}\,\delta\,(m_{\mathrm{d}} - m_{\mathrm{u}}), \qquad c = \frac{3}{8}\gamma\,(m_{\mathrm{d}} - m_{\mathrm{u}}). \tag{46}$$

In Figs 2 and 3 we plot the splitting constants of Eqs(45, 46) as functions of m_s and $m_{\mathrm{d}} - m_{\mathrm{u}}$ respectively, together with the error bars corresponding to Eqs(6, 11). For $m_s \approx 175$ MeV the N–Ξ splitting (*i.e.* constant F) is reproduced. For the same value of m_s constant D corresponding to Σ–Λ splitting is overestimated by 35 MeV,

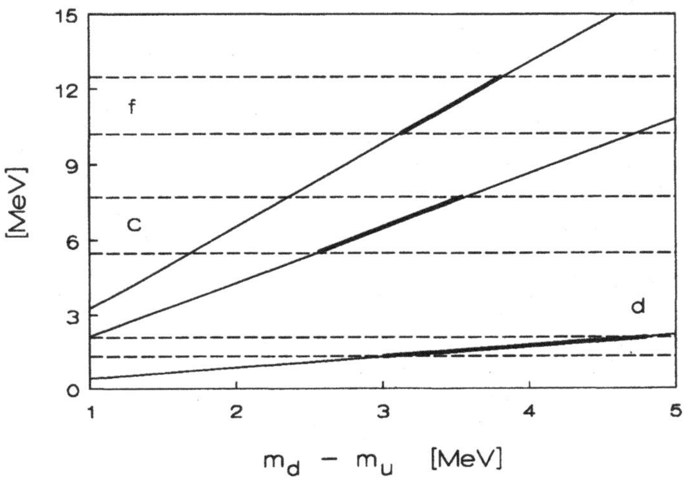

Fig. 3. Constants f, d and c. Solid lines represent model predictions as functions of $m_d - m_u$ mass difference. Dashed line correspond to the error bars of Eq.(11).

whereas constant C is underestimated by 60 MeV. The resulting spectrum is shown in column 4 of Fig.1. On the other hand, the isospin breaking constants f, d and c are reproduced within experimental errors for $m_{\rm d} - m_{\rm u} \approx 3.5$ MeV (see Fig.3).

The present model makes specific prediction for the ratio of hadronic to isospin breaking constants:

$$\frac{f}{F}\,(2.99 \pm 0.3) = \sqrt{5}\,\frac{d}{D}\,(4.90 \pm 2.15) = \sqrt{2}\,\frac{c}{C}\,(2.25 \pm 0.42) \quad (\times 10^{-2}), \tag{47}$$

where the numbers in brackets correspond to the experimental values of Eqs(6, 11). Certainly the central values are fairly scattered. We would like to offer the following explanation of this discrepancy. The isospin splittings are proportional to a tiny parameter, namely $m_{\rm d} - m_{\rm u}$, and therefore the first order of the perturbation theory is legitimate. On the contrary, for the hyperon splittings which are proportional to the much larger parameter, namely strange quark mass, one may expect some corrections from the higher order of the perturbative expansion in m_s. Indeed, already the second order brings the splittings to their experimental values with an accuracy of a few MeV. Last column in Fig.1 shows the spectrum obtained by applying the Yabu Ando procedure to the present model.

To summarize: we have studied the symmetry breaking effects due to the quark masses in the Skyrme model and in the solitonic sector of the semibosonised NJL model. In the simplest version of the Skyrme model the splitting operator is too simple to account for hyperon splittings. On the contrary, in the simplest version of the NJL model, a satisfactory description of the hadronic mass spectrum including both the hyperon and the isospin splittings was found. The new terms in the splitting operator, which are not present in the Skyrme model, are proportional to the *anomalous* moments of inertia which get the main contribution from the valence level. The absolute masses are too big, but there exist several mechanisms which may bring them down, namely gluon corrections,[24] rotational and translational band subtraction[7] and Casimir energies of quantum fluctuations.[25,26]

Acknowledgements

I would like to thank the organizers for very nice and stimulating atmosphere during the Workshop. This work summarizes a series of papers written in collaboration with A. Blotz, D.I. Diakonov, Z. Duliński, K. Goeke, M.A. Nowak, P.O. Mazur, P.V. Pobylitsa, and P. Sieber whom I would like to thank for many discussions.

This work was partly sponsored by *Polish Research Grant 2.0091.91.01.*

References

1. H. Reinhardt, "Effective Chiral Meson Theory of QCD", talk at this conference and references therein.
2. E. Guadagnini, *Nucl. Phys.* **B236**, 35 (1984).
3. S. Jain and S. R. Wadia, *Nucl. Phys.* **B258**, 713 (1985).
4. P. O. Mazur, M. A. Nowak, and M. Praszałowicz, *Phys. Lett.* **147B**, 137 (1984).
5. M. Praszałowicz, *Phys. Lett.* **158B**, 264 (1985).
6. A. Blotz, D. I. Diakonov, K. Goeke, N. W. Park, V. Petrov and P. V. Pobylitsa, "Strange Baryons in the Solitonic Sector of the Nambu–Jona-Lasinio model", Bochum Univ. preprint RUB-TPII-16/92, 1992, *Phys. Lett.* **B** in print.
7. A. Blotz, D. I. Diakonov, K. Goeke, N. W. Park, V. Petrov and P. V. Pobylitsa, "The SU(3)-Nambu–Jona-Lasinio Soliton in the Collective Quantization Formulation", Bochum Univ. preprint RUB-TPII-27/92, 1992, submitted to *Nucl. Phys.* **A**.
8. H. Weigel, R. Alkofer, and H. Reinhardt, *Phys. Lett.* **B284**, 296 (1992).
9. M. Praszałowicz, A. Blotz, and K. Goeke, "Isospin Baryon Mass Differences in SU(3) Nambu–Jona-Lasinio Model", Bochum Univ. preprint RUB-TP2-20/92, 1992.
10. H. Yabu and K. Ando, *Nucl.Phys.* **B301**, 601 (1988).
11. Th. Meissner, F. Grümmer, and K. Goeke, *Ann. of Phys (NY)* **202**, 297 (1990).
12. R. Alkofer, "Chiral Soliton of the Nambu–Jona-Lasinio Model", talk at this conference and references therein.
13. M. Gell-Mann, *Phys. Rev.* **125**, 1067 (1962).
14. S. Okubo, *Progr. Theor. Phys.*, **27**, 349 (1962).
15. J. J. de Swart, *Rev. Mod. Phys.* **35**, 916 (1963).
16. J. Gasser and H. Leutwyler, *Phys. Rep.* **87**, 77 (1982).
17. T. H. R. Skyrme, *Proc. Royal Soc.* **A260**, 127 (1961).

18. T. H. R. Skyrme, *Nucl. Phys.* **31**, 556 (1962).
19. E. Witten, *Nucl. Phys.* **B223**, 422 (1983).
20. E. Witten, *Nucl. Phys.* **B223**, 433 (1983).
21. C. Callan, C. Coleman, J. Wess and B. Zumino, *Phys. Rev.* **177**, 2247 (1969).
22. J. Wess and B. Zumino, *Phys. Lett.* **37B**, 95 (1971).
23. M. Praszałowicz, *Acta. Phys. Pol.* **B22**, 525 (1991).
24. D. I. Diakonov, J. Jaenicke and M. V. Polyakov, "Gluon Exchange Corrections to the Nucleon Mass in the Chiral Theory", St. Petersburg preprint 1738, 1991.
25. B. Moussalam and D. Kalafatis, *Phys. Lett.* **B272**, 196 (1991).
26. B. Moussalam, "The Casimir Energy of a Skyrmion", talk at this conference.
27. C. G. Callan and I. Klebanov, *Nucl. Phys* **B262**, 365 (1985).
28. C. G. Callan, K. Hornbostel and I. Klebanov, *Phys. Lett.* **202B**, 269 (1988).
29. Z. Duliński, M. Praszałowicz and P. Sieber, "H Dibaryon in Chiral Quark Model", Bochum Univ. preprint RUB-TP2-19/92, 1992.

QUARK SOLITONS AS CONSTITUENTS OF HADRONS *

JOHN ELLIS
CERN - Geneva

YITZHAK FRISHMAN and AMIHAY HANANY
Department of Physics, Weizmann Institute of Science
76100 Rehovot, Israel

and

MAREK KARLINER
Raymond and Beverly Sackler Faculty of Exact Sciences
School of Physics and Astronomy
Tel-Aviv University, 69978 Tel-Aviv, Israel

ABSTRACT

We exhibit static solutions of multi-flavour QCD in two dimensions that have the quantum numbers of baryons and mesons, constructed out of quark and anti-quark solitons. In isolation the latter solitons have infinite energy, corresponding to the presence of a string carrying the non-singlet colour flux off to spatial infinity. When N_c solitons of this type are combined, a static, finite-energy, colour singlet solution is formed, corresponding to a baryon. Similarly, static meson solutions are formed out of a soliton and an anti-soliton of different flavours. The stability of the mesons against annihilation is ensured by flavour conservation. The static solutions exist only when the fundamental fields of the bosonized Lagrangian belong to $U(N_c \times N_f)$ rather than to $SU(N_c) \times U(N_f)$. Discussion of flavour symmetry breaking requires a careful treatment of the normal ordering ambiguity. Our results can be viewed as a derivation of the constituent quark model in QCD_2, allowing a detailed study of constituent mass generation and of the heavy quark symmetry.

1. Introduction

One of the key outstanding problems in strong interaction physics is the derivation of hadron spectroscopy from QCD, the underlying theory. Quarks were first postulated as constituents of hadrons to describe qualitatively the spectroscopy of mesons and baryons containing the three lightest u, d and s quark flavours[1]. Subsequently, it was realized that the short-distance properties of strongly-interacting matter could be described exactly in terms of current quarks and the asymptotic freedom of QCD. The phenomenological successes of current algebra[2] and chiral symmetry[3] implied

*presented by M. Karliner. Publ. previously in *Nucl.Phys.*B **382**(1992)189 that the current light (u, d, s) quarks must be much lighter than the original constituent quarks, and the relation between current and constituent light quarks awaits clarification. [4-8] The distinction between current and constituent heavy quarks (c, b, t) is also not clear, but is not so crucial. QCD-inspired potential models work well for mesons made out of heavy quarks, and lattice techniques provide fair understanding of the effective heavy quark–anti-quark potential.

The mystery of the relationship between light current and apparently the heavier constituent quarks is only deepened by the successes of calculations of baryon properties made using the Skyrme model[9,10], a soliton in the low-energy chiral approximation[11] to QCD in terms of bosonic matrix variables. Constituent quarks do not appear in the Skyrme model, their rôles being usurped by coherent states of current quarks.

In $1+1$ dimensions, the spectrum and interactions of mesons in QCD_2 were first discussed in the framework of the large-N_c expansion.[12-15] For baryons, non-Abelian bosonization methods[16] applied to QCD_2[17] have made it possible to obtain the low-lying spectrum[18-21] in the case of an unbroken light flavour symmetry, again without any reference to the idea of constituent quarks. [a] More recently, explicit asymptotic static soliton solutions of the bosonized heavy quark theory have been exhibited.[8] These have the quantum numbers of quarks and an infinite energy associated with a colour flux tube of infinite length. [b]

In this paper we extend this approach by exhibiting static soliton solutions of QCD_2 that have the quantum numbers of baryons and mesons. These new solutions are colour singlets and have finite energy. The solutions with baryon number zero are bound states of the quark and anti-quark solitons, while those with non-zero baryon number are bound states of N_c quark solitons, corresponding to mesons and baryons, respectively. They provide a theoretical laboratory in which the concept of a constituent quark can be dissected. They also provide insight into the QCD description of heavy-light $Q\bar{q}$ mesons such as the D and B, and baryons with one or two heavy quarks, to which the previous heavy-quark effective potential and light-quark chiral approaches have not been applicable. We show that the D and B mesons are likely to contain OZI-evading densities of quark–anti-quark pairs that are absent in the naïve constituent quark description, and could play observable rôles in their dynamics and decays.

The paper is organized as follows. In section 2 we show how mesons emerge as electron-positron solitons in QED_2, developing intuition for the QCD_2 case discussed in section 3. Section 4 discusses explicit meson and baryon solutions in QCD_2, in terms of quark and anti-quark solitons. Such solutions turn out to exist only when the fundamental fields of the bosonized Lagrangian belong to $U(N_c \times N_f)$ rather than to $SU(N_c) \times U(N_f)$. Section 5 discusses their semi-classical quantization. Section 6 contains comments on D- and B-meson physics, and section 7 is a summary and out-

---

[a]The one-flavour case in the discretized light-cone formalism is discussed in ref. [22].

[b]There are also qualitative and group-theoretical indications that such a mechanism could be responsible for appearance of constituent quarks in QCD_4, but the relevant dynamics is as yet unknown.[5]

look. Formal aspects of mass splitting and normal ordering ambiguities are discussed in an Appendix.

2. Mesons from solitons in QED_2

Some of the interesting nonperturbative phenomena in QCD_2 have close analogues in QED_2 and are easy to derive, once the bosonized form of the Lagrangian is known. In this section we present a rederivation of the relevant results obtained long ago, via Abelian bosonization, by Coleman[23], and add some new results of our own, namely explicit solutions in the case of broken flavour symmetry. We believe the reader will find this section useful for developing physical intuition for the discussion of QCD_2 to follow in the next section.

The Lagrangian of multi-flavour massive QED in two dimensions is

$$\begin{aligned} \mathcal{L} &= \sum_k \bar{\psi}_k (i\not{D} - m_k)\psi_k - \frac{1}{4}F_{\mu\nu}F^{\mu\nu} \\ iD_\mu &= i\partial_\mu - eA_\mu \end{aligned} \tag{1}$$

where k is the flavour index, A_μ is the gauge potential, and

$$F_{\mu\nu} = \partial_\mu A_\nu - \partial_\nu A_\mu \equiv \epsilon_{\mu\nu}F \tag{2}$$

The bosonized version of (1) reads[23]

$$\begin{aligned} \mathcal{L} &= \frac{1}{2}\sum_k(\partial_\mu\chi_k)^2 + \frac{1}{2}F^2 + \frac{e}{\sqrt{\pi}}F\sum_k\chi_k + \sum_k\frac{C}{\pi}m_k\mu_k N_{\mu_k}\cos\sqrt{4\pi}\chi_k \\ \bar{\psi}_k\gamma_\mu\psi_k &= -\frac{1}{\sqrt{\pi}}\epsilon_{\mu\nu}\partial^\nu\chi_k \\ \bar{\psi}_k\psi_k &= -\frac{C}{\pi}\mu_k N_{\mu_k}\cos\sqrt{4\pi}\chi_k \\ Q = \int_{-\infty}^{\infty}dx\sum_k\psi_k^\dagger\psi_k &= \frac{1}{\sqrt{\pi}}\sum_k[\chi_k(\infty) - \chi_k(-\infty)] \end{aligned} \tag{3}$$

where Q is the total electric charge, $C = 1/2\, e^\gamma \simeq 0.891$ (γ is Euler's constant), N_{μ_k} stands for normal ordering [c] with respect to a scale μ_k, and we take the θ vacuum angle[23] to be zero, for simplicity.

We integrate out F and take $\mu_k = m_k$, and then "absorb" a factor of C/π in a redefinition of m_k^2. Then the Lagrangian (3) becomes

$$\mathcal{L} = \frac{1}{2}\sum_k(\partial_\mu\chi_k)^2 - \frac{e^2}{2\pi}(\sum_k\chi_k)^2 + \sum_k m_k^2\cos\sqrt{4\pi}\chi_k \tag{4}$$

[c]For a detailed discussion of normal ordering in the presence of different masses, see the Appendix.

The equations of motion in the static case read

$$\chi_k'' - 4\alpha(\sum_l \chi_l) - \sqrt{4\pi} m_k^2 \sin\sqrt{4\pi}\chi_k = 0 \tag{5}$$

where $\alpha \equiv \frac{e^2}{4\pi}$. The static potential is given by

$$V = 2\alpha(\sum_l \chi_l)^2 + \sum_k m_k^2(1 - \cos\sqrt{4\pi}\chi_k) \tag{6}$$

Multiplying (5) by χ_k', summing over k and integrating, we get

$$\frac{1}{2}\sum_k {\chi_k'}^2 - V(x) = (\text{const.}) \tag{7}$$

We are looking for finite energy static solutions, therefore the gradient terms must vanish asymptotically,

$$\chi_k'(\pm\infty) = 0 \tag{8}$$

so that

$$V(\infty) = V(-\infty) \tag{9}$$

Taking $\chi_k(-\infty) = 0$, we see that $V(-\infty) = 0$, and hence $V(\infty) = 0$. Thus

$$\begin{aligned} \sum_l \chi_l(\infty) &= 0 \\ \cos\sqrt{4\pi}\chi_k(\infty) &= 1 \end{aligned} \tag{10}$$

We thus see that only states with zero total charge Q are allowed, which is what one expects, since QED_2 is confining. From (10) it follows

$$\chi_k(\infty) = \sqrt{\pi}\, n_k, \qquad n_k = 0, \pm1, \pm2, \ldots \tag{11}$$

For two flavours, from (10) and (11), taking the lowest non-trivial n_k's, namely $n_1 = 1, n_2 = -1$, we obtain

$$\begin{aligned} \chi_1(\infty) &= \sqrt{\pi} \\ \chi_2(\infty) &= -\sqrt{\pi} \end{aligned} \tag{12}$$

The boundary conditions (12) correspond to a meson built out of a soliton and an anti-soliton. Equations (5) with boundary conditions (12) can be solved explicitly. When $m_1 = m_2$ it is easy to see that $\chi_1(x) = -\chi_2(x)$ and the "string tension" term proportional to α in (5) vanishes, leading to two "mirror" decoupled sine-Gordon equations for χ_1, χ_2. When $m_1 \neq m_2$, a solution can be found numerically. It is particularly interesting to examine the solutions for widely unequal masses. Some representative solutions are shown in Fig. 1. It is worthwhile pointing out that for $\alpha \ll m_1^2, m_2^2$, the widths of χ_1 and χ_2 are governed by $1/m_1$ and $1/m_2$, respectively. The lighter "quark", say χ_2, will be more influenced by the string tension, getting

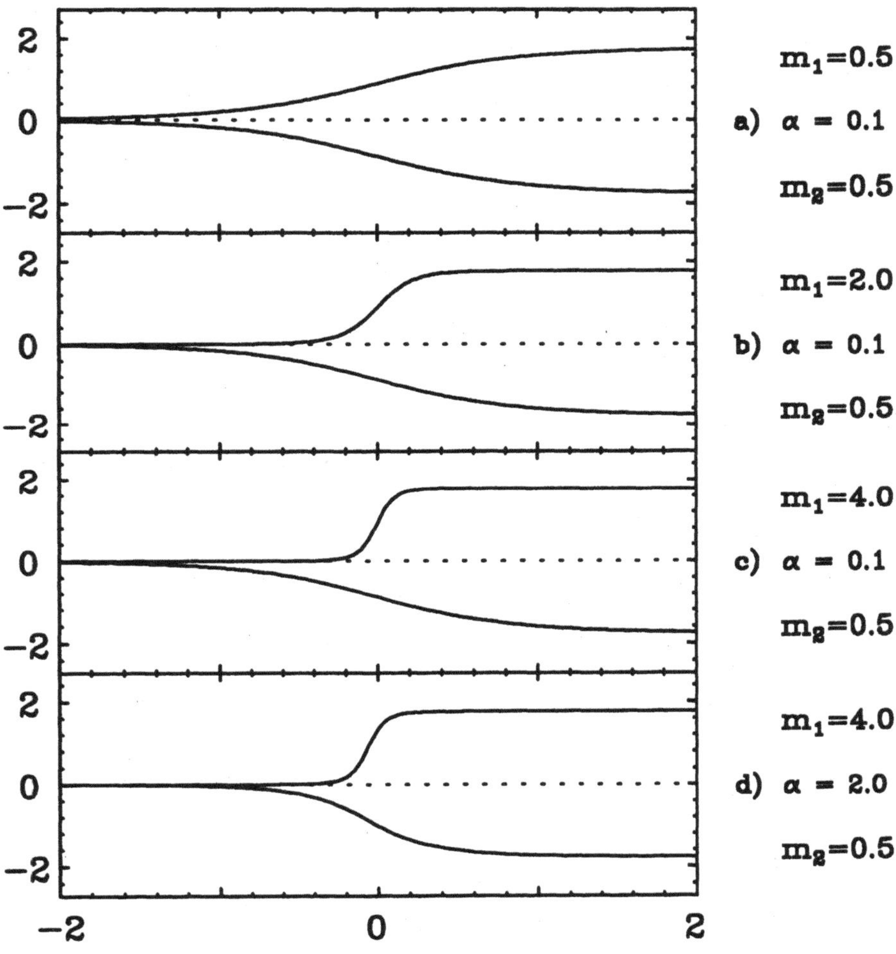

Fig. 1. Mesons in QED_2 with two flavours, solutions of eqs. (5), (12).
(a) $\alpha = 0.1$, $m_1 = 0.5$, $m_2 = 0.5$;
(b) $\alpha = 0.1$, $m_1 = 2.0$, $m_2 = 0.5$;
(c) $\alpha = 0.1$, $m_1 = 4.0$, $m_2 = 0.5$;
(d) $\alpha = 2.0$, $m_1 = 4.0$, $m_2 = 0.5$.
In all four plots the upper line corresponds to χ_1, and the lower line to χ_2. (a), (b) and (c) show how, with increasing m_1, the χ_1 profile converges to a step function, while the light flavour profile, χ_2, remains almost unaffected. The α dependence can be inferred from the difference between (c) and (d): χ_2 gets heavier with increasing α/m_2^2. The effect is not a sharp one, because part of the α dependence is already included in the value of m_2, through the normal ordering (see Appendix).

heavier with increasing α/m_2^2. This is just the intuitive picture we usually associate with a light current quark getting a constituent mass of the order of the gauge scale. Let us now consider the case $m_1 \to \infty$. Then, in (5) with $k = 1$, we can neglect the α term and get a free soliton of mass m_1, which in the limiting case tends to a theta function

$$\chi_1(x) \underset{m_1 \to \infty}{\longrightarrow} \sqrt{\pi}\,\theta(x) \tag{13}$$

as follows from the explicit form of the sine-Gordon solution,

$$\chi_1 = \frac{2}{\sqrt{\pi}} \tan^{-1} \exp\left(\sqrt{4\pi} m_1 x\right) \tag{14}$$

Then, for the light flavour, $k = 2$,

$$\chi_2'' - 4\alpha\chi_2 - \sqrt{4\pi} m_2^2 \sin\sqrt{4\pi}\chi_2 = 4\alpha\sqrt{\pi}\theta(x)\ . \tag{15}$$

We see that the light "anti-quark" field χ_2 feels a point-like "source" term due to the heavy "quark" χ_1. In the following sections we shall demonstrate that a very similar phenomenon occurs in QCD_2.

3. Hadronic Solitons

Two non-Abelian bosonizations of QCD_2 have been developed, one in terms of $SU(N_c) \times U(N_f)$ bosonic variables[18] where N_c is the number of colours and N_f is the number of flavours, and the other[19] in terms of $U(N_c \times N_f)$ bosonic variables. [d] We will now show that in the $U(N_c \times N_f)$ scheme, but not in the $SU(N_c) \times U(N_f)$ scheme, there are static solutions that have the quantum numbers of baryons and mesons, constructed out of quark and anti-quark solitons. In isolation the latter solitons have infinite energy, corresponding to the presence of a string carrying the non-singlet colour flux off to spatial infinity. When N_c solitons of this type are combined, a static, finite-energy, colour singlet solution is formed, corresponding to a baryon. Similarly, static meson solutions are formed out of a soliton and an anti-soliton of different flavours. The stability of the mesons against annihilation is ensured by flavour conservation.

The effective exact $U(N_c \times N_f)$ flavour symmetric bosonic action with gauge fields integrated out reads[19]

$$S_{\mathit{eff}}[u] = S_0[u] + \frac{e_c^2 N_f}{8\pi^2} \int d^2x \,\mathrm{Tr}\, \left[\partial_-^{-1}\left(u\,\partial_- u^\dagger\right)_c\right]^2 + m'^2 N_{\tilde{m}} \int d^2x \,\mathrm{Tr}\, \left(u + u^\dagger\right) \tag{16}$$

where[16]

$$S_0[u] = \frac{1}{8\pi} \int d^2x \,\mathrm{Tr}\,(\partial_\mu u)(\partial^\mu u^\dagger) + \frac{1}{12\pi} \int_B d^3x \,\mathrm{Tr}\, \epsilon^{ijk}(u^\dagger\partial_i u)(u^\dagger\partial_j u)(u^\dagger\partial_k u) \tag{17}$$

[d] The specific case of $SU(N_c)$, $N_f = 2$ has also been considered in a mixed Abelian – non-Abelian formalism in Ref. [24].

e_c is the strong coupling constant (having dimensions of mass in 1+1 space time), u is an element in $U(N_c \times N_f)$, the subscript c denotes projection onto the colour part, i.e. averaging over flavour and subtracting the $U(1)$ part, and

$$m'^2 = m_q \tilde{m} \frac{C}{2\pi} . \tag{18}$$

where $\tilde{m}$ is a normal ordering scale, and $N_{\tilde{m}}$ stands for normal ordering with respect to $\tilde{m}$ (see the discussion in the Appendix). We shall look for solutions with u in diagonal form (see the discussion in ref. [18]),

$$\begin{aligned} u_{\alpha\alpha' jj'} &= \delta_{\alpha\alpha'} \delta_{jj'} e^{-i\sqrt{4\pi}\chi_{\alpha j}} \\ \alpha &= 1, \ldots, N_c \\ j &= 1, \ldots, N_f \end{aligned} \tag{19}$$

The Wess-Zumino term vanishes for either static or diagonal solutions. In general, for non-equal masses, in the static case,

$$\begin{aligned} S_{eff}[u] = &-\frac{1}{2}\sum_{\alpha j} \int \left[\chi'_{\alpha j}(x)\right]^2 - 2\alpha_c \sum_{\alpha} \int \left[\sum_l \chi_{\alpha l} - \frac{1}{N_c}\sum_{\beta l} \chi_{\beta l}\right]^2 \\ &+ \sum_{\alpha j} \int m_j^2 \cos \sqrt{4\pi}\chi_{\alpha j} \end{aligned} \tag{20}$$

where $\alpha_c = e_c^2/4\pi$ and m_j stands for the j-th mass. The relation of m_j's to the "current" quark masses and to α_c, as obtained through the normal ordering, is discussed in the Appendix.

The Hamiltonian density is

$$\begin{aligned} H &= \frac{1}{2}\sum_{\alpha j} \left(\chi'_{\alpha j}\right)^2 + V, \\ V &= 2\alpha_c \sum_{\alpha} \left[\sum_l \chi_{\alpha l} - \frac{1}{N_c}\sum_{\beta l} \chi_{\beta l}\right]^2 + \sum_{\alpha j} m_j^2 \left[1 - \cos \sqrt{4\pi}\chi_{\alpha j}\right] \end{aligned} \tag{21}$$

where we have added a constant to V, to make V vanish for $\chi_{\alpha j} = 0$. The equations of motion are

$$\chi''_{\alpha j} - 4\alpha_c \left[\sum_l \chi_{\alpha l} - \frac{1}{N_c}\sum_{\beta l} \chi_{\beta l}\right] - \sqrt{4\pi} m_j^2 \sin \sqrt{4\pi}\chi_{\alpha j} = 0 \tag{22}$$

To obtain an integral of motion, we multiply by χ'_{α_j} and sum over α, j, to obtain in the static case

$$\sum_{\alpha j} -\frac{1}{2}\left(\chi'_{\alpha j}\right)^2 + 2\alpha_c \sum_{\alpha} \left[\sum_l \chi_{\alpha l} - \frac{1}{N_c}\sum_{\beta l} \chi_{\beta l}\right]^2 + \sum_{\alpha j} m_j^2 \left[1 - \cos \sqrt{4\pi}\chi_{\alpha j}\right] = (\text{const.}) \tag{23}$$

Namely, the left-hand side of (23) is independent of x. For static solutions, the existence of such an x-independent integral of motion is the analogue of energy conservation for solutions evolving in time.

If we choose the boundary condition $\chi_{\alpha j}(-\infty) = 0$, then the constant on the right-hand side of (23) vanishes, and hence also for $\chi_{\alpha j}(+\infty)$ (denoted hereafter simply by $\chi_{\alpha j}$), we must have

$$2\alpha_c \sum_\alpha \left[\sum_l \chi_{\alpha l} - \frac{1}{N_c}\sum_{\beta l}\chi_{\beta l}\right]^2 + \sum_{\alpha j} m_j^2 \left[1 - \cos\sqrt{4\pi}\chi_{\alpha j}\right]\Big|_{+\infty} = 0 \tag{24}$$

Note that this condition is also obtained by requiring finite energy, as follows from eq. (21). We infer from (24) that

$$\begin{aligned} &\frac{1}{\sqrt{\pi}}\chi_{\alpha j} = n_{\alpha j} \qquad \text{integers} \\ &\sum_l \chi_{\alpha l} = \sqrt{\pi}\sum_l n_{\alpha l} = \sqrt{\pi} n \qquad \text{independent of } \alpha \end{aligned} \tag{25}$$

The baryon number of any given flavour l is given by

$$B_l = \sum_\alpha n_{\alpha l} \tag{26}$$

Combining eqs. (25) and (26) we get the total baryon number

$$B = \sum_l B_l = nN_c \tag{27}$$

which clearly is an integer multiple of N_c.

4. Explicit solutions and bosonization schemes

For the mesonic solutions, we need $B = 0$ and hence $n = 0$. Let us consider the case of a meson containing a quark of flavour $l = 1$ and an anti-quark of flavour $l = 2$, and no other constituents. Thus, we need

$$\begin{aligned} &\sum_l n_{\alpha l} = 0 \\ &\sum_\alpha n_{\alpha l} = \begin{cases} 1 & l = 1 \\ -1 & l = 2 \\ 0 & l \geq 3 \end{cases} \end{aligned} \tag{28}$$

Note that in the product scheme, i.e. with h in $SU(N_c)$ and g in $U(N_f)$, we would have

$$\begin{aligned} &n_{\alpha l} = p_\alpha + q_l \\ &\sum_\alpha p_\alpha = 0 \end{aligned} \tag{29}$$

Then

$$N_f p_\alpha + \sum_l q_l = 0 \tag{30}$$

which implies that p_α is independent of α and hence must be zero. Thus

$$n_{\alpha l} = q_l \qquad \text{integers} \tag{31}$$

$$\sum_l q_l = 0$$

$$N_c\, q_l = \begin{cases} 1 & l = 1 \\ -1 & l = 2 \\ 0 & l \geq 3 \end{cases} \tag{32}$$

which is impossible. Thus there are no solutions with mesons as quark–anti-quark states in the "product scheme" of $SU(N_c) \times U(N_f)$.

Let us now give an example of a solution in the $U(N_c \times N_f)$ scheme. In view of the discussion above, such a solution will obviously have components in what we called l in ref. [19], namely a non-factorizable part in the colour-flavour space.

As obtained above, the asymptotic boundary conditions are:

$$\begin{aligned} \chi_{\alpha j}(-\infty) &= 0 \\ \chi_{\alpha j}(+\infty) &= \sqrt{\pi} n_{\alpha j} \end{aligned} \tag{33}$$

where the set $\{n_{\alpha j}\}$ must satisfy the constraints (28). A possible solution is

$$\begin{aligned} n_{11} &= \ \ 1 \\ n_{12} &= -1 \end{aligned} \tag{34}$$

with all other $n_{\alpha j}$ being zero. Having specified the asymptotic boundary conditions at $x \to \pm\infty$, we must now see whether a solution exists for all x. In general such solutions can only be found numerically.[e] The case of an exact $SU(N_f)$ symmetry, i.e. equal quark masses, is an exception where an explicit analytical solution can be found:

$$\chi_{11}(x) = -\chi_{12}(x) = \frac{2}{\sqrt{\pi}} \tan^{-1} \left[\exp\left(\sqrt{4\pi} m\, x \right) \right], \tag{35}$$

with all others vanishing identically. In order to show that this is a solution, we note that with (35) the string tension term proportional to α_c in (22) vanishes identically, and the equations for the various $\chi_{\alpha j}$ decouple, yielding a set of independent sine-Gordon equations with equal masses. It is then obvious that with the boundary conditions (33), (34), $\chi_{11}(x)$ is the usual sine-Gordon soliton.

Recently there has been much discussion in the literature of the so-called heavy quark symmetry[26-31]. Here this symmetry manifests itself in a rather clear way. Consider a $Q\bar{q}$ meson made out of a heavy quark Q and a light anti-quark $\bar{q}$, such as

---

[e]We have used the subroutine package COLSYS.[25]

the D- or B-mesons. When m_Q is much larger than the scale of the theory, $m_Q \gg e_c$, its profile tends to a theta function,

$$\chi_{1Q} \underset{m_Q/e_c \to \infty}{\longrightarrow} \sqrt{\pi}\,\theta(x) \tag{36}$$

as in the QED_2 case, while the baryonic current of the heavy quark tends to a delta function,

$$J_Q^B = \frac{1}{\sqrt{\pi}} \sum_\alpha \partial_x \chi_{\alpha Q} \underset{m_Q/e_c \to \infty}{\longrightarrow} \delta(x) \tag{37}$$

Thus the heavy quark acts as a static colour source, while the profile $\chi_{\alpha q}$ of the light quark becomes independent of m_Q, as shown in Fig. 2. The physics of B and D mesons in the context of the heavy quark symmetry will be discussed in more detail in section 6. The presence of the static colour source (37) makes the total energy of the $Q\bar{q}$ system finite. From (21), (26), (33) and (34) we see that had there been no other colour source, the light flavour profile $\chi_{1\bar{q}}$ on its own would correspond to a configuration with a net baryon number -1, one "unit" of colour charge and a finite energy density per unit length, associated with a colour flux tube of infinite length,

$$\begin{aligned} \chi_{1\bar{q}} &\underset{x \to \infty}{\longrightarrow} -\sqrt{\pi} \\ V_{\bar{q}}(x) &\underset{x \to \infty}{\longrightarrow} 2\pi\alpha_c \left(1 - \frac{1}{N_c}\right)^2 \end{aligned} \tag{38}$$

resulting in the total energy being infinite. This is reminiscent of the single quark solution discussed in ref. [8]. It gives precise meaning to the intuitive concept of quark confinement: isolated quarks have infinite energy because flux conservation forces them to emit a flux tube which has no sink to absorb it.

Multi-quark baryonic solutions can be obtained in a similar way to the meson solutions. For example, taking $N_c = 3$ and $B = 3$ (a 3-quark state[f]) we find $n = 1$ (*cf.* eq. (27)). One possible solution is

$$n_{11} = n_{21} = n_{31} = 1 \tag{39}$$

Corresponding to a "*uuu*"-like baryon, i.e. a "Δ^{++}". When quark masses are equal, $m_1 = m_2 = \ldots = m_{N_f}$, this solution coincides with one found earlier in the "product scheme" [18], in which the colour part is "frozen", i.e. the string tension vanishes identically. The vanishing of string tension in (39) is caused by a mechanism similar to the one occurring in the meson with equal quark masses described above, leading to

$$\chi_{11}(x) = \chi_{21}(x) = \chi_{31}(x) = \frac{2}{\sqrt{\pi}} \tan^{-1}\left[\exp\left(\sqrt{4\pi}m\,x\right)\right], \tag{40}$$

This baryon solution is plotted in Fig. 3(a). There is an important difference, however, between the meson and the baryon cases. The vanishing of the string tension term in

[f] In our normalization a single quark carries one unit of baryon number.

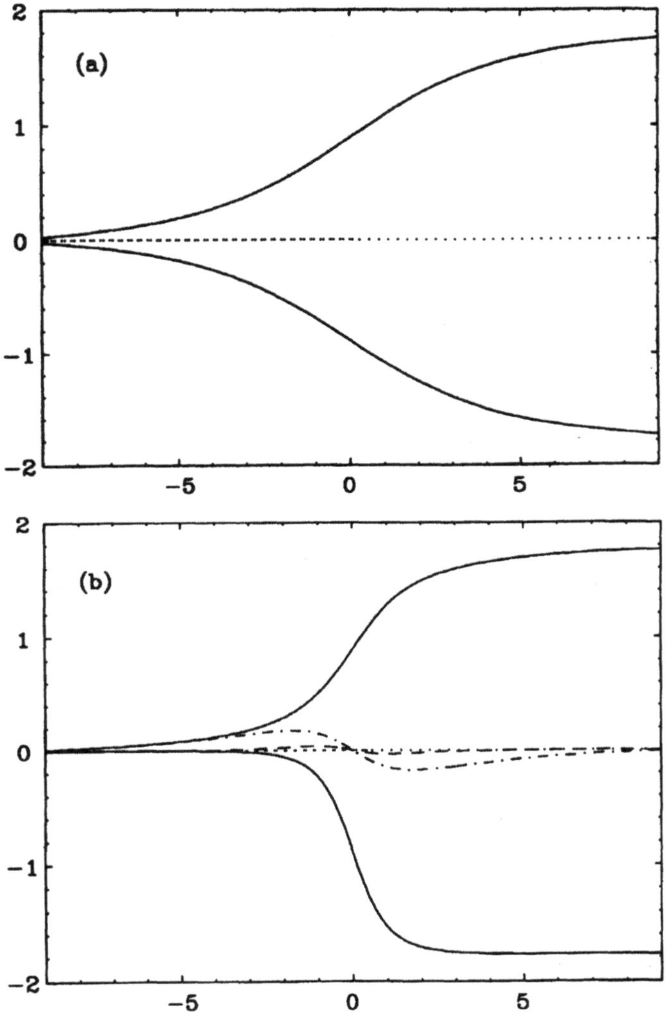

Fig. 2. Mesons in QCD_2 with $N_c = 2$, $N_f = 2$, solutions of eqs. (22), (34).
(a) $\alpha_c = 1.0$, $m_1 = 0.1$, $m_2 = 0.1$;
(b) $\alpha_c = 1.0$, $m_1 = 0.1$, $m_2 = 0.5$;
(c) $\alpha_c = 1.0$, $m_1 = 0.1$, $m_2 = 2.0$;
(d) $\alpha_c = 10.0$, $m_1 = 0.1$, $m_2 = 2.0$.
χ_{11} and χ_{12} are the upper and lower continuous curves, respectively. The "non-valence" components χ_{21} and χ_{22} are denoted by dot-dashed and dashed lines, respectively, except for (a), where they are exactly zero.

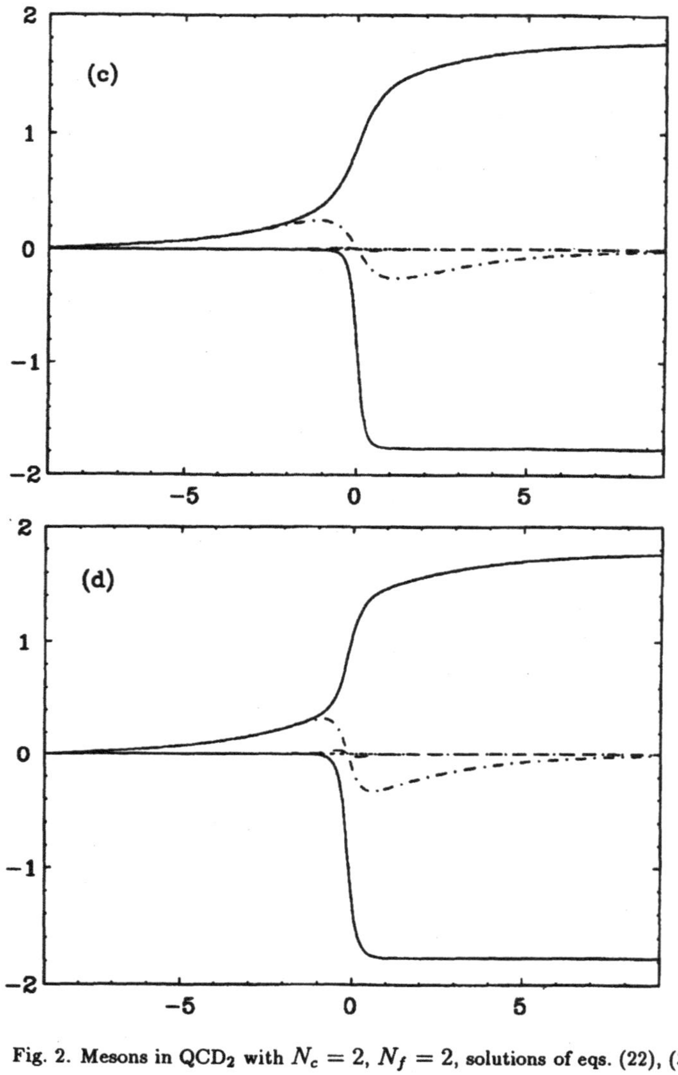

Fig. 2. Mesons in QCD_2 with $N_c = 2$, $N_f = 2$, solutions of eqs. (22), (34).

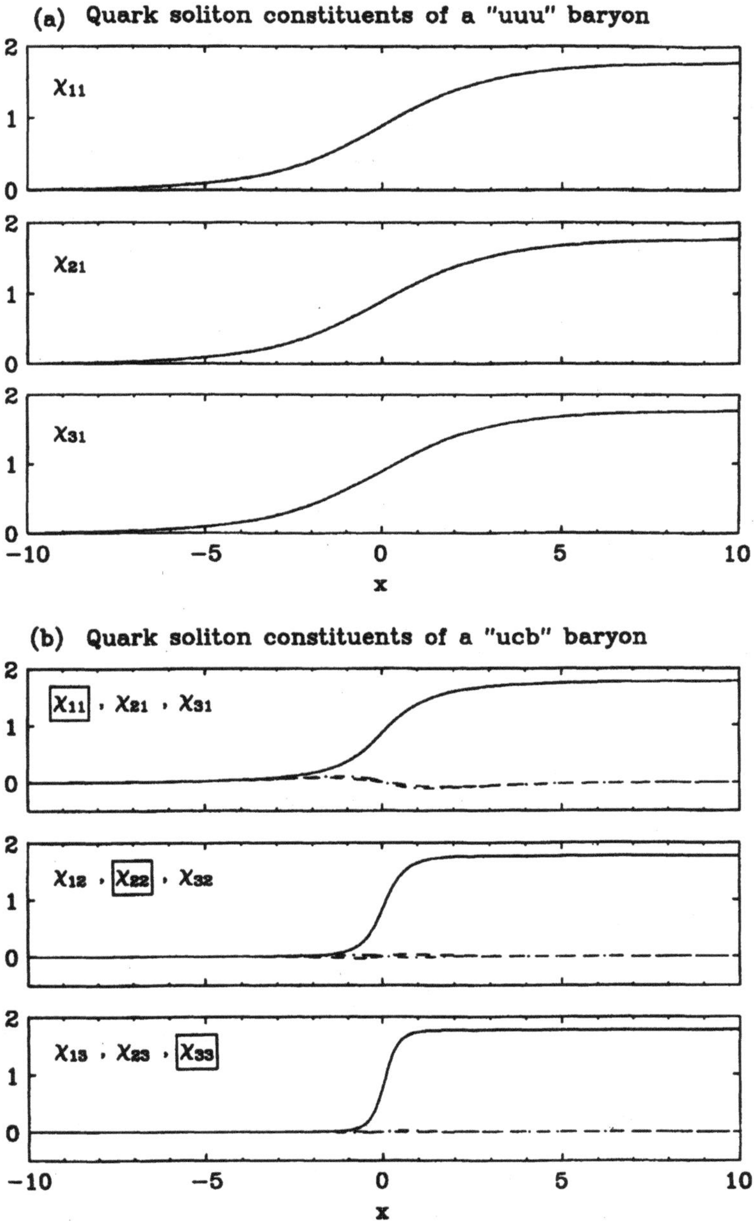

Fig. 3. Baryons in QCD_2 with N_c=3, N_f=3, solutions of eq. (22).

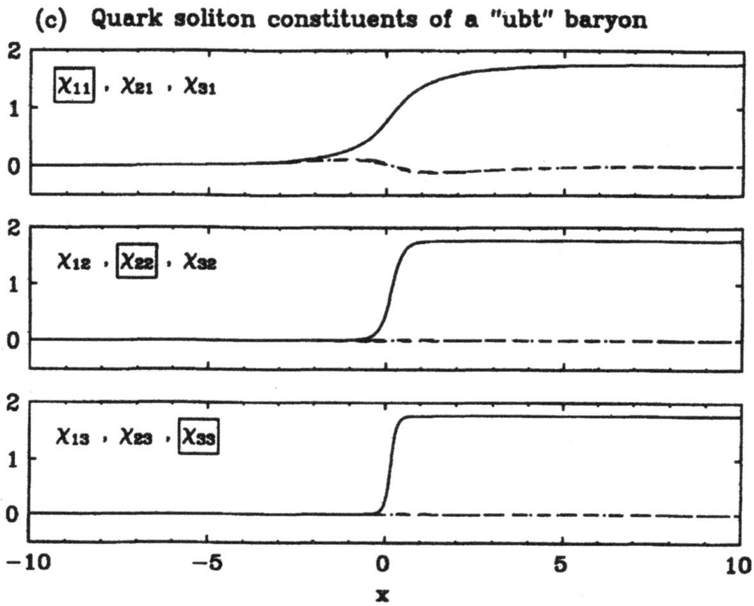

Fig. 3. Baryons in QCD_2 with $N_c=3$, $N_f=3$, solutions of eq. (22).
(a) a "uuu"-like baryon, eq. (39), with $\alpha_c=1.0$, $m_1=m_2=m_3=0.1$;
(b) a "ucb"-like baryon, eq. (41), with $\alpha_c=1.0$, $m_1=0.1$, $m_2=0.5$, $m_3=0.8$;
(c) a "$ub\,t$" -like baryon, eq. (41), with $\alpha_c=1.0$, $m_1=0.1$, $m_2=1.0$, $m_3=2$.
In (a) the "non-valence" components, χ_{21}, χ_{31}, etc. are exactly zero. In (b) and (c) the continuous lines denote the "valence" components, while the dashed and dot-dashed lines denote the "non-valence" components. The "non-valence" components of the heavy flavours are clearly much more suppressed than those of the light flavour.

the meson is a manifestation of the fact that the chromoelectric fluxes of the quark and anti-quark cancel each other. This phenomenon has its counterpart in QED_2[23], as discussed in section 2. The cancellation of fluxes of N_c quarks has no such counterpart and can only occur in a non-Abelian theory. Another difference between the baryonic and mesonic solitons is that the latter do not exist in the "product scheme".

When the boundary conditions (39) are taken together with non-equal quark masses, we still obtain a "uuu"-like baryon, with the classical solution (40), as in the "product scheme". The quantum fluctuations, however, must be treated differently, due to flavour symmetry breaking (see the Appendix for additional discussion).

An intrinsically new 3-quark solution occurs when more than one flavour appears in the nontrivial solution, for example,

$$n_{11} = n_{22} = n_{33} = 1 \tag{41}$$

This corresponds to a baryon in which each quark has a different flavour. Such a solution is particularly interesting when quark masses are taken to be non-equal, corresponding to a "uds"-like baryon, or to a baryon in which one quark is much heavier than the QCD scale, such as the Σ_c or Σ_b, [g] again serving as a theoretical laboratory for the study of the heavy quark symmetry discussed earlier. [h]

One can also study baryons containing two heavy quarks, as shown in Fig. 3(b). For $N_c = 3$, the light quark distribution in such a baryon is the same as in a $\bar{Q}q$ meson. The physical reason for this is that the two heavy quarks are essentially at rest and act as a static colour source. They are colour triplets and combine to a colour anti-triplet, $\mathbf{3} \otimes \mathbf{3} \to \mathbf{3}^*$, i.e. the effective field seen by the light quark q in a QQq baryon is that of a static anti-quark. This gives precise meaning to the concept of constituent quark in QCD_2. The QQq case is particularly clear, since this type of baryon contains only one light constituent quark, while in Qqq or qqq baryons there are two or three such objects, superimposed nonlinearly. A solution of the QQq type with $m_1 \ll m_2 \ll m_3$ is shown in Fig. 3(c).

5. Semi-classical quantization

As has already been mentioned, one of the specific applications which motivated this study was that to mesons containing just one heavy quark Q. In the constituent quark language, these would be described as $Q\bar{q}$ mesons, where $\bar{q}$ represents some light constituent quark (u, d or s). Examples include the D, D_s and B mesons. To describe such mesons within our QCD_2 approach, we need to consider a quark mass matrix with one heavy eigenvalue $M \gg e_c$, and $N_f - 1$ (typically three) light eigenvalues $m \ll e_c$. In such a case, the light quark degrees of freedom should be quantized semi-classically, as was already done for baryons made out of light current quarks.[18] The resulting lump is the best QCD model we can derive for the concept of a light constituent quark. However, clearly it is a coherent state containing in some sense an infinite number of current quarks, at least in the massless limit.

When discussing quantum fluctuations around the classical solution $u_c(x)$, we write

$$U = A u_c(x) A^{-1} \tag{42}$$

with A unitary. Now, we take A to be a product

$$A = A_c A_f \tag{43}$$

---

[g] In 1+1 dimensions there is no spin and therefore baryons must be in the completely symmetric representation of $SU(N_f)$, as the space part of the wave function is symmetric for lowest energy states. Thus with $N_c = 3$ and three light flavours there is only decuplet and no octet, and there is no analogue of Λ_c.

[h] For a different approach to heavy-quark baryons in the chiral soliton framework, see ref. [32].

where A_c acts only in the colour space and A_f in flavour space. This factorization is implied by the fact that although the bosonic variables $u(x)$ belong to $U(N_c \times N_f)$, the full theory is *not* $U(N_c \times N_f)$ invariant. Only the free-fermion theory has the full $U(N_c \times N_f)$ symmetry, and gauging of the colour part results in this symmetry being broken down to $SU(N_c) \times U(N_f) \times U(1)$. If we take A_c to be a function of x_- only, we see that it will not change the interaction term in $S_{eff}[u]$ of eq. (16), thus having no α_c effects. Thus we use "light-cone quantization" for the colour part. For A_f we consider a function of t only, thus reducing the treatment of quantum fluctuations to that which has already been treated in ref. [18], for the equal mass case. The unequal mass case is discussed in the Appendix.

Taking the colour part A_c to be a function of x_- , while the flavour part A_f is taken to be a function of t, might turn out to be problematic. A more natural choice would have been to take both A_c and A_f as functions of t, since we are performing the quantization around a static classical solution, $u_c = u_c(x)$.

In any case, we expect that for equal quark masses the fluctuations in the colour space will not influence ratios such as flavour content of hadrons (sec. 6), since the physical states are colour singlets. The situation might change, though, when masses are different. The flavour content will then in general depend on the mass ratios, and the different masses could in general be influenced differently by the colour fluctuations. This is because the flavours which are light compared with the typical scale of the theory are in the strong-coupling regime, and will be more affected than the flavours which are much heavier than the dynamical scale of the theory. In our case, we will be interested in the u, d and s contents in the approximation where they are all very light.

6. Comments on D and B physics

In the previous sections we have described QCD_2 solitons which could serve as models for $Q\bar{q}$ mesons such as the D or B. In addition to giving some insight into the concept of a constituent quark from the point of QCD, this study may also give some new insights into the dynamics and weak decays of D and B mesons. [i] In particular, we would like to comment on the existence and possible phenomenological rôle of non-valence quarks in the D and B meson wave functions.

There are various phenomenological indications that the proton wave function contains a significant density of non-valence $\bar{s}s$ quarks. Moreover, their abundance relative to $\bar{u}u$ and $\bar{d}d$ quarks is qualitatively reproduced by Skyrme model soliton calculations. The relative abundances of $\bar{s}s, \bar{u}u$ and $\bar{d}d$ have also been calculated in baryonic QCD_2 solitons,[21] and found to be qualitatively similar to the QCD_4 results.[33-35] In this section we invert the logic: QCD_2 mesonic solitons contain calculable non-valence quark densities, and we would expect QCD_4 mesonic solitons and hence physical D and B mesons to contain similar non-valence quark densities.

The calculation of the different light quark densities in a QCD_2 mesonic soliton

[i] $Q\bar{q}$ mesons in QCD_2 have also been recently studied in the large-N_c limit.[36]

parallels very closely that in a baryonic soliton:[21]

$$\frac{\langle M|\bar{q}_i q_i|M\rangle}{\langle M|\bar{q}_j q_j|M\rangle} = \frac{\ll z_i^* z_i \gg}{\ll z_j^* z_j \gg} \tag{44}$$

where $\ll\ \gg$ denotes an average over the collective coordinate representation of the mesonic soliton. In the physical case with one heavy and three light quarks we find for a $Q\bar{q}_1$ meson:

$$\langle M|\bar{q}_1 q_1|M\rangle : \langle M|\bar{q}_2 q_2, \bar{q}_3 q_3|M\rangle = 2:1 \tag{45}$$

This parallels the previous result [21] for the sea- and valence-quark content of baryons in the semi-classical quantization of QCD_2 with N_l light flavours and N_c colours. For a baryon B containing k light valence quarks of flavour v

$$\langle \bar{v}v \rangle_B = \frac{k+1}{N_l + N_c} \tag{46}$$

and

$$\langle (\bar{q}q)_{sea} \rangle_B = \frac{1}{N_l + N_c}, \tag{47}$$

where $(\bar{q}q)_{sea}$ refers to the non-valence quarks in the baryon B, and the total flavour content is normalized to 1. Equations (46) and (47) suggest an "equipartition" for valence and sea, each with a content of $1/(N_l + N_c)$. Since the semi-classical quantization of light flavours in $\bar{Q}q$ mesons parallels that in the baryons, the "equipartition" result applies to the light flavour content of $\bar{Q}q$ mesons as well. This implies, for example, that

$$\langle D|\bar{s}s|D\rangle / \langle D|\bar{u}u|D\rangle = \frac{1}{2} \tag{48}$$

and similarly for B mesons. We might expect the ratio (48) to be qualitatively similar, though possibly smaller by about a factor 2, [21] for the realistic case of QCD_4 with light flavour $SU(3)$ breaking.[33-35]

The presence of significant amounts of $\bar{s}s$ quarks in the D and B mesons implies that the naïve OZI rule forbidding disconnected quark diagrams can be evaded.[37] This might also have implications for D and B production dynamics, but here we only emphasize some possible implications for D and B decays.

- 1. Annihilation diagrams: $c\bar{s} \to u\bar{d}$ could be more important for D^0 and D^+ mesons than is normally supposed when only the D_s wave function is assumed to contain strange quarks.

- 2. The final states from D and B decays could contain more strange particles than is normally supposed when $\bar{s}s$ pairs need to pop out of the QCD vacuum or be created at the weak vertex. This could help explain the surprisingly large [38,39,40] branching ratios for $D^0 \to \phi K^0$ and $D^+ \to \phi K^+$. This observation could also have implications for attempts to estimate the ratio$|V_{cs}/V_{cd}|$ of Kobayashi-Maskawa matrix elements on the basis of strange final states. [41] It might also have implications for the ratios of $D \to K\bar{K}$ and $\pi\pi$ final states.

- 3. The $\bar{s}s$ pairs could provide an additional source for $B \to \phi + X$. At this time it is premature to compare this with the data.

Detailed investigations of these possibilities should await more realistic calculations in QCD_4, however.

7. Summary and outlook

We have shown in this paper that the spectrum of QCD_2 includes finite-energy mesonic and baryonic solitons which contain one heavy quark. These solitons can be regarded as bound states of the infinite-energy single-quark solitons that we found previously. They provide meaning for the previously fuzzy concept of a constituent quark, at least in QCD_2. A particularly interesting application is to the study of mesons and baryons containing both heavy and light quarks. A constituent light quark is seen to be a semi-classical coherent state containing an indefinite number of light $\bar{q}q$ pairs, among which non-valence flavours have as much as one half of the abundance of the valence flavour. This observation could have phenomenological implications for the dynamics and weak decays of D and B mesons.

The next step in the programme of developing accurate QCD descriptions of constituent quarks and $Q\bar{q}$ mesons is to extend the analysis of this paper to the realistic case of four dimensions. This may be possible for the lowest-lying $Q\bar{q}$ mesons if they are describable by spherically symmetric wave functions, which could be analyzed using an effective two-dimensional field theory in (r, t) coordinates. We are now investigating this possibility.

8. Acknowledgements

One of us (J.E.) thanks the Minerva Foundation, the Weizmann Institute and Tel Aviv University for their kind hospitality during the early stages of the work described here. The research of M.K. was supported in part by the Einstein Center at the Weizmann Institute and by the Basic Research Foundation administered by the Israel Academy of Sciences and Humanities and by grant No. 90-00342 from the United States-Israel Binational Science Foundation(BSF), Jerusalem, Israel. M.K. would like to thank Shimon Yankielowicz for useful discussions.

9. References

1. For a review see: J. J. J. Kokkedee, *The Quark Model*, W. A. Benjamin, New York, 1969.
2. For a detailed exposition of current algebra see:
   S. L. Adler and R.F. Dashen, *Current Algebras and Applications to Particle Physics*, W. A. Benjamin, New York, 1968;
   S.B. Treiman, R. Jackiw and D.J. Gross, *Lectures on Current Algebra and its Applications*, Princeton Univ. Press, 1972;
   S.B. Treiman, R. Jackiw, B. Zumino and E. Witten, *Current Algebra and Anomalies*, Princeton Univ. Press, 1985.

3. M. Gell-Mann, R. J. Oakes and B. Renner, *Phys. Rev.* **175**(68)2195 and references therein.
4. H. Georgi and A. Manohar, *Nucl. Phys.* **B310**(88)527.
5. D. B. Kaplan, *Phys. Lett.* **B235**(90)163; *Nucl. Phys.* **B351**(1991), 137.
6. S. Weinberg, *Phys. Rev. Lett.* **65**(90)1181.
7. U. Ellwanger and B. Stech, *Z. Phys.* **C49**(91)683.
8. J. Ellis, Y. Frishman and M. Karliner, *Phys. Lett.* **B272**(91)333.
9. T.H.R. Skyrme, *Proc. Roy. Soc. London* **A260**(1961)127.
10. E. Witten, *Nucl. Phys.* **B223**(1983)422, *ibid* 433; G. Adkins, C. Nappi and E. Witten, *Nucl. Phys.* **B228**(1983)433; for the 3 flavour extension of the model see: E. Guadagnini, *Nucl. Phys.* **236**(1984)35; P. O. Mazur, M. A. Nowak and M. Praszałowicz, *Phys. Lett.* **147B**(1984), 137.
11. E. Witten in Lewes Workshop Proc.; A. Chodos *et al.*, Eds; Singapore, World Scientific, 1984.
12. G. 't Hooft, *Nucl. Phys.* **B72**(74)461; **B75**(1974), 461.
13. C. G. Callan, N. Coote and D. J. Gross, *Phys. Rev.* **D13**(76)1649.
14. M. Einhorn, *Phys. Rev.* **D14**(76)3451.
15. R.C. Brower, J. Ellis, M.G. Schmidt and J.H. Weis, *Nucl. Phys.* **B128**(77)131, **B128**(77)175.
16. E. Witten, *Comm. Math. Phys.* **92**(1984)455.
17. D. Gonzales and A.N. Redlich, *Nucl. Phys.* **B256**(1985)621.
18. G.D. Date, Y. Frishman and J. Sonnenschein, *Nucl. Phys.* **B283**(1987), 365.
19. Y. Frishman and J. Sonnenschein, *Nucl. Phys.* **B294**(1987), 801.
20. Y. Frishman and W. J. Zakrzewski, *Nucl. Phys.* **B331**(1990)781 and in *Proc. of the 25-th Int. High Energy Physics Conference*, Singapore(1990).
21. Y. Frishman and M. Karliner, *Nucl. Phys.* **B334**(1990), 339.
22. K. Hornbostel, S. J. Brodsky and H.C. Pauli, *Phys. Rev.* **D41**(1990)3814.
23. S. Coleman, *Ann. Phys.* **101**(76)239.
24. D. Gepner, *Nucl. Phys.* **B252**(1985)481.
25. U. Ascher, J. Christiansen and R.D. Russel, *ACM Trans. Math. Soft.* **7**(81)209.
26. N. Isgur and M.B. Wise, *Phys. Lett.* **B232**(1989)113; **B237**(1990), 527.
27. M.B. Voloshin and M.A. Shifman, *Yad. Fiz.* **45** (1987)463 (*Sov. J. Nucl. Phys.* **47**(1988), 511).
28. H.D. Politzer and M.B. Wise, *Phys. Lett.* **B206**(88)68,, **B208**(88)504.
29. S. Nussinov and W. Wetzel, *Phys. Rev.* **D36**(87)130.
30. A. Falk, H. Georgi, Grinstein and M.B. Wise, *Nucl. Phys.* **B343**(1990)1.
31. For recent reviews, see B. Grinstein, *Lectures on Heavy Quark Effective Theory*, SSCL-PREPRINT-17, Dec 1991, to be published in Workshop on High Energy Phenomenology, Mexico City Mexico, Jul 1-10, 1991; H. Georgi, *Heavy Quark Effective Field Theory*, HUTP-91-A039; M.B. Wise, *New Symmetries of Strong Interactions*, Lake Louise Lectures, CALT-68-1721, Feb. 1991.
32. M. Rho, D.O. Riska and N.N. Scoccola, *Phys. Lett.* **B251**(90)597.
33. J. Donoghue and C. Nappi, *Phys. Lett.* **B168**(1986)105; J. Donoghue, in Proc. of II-nd Int. Conf. on πN Physics.

34. H. Yabu, *Phys. Lett.* **B218**(89)124; D. B. Kaplan and I. Klebanov, *Nucl. Phys.* **B335**(90)45; for a review see I. Klebanov in Proc. NATO ASI on Hadron and Hadronic Matter, Cargese, 1989.
35. H. Weigel, J. Schechter, N.W. Park and U.-G. Meißner, *Phys. Rev.* **D42** (1990)3177.
36. B. Grinstein and P.F. Mende, *Heavy Mesons in Two Dimensions*, SSCL-Preprint-64/BROWN HET-851, submitted to *Phys. Rev. Lett.*
37. J. Ellis, E. Gabathuler and M. Karliner, *Phys. Lett.* **B217**(88)173; for a recent review of the evidence for this, see R. Decker, M. Nowakowski and U. Wiedner, CERN preprint PPE 92-010 (1992).
38. J. F. Donoghue, *Phys. Rev.* **D33**(86)1516; M. Bauer, B. Stech and M. Wirbel, *Z. Phys* **C34**(1987)103.
39. CLEO Collaboration (C. Bebek *et al.*), *Phys. Rev. Lett.* **56**(1986)1893; ARGUS Collaboration (H. Albrecht, *et al.*), *Z. Phys.* **C33**(1987)359; ACCMOR Collaboration (S. Barlag *et al.*), *Phys. Lett.* **B232**(89)561.
40. Tagged Photon Spectrometer Collaboration (J.C. Anjos *et al.*), *Study of the doubly Cabibbo suppressed decay $D^+ \to \phi K^+$ and the singly Cabibbo suppressed decay $D_s^+ \to \phi K^+$*, FERMILAB-PUB-91-331.
41. WA82 Collaboration (M. Adamovich *et al.*), *Measurement of Relative Branching Fractions of D^0 Cabibbo Suppressed Decays*, CERN preprint PPE-92-015 and references therein.

Appendix

We discuss here the more formal aspects of mass splitting. In the following we will focus mainly on the bosonization in the product scheme, since the technicalities are simpler there, while the conclusions are the same as in the $U(N_c \times N_f)$ scheme. We recall the initial bosonic action in the product scheme[18]

$$\begin{aligned}\tilde{S}\,[g,h,A_\mu] = N_c\, S\,[g] + N_f S\,[h,A_\mu] - \frac{1}{2e_c^2}\int d^2x \,\mathrm{Tr}\, F_{\mu\nu}F^{\mu\nu} \\ + C\tilde{m}N_{\tilde{m}}\int d^2x\,[\,\mathrm{Tr}\,(M\,g)\,\mathrm{Tr}\,h + h.c.]\end{aligned} \tag{A.1}$$

where $g \in U(N_f)$, $h \in SU(N_c)$, A_μ is colour gauge field, $M = \mathrm{diag}(m_1, \ldots, m_{N_f})$ is the mass matrix, m_i are the quark masses, $N_{\tilde{m}}$ stands for normal ordering with respect to $\tilde{m}$, and $F_{\mu\nu}$ is the usual non-Abelian field-strength tensor. Now define the "dimensionless mass matrix" D as

$$D = \frac{1}{m_0} M \tag{A.2}$$

where m_0 is an arbitrary mass parameter. In the gauge $A_- = 0$, integrating out A_+, and then taking the strong-coupling limit, $e_c \gg$ all m_i, we get [18]

$$\tilde{S}\,[g] = N_c S\,[g] + m^2 N_m \int d^2x\, Tr\left[D(g + g^\dagger)\right] \tag{A.3}$$

$$m = \left[C N_c m_0 \left(\frac{e_c^2 N_f}{2\pi} \right)^{\frac{1}{2}p} \right]^{\frac{1}{p+1}} \tag{A.4}$$

$$p = \frac{N_c^2 - 1}{N_c(N_c + N_f)} \tag{A.5}$$

In the $U(N_c \times N_f)$ bosonization scheme the mass term, denoted M.T., has the form

$$M.T. = m_0 C \tilde{m} N_{\tilde{m}} \operatorname{Tr} \left[\hat{D}(u + u^\dagger) \right] \tag{A.6}$$

where $u \in U(N_c \times N_f)$ and $\hat{D}$ is the dimensionless mass matrix (in flavour only). In the strong-coupling limit the mass term (A.6) exactly coincides with that of (A.3).[19] The classical low-lying solutions to the action (A.3) are given by the time-independent form

$$g = \operatorname{diag} \left[\exp\left(-i\sqrt{\frac{4\pi}{N_c}} \phi_1(x) \right), \ldots, \exp\left(-i\sqrt{\frac{4\pi}{N_c}} \phi_{N_f}(x) \right) \right] \tag{A.7}$$

yielding an action density

$$\tilde{S}_d[g] = -\int dx \sum_{i=1}^{N_f} \left[\frac{1}{2} \left(\frac{d\phi_i}{dx} \right)^2 - 2\tilde{m}_i^2 \left(\cos\sqrt{\frac{4\pi}{N_c}} \phi_i - 1 \right) \right] \tag{A.8}$$

with mass parameters

$$\tilde{m}_i^2 = \frac{m_i}{m_0} m^2 \tag{A.9}$$

Each ϕ_i has a classical solution

$$\phi_i(x) = \sqrt{\frac{4N_c}{\pi}} \arctan \left[\exp\left(\sqrt{\frac{8\pi}{N_c}} \tilde{m}_i x \right) \right] \tag{A.10}$$

with an energy

$$E_i = 4\tilde{m}_i \sqrt{\frac{2N_c}{\pi}} \tag{A.11}$$

The minimum classical energy solution is given by a solution with exactly one non-trivial entry,

$$g_0(x) = \operatorname{diag} \left[1, 1, \ldots, \exp\left(-i\sqrt{\frac{4\pi}{N_c}} \phi_{N_f}(x) \right) \right] \tag{A.12}$$

with m_{N_f} chosen to be the smallest mass.

In order to quantize the system we introduce the collective coordinates $A(t)$,

$$g(x,t) = A(t)\, g_0(x)\, A^\dagger(t), \qquad A(t) \in U(N_f) \tag{A.13}$$

Writing $A = \left(A_1, \ldots, A_{N_f-1}, z\right)$ where $A_1, \ldots, A_{N_f-1}, z$ are column vectors, we get

$$\begin{aligned}\tilde{S}[g] - \tilde{S}[g_0] &= \frac{1}{2M_{N_f}} \int dt (Dz)_i^\dagger (Dz)_i + iN_c \int dt \dot{z}_i^\dagger z_i \\ &\quad - \frac{2\pi}{M_{N_f} N_c} \int dt \sum_{i=1}^{N_f} (\tilde{m}_i^2 - \tilde{m}_{N_f}^2) |z_i|^2\end{aligned} \tag{A.14}$$

$$Dz = \dot{z} - z(z^\dagger \dot{z}) \tag{A.15}$$

$$\frac{1}{2M_i} = \frac{N_c}{2\pi} \int_{-\infty}^{\infty} \left(1 - \cos\sqrt{\frac{4\pi}{N_c}}\phi_i\right) dx = \frac{\sqrt{2}}{\tilde{m}_i} \left(\frac{N_c}{\pi}\right)^{\frac{3}{2}}, \qquad i = 1 \ldots N_f. \tag{A.16}$$

The resulting Hamiltonian is

$$H = 4\tilde{m}_{N_f} \left(\frac{2N_c}{\pi}\right)^{\frac{1}{2}} \left\{1 + \left(\frac{\pi}{2N_c}\right)^2 \left[C_2(R) - \frac{N_c^2}{2N_f}(N_f - 1)\right] + \sum_{i=1}^{N_f} \frac{m_i - m_{N_f}}{m_{N_f}} |z_i|^2 \right\} \tag{A.17}$$

The Hamiltonian depends on m_0 only through $\tilde{m}_{N_f}$, therefore the overall mass scale is undetermined, and only mass ratios are meaningful. The quantity $|z_i|^2$ is proportional to the $\bar{q}_i q_i$ operator. Therefore the last term in ; due to $m_i \neq M_{N_f}$, is proportional to the weighted average over flavours of the $\bar{q}q$ operators, the weights being proportional to the mass differences. Since this term comes from quantum fluctuations around the classical solution, consistency with the semi-classical approximation requires that it be very small compared to one.

Generalizing eq. (A.2), by introducing an extra undetermined mass parameter m_{i0} for each flavour, we define $\rho_i = m_i/m_{i0}$. The mass term has the form

$$M.T. = C\tilde{m}N_{\tilde{m}} \operatorname{Tr}(M\, g) = C \sum_{i=1}^{N_f} m_{i0} \tilde{m} N_{\tilde{m}} \rho_i g_{ii} \tag{A.18}$$

By integrating out the gauge fields and taking the strong-coupling limit, we get

$$\tilde{S}[g] = N_c S[g] + \sum_{i=1}^{N_f} \hat{m}_i^2 N_{\hat{m}_i} \int d^2x \frac{m_i}{m_{i0}} (g_{ii} + g_{ii}^*) \tag{A.19}$$

$$\hat{m}_i = \left[CN_c m_{i0} \left(\frac{e_c^2 N_f}{2\pi}\right)^{\frac{1}{2}p}\right]^{\frac{1}{p+1}} \tag{A.20}$$

The classical low mass solutions to the action (A.19) are given by (A.7) with an action density (A.8), but now with mass parameters

$$\tilde{m}_i^2 = \rho_i \hat{m}_i^2 = \frac{m_i}{m_{i0}} \hat{m}_i^2 \tag{A.21}$$

Each ϕ_i has a classical solution (A.10) with energy (A.11). The minimum classical energy solution is given by the ansatz (A.12), with $\tilde{m}_{N_f}$ chosen such that it is the smallest mass. Define now $A(t)$ as in (A.13), resulting in an effective action like in (A.14), and a Hamiltonian

$$H = M\left\{1+\left(\frac{\pi}{2N_c}\right)^2\left[C_2(R)-\frac{N_c^2}{2N_f}(N_f-1)\right]+\sum_{i=1}^{N_f}\left[\frac{m_i}{m_{N_f}}\left(\frac{m_{i0}}{m_{N0}}\right)^{\frac{1-p}{1+p}}-1\right]|z_i|^2\right\} \tag{A.22}$$

$$M = 4\tilde{m}_{N_f}\left(\frac{2N_c}{\pi}\right)^{\frac{1}{2}}. \tag{A.23}$$

There are several problems connected with these arbitrary mass parameters. Since the choice of the parameters m_{i0} is arbitrary, we can choose

$$m_{i0} = m_{N0}\left(\frac{m_i}{m_{N_f}}\right)^{\frac{p+1}{p-1}}, \tag{A.24}$$

in which case H has no explicit contribution due to the mass differences, which is unacceptable.

The question now arises at what stage should one apply the semi-classical approximation.[20] The classical solution is governed by $\tilde{m}_i$ but the lightest $\tilde{m}_i$ is not necessarily the lightest m_i. In fact, the normal ordering ambiguity is present even in a non-interacting theory. To see this, we formally set $N_c = 1$, i.e. we have the field g with level 1 which means free quarks. There is no gauge group, so the dependence on e_c should vanish and indeed $p = 0$, $\hat{m}_i = Cm_{i0}$ and $M = 4C\left(\frac{2m_{N_f}m_{N0}}{\pi}\right)^{\frac{1}{2}}$. The lowest multiplet corresponds to a Young tableaux of one box and $C_2(R) = \frac{N_f^2-1}{2N_f}$, therefore

$$H = M\left[\left(\frac{\pi}{2}\right)^2\frac{N_f-1}{2}+\sum_{i=1}^{N_f}\frac{m_i m_{i0}}{m_{N_f}m_{N0}}|z_i|^2\right] \tag{A.25}$$

which in general does not look like a non-interacting theory, unless one takes $m_{i0} = m_{N0}$ for all i.

One could argue that the above example is not relevant for a generic case, since for small N_c the semi-classical approximation is not justified. This can be seen from (A.17), as follows. Consistency with the semi-classical approximation requires the quantum corrections to the energy to be much smaller than the classical contribution. In the case of one box this means $\left(\frac{\pi}{2}\right)^2\frac{N_f-1}{2} \ll 1$, which for any $N_f \geq 2$ is not correct.

Nevertheless, since the choice of the m_{i0}'s is *a priori* arbitrary, we see that there is an ambiguity. In the non-interacting case it is obvious what is the right choice of the m_{i0}'s, but in general we do not know what is the choice leading to the optimal semi-classical approximation, as is best illustrated by the case of equal quark masses, which can be made to appear unequal by a suitable choice of the m_{i0}'s.

One can summarize the situation as follows. When all masses are equal, the treatment of ref. [18], in the strong coupling limit, yields an effective Lagrangian in terms of flavour degrees of freedom only, with a mass scale that involves the coupling α_c, as in eq. (A.4). When the current mass M of some quark is heavy, $M^2 \gg \alpha_c$, we expect the constituent mass of that quark to be heavy and close to the current mass. The problems arise in the intermediate cases, when we do not know what is the "best" starting point for the normal ordering scale, before going to the semi-classical approximation. Of course, if one were able to sum all corrections, one would obtain the full result, regardless of the starting point.

SKYRMIONS IN THE WEAK INTERACTIONS

YVES BRIHAYE
Department of Mathematics and Physics, Mons University, Av. Maistriau
B-7000 Mons, Belgium

and

JUTTA KUNZ *
Institute for Theoretical Physics, University of Utrecht
NL-3508 TA Utrecht, the Netherlands

ABSTRACT
The Weinberg-Salam theory possesses an approximate chiral symmetry and becomes equivalent to a gauged non-linear σ-model in the infinite Higgs mass limit. When supplemented with a gauged stabilizing term, the effective lagrangian for the weak interactions allows for classical skyrmion solutions for large values of the Higgs mass. We discuss the classical solutions and their properties, and we perform a collective quantization to obtain the quantum spectra of the "weak baryons".

1. Introduction

Perturbative calculations performed in the Weinberg-Salam theory are in impressive agreement with a large amount of experimental data. While there are now strong experimental bounds on the mass of the top quark, the mass of the Higgs particle remains much less restricted. In fact, the whole interpretation of the Higgs sector remains problematic, since the predicted Higgs boson has not yet been observed. It appears quite possible that the Weinberg-Salam theory only represents a low-energy effective model of the weak interactions, and that at high energies, e.g. in the TeV region, new weak interaction phenomena wait to be discovered by future experiments. Here we consider possible non-perturbative phenomena present in the weak interactions in this energy range, with particular emphasis on weak skyrmions.

In the strong interactions baryons emerge as topological solitons of an effective chiral lagrangian[1-3]. The electroweak interactions also possess an approximate $SU(2)_L \times SU(2)_R$ chiral symmetry, broken slightly by the non-vanishing mixing angle θ_w. In the infinite Higgs mass limit ($M_H \to \infty$, $\theta_w = 0$) the bosonic sector of the Weinberg-Salam theory simply represents a non-linear σ-model with gauged $SU(2)_L$. Therefore, the possible existence of chiral soliton solutions, weak skyrmions, has been predicted[4].

In the pure Weinberg-Salam theory only classically unstable solutions, sphalerons, are known to exist[5-8]. As in the strong interactions, to obtain classically stable solutions, weak skyrmions, a stabilizing term must be added[4]. The presence of such a Skyrme term in the effective lagrangian of the weak interactions has been suggested from two different points of view. On the one hand, Gipson and Tze[4] have invoked the one-loop effective lagrangian for a strongly interacting Higgs sector ($M_H \geq 1$

*Department of Physics, University of Oldenburg, W-2900 Oldenburg, Germany

TeV), obtained by Appelquist and Bernard[9]. On the other hand, Rubakov[10,11] has introduced a Skyrme term by assuming an underlying technicolour theory for the weak interactions[12,13], much in the spirit of the large N_c-expansion of the strong interactions.

Neglecting the gauge fields, Gipson and Tze[4] have made first estimates for the weak skyrmions. Subsequently, Ambjørn and Rubakov[11] and Eilam, Klabucar and Stern[14,15] have obtained and analyzed the skyrmion solution of the effective weak lagrangian with a gauged Skyrme term in the infinite Higgs mass limit. Due to the coupling to the gauge fields classical weak skyrmion solutions exist only beyond a critical value of the Skyrme coupling constant e. At the critical value a bifurcation occurs and a second classically unstable branch of solutions emerges. We have shown, that many more classical solutions of this effective lagrangian exist[16–17].

Analogously to the skyrmions of the strong interactions[1–3], the classically stable weak skyrmions can be quantized by the method of collective coordinates, giving an idea about the rotational and vibrational spectra of the "weak baryons" or "technibaryons"[15,18].

In section 2 we briefly review the effective weak lagrangian for large finite and for infinite Higgs mass. In section 3 we describe the general spherically symmetric ansatz for the classical solutions. We discuss the classical solutions and their properties in section 4. We perform the collective quantization of the skyrmion solutions in section 5 and give our conlusions in section 6.

2. The Lagrangian

Let us consider the bosonic sector of the Weinberg-Salam theory in the limit of vanishing mixing angle. In this limit the U(1) field decouples from the SU(2) fields and may consistently be set to zero. The model then reduces to an SU(2) gauge theory coupled to a complex Higgs doublet with lagrangian density

$$\mathcal{L} = -\frac{1}{2g^2}\mathrm{Tr}(F_{\mu\nu}F^{\mu\nu}) + (D_\mu\Phi)^\dagger(D^\mu\Phi) - \lambda(\Phi^\dagger\Phi - \frac{v^2}{2})^2 , \tag{1}$$

the SU(2) field strength tensor

$$F_{\mu\nu} = \partial_\mu V_\nu - \partial_\nu V_\mu - i[V_\mu, V_\nu] , \tag{2}$$

and the covariant derivative for the Higgs field

$$D_\mu\Phi = (\partial_\mu - iV_\mu)\Phi . \tag{3}$$

The gauge symmetry is spontaneously broken via the Higgs potential, leading to a nonvanishing vacuum expectation value for the Higgs field

$$\Phi = \begin{pmatrix} \phi_1 \\ \phi_2 \end{pmatrix} , \qquad \langle\Phi\rangle = \frac{v}{\sqrt{2}}\begin{pmatrix} 0 \\ 1 \end{pmatrix} \tag{4}$$

and to masses for the gauge and Higgs bosons

$$M_W = M_Z = \frac{1}{2}gv \ , \qquad M_H = v\sqrt{2\lambda} \ . \tag{5}$$

To exhibit the global $SU(2)_L \times SU(2)_R$ symmetry of the Higgs sector let us rewrite the lagrangian density in terms of the complex 2×2 matrix M for the Higgs field

$$M = \begin{pmatrix} \phi_2^* & \phi_1 \\ -\phi_1^* & \phi_2 \end{pmatrix} . \tag{6}$$

The lagrangian density (1) then reads

$$\mathcal{L} = -\frac{1}{2g^2}\mathrm{Tr}(F_{\mu\nu}F^{\mu\nu}) + \frac{1}{2}\mathrm{Tr}(D_\mu M)^\dagger(D_\mu M) - \lambda(\frac{1}{2}\mathrm{Tr}(M^\dagger M) - \frac{v^2}{2})^2 \ , \tag{7}$$

with the covariant derivative D_μ

$$D_\mu M = (\partial_\mu - iV_\mu)M \ , \tag{8}$$

where the $SU(2)_L$ symmetry is local due to the presence of the gauge fields.

We now consider the polar decomposition of M,

$$M = \frac{h}{\sqrt{2}}U \ , \qquad h^2 = \mathrm{Tr}(M^\dagger M) \ , \tag{9}$$

where the unitary matrix U constitutes the "angular" part and h the "radial" part of M. In the infinite Higgs mass limit the length of the Higgs field is frozen to its vacuum expectation value, $h = v$. Thus in the infinite Higgs mass limit, the lagrangian density (7) reduces to a gauged non-linear σ-model.

To allow for the existence of weak skyrmions, a stabilizing term needs to be added to the lagrangian density (7). For finite Higgs mass we consider the stabilizing term[19,20]

$$\mathcal{L}_I = \frac{1}{8e^2v^4}\mathrm{Tr}((D_\mu M)^\dagger(D_\nu M) - (\mu \to \nu))^2 \ , \tag{10}$$

which is a polynomial in M. In the infinite Higgs mass limit this term coincides with the gauged Skyrme term[10–11,14–18]

$$\mathcal{L}_{II} = \frac{1}{32e^2}\mathrm{Tr}[D_\mu UU^\dagger, D_\nu UU^\dagger]^2 \ . \tag{11}$$

3. Ansatz and Energy Functional

To construct classical solutions of the equations of motion of the effective weak lagrangians $\mathcal{L} + \mathcal{L}_I$ and $\mathcal{L} + \mathcal{L}_{II}$, we restrict ourselves to static, spherically symmetric configurations of the fields. The corresponding ansatz for the fields[10–11,14–21]

$$M = \frac{v}{\sqrt{2}}L(r)\Big(\cos F(r) + i\hat{r}_a\tau_a \sin F(r)\Big) \ , \tag{12}$$

$$V_i = \frac{G(r)}{2r}\epsilon_{iab}\hat{r}_a\tau_b + \frac{H(r)}{2r}(\delta_{ia} - \hat{r}_i\hat{r}_a)\tau_a + \frac{K(r)}{2r}\hat{r}_i\hat{r}_a\tau_a \ , \tag{13}$$

$$V_0 = 0 \ , \tag{14}$$

involves the five radial functions $L(r)$, $F(r)$, $G(r)$, $H(r)$ and $K(r)$

The spherically symmetric energy functionals associated with the lagrangian terms (7), (10) and (11) read, respectively,

$$E = \frac{2\pi M_W}{g^2} \int dx \, [\frac{((G-1)^2 + H^2 - 1)^2}{x^2} + 2(G' - \frac{HK}{x})^2 + 2(H' + \frac{(G-1)K}{x})^2$$

$$+ L^2(2P + (2xF' - K)^2) + 4x^2L'^2 + 2\epsilon x^2(L^2-1)^2] \, , \tag{15}$$

$$E_I = \frac{2\pi M_W}{g^2} \frac{\xi}{16} \int dx \; L^2 P \left[\frac{L^2[P + 2(2xF' - K)^2]}{x^2} + 8L'^2 \right] \, , \tag{16}$$

and

$$E_{II} = \frac{2\pi M_W}{g^2} \frac{\xi}{16} \int dx \; \frac{P[P + 2(2xF' - K)^2]}{x^2} \tag{17}$$

with

$$P = (G - 1 + \cos 2F)^2 + (H - \sin 2F)^2 \, , \tag{18}$$

with the dimensionless coordinate

$$x = M_W r \, , \tag{19}$$

the prime means derivative with respect to x, and with the parameters

$$\epsilon = \frac{4\lambda}{g^2} = \frac{1}{2}\left(\frac{M_H}{M_W}\right)^2 , \qquad \xi = \frac{g^2}{e^2} \, . \tag{20}$$

The energy functionals (15)-(17) contain a residual U(1) gauge symmetry. Under this U(1) symmetry the functions $G(x)$, $H(x)$, $K(x)$ and $F(x)$ transform as

$$\begin{aligned} (\tilde{G}(x) - 1) &= (G(x) - 1)\cos(2\chi(x)) + H(x)\sin(2\chi(x)) \, , \\ \tilde{H}(x) &= -(G(x) - 1)\sin(2\chi(x)) + H(x)\cos(2\chi(x)) \, , \\ \tilde{K}(x) &= K + 2x\chi'(x) \, , \\ \tilde{F}(x) &= F(x) + \chi(x) \, , \end{aligned} \tag{21}$$

while the length function of the Higgs field $L(x)$ is invariant. By making a choice of gauge, one of the radial functions (except for $L(x)$) can be eliminated.

Let us now consider the "topological charge" associated with the effective weak lagrangian. The non-linear σ-model possesses a conserved topological current

$$T_\mu = \frac{1}{24\pi^2} \varepsilon_{\mu\nu\rho\sigma} \mathrm{Tr}(\partial^\nu U U^\dagger \partial^\rho U U^\dagger \partial^\sigma U U^\dagger) \, , \tag{22}$$

normalized such, that the total topological charge

$$T = \int d^3 r T^0 \, , \tag{23}$$

takes on integer values. However, the current T_μ is not gauge invariant. In contrast, the current

$$t_\mu = \frac{1}{24\pi^2}\varepsilon_{\mu\nu\rho\sigma}\mathrm{Tr}(D^\nu U U^\dagger D^\rho U U^\dagger D^\sigma U U^\dagger + \frac{3i}{2} D^\nu U U^\dagger F^{\rho\sigma}) , \tag{24}$$

is gauge invariant, but not conserved

$$\partial^\mu t_\mu = \frac{1}{32\pi^2}\varepsilon_{\mu\nu\rho\sigma}\mathrm{Tr}(F^{\mu\nu}F^{\rho\sigma}) . \tag{25}$$

In the unitary gauge $U = 1$ this current is the familiar Chern-Simons current of the gauge fields

$$t_\mu = \frac{1}{24\pi^2}\varepsilon_{\mu\nu\rho\sigma}\mathrm{Tr}(iV^\nu V^\rho V^\sigma + \frac{3}{2} V^\nu F^{\rho\sigma}) , \tag{26}$$

whose divergence represents the U(1) anomaly. The Chern-Simons charge or "topological charge" of a configuration is then given by

$$t = \int d^3 r t^0 . \tag{27}$$

4. Classical Solutions

Having fixed a gauge, one obtains from the variational principle a set of equations of motion for the dynamical degrees of freedom of the system. (Note, that for the function $K(x)$ one obtains only constraint equations, because $K'(x)$ does not occur in the energy functionals.)

Regular, finite energy solutions of the equations of motion require certain boundary conditions. Subject to these boundary conditions the equations of motion are then solved numerically. The numerical results presented in the following are based on the parameter choice $g = 0.67$ and $M_W = 83$ GeV.

4.1. *Classical solutions at infinite Higgs mass*

Ambjørn and Rubakov[11] have studied the classical weak skyrmions in the unitary gauge, where the chiral angle vanishes, $F(r) = 0$. In contrast, Eilam, Klabucar and Stern[14,15] have used the gauge $H(r) = 0$. This gauge we adopt in the following.

To obtain skyrmion-like solutions we choose the boundary conditions

$$G(0) = G(\infty) = 0 , \tag{28}$$

$$F(0) = 0 , \quad F(\infty) = N\pi \ (N \text{ integer}) . \tag{29}$$

The classical solutions are characterized by their energy E and their "topological number" t. In the present gauge, the latter is given by

$$t = N - \frac{1}{2\pi}\int_0^\infty dx \Big((G(x) - 1)^2 - 1 \Big) \frac{K(x)}{x} . \tag{30}$$

4.1.1. *Weak skyrmions*

The weak skyrmion solutions are obtained for $N = 1$. The function $F(x)$ plays the part of the chiral angle of the skyrmion, and the gauge field function $G(x)$ is close to its vacuum value of zero for small values of ξ. The solution is shown in Fig. 1 for $\xi = 0.38$. The energy of the skyrmion solutions is exhibited in Fig. 2, where it corresponds to the lowest branch (labelled $N = n = 1$). Interestingly, the classical skyrmion solutions exist only within a certain range of the parameter ξ, $0 \leq \xi \leq \xi_{cr}$[14]. At the critical value ξ_{cr} a cusp is encountered and a second branch of classical solutions emerges for $0 \leq \xi \leq \xi_{cr}$[14]. The solution for $\xi = 0$ on the upper branch is also shown in Fig. 1. (This solution is in fact a sphaleron of the Weinberg-Salam theory in the infinite Higgs mass limit[7,8].) The "topological number" of the skyrmion solutions is shown in Fig. 3, where it corresponds to the upper branch (labelled $N = n = 1$). Note, that the "topological number" changes smoothly at the cusp, while its derivative becomes infinite.

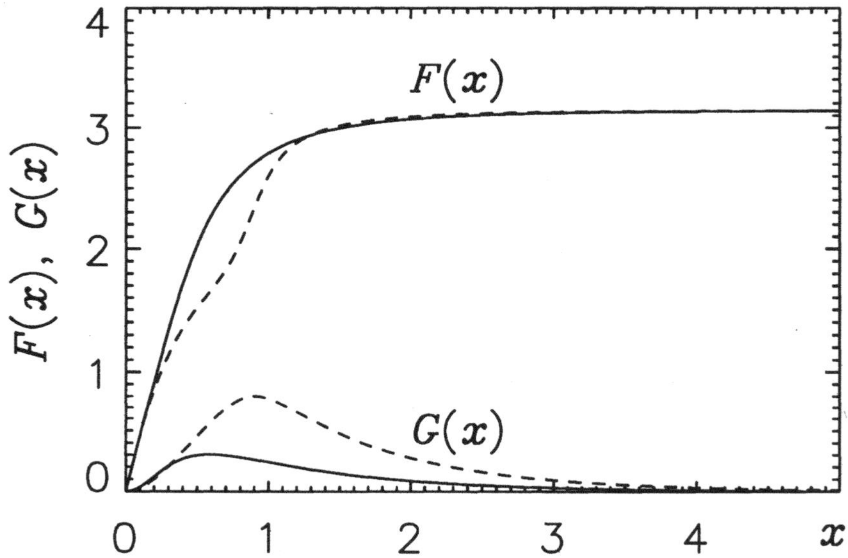

Fig. 1. Functions $F(x)$ and $G(x)$ at the cusp (solid) and at $\xi = 0$ on the upper branch (dashed).

4.1.2. *Weak multiskyrmions*

Analogously to the skyrmion solutions one can construct solutions for higher values of N[16]. The energy and the "topological number" of the solutions for $N = 2$ are also shown in Fig. 2 and Fig. 3, respectively[16]. For a given N, there exist two branches of solutions, which merge and end in a cusp at some critical value ξ_{cr}. The lower branch

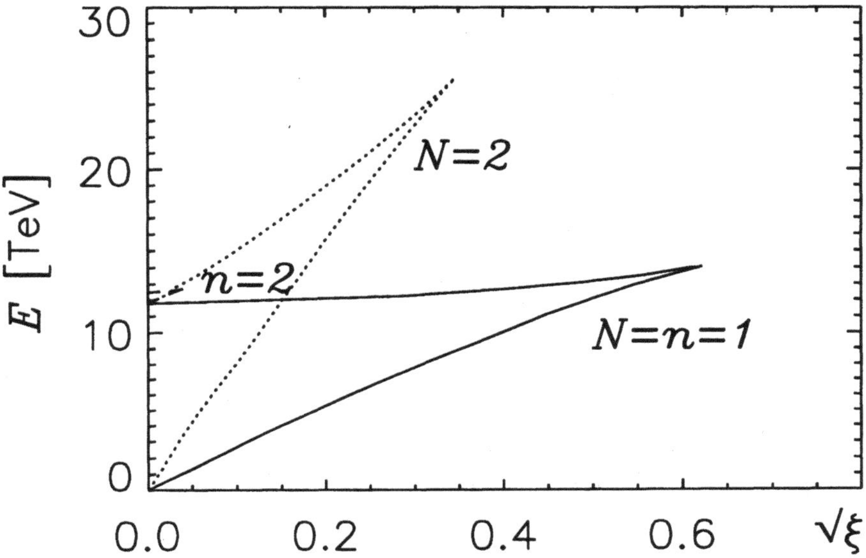

Fig. 2. Energy in TeV of the solutions with $N = n = 1$ (solid), $N = 2$ (dotted), $n = 2$ (dashed) .

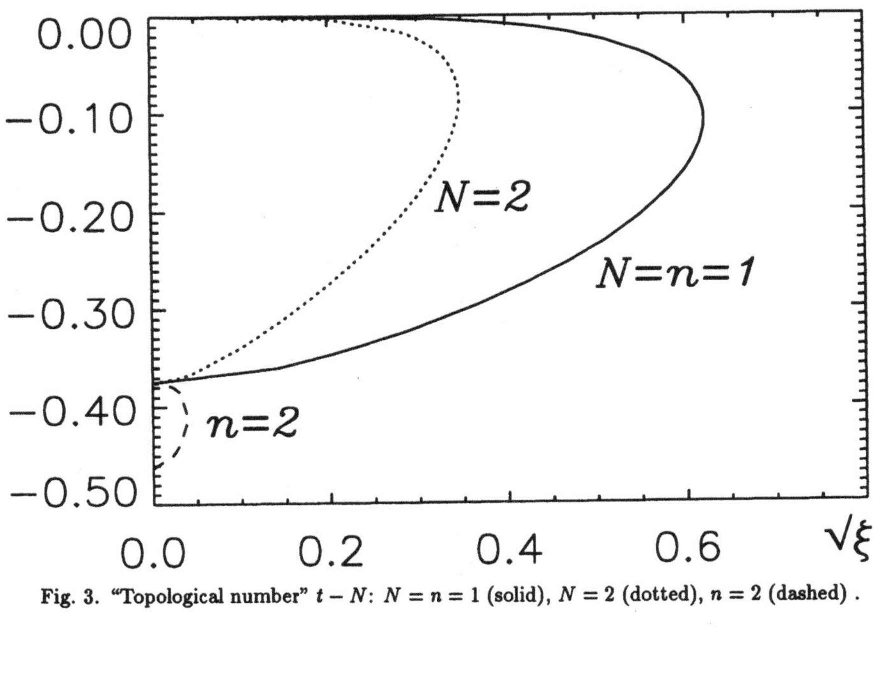

Fig. 3. "Topological number" $t - N$: $N = n = 1$ (solid), $N = 2$ (dotted), $n = 2$ (dashed) .

corresponds to spherically symmetric (multi-)skyrmion solutions. The upper branch is associated with sphaleron solutions[7,17]. The critical values ξ_{cr} decrease as a function of N. For $N = 1, 2$ and 3 they are $\xi_{cr} = 0.386, 0.121$ and 0.057, respectively. At $\xi = 0$, all upper branches converge to the same value of the energy and of the "topological charge". We conjecture, however, that in analogy to the strong interactions the energetically lowest multiskyrmion solutions are deformed[22].

4.1.3. *Radial excitations*

With the boundary conditions

$$G(0) = G(\infty) = 0 \,, \tag{31}$$

$$F(0) = \left(\frac{1+(-1)^n}{2}\right)\pi \,, \quad F(\infty) = \pi \tag{32}$$

another sequence of classical solutions can be constructed[7,17] based on the upper branch of the skyrmion solution $n = 1$. With increasing n the functions $F_n(x)$ develop a number of oscillations and approach two plateaus, $F_n(x) \approx \pi/2$ for $0 \le x \le x_{cr}$ and $F_n(x) \approx \pi$ for $x_{cr} \le x \le \infty$. At the same time, the functions $G_n(x)$ approach the corresponding sphaleron configurations[5].

The energy and the "tolopogical number" of the solutions with $n = 2$ are shown in Fig. 2 and Fig. 3, respectively, as functions of the parameter ξ.

4.1.4. *Classical stability*

To determine the classical stability of the classical solutions one has to consider small oscillations about these solutions. The resulting eigenvalue equations have so far not been solved for the effective weak lagrangian with stabilizing term. But the stability equations have been solved for the classical solutions of the Weinberg-Salam theory, the sphalerons[8,23].

At $\xi = 0$ the solutions $F_n(x)$, $G_n(x)$ on the upper branches coincide with the corresponding sphaleron solutions[7,8]. For these solutions it is known, that they are classically unstable, with n unstable modes[8,23]. The eigenvalues of these unstable modes are $\omega^2 = -3.94M_W^2$ for $n = 1$, and $\omega^2 = -4.25M_W^2$ and $\omega^2 = -185.7M_W^2$ for $n = 2$. Assuming continuity for the solutions when ξ is increased from zero, the solutions $F_n(x)$, $G_n(x)$ have n unstable modes along their upper branches.

Lacking a complete numerical stability analysis, we invoke an argument from catastrophe theory for analyzing the stability along the lower branches[14,15]. According to catastrophe theory, a transition from an unstable to a stable configuration occurs precisely at the critical value ξ_{cr} of a cusp, where an upper and a lower branch merge. At the cusp the curvature eigenvalue changes sign. (This argument is indeed true for the classical solutions of a similar effective lagrangian for the strong interactions[24].)

We now apply this argument to the classical skyrmion solutions ($N = n = 1$). On the upper branch the classical solutions have a single unstable mode, whose eigenvalue decreases in size as the critical value ξ_{cr} is approached. At the cusp this mode has a zero eigenvalue. On the lower branch, the skyrmion branch, there is then no more unstable mode. We thus conjecture that the weak skyrmion solutions are classically stable. (For the other solutions ($N > 1$, $n > 1$) we conclude, that they are unstable also on their lower branches, but with one less unstable mode than on their upper branches.)

4.2. *Classical solutions at finite Higgs mass*

Let us now consider the case of a large but finite Higgs mass. The interesting question then is: Do classically stable branches of solutions exist also for finite values of the Higgs mass, which could be identified with weak skyrmions?

In the infinite Higgs mass limit the relevant set of solutions starts from a sphaleron at $\xi = 0$, approaches a cusp, and then forms the stable skyrmions branch. This sphaleron solution exists for large, but finite Higgs mass (for $M_H \geq 12M_W$)[7,8]. By continuity, one can construct a branch of solutions starting from this sphaleron for $\xi \geq 0$ and for $M_H \geq 12M_W$. However, only for very large values of the Higgs mass this branch leads to a skyrmion branch. A typical example is shown in Fig. 4, where the energy of the set of classical solutions is exhibited for $M_H = 136M_W$. The upper branch starting at $\xi = 0$ indeed ends in a cusp at a critical value ξ^1_{cr}, where a lower branch emerges. The lower branch, however, also ends in a cusp at a second critical value ξ^2_{cr}, it does not extend all the way to $\xi = 0$. From the second critical value ξ^2_{cr} a second upper branch emerges.

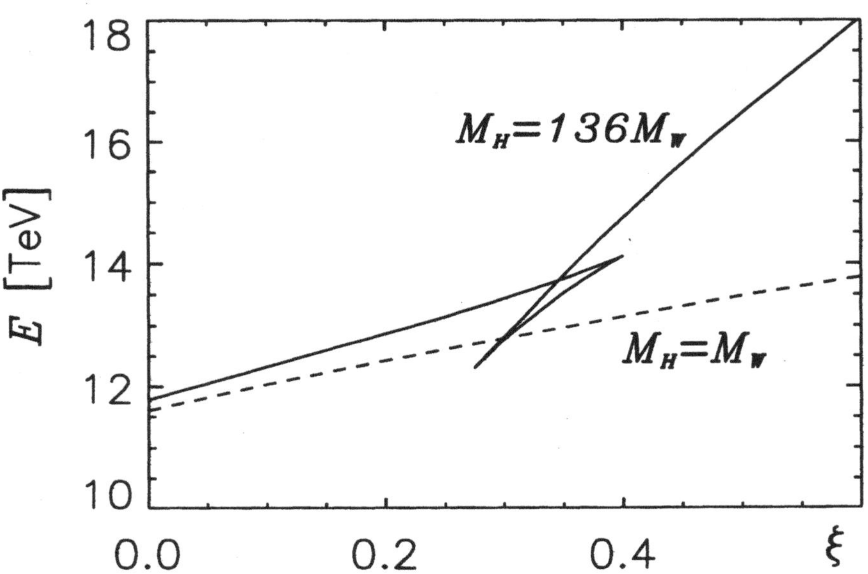

Fig. 4. Energy in TeV of the classical solutions for $M_H = 136M_W$ (solid), $M_H = M_W$ (dashed) .

To discuss the stability of the classical solutions we again invoke the argument from catastrophe theory and the known instability of the sphaleron solution. At $\xi = 0$ on the first upper branch the classical solution has a single unstable direction. The stability doesn't change along the upper branch until the first cusp is reached. Here the unstable mode becomes stable. The lower branch is thus stable and corresponds to the skyrmion branch. At the second cusp the stability changes again, and the second upper branch is unstable again, with a single unstable mode.

Repeating the numerical analysis for various values of M_H we find that the domain of existence of the lower stable branch increases, when M_H increases and that $\xi^2_{\rm cr}$ approaches zero in the infinite Higgs mass limit (i.e. the set of solutions approaches the one of Fig. 2). On the other hand, when M_H is decreased, the domain of existence of the lower stable branch diminishes and it disappears for $M_H \approx 100 M_W$. As an example the single branch of solutions for $M_H = M_W$ is also shown in Fig. 4.

5. Collective Quantization

Classical solutions which represent local minima of the energy functional (apart from zero modes associated with continuous symmetries broken by the solution) may be quantized by standard techniques to obtain the corresponding physical particle spectrum. To quantize the classical weak skyrmions along the lower, stable branch we proceed analogously to the quantization of skyrmions in the strong interactions[1-3]. The zero modes are identified, and the variables associated with these zero modes are promoted to dynamical variables, the collective variables. Then the standard commutators are imposed for the collective coordinates and their canonically conjugate momenta.

In the following we restrict the discussion to the infinite Higgs mass case, since the finite Higgs mass case has not yet been treated.

5.1. *Rotational excitations*

The relevant zero modes of the weak skyrmion are associated with spatial rotations and with weak "vector" isospin rotations. Due to the special choice of the classical ansatz, i. e. the generalized spherical symmetry, both are related. We parametrize the collective variables by the SU(2) matrix $A(t)$ and neglect - consistent with the adiabatic approximation - terms of higher order in time derivatives. This leads to the time dependent ansatz

$$U(x_i, t) = A(t) U^{\rm cl}(x_i) A(t)^\dagger \ , \tag{33}$$

$$V_k(x_i, t) = A(t) V_k^{\rm cl}(x_i) A(t)^\dagger \ , \tag{34}$$

(the superscript cl denotes the classical ansatz (12)-(14)) and

$$V_0(x_i, t) = \frac{\bar{G}(x)}{2} \epsilon_{abc} \hat{x}_a \tau_b k_c + \frac{(\bar{H}(x) - 2)}{2} (\delta_{ab} - \hat{x}_a \hat{x}_b) \tau_a k_b + \frac{\bar{K}(x)}{2} \hat{x}_a \hat{x}_b \tau_a k_b \ , \tag{35}$$

with the isovector $\vec{k}$

$$i\, k_j \tau_j = A^\dagger \dot{A} \ . \tag{36}$$

Note, that the time component of the gauge field is no longer zero. This is essential to insure invariance of the lagrangian density under the residual gauge transformation (21). Under this transformation, the induced functions $\bar{G}(x)$ and $\bar{H}(x)$ transform as a doublet analogously to $(G(x)-1)$ and $H(x)$, while the induced function $\bar{K}(x)$ is invariant.

Substituting the ansatz (33)-(35) into the total lagrangian $\mathcal{L}+\mathcal{L}_{II}$ we obtain

$$L = -\frac{1}{2}a[U^{\rm cl}, V_i^{\rm cl}, V_0]\, Tr(A^\dagger \dot{A})^2 - E[U^{\rm cl}, V_i^{\rm cl}] \,, \tag{37}$$

where the moment of inertia $a[U^{\rm cl}, V_i^{\rm cl}, V_0]$ is a gauge invariant functional of the fields $G(x), H(x), K(x)$ and $F(x)$ and of the induced components $\bar{G}(x), \bar{H}(x), \bar{K}(x)$[18]. By minimizing the functional $a[U^{\rm cl}, V_i^{\rm cl}, V_0]$, with respect to the induced functions $\bar{G}(x)$, $\bar{H}(x)$ and $\bar{K}(x)$ we obtain a set of linear differential equations for these functions, which are solved numerically[18]

For the weak skyrmions we observe that for small values of ξ the induced functions $\bar{G}(x)$, $\bar{H}(x)$ and $\bar{K}(x)$ deviate very little from their trivial values corresponding to $V_0 = 0$. In contrast, for larger values of ξ these deviations become substantial.

Standard quantization of the collective variables, parametrized by the matrix $V(t)$, leads to the mass spectrum for the weak skyrmions, which takes the form of a rotator

$$M_l = E[U^{\rm cl}, V_i^{\rm cl}] + \frac{2l(l+1)}{a[U^{\rm cl}, V_i^{\rm cl}, V_0]} \,. \tag{38}$$

When the solitons are quantized as bosons[14,15], and - unlike the strong interactions - there is no anomalous term requiring the quantization of the solitons as fermions, the lowest state has $l=0$ and its energy equals the energy of the classical solution (up to an unknown constant shift due to zero point oscillations[25]). The energy spacing within the collective rotational band is determined by the moment of inertia $a[U^{\rm cl}, V_i^{\rm cl}, V_0]$.

Several values of the classical energy $E^{\rm cl} = E[U^{\rm cl}, V_i^{\rm cl}]$, the moment of inertia $a = a[U^{\rm cl}, V_i^{\rm cl}, V_0]$ and the rotational energy $E^{rot}(l=1)$ for the weak skyrmions are shown in Table 1[18]. For intermediate values of the parameter ξ the moment of inertia for the weak skyrmions $a[U^{\rm cl}, V_i^{\rm cl}, V_0]$ is very big, leading to a very closely spaced rotational spectrum, where the excited quantum states of the weak skyrmion are almost degenerate with the ground state. In this region the collective quantization procedure appears well justified. Close to the critical value of ξ the classical minimum becomes very shallow and finally disappears at the cusp. Therefore the semi-classical quantization procedure becomes inadequate close to the cusp. For small values of ξ on the other hand the classical energy of the weak skyrmions appoaches zero while the rotational energy diverges. Thus for small values of ξ the quantum corrections to the energy exceed the classical energy by far, invalidating the collective quantization procedure for small values of ξ. We therefore conclude that the model does not allow for arbitrarily light and small solitonic states[18].

Table 1. Classical energy, rotational energy ($l = 1$) and vibrational energy of weak skyrmions

$\xi^{1/4}$	$E^{\rm cl}$(TeV)	$E^{\rm rot}$(TeV)	$\omega^{\rm br}$(TeV)
0.1	0.289	2792.	9.9
0.2	1.078	43.6	2.5
0.3	2.419	3.82	1.1
0.4	4.269	0.68	0.63
0.5	6.569	0.18	0.41
0.6	9.196	0.059	0.28
0.7	11.92	0.023	0.20
0.785	13.98	0.009	0.14

5.2. *Vibrational excitations*

In the Skyrme model of the nucleon the Roper resonance has been described as a vibrationally excited nucleon, performing monopole vibrations. We now consider the corresponding vibrational excitations for the weak skyrmions and determine the excitation energies.

For the monopole vibration of the weak solitons we consider the time-dependent scale factor $\lambda(t)$[26]. The fields then scale according to

$$U^{\rm cl}(x_i) \rightarrow U^{\rm cl}(\lambda x_i) \ , \tag{39}$$

$$V_k^{\rm cl}(x_i) \rightarrow \lambda V_k^{\rm cl}(\lambda x_i) \ . \tag{40}$$

Again the time component of the gauge field acquires a non-trivial value which we parametrize as[18]

$$V_0(x_i, t) = \frac{\dot{\lambda}}{\lambda} P(\lambda x)(\hat{x}_a \tau_a) \ . \tag{41}$$

After inserting the scaled fields into the effective weak lagrangian and making use of the virial theorem we obtain[18]

$$L(\lambda, \dot{\lambda}) = \frac{1}{2}\left(\frac{C_2}{\lambda^2} + C_4\right)\frac{\dot{\lambda}^2}{\lambda^3} - \frac{1}{2}\left(\lambda + \frac{1}{\lambda}\right)E^{\rm cl} \ . \tag{42}$$

The coefficients C_2 and C_4 are functionals depending on the induced function $P(x)$ and on the classical functions $G(x)$, $H(x)$, and $K(x)$[18]. They are invariant under the gauge transformation (21), when the induced function $P(x)$ transforms analogously to the function $K(x)$.

For small oscillations about the classical minimum at $\lambda = 1$ the lagrangian (42) reduces to the one of a harmonic oscillator. The vibrational excitation energy $\omega^{\rm br}$ is then given by

$$\omega^{\rm br} = \sqrt{\frac{E^{\rm cl}}{C_2 + C_4}} \ . \tag{43}$$

Minimizing the functional C_2+C_4 leads to a differential equation for the new function $P(x)$ to be solved. Table 1 contains the excitation energy $\omega^{\rm br}$ for the weak skyrmions for several values of ξ along the lower branch.

6. Discussion

The Weinberg-Salam theory describes the low energy regime of the weak interactions excellently. However, at energies in the TeV region modifications of the electroweak lagrangian may become necessary. This may be the case, if there is an underlying non-abelian gauge theory, such as a technicolour theory, for which non-perturbative effects will play a central role. But also a very heavy, strongly interacting Higgs particle will require modifications of the electroweak lagrangian and may lead to new non-perturbative effects.

Based on this reasoning and lacking the knowledge of the true effective weak interaction lagrangian, we have studied as model lagrangians Skyrme-type models with gauged $SU(2)_L$, where the gauge fields describe the weak vector bosons (and the U(1) field decouples). We have put the main emphasis on an effective weak lagrangian with infinite Higgs mass, but we have also considered a large but finite Higgs mass, $M_H \geq 1\text{TeV}$.

6.1. *Classical skyrmion solutions*

At infinite Higgs mass the effective weak lagrangian allows for a plethora of classical solutions. We have argued that the energetically lowest branch of solutions is classically stable, and that it describes the weak skyrmions or techni-skyrmions. Also at finite Higgs mass a skyrmion branch exists if the Higgs mass is sufficiently large. (The lower bound on the Higgs mass found here may decrease, when the full effective weak lagrangian[9] is considered.) Analogy with the strong interactions then suggests, that even though being crude, the model predicts the properties of the weak soliton states at least with the right order of magnitude, and that a rich set of quantum states, particles and resonances, resides in the TeV region[27].

6.2. *Quantization and spectrum*

To obtain the quantum spectrum of the weak skyrmions, we have semi-classically quantized the stable classical solutions, using collective coordinates. The zero modes associated with spatial and isospin rotations give rise to a rotational quantum spectrum. We have quantized the solitons as bosons. (The consistency requirements and implications of quantizing the solitons as fermions have been discussed by Hung and Tze[28].) The energy of the ground state then agrees with the classical energy (up to an unknown shift due to zero point oscillations[25]). The excitation energy of the higher angular momentum states is determined by the moment of inertia a. We expect the ground state energy of the weak quantum soliton roughly between 3 and 10 TeV and the excitation energy of the first rotationally excited state to be small, ranging between several hundred and several tens of GeV, resulting in a closely spaced rotational band[18]. We expect the excitation energy of the collective vibrational state to be on the order of a few hundred GeV[18].

6.3. *Electromagnetic properties*

Perturbation theory in the weak mixing angle θ_w is an excellent approximation for the electroweak sphalerons[5,6]. Using perturbation theory in the weak mixing angle θ_w also for the weak skrymions, Eilam and Stern[15] have determined the electric charges of the lowest states: the ground state $l = 0$ is neutral, while the isotriplet $l = 1$ has charges 0 and ± 1. Sphalerons have a very big classical magnetic dipole moment, therefore the electromagnetic properties of the weak skyrmions may also hold surprises.

7. References

1. T. H. R. Skyrme, *Proc. Roy. Soc.* **A260**, 127 (1961); *Nucl. Phys.* **31**, 556 (1962).
2. E. Witten, *Nucl. Phys.* **B223**, 422, 433 (1983).
3. G. S. Adkins, C. R. Nappi and E. Witten, *Nucl. Phys.* **B228**, 552 (1983).
4. J. M. Gipson and H. C. Tze, *Nucl. Phys.* **B183**, 524 (1981); J. M. Gipson, *Nucl. Phys.* **B231**, 365 (1984).
5. F. R. Klinkhamer and N. S. Manton, *Phys. Rev.* **D30**, 2212 (1984).
6. B. Kleihaus, J. Kunz and Y. Brihaye, *Phys. Lett.* **B273**, 100 (1991).
7. J. Kunz and Y. Brihaye, *Phys. Lett.* **B216**, 353 (1989).
8. L. G. Yaffe, *Phys. Rev.* **D40**, 3463 (1989).
9. T. Appelquist and C. Bernard, *Phys. Rev.* **D22**, 200 (1980).
10. V. A. Rubakov, *Nucl. Phys.* **B256**, 509 (1985).
11. J. Ambjørn and V. A. Rubakov, *Nucl. Phys.* **B256**, 434 (1985).
12. S. Weinberg, *Phys. Rev.* **D13**, 974 (1976).
13. L. Susskind, *Phys. Rev.* **D20**, 2619 (1979).
14. G. Eilam, D. Klabucar and A. Stern, *Phys. Rev. Lett.* **56** 1331 (1986).
15. G. Eilam and A. Stern, *Nucl. Phys.* **B294**, 775 (1987).
16. Y. Brihaye and J. Kunz, *Z. Phys.* **C41**, 663 (1989).
17. Y. Brihaye and J. Kunz, *Mod. Phys. Lett.* **A4**, 2723 (1989).
18. Y. Brihaye, B. Kleihaus and J. Kunz, *Phys. Lett.* **B236**, 54 (1990).
19. L. Carson, "Are there electroweak skyrmions?", UAHEP9012 preprint, December 1990.
20. Y. Brihaye, F. Mousset and J. Kunz, *Z. Phys.* **C**, in press.
21. J. Baacke, G. Eilam and H. Lange, *Phys. Lett.* **B199**, 234 (1987).
22. J. J. M. Verbaarschot, *Phys. Lett.* **B195**, 235 (1987).
23. Y. Brihaye and J. Kunz, *Phys. Lett.* **B249**, 90 (1990); *Acta Physica Polonica* **B23**, 513 (1992).
24. Y. Brihaye, C. Semay and J. Kunz, *Phys. Rev.* **D44**, 250 (1991).
25. G. Holzwarth, *Phys. Lett.* **B291**, 218 (1992).
26. C. Hajduk and B. Schwesinger, *Phys. Lett.* **B140**, 172 (1984).
27. A. Dobado and J. Terron, *Phys. Rev.* **D45**, 3090 (1992).
28. P. Q. Hung and C.-H. Tze, *Z. Phys.* **C39**, 371 (1989).

II. Weak and Electromagntic Properties of Baryons

CHIRAL SYMMETRY AND PARITY–VIOLATING MESON–NUCLEON VERTICES*

Ulf-G. Meißner[†]

Universität Bern
Inst. für Theoret. Physik
Sidlerstr. 5
CH-3012 Bern Switzerland

ABSTRACT

In this lecture, I review progress made in the calculations of the parity-violating meson-nucleon interaction regions. The underlying framework is the topological chiral soliton model of the nucleon. Emphasis is put on the computation of theoretically and experimentally accessible nuclear parity violating observables. I stress the importance of the interplay of strong and weak interactions which makes this field interesting and challenging. I also discuss recent developments pointing towards the importance of strange quark admixtures in the proton wave function.

INTRODUCTION

Our understanding of the hadronic weak interactions has progressed considerably in the last two decades. Still, the almost unique tool to study the non-leptonic, strangeness conserving part of the weak Hamiltonian are few-nucleon systems. In general, nuclear parity-violating (pv) observables cannot be calculated reliably enough so that we could deduce stringent limits on the standard model from them. Stated in another way: We are still far away from extracting *e.g.* the Weinberg angle to some decent precision from nuclear parity violation. On the other hand, there are now very sophisticated parametrizations of the strong force between two nucleons available which allow us to test our understanding of the hadronic weak interactions in terms of meson exchanges. Direct W- or Z-exchange between nucleons is wiped out by the hard core of the NN-force, but there still remains a long-range component

* Work supported in part by Deutsche Forschungsgemeinschaft and by Schweizerischer Nationalfonds.

† Heisenberg fellow

of the weak interactions between nucleons, which can be parametrized in terms of pv meson-nucleon interaction vertices. One way to calculate these pv couplings is to make use of the quark model.[1] There is, however, a considerable uncertainty in these calculations which stems from the fact that the pertinent multiquark operators have to be calculated at low energies ($E \ll M_{W,Z}$). Gluonic corrections arise, and unavoidably one enters the non-perturbative regime where the strong coupling constant $\alpha_s(q^2)$ becomes larger than unity. These problems are most pronounced in the case of the pion which dominates the long-range part of the pv potential. The Goldstone-boson-like character has always posed problems for quark-model practitioners. Quite contrary, the recently popular topological soliton models of the nucleon like the Skyrme model[2] and generalizations thereof[3] naturally incorporate the pseudoscalar as well as the low-lying vector multiplets. Here we have reached our starting point for a calculation of the parity-violating meson-nucleon couplings and form factors.[4,5,6,7] The soliton approach to the nucleon is far from being perfect, but it has the conceptual advantage that it allows for a simultaneous calculation of the strong and weak interaction regions, a point which is generally overlooked by quark model enthusiasts. Furthermore, nuclear parity violation can also be used as a testing ground to find out the limitations of the soliton scenario — often more can be learned from the failures of a model than from its successes.

Another interesting aspect of nuclear parity violation is the quest for finding few-nucleon systems which can be calculated with some reliability and where the experimenters have a change of detecting a clear signal. Here, I will focus on two rather different systems. In proton-proton scattering, one can observe longitudinal asymmetries of the order 10^{-7}, which appear to be awfully small. However, progress in experimental techniques now allows for experiments with an accuracy of $\delta A_L \simeq \pm 1.0 \cdot 10^{-8}$ and therefore a fairly sensitive test of the meson-exchange picture underlying the theoretical description of this process. A very different system is the nucleus ^{18}F, in which nuclear amplification takes place and the observed circular polarization of emitted γ-rays is of the order $2 \cdot 10^{-4}$. Luckily for the theorists, the β-decay of the daughter nucleus ^{18}Ne allows one to gauge the rather involved shell model calculations,[8] although sceptical minds tend to look at these calculations with a certain dose of disbelieve. As we will see, too few "good" nuclear systems are considered at present and therefore the restrictions on the pv meson-nucleon couplings are by far too soft.

Finally, during the last year, effective field theory methods have been used to gain further insight into the strength of the pv meson–nucleon couplings.[9] These results seem to indicate a large enhancement from operators involving strange quarks to various coupling constants. Furthermore, some couplings not considered so far (like *e.g.* the pv $\pi N\gamma$ vertex) might be of importance. I will discuss these topics in the end of this lecture.

PV MESON-NUCLEON INTERACTION REGIONS

In the meson-exchange parametrization of the weak nuclear force, one usually only considers the exchange of charged pions and the vector mesons ρ and ω. CP invariance does not allow for the coupling of neutral scalar or pseudoscalar mesons to nucleons, eliminating the infamous scalar mesons, the η, η' and the π^0 (the $\delta^{\pm}$ are considered a form factor corrections to the $\pi^{\pm}$-exchange). Then there remains the $\phi(1020)$ — its coupling to the nucleon is generally supposed to be OZI-suppressed and not considered.[10] This might, however, be a too simplistic approach in light of the discussion surrounding the admixture of strange operators into the proton's wavefunction. At present no final conclusion can be drawn and I will make life easy on us and neglect the ϕ for the time being. I will pick up this theme in the final section.

Unavoidably I will have to define the basic couplings which parametrize the pv nuclear potential. For the pion, there is only a $\Delta I = 1$ (isovector) coupling (to first order in the pion field)

$$\mathcal{L}^{\text{pv}}_{\pi N} = -\frac{G_\pi(q^2)}{\sqrt{2}}\frac{E}{M_N}\chi_f^\dagger\,(\vec{\tau}\times\vec{\pi})_3\,\chi_i \tag{1}$$

with $\chi_{i,f}$ denoting nucleon spinors, q^2 the invariant momentum transfer squared at the πN-vertex and I have (for simplicity) given the non-relativistic reduction of this vertex. In the case of the ω, $\Delta I = 0$ and $\Delta I = 1$ couplings are possible,

$$\mathcal{L}^{\text{pv}}_{\omega N} = \chi_f^\dagger\left[h^0_\omega(q^2) + h^1_\omega(q^2)\tau^3\right]\left[\frac{E}{M_N}\vec{\sigma}_T + \vec{\sigma}_L\right]\cdot\vec{\omega}\chi_i \tag{2}$$

with $E = \left(M_N^2 + \vec{q}^{\,2}/4\right)^{1/2}$ and $\vec{\sigma}_{L,T}$ the longitudinal and transverse spin-operator, respectively. For the ρ, one has isoscalar, isovector and isotensor vertices

$$\begin{aligned}\mathcal{L}^{\text{pv}}_{\rho N} = \chi_f^\dagger&\left[h^0_\rho(q^2)\tau^a + h^1_\rho(q^2)\delta^{a3} + \frac{h^2_\rho(q^2)}{2\sqrt{6}}\left(3\tau^3\delta^{a3} - \tau^a\right)\right]\\ \times&\left[\frac{E}{M_N}\vec{\sigma}_T + \vec{\sigma}_L\right]\cdot\vec{\rho}^{\,a}\chi_i - \frac{iE}{2M_N^2}h'^1_\rho(q^2)\chi_f^\dagger\vec{\sigma}\cdot\vec{q}\left(\vec{\tau}\times\vec{\rho}^{\,0}\right)_3\chi_i\;.\end{aligned} \tag{3}$$

Generally, the coupling $h_\rho^{'1}$ is neglected,[11] but I will not follow this historical path here. From Eqs. (1) – (3) it becomes obvious that the pv interaction regions are characterized by coupling constants $h_M = h_M(q^2 = 0)$ and form factors $F_M^{\rm pv}(q^2) = h_M(q^2)/h_M$ (in case of the pion, I use $G_\pi \equiv h_\pi$).

In the topological chiral soliton model[12] underlying the calculation of the pv vertices, nucleons arise as solitons of a non-linear meson theory. This non-linear meson theory is constructed in harmony with chiral symmetry and anomaly constraints and all its parameters are fixed from mesonic reactions like *e.g.* ($\rho^0 \to \pi^+\pi^-$, $\omega \to \pi^+\pi^-\pi^0$, $\omega \to \pi^0\gamma, \ldots$. The Lagrangian and its parameters are completely determined in the meson sector, and the calculation of nucleon properties proceeds without any new parameters, *i.e.* no fudging is possible! That is certainly an appealing aspect of the soliton approach to the nucleon and it poses several restrictions. Of course, the model does not perfectly predict all nucleon properties.

Now: How can we calculate the pv couplings appearing in Eqs. (1) – (3)? For that, we consider the current × current form of the weak Hamiltonian with the currents being of (V–A)-type. To pick out the pv pieces, consider the $\Delta I = 0, 1$ or 2 components of products like $V_\mu A^\mu$ and $I_\mu A^\mu$, with V_μ the vector, I_μ the isoscalar and A_μ the axial current. These currents are already given in terms of the meson fields which make up the soliton, and their explicit expressions can be found *e.g.* in Ref. [5]. One then makes use of the "background-scattering" method, which amounts to an expansion of the meson fields around the soliton background. For the pion, we write[13]

$$\vec{\pi} = \vec{\pi}_S + \vec{\pi}_f \tag{4}$$

and similarly for ρ and ω. $\vec{\pi}_S$ is the "hard" component of the pion field making up the soliton and $\vec{\pi}_f$ a small pionic fluctuation ("soft" component). Inserting the expressions (4) into the soliton currents and these into the weak Hamiltonian, all one has to do is to find the terms linear in $\vec{\pi}_f$ (or $\vec{\rho}_f$ or ω_f). Quantizing the respective operators which are given in terms of the collective variables $(A, \dot{A})$,[2,3] one can immediately read off the coupling constants and form factors for the meson under consideration. In particular, one cannot only construct pv meson-nucleon vertices, but also the equivalent pv $N\Delta$-transition couplings. I will come back to this point later on. For details, the interested reader should consult Refs. [5,6]. I will not give any explicit formula here, but rather make a few comments on the results of Ref. [5]. First, the pv πN coupling is completely dominated by the neutral current contribution.

	KM	DDH	DZ	AH	RR
$\tilde{F}_\pi$	0.6	10.8	3.1	5.0	0.0→27.1
F_0	5.9	15.9	11.5	8.0	−15.9→43.0
F_1	0.2	0.3	−0.5	0.3	−0.1→0.6
F_2	5.3	13.3	9.3	9.8	10.6→ 15.3
H_1	1.7	0.5	0.0	0.0	—
G_0	24.5	8.0	16.3	27.0	−23.9→ 43.1
G_1	4.1	4.8	9.2	10.0	−3.3→ 8.0

Table: Effective weak meson-nucleon coupling constants in units of 10^{-7}. We present the result of the soliton model calculation of Kaiser and Meißner(KM) [5,6] together with the quark model results of Desplanques *et al.*(DDH) [1] as well as Dubovik and Zenkin(DZ) [16]. The value for $h_\rho^{'1}$ in the column DDH is taken from Holstein's calculation in ref.[11]. The "reasonable ranges"(RR) defined by DDH are also given. The column AH gives the best fit values of Adelberger and Haxton.[17]

The charged current contribution can be estimated in the factorization approximation, $G_\pi^{CC} = \cos^2\theta_c < \pi|A_\mu|0 >< p|V_\mu|n >= G_F \cos^2\theta_c f_\pi(M_n - M_p)$. The electromagnetic mass difference of the neutron and the proton is well-reproduced in the model,[4] whereas a strong part of $M_n - M_p$ is somewhat underestimated.[14]. Taking as an upper limit the empirical value $M_n - M_p \simeq 1.3$ MeV, we find $A = G_\pi^{NC}/G^{CC} \sim 13.5$, consistent with previous estimates[15] and the quark model calculations of Ref. [1] ($A \simeq 24$). The numerical value for the effective pion-exchange coupling, $\tilde{F}_\pi = g_{\pi NN} G_\pi/\sqrt{32}$ with $g_{\pi NN}$ the strong πN coupling, is considerably smaller in the soliton model than in the quark model, we find $\tilde{F}_\pi^{sol} = 0.6$ ve rsus $\tilde{F}_\pi^q = 10.8$,[8] or $\tilde{F}_\pi^q = 3.1$ [16] (in units of 10^{-7}).

For the vector meson couplings, the results are less different. Using the standard definitions $F_i = -g_{\rho NN} h_\rho^i/2$, and $G_i = -g_{\omega NN} h_\omega^i/2$ and $H_1 = -g_{\rho NN} h_\rho^{'1}/4$, we find that the soliton and the quark model predict the following pattern for the ρ-couplings: $F_0 \gtrsim F_2 \gg F_1$. The absolute values of the constant s F_i, are, however, reduced in the soliton approach. For H_1, the soliton model predicts a value three times as large as the quark model.[11] In the case of the ω, all calculations give $G_0 > G_1$, but G_0 is considerably enhanced in the soliton approach and close to the "best-fit" estimate of Adelberger and Haxton.[17] These results are summarized in the table.

As already stated, the calculation of pv $N\Delta M$ $(M = \pi, \rho, \omega)$ transition vertices proceeds along the same lines and only differs at the step when one quantizes the collective coordinates. In our model, the pv $\pi N\Delta$ coupling has $\Delta I = 0$ and 1 components and reads non-relativistically

$$\mathcal{L}^{\rm pv}_{\Delta N\pi} = \left(\frac{E}{4M_N^3}\right) \left[\delta^{ab} h^0_\Delta(q^2) + \epsilon^{3ab} h^1_\Delta(q^2)\right] \chi^\dagger_\Delta \vec{S} \cdot \vec{q}\vec{\sigma} \cdot \vec{q} T^a \chi_N \pi^b + \text{h.c.} \tag{5}$$

with $\vec{S}$ and $\vec{T}$ the conventional $N\Delta$ transition spin and isospin operators, respectively. It is easy to convince oneself that $h^0_\Delta(q^2) = 0$ in this model,[6] naively, a non-vanishing isoscalar $\pi\Delta N$-vertex would lead to a non-zero CP-violating πN-vertex (a more fool-proof argument is given in Ref. [6]). The isovector vertex does not vanish, and for the "minimal" model[18] we find

$$h^1_\Delta(0)/G_\pi = 1.10 \ . \tag{6}$$

The presence of $\pi\rho\omega$-correlations in the effective action tends, however to decrease this ratio. For the $V\Delta N$-couplings, we find ($V = \rho$ or ω):

$$h^i_{\rho\Delta N}(q^2) = \frac{3}{\sqrt{2}} h^i_\rho(q^2) \qquad (i = 0,' 1, 2) \ , \qquad h^1_{\omega\Delta N}(q^2) = \frac{3}{\sqrt{2}} h^1_\omega(q^2) \tag{7}$$

and $h^1_{\rho\Delta N}(q^2) = h^0_{\omega\Delta N}(q^2) \equiv 0$. These predictions are insofar interesting since in the seventies it was argued that *e.g.* the $\rho\Delta N$-couplings are negligible[19] — quite in contrast to our results. A recently performed quark model calculation by Feldman *et al.*[29] along the lines of DDH gives results rather different from the soliton model predictions. The source of these discrepancies is not yet understood. In a similar fashion, one easily derive the corresponding $M\Delta\Delta$ vertices,

$$\begin{aligned} &G_{\pi\Delta\Delta}(q^2) = G_\pi(q^2) \\ &h^i_{\rho\Delta\Delta}(q^2) = \frac{1}{5} h^i_\rho(q^2) \ (i = 0,' 1, 2) \ ; \ h^1_{\rho\Delta\Delta}(q^2) = h^1_\rho(q^2) \\ &h^0_{\omega\Delta\Delta}(q^2) = h^0_\omega(q^2) \ ; \ h^1_{\omega\Delta\Delta}(q^2) = \frac{1}{5} h^1_\omega(q^2) \end{aligned} \tag{8}$$

The calculation of the associated weak form factors proceeds in a straightforward way.[6] In Fig. 1 we show the weak πN form factor $G_\pi(q^2)$ in comparison with the equivalent strong form factor $G_{\pi NN}(q^2)$ as well as the monopole with cut-off $\Lambda =$ 1 GeV. As it turns out, all form factors can be fitted by monopoles at low $\vec{q}^{\,2}$,

$$F^M(\vec{q}^{\,2}) = h_M \frac{\Lambda^2}{\Lambda^2 + \vec{q}^{\,2}} \tag{9}$$

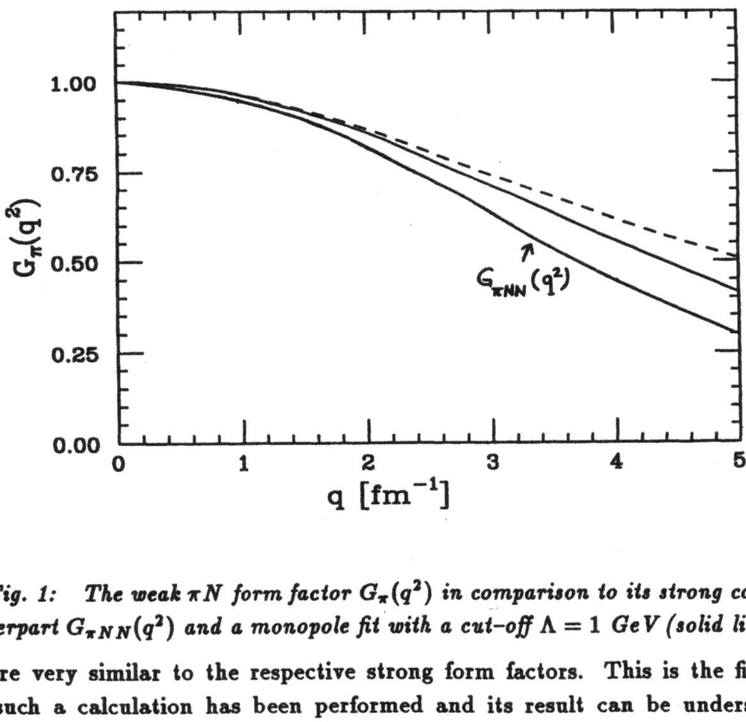

Fig. 1: The weak πN form factor $G_\pi(q^2)$ in comparison to its strong counterpart $G_{\pi NN}(q^2)$ and a monopole fit with a cut-off $\Lambda = 1$ GeV (solid line).

and are very similar to the respective strong form factors. This is the first time that such a calculation has been performed and its result can be understood as follows: The intrinsic scale of the meson-nucleon interaction regions is set by the topological baryon charge radius, $r_B \simeq 0.5$ fm. From that, one can deduce a cut-off scale $\Lambda \simeq \sqrt{6}/r_B \simeq 1$ GeV. It is, however, not that simple because the dynamical treatment of the vector mesons modifies this result. Defining by R_M the ratio of the (averaged) weak to strong MN cut-offs (all form factors of monopole type), we find

$$R_\pi = 1.15 \quad , \quad R_\rho = 0.91 \quad , \quad R_\omega = 0.77 \tag{10}$$

which justifies within the accuracy of the model the assumption of taking the same form factors for the strong and weak vertices as it was done *e.g.* by Driscoll and Miller[20] in their study of the pv *pp*-interaction.

Another topic which can be discussed in the framework of the chiral soliton model are the corrections of pv two–pion exchange. One motivation to do this is that correlated 2π–exchange gives rise to the intermediate range attraction of the parity–conserving NN interaction. Furthermore, recent investigations point towards the importance of pv 2π exchange even below production threshold.[31] Using the soliton model, Norbert Kaiser and I have shown that the inclusion of pion loops gives the intermediate range attraction with just the right strength as compared to the Paris potential.[32] Similarly, we have worked out corrections to the various pv ρN couplings.[33] The effects of irreducible two–pion corrections are generally small, of the order of $10\ldots20\%$. This is in agreement with older dispersion–theoretical investigations.[34] So we finally have all the tools at hand to make contact to experiment.

PARITY-VIOLATION IN PROTON-PROTON SCATTERING

The simplest system in which one can probe certain components of the weak pv inter-nucleon force is the two nucleon system. By scattering polarized protons off a hydrogen target, parity violation shows itself in a non-vanishing longitudinal asymmetry,

$$A_L = \frac{\sigma^+ - \sigma^-}{\sigma^+ + \sigma^-} \tag{11}$$

assuming a 100% longitudinal polarization of the beam and having taken care of the Coulomb-corrections $\sigma^\pm$ are the cross-sections for scattering positive/negative helicity protons from an unpolarized target. The calculation of this process in the DWBA as pioneered by Brown *et al.*[21] goes as follows. One splits the total scattering amplitude $\mathcal{F}_{ss'}$ into a strong and weak part

$$\mathcal{F}_{ss'} = F_{ss'} + f_{ss'} \tag{12}$$

for total spins s and s'. Now it is of utmost importance to take into account the strong distortions, *i.e.* calculating the weak scattering amplitude with distorted waves $\psi_s^{(-)}$ and $\psi_s^{(+)}$, *i.e.* $f_{ss'} = < \psi_s^{(-)} | V_{\rm pv} | \psi_s^{(+)} >$ with $V_{\rm pv}$ the pv one-meson-exchange potential. It should be pointed out that the strong distortions govern the energy-dependence of the analyzing powers A_L. Recently, Driscoll and Miller[20] have done the most complete calculation based on the Bonn-Potential[21] for the strong force and an equivalently constructed weak potential with the pv couplings taken from the quark model[1] and using the same vertex functions for the weak and strong form factors. I should point out here that for obvious reasons there is no pion contribution to this process and one essentially tests the vector-meson couplings $h_\rho^{pp} = h_\rho^0 + h_\rho^1 + h_\rho^2/\sqrt{6}$ and $h_\omega^{pp} = h_\omega^0 + h_\omega^1$.

Recently, Doug Driscoll and I have repeated this calculation[22] by including the soliton model predictions $h_\rho^{pp} = -5.15 \cdot 10^{-7}$ and $h_\omega^{pp} = -8.20 \cdot 10^{-7}$. The resulting curve for A_L is shown in Fig. 2, for the quark[1,16] and the soliton model.[22] The shape of the curve as predicted by the soliton model follows closer the empirical trend suggested by the low-energy data.[23] In fact, a χ^2 calculation for the three curves shown in fig.2 gives $\chi^2 = 34/3$ (DDH), $\chi^2 = 26/3$ (DZ) and $\chi^2 = 8/3$ (KM) as disussed in ref.26 (at that time, the Bonn result was not available). Also, the maximum at $p_{\rm lab} = 0.95$ GeV/c is flatter than in the calculation using the quark model parameter. Furthermore, the energy at which the asymmetry changes sign is larger than the quark model predicts, which can be traced back to the fact that in the soliton model $h_\omega^{pp} > h_\rho^{pp}$, in contrast to the quark model with $h_\omega^{pp} < h_\rho^{pp}$. Of particular interest is the value of A_L at 222 MeV. This is the energy selected for an upcoming pp parity violation measurement at TRIUMF because $\delta(^1S_0) + \delta(^3P_0) = 0$ at this energy and the $j = 0$ contribution to the analyzing power consequently vanishes. The measurement of the dominant $j = 2$ contribution gives a different combination of h_ρ^{pp} and h_ω^{pp} than the $j = 0$ contribution to A_L, which is already measured at 15 and 45 MeV.[23]. The predictions using the quark[1] and soliton model[5] weak parameters, respectively, differ by $\Delta A_L = 4.6 \cdot 10^{-8}$. To be more precise, the various predictions are:

$$A_L(\mathrm{DDH}) = 5.0 \cdot 10^{-8}\ ,\ A_L(\mathrm{DZ}) = 2.6 \cdot 10^{-8}\ ,\ A_L(\mathrm{KM}) = 3.7 \cdot 10^{-9} \tag{13}$$

The projected long-term accuracy of the upcoming TRIUMF experiment is $(\delta A_L)_{\rm stat} \simeq \pm 1 \cdot 10^{-8}$,[24] which should be sufficient to discriminate between these two predictions. Notice that a similar experiment is also planned at COSY.[30] This experiment should set rather stringent limits on some combinations of the pv ρN and ωM couplings. To stress it again, the pp system is a particularly good example of the interplay of weak and strong interactions and it is therefore mandatory to treat both of them consistently (for further discussion, see Refs. [17,20]). A possible loophole to all of this will be discussed in the last section.

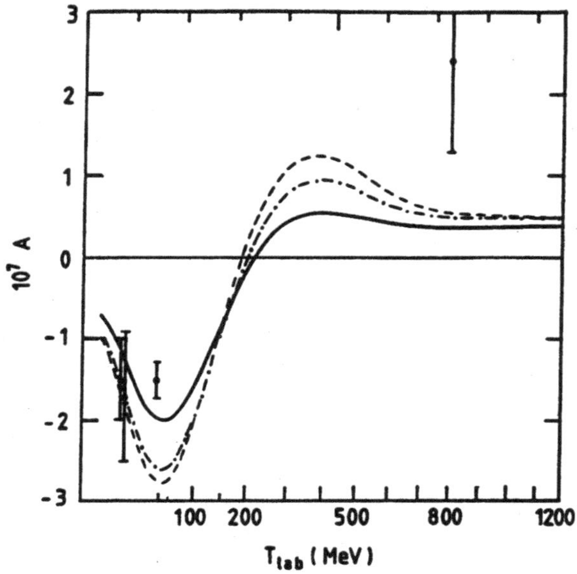

Fig. 2: Parity-violating asymmetry in pp-scattering. The solid line gives the prediction based on the weak couplings as given by the soliton model,[22] whereas the dashed and dashed-dotted lines are based on the quark model calculations of refs.1 and 16, respectively.

PARITY VIOLATION IN ^{18}F AND DEUTERON PHOTODISINTEGRATION

The nucleus ^{18}F is what I called a "good system" before. It exhibits "nuclear amplification" in that it has two close-by levels of opposite parity which are separated by only 39 keV (the next level which could mix with these is approximately 2 MeV away) and the dominant E1-transition from the level at 1.081 MeV to the ground state is suppressed, which leads to $|M1/E2| \simeq 112$. The M1-transition is, of course, only possible because of the mixing of the opposite parity-levels. Altogether, this amounts to an amplification of approximately $(2/0.039) * 112 \simeq 6 \cdot 10^3$ (for further details, see Ref. [17]). Theoretically, one can calibrate the shell-model calculation to extract the pv circular polarization from the β-decay of ^{18}Ne, because the pion-exchange of this β-decay up to an overall isospin rotation,[25] and therefore calculation and measurement of ^{18}Ne $(0^{+}1) \longrightarrow$ ^{18}F $(0^{-}0)$ β-decay serves as a "gauge" for the accuracy and amounts effectively to a large model-independent limit on the weak pion decay constant. The latter dominates completely this $\Delta I = 1$ pv observable, and one can dedūce a limit on $\tilde{F}_\pi$, $\tilde{F}_\pi \leq 3.4 \cdot 10^{-7}$. Here, we have used the experimental circular polarization, $|P_\gamma(^{18}\mathrm{F})| = (0.17 \pm 0.58) \cdot 10^{-3}$. The quark model prediction of Ref. [1], $\tilde{F}_\pi = 10.8 \cdot 10^{-7}$, is clearly in contradiction to this result.

What does the soliton model give? Of course, $\tilde{F}_\pi$ is considerably reduced, so we expect a smaller asymmetry. The vector meson contribution is enhanced, and taking nuclear structure calculation from Ref. [17], we predict $P_\gamma(^{18}\mathrm{F}) = 2.2 \cdot 10^{-4}$, not far from the central value of the experiment. We should, however, not put too much emphasis on this closeness of the experimental and theoretical number, but rather state that the strength of the pv πN coupling should still be considered as the main theoretical puzzle. I am sure that $\tilde{F}_\pi$ should come out smaller than in Ref. [1], but whether it is as small as predicted by the soliton model can only be checked if more theoretical and experimental information on the $\Delta I = 1$ part of the pv nuclear force are available. One particularly interesting candidate to study in more detail would be the reaction $\vec{n} + p \rightarrow d + \gamma$ or the inverse process $\vec{\gamma} + d \rightarrow n + p$. A calculation of the circular asymmetry as a function of the photon energy has been performed some time ago by Oka.[27] I have used this calculation in Ref.28 to investigate the sensitivity of the circular asymmetry A_L to the various pv couplings. Considering photon energies below 30 MeV, $A_L(\omega)$ increases linearly with energy when one uses the quark model couplings of DDH or DZ, with the slope determined by the strength of the pv πN coupling. For the DDH-case the pion contribution is completely dominant for all energies, whereas for the DZ-parameters the reduced πNN strength leads to an overall decrease of $A_L(\omega)$. For the soliton model, however, things are significantly different. First, between 1 and $20\,MeV$, $A_L(\omega)$ shows a flat minimum at about $\omega_L \approx 12\,MeV$ and only after $\omega_L \geq 20\,MeV$ a gradual rise in $A_L(\omega)$ sets in. Also, the overall magnitude of the effect is an order of magnitude smaller for the weak parameters predicted by the soliton model. It would be worthwile to measure the asymmetry say at 10 and $20\,MeV$ incident energy, although the effect is small, the tremendously different slope of $A_L(\omega)$ should be detectable in an dedicated experiment. Of course, as already mentioned, a more thorough theoretical study has also to be done. First, a more consistent calculation employing *e.g.* the Bonn-potential and the equivalently constructed weak potential should be performed. Second, the effects of meson-exchange curents, which play an important role in the accurate description of the deuteron properties have to be included. Therefore, these results should only be considered as a guide, but the trends exhibited will certainly not be wiped out by a more elaborate calculation. A more detailed discussion is given in Ref.28.

In the last section, I will discuss some medium renormalization effects which might come to the rescue of the large value for the pv pion–nucleon coupling as predicted by DDH. However, for the deuteron photodisintegration process just discussed, such a renormalization cannot be operative since the deuteron is essentially an ensemble of two free nucleons.

THE NUCLEON ANAPOLE MOMENT

Apart from the electric dipole moment, there is one other pv coupling of the photon to nucleons (spin-1/2 fields), the so–called anapole moment. It has recently attracted new interest[35] since its contribution might be enhanced considerably in nuclei, similar to the case of ^{18}F just discussed. For on–shell nucleons, current conservation and Lorentz invariance require that pv corrections to matrix elements of the electromagnetic current take the form

$$< N(p')|J^{em}_{\mu,pv}(0)|N(p) >= \frac{a(q^2)}{M_N^2}\bar{u}(p')[\gamma_\nu q^\nu q_\mu - q^2\gamma_\mu]\gamma^5 u(p) \tag{14}$$

with $q^2 = (p' - p)^2$. In the Breit frame, where the photon transfers no energy, this matrix element reads

$$< N(\vec{q}/2)|J^{em}_{\mu,pv}|N(-\vec{q}/2) >= \frac{E\vec{q}^{\,2}}{M_N^3}a(\vec{q}^{\,2})\chi_f^\dagger \vec{\sigma}_T\chi_i \tag{15}$$

with $\vec{\sigma}_T = \vec{\sigma} - \hat{q}\vec{\sigma}\cdot\hat{q}$ the transverse spin operator and $a(q^2)$ the nucleon anapole form factor. The anapole moment has isoscalar and isovector components,

$$a(0) = a_S(0) + a_V(0)\tau_3 \tag{16}$$

In the soliton model, one can easily calculate the anapol moment and form factor.[36] For that, one identifies the matrix element in (15) with the Fourier transform of the pv electromagnetic current. For the usual hedgehog ansätze, its has the general form

$$\vec{J}_{pv}(\vec{r}) = \Gamma_1(r)\vec{\sigma} + \Gamma_2(r)\hat{r}\vec{\sigma}\cdot\vec{r} \tag{17}$$

with $\Gamma_{1,2}(r)$ functions of the various meson profiles whose explicit form we do not need here. However, one immediately encounters a difficulty. Current conservation demands $\Gamma_1'(r) + \Gamma_2'(r) + 2\Gamma_2(r)/r = 0$, where the prime denotes differentiation with respect to r. This condition is not met. Interestingly, if one switches off the ρ–meson fields and considers the so–called ω–stabilized Skyrmion[37], then the divergence condition is fulfilled. This peculiar behaviour might be traced back to the fact that in the isoscalar channel one has exact vector meson dominance (VMD) but not in the isovector one (compare the discussion of Hwang and Nigoyi[38] of VMD and gauge invariance for pv photon–nucleon couplings). To get an idea of the size of the anapole moment, let me crudely rstore gauge invariance by subtracting the pieces which violate current conservation. In that case, the "minimal" model gives $a_s(0) = 4\cdot 10^{-8}$, and the extension of the pv γN vertex is given by a mean square radius of about 0.4 fm corresponding to a monopole form factor with a cut–off of Λ = 1.23 GeV. At present, I can not offer a solution to the problem concerning the violation of current conservation, but I suspect that is it related to the rather crude quantisation procedure used (which is known to do harm to e.g. the chiral algebra of the charges[38]).

RECENT DEVELOPMENTS

There are some recent developments (partly outside the soliton model) which indicate some interesting new effects and might lead to a reconsideration of some topics discussed so far. The first one is due to a calculation of Dai, Savage, Liu and Springer.[9] They calculate an effective Hamiltonian for $\Delta I = 1$ nuclear parity violation, including the effects of the heavy quarks s, c and b. At the scale of the W-boson mass, the pv $\Delta I = 1$ Hamiltonian is, of course, well known and given in terms of eight four–quark operators with known Wilson coefficients. Integrating out the b and the c quark successively, one has a tower of effective theories. For each of these, the anomalous dimension matrix is calculated to one loop in the QCD corrections and the effective field theories are matched. By this procedure, one can finally go down to the hadronic scale of Λ_χ = 1 GeV and compare the Wilson coefficients $C_i(\Lambda_\chi)$ with the original ones, $C_i(M_W)$. The important observation made in ref.9 is that the operators involving strange quarks are substantially larger than the ones involving only the up and down quarks, approximately

$$\frac{C_i^{\mathrm{strange}}(\Lambda)}{C_i^{\mathrm{non-strange}}(\Lambda)} \sim \frac{1}{\sin^2\Theta_W} \sim 5 \tag{18}$$

The authors of ref.9 did not compute hadronic matrix elements at the scale Λ_χ, but resorted to the meson–exchange picture and large N_c arguments. In that case ($N_c \to \infty$), factorization can be justified and one finds for the $\rho^0 N$ pv matrix element

$$\begin{aligned} < \rho^0 N | H_{pv}^{\Delta I=1} | N > = \frac{1}{3} G_F \sin^2\Theta_W f_\rho \epsilon_\rho^{*\mu} \{ & -0.95 < N | \bar{u}\gamma_\mu\gamma_5 u + \bar{d}\gamma_\mu\gamma_5 d | N > \\ & + 13.4 < N | \bar{s}\gamma_\mu\gamma_5 s | N > \} \end{aligned} \tag{19}$$

Consider first the case were the strange matrix element vanishes (like in DDH). In that case, one finds $h_\rho^{(1)} = -1.9 \cdot 10^{-8}$, quite consistent with the DDH value. However, the large relative factor in front of the new, un–colored strange contribution can easily alter this result by an order of magnitude. Combining the EMC–data and hyperon decay rates, one has $< N|\bar{u}\gamma_\mu\gamma_5 u + \bar{d}\gamma_\mu\gamma_5 d|N > \simeq - < N|\bar{s}\gamma_\mu\gamma_5 s|N > \simeq -(0.2 \pm 0.1)S_\mu$ with S_μ the nucleon spin vector. In this case, $h_\rho^{(1)} = -2.9 \cdot 10^{-7}$, which is an enhancement of a factor 15. If that were true, all previous estimates of pv meson–nucleon couplings can be off the mark by large factors. However, we should not forget that in last years many of the matrix elements which indicated a large contribution of the stramge quark sea to various nucleon properties have been tamed, the prime example being the famous πN Σ term.

In a similar fashion, Kaplan and Savage[40] have recently reanalyzed the pv pion–nucleon couplings making use of baryon chiral perturbation theory. They have derived the most general pv and CP–conserving effective pion–nucleon–photon Lagrangian to first order in derivatives and first order in the photon field and to all orders in the pion field. This effective Lagrangian is parametrized by a few coupling constants, which are labelled $h_V^{0,1,2}$, $h_A^{1,2}$ and $h_{\pi NN}^1 = G_\pi$. Apart from the standard pv pion–nucleon coupling (discussed before), the authors of ref.40 mainly concentrate on the novel pv $\gamma\pi NN$ and the pv $\pi\pi NN$ vertices (the latter one has been already been considered by nuclear theorists in the seventies). Three different methods are used in ref.40 to estimate the strength of these coupling constants, namely factorization, dimensional analysis and relations to $\Delta S = 1$ hyperon decay matrix elements. From these methods, the dimensional analysis is considered most reliable. The most interesting results of this are 1) a large contribution of the strange quarks to G_π (together with a large value for this coupling), 2) a sizeable strangeness enhancement for the pv $\pi\pi NN$ coupling h_A^1 and 3) a large value for the strength of the pv $\gamma\pi NN$ coupling. Taking these estimates face value, drastic consequences would arise. First, in the case of the ^{18}F experiment, interference between the one–pion exchange (considered so far) and the novel $\gamma\pi NN$ vertex might complicate the analysis of the data and ultimately relax the bound on G_π. Similarly, for the planned TRIUMF and COSY experiments measuring parity violation in pp scattering at 230 MeV, one would have to consider two–pion exchange, not only the conventional one arising from e.g. intermediate Δ resonances, but also the one due to the large pv $\pi\pi NN$ coupling. However, before jumping too far, one should not forget that the results of ref.40 should be considered indicative – more elaborate calculations of the hadronic matrix elements are necessary (using e.g. lattice methods) and also more complex nuclear structure calculations involving these novel couplings have to be performed before one can draw a final conclusion. For more details on these topics, please consult ref.40.

MEDIUM RENORMALIZATION OF G_π ?

There exist ample evidence that suggests scale changes of fundamenral properties of nucleons in nuclei. Some pertinent examples are the first EMC effect, the quenching of the axial–vector coupling constant g_A in nuclear β–decay or the behaviour of the longitudinal and transverse strength functions in quasi–elastic electron scattering off nuclei. These effects are there and they are important, but thier origin still remains to be explained in a consistent treatment of many–body effects and fundamental scale changes of the nucleon properties. The chiral soliton model allows to systematically investigate the constraints from chiral symmetry on such possible medium modifications.[41,42] The basic idea is the following: In the soliton model, baryon properties are fixed once the mesonic input is determined. We know, however, that meson masses and coupling constants change in the baryon–rich environment.[43] This immediately leads to density or temperature–dependent nucleon properties.[44] For the meson sector, I will use here results from the Nambu–Jona-Lasinio model which have been obtained in collaboration with Véronique Bernard.[42] For not too large densities ρ, one finds for the pion decay constant and the vector and scalar meson masses (all other quantities are essentially unaffected)

$$F_\pi^* = F_\pi(0)[1 - R_\pi \frac{\rho}{\rho_0}] \, , \; m_V^* = m_V(0)[1 - R_V \frac{\rho}{\rho_0}] \, , \; m_\sigma^* = m_\sigma(0)[1 - R_\sigma \frac{\rho}{\rho_0}] \qquad (20)$$

where the '*' denotes quantities in the medium and ρ_0 is the nuclear matter density. The range of values for $R_{\pi,V,\sigma}$ is discussed in ref.42. For simplicity, let me take an universal and equal value, $R_\pi = R_V = R_\sigma = R$. This is not a direct consequence of the NJL model but compatible with it. For the sake of the argument I will make here, this simplification is justified. In ref.42, which is a widely overlooked paper, I have shown that most of the pv meson–nucleon couplings are very sensitive to such medium effects, quite in contrast to their strong counterparts. In particular, the most important pion–nucleon couplings show the following medium renormalization (for R = 0.2 and at nuclear matter density)

$$\frac{G_\pi^*(\rho_0)}{G_\pi(0)} = 0.65 \quad , \quad \frac{g_{\pi NN}^*(\rho_0)}{g_{\pi NN}(0)} = 0.99 \qquad (21)$$

and similar results for the vector meson couplings. One can understand this very different behaviour if one takes a closer look at the expressions for the various coupling constants. Using the dimensionless variable $x = gF_\pi r$, with g the universal vector-meson–pion coupling, one notices that the weak couplings depend one much higher powers of F_π than their strong counterparts and thus are more sensitive to medium modifications. A more detailed account of this can be found in ref.42.

Finally, let me point out some recent work by Grach and Shmatikov[45] which concerns yet another mechanism to bring down the value of the pv πN coupling in the medium. The basic idea of thier work is that the rescattering of emitted pions leads to a strong suppression of G_π (the basic Feynman diagrams are shown in fig.3). Using monopole form factors with a cut off $\Lambda \simeq 7M_\pi \simeq 1$ GeV to regulate the diverging loop integrals, they find

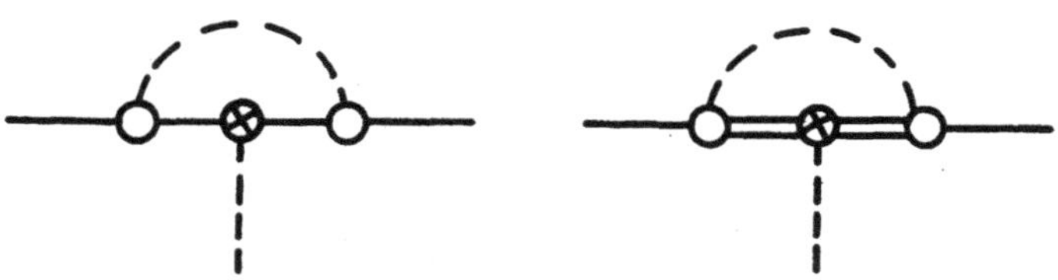

Fig. 3: Strong pion rescattering in the medium.[45] The solid, double and dashed lines denote nucleons, Δ's and pions, in order. Strong meson–nucleon vertices are depicted by open circles and weak vertices by the crossed circles.

$$\begin{aligned} G_\pi^{(r)} &= G_\pi\left(1 + \frac{g_{\pi NN}^2}{8\pi^2} I_1 + \frac{g_{\pi\Delta N}^2}{8\pi^2}\frac{14}{27} I_2\right) \\ &= G_\pi(1 - 0.76 + 0.01) \end{aligned} \tag{22}$$

which leads to

$$\tilde{F}_\pi^{(r)} = G_\pi^{(r)} g_{\pi NN}/\sqrt{32} \simeq 2.9 \cdot 10^{-7} \tag{23}$$

which is below the bound from the ^{18}F experiment. This is an interesting suggestion, but it definitively needs a better treatment (better regularisation procedure) and should also be applied to the other pv meson–nucleon couplings. Also, one should understand the relation to the soliton model results discussed before.

OPEN PROBLEMS

Instead of rephrasing what I have said so far, let me just mention the two salient problems which have to be addressed in the framework of the chiral soliton model to allow for a deeper understanding of the pv meson–nucleon interaction regions.

- Realistic versions of the three flavor Skyrme model are now available. They do not indicate a large strange component in proton wave function. It would be worthwhile and necessary to extend the analysis of the pv interaction region discussed here. This would also allow to addres such questions like the strength of the ϕ–couplings and the relation to the $\Delta S = 1$ hyperon decay matrix elements. Ultimately, such calculations will shed some light on the recent developments concerning the possible enhancement of various weak couplings due to the strange color–singlet operators.
- An old problem is whether the soliton model calculations should be supplemented by strong interaction enhancement factors or whether these are already contained in the non–perturbative soliton currents. This question was to some extent addressed in ref.7, where it was argued that the inclusion/omission of these factors would at most lead to uncertainties of the order of 30 per cent, i.e. lead to corrections within the accuracy of the model. To my opinion, this question is not yet settled. Its resolution will also bring about the answer to the question of including operators which are not of the canonical $V_\mu A^\mu$–type.

ACKNOWLEDGEMENTS

It is my pleasure to thank the organizers for their excellent work which made this meeting truely memorable. I would like to thank V. Bernard, D. Driscoll and N. Kaiser for many enjoyable collaborations. I have also profited from various discussions with B. Desplanques.

REFERENCES

1. B. Desplanques, J. F. Donoghue and B. R. Holstein, *Ann. Phys.* (NY) **124** (1980) 449.
2. T. H. R. Skyrme, *Nucl. Phys.* **31** (1962) 556; U.-G. Meißner and I. Zahed, *Adv. Nucl. Phys.* **17** (1986) 143.
3. U.-G. Meißner, *Phys. Rep.* **161** (1988) 213.
4. N. Kaiser and U.-G. Meißner, *Nucl. Phys.* **A489** (1988) 671.

5. N. Kaiser and U.-G. Meißner, *Nucl. Phys.* **A499** (1989) 699.
6. N. Kaiser and U.-G. Meißner, *Nucl. Phys.* **A510** (1990) 759.
7. M. Shmatikov, Kurchatov Institute of Atomic Energy preprint, IAE–4636/2 (1988); I. L. Grach and M. Shmatikov, *Z.Phys.* **C44** (1989) 393.
8. E. G. Adelberger *et al. Phys. Rev.* **C27** (1988) 2833.
9. J. Dai, M. J. Savage, J. Liu and R. Springer, *Phys. Lett.* **B271** (1991) 403.
10. H. Genz and G. Höhler, *Phys. Lett.* **B61** (1976) 389.
11. B. R. Holstein, *Phys. Rev.* **D23** (1981) 1613.
12. P. Jain, R. Johnson, U.-G. Meißner, N. W. Park and J. Schechter, *Phys. Rev.* **D37** (1988) 3252; U.-G. Meißner, N. Kaiser, H. Weigel and J. Schechter, *Phys. Rev.* **D39** (1989) 1956, (E) **D40** (1989) 262.
13. H. Schnitzer, *Phys. Lett.* **B139** (1984) 217.
14. P. Jain *et al., Phys. Rev.* **D40** (1989) 855.
15. M. Gari and J. H. Reid, *Phys. Lett.* **53B** (1974) 237.
16. V. N. Dubovik and S. V. Zenkin, *Ann. Phys.* (NY) **172** (1986) 100.
17. E. G. Adelberger and W. C. Haxton, *Ann. Rev. Nucl. Part. Sci.* **35** (1985) 501.
18. U.-G. Meißner, N. Kaiser and W. Weise, *Nucl. Phys.* **A466** (1987) 685.
19. G. N. Epstein, *Phys. Lett.* **B71** (1977) 267; B. H. J. McKellar and P. Pick, *Nucl. Phys.* **B22** (1970) 265.
20. D. E. Driscoll and G. E. Miller, *Phys. Rev.* **C39** (1989) 1951; **C40** (1989) 2159.
21. R. Machleidt, K. Holinde and C. Elster, *Phys. Rep.* **149** (1987)1.
22. D. E. Driscoll and U.-G. Meißner, *Phys. Rev.* **C41** (1990) 1303.
23. D. E. Nagle *et al.*, AIP Conference Proceedings No. 51, p. 224; S. Kistryn *et al. Phys. Rev. Lett.* **58** (1987) 1616; P. D. Eversheim *et al., Phys. Lett.* **B256** (1991) 11.
24. S. Page, in TRIUMF brown report 89–5, 1989.
25. W. C. Haxton, *Phys. Rev. Lett.* **46** (1981) 698; C. Bennett, M. M. Lowry and K. Krien, *Bull. Am. Phys. Soc.* **25** (1980) 486.
26. D. E. Driscoll, PhD Thesis, University of Washington, 1990, unpublished.
27. T. Oka, *Phys. Rev.* **D27** (1983) 523.
28. U.-G. Meißner, *Mod. Phys. Lett.* **A5** (1990) 1703.

29. G. B. Feldman, C. A. Crawford, J. Dubach and B. R. Holstein, *Phys. Rev.* **C43** (1991) 863.
30. D. Eversheim, private communication.
31. R. R. Silver, W. M. Kloet, L. S. Kisslinger and J. Dubach, *Phys. Rev.* **C40** (1989) 2218; M. M. Musakhanov and Y. Y. Podgornov, *Yad. Fiz.* **44** (1986) 709.
32. N. Kaiser and U.-G. Meißner, *Nucl. Phys.* **A506** (1990) 417.
33. N. Kaiser and U.-G. Meißner, *Nucl. Phys.* **A515** (1990) 648.
34. M. Chemtob and B. Desplanques, *Nucl. Phys.* **B78** (1974) 139.
35. W. C. Haxton, E. M. Henley and M. J. Musolf, *Phys. Rev. Lett.* **63** (1989) 949.
36. N. Kaiser and U.-G. Meißner, unpublished.
37. G. S. Adkins and C. R. Nappi, *Phys. Lett.* **B137** (1984) 251.
38. W.-Y. P. Hwang and K. Nigoyi, *Phys. Rev.* **D23** (1981) 1604.
39. N. Kaiser, U. Vogl and W. Weise, *Nucl. Phys.* **A490** (1988) 602.
40. D. B. Kaplan and M. J. Savage, preprint UCSD/PTH 92–04, 1992.
41. U.-G. Meißner, *Phys. Rev. Lett.* **62** (1989) 1013; *Phys. Lett.* **B220** (1989) 1.
42. U.-G. Meißner, *Nucl. Phys.* **A503** (1989) 801.
43. V. Bernard, U.-G. Meißner and I. Zahed, *Phys. Rev. Lett.* **59** (1987) 966; V. Bernard and U.-G. Meißner, *Nucl. Phys.* **A489** (1988) 647.
44. V. Bernard and U.-G. Meißner, *Ann. Phys.* (NY) **206** (1991) 50.
45. I. Grach and M. Shmatikov, *Nucl. Phys.* **A536** (1992) 509.

RIGOROUS CHIRAL CONSTRAINTS AND THE SOLITON MODEL OF THE NUCLEON

NORBERT KAISER
Physik Department T30, Technische Universität München,
James Franck Straße, W-8046 Garching, Germany

ABSTRACT
I consider several nucleon observables like the isovector electromagnetic and scalar radii and the electromagnetic polarizabilities in the Skyrme soliton model. The behaviour of these quantities for the pion mass tending to zero is studied and compared to rigorous results from Chiral Perturbation Theory. One finds that the present treatment of the Skyrme soliton does not reproduce all constraints following from chiral QCD. In many cases this failure can be traced back to the non-commutativity of the chiral limit and the large N_c limit. Improved treatments of nucleons as Skyrme solitons are therefore necessary in order to fullfil all QCD constraints.

1. Introduction

Over the last decade a radically new picture of the nucleon and other baryons has been developed. Baryons are viewed as topological solitons of effective meson theories and all baryon properties can be predicted from few mesonic input parameters[1]. Such a scenario is based on the large N_c expansion of QCD by 't Hooft and Witten[2]. The starting point in any of the effective meson lagrangians is the non-linear σ-model

$$\mathcal{L}_\sigma = \frac{f_\pi^2}{4} Tr(\partial_\mu U \partial^\mu U^\dagger) + \frac{f_\pi^2 m_\pi^2}{4} Tr(U + U^\dagger - 2) \tag{1}$$

with $f_\pi = 93$ MeV the weak pion decay ($\pi \to \mu\nu_\mu$) constant, $m_\pi = 138$ MeV the pion mass and $U = \exp(i\vec{\tau}\cdot\vec{\pi}/f_\pi)$ incorporates the pseudoscalar pion fields. The effective lagrangian eq.(1) encodes one of the crucial features of low energy strong interaction physics, *i.e.* the spontaneous breakdown of chiral symmetry in QCD and the appearance of pions as (approximate) Goldstone bosons. Furthermore, the non-linear σ-model describes the interaction of the Goldstone bosons ($\pi\pi$ - scattering) in accordance with low energy theorems based on current algebra. Expanding eq.(1) up to order $\vec{\pi}^4$, one finds the famous Weinberg result[3] for the invariant $\pi\pi$ - scattering amplitude

$$A(s,t,u) = \frac{s - m_\pi^2}{f_\pi^2} \,. \tag{2}$$

The effective meson lagrangian eq.(1) does not possess stable topological solitons unless higher derivative terms (like Skyrme's fourth order term) or explicit vector mesons are added. The developments of the last years have shown that the introduction of vector mesons (ρ and ω) is more or less unique if chiral symmetry is used as a guiding principle[1]. For the system of coupled pions, ρ and ω mesons the so-called "Realistic-Pseudoscalar-Vector-Lagrangian" emerged as the most general candidate of an effective meson lagrangian consistent with chiral symmetry and phenomenological requirements[4]. A detailed study of various nucleon properties like electromagnetic, axial and strong meson-nucleon form factors has been performed in refs.[5,6] and it was found that this model gives results which are in rather good agreement with the data. The generic problem of all soliton models, *i.e.* a too large nucleon mass $M_N \simeq 1.5$ GeV and a too small value of the axial vector coupling $g_A < 1$, however, remains still unsolved. I will not further go into the details of this model here (which was part of my talk at the Siegen workshop) but refer the interested reader to the literature[4,5,6]. (See also the contributions of J. Schechter and H. Weigel to these proceedings).

Comparing the results of the Skyrme soliton model directly with experimental data is not the only possibility to check the quality of such a model. One can furthermore use rigorous and model independent constraints on certain nucleon observables which follow directly from the chiral structure of QCD. Well-known examples are the Weinberg-Tomozawa relation for the S-wave πN scattering lengths[3,7] and the Kroll-Ruderman theorem for the electric dipole amplitudes in pion photoproduction[8]. These low energy theorems first derived in the fifties and sixties using current algebra methods uniquely fix the leading terms in the pion mass or quark mass expansion of the corresponding amplitudes. Quark mass and pion mass expansion are equivalent since the square of the pion mass is proportional to the light quark mass by the Gell-Mann-Oakes-Renner relation.

Recently, considerable progress has been made in systematically exploring the consequences of chiral symmetry in QCD. The method to do this is the so-called Chiral Perturbation Theory which allows to expand the QCD Green's functions systematically in external momenta and quark masses [9,10,11]. In order to solve the chiral Ward identities of QCD one makes use of an effective field theory for the asymptotically observed fields (pions and nucleons) which consists of a string of terms with increasing number of derivatives and quark mass insertions. To lowest order in the chiral expansion this effective theory is the non-linear σ-model eq.(1) (coupled to external vector or scalar sources) for the pure meson sector. The couplings of the pions (Goldstone bosons) to the nucleons are severely constrained by chiral symmetry. To leading order the effective πN lagrangian involves a conventional pseudovector πNN coupling with the πNN coupling constant given by the Goldberger-Treiman relation $g_{\pi NN} = g_A\, M/f_\pi$ and the $\pi\pi NN$ and $\gamma\pi NN$ contact vertices of Weinberg and Kroll-Ruderman type. These contact terms are a consequence of chiral symmetry and they reproduce the beforementioned current algebra results for πN scattering and pion photoproduction as tree level results. The chiral expansion is then systematically improved by working out (pion)-loop diagrams generated by the effective vertices and by including chirally symmetric higher order counter terms. The latter are necessary for a consistent renormalization of the divergences generated by the loops.

2. Rigorous results from CHPT

Let me now come to the pertinent results from CHPT. For the isovector electric and magnetic radii of the nucleon CHPT sets the following constraint[10,11,17]

$$\begin{aligned} 6\,F_1^{v\prime}(0) &= -\frac{5g_A^2+1}{8\pi^2 f_\pi^2}\ln\frac{m_\pi}{\lambda}+\ldots \\ \frac{6}{M}\,F_2^{v\prime}(0) &= \frac{g_A^2}{8\pi f_\pi^2}\frac{1}{m_\pi}+\ldots \end{aligned} \tag{3}$$

The dots in eq.(3) stand for terms depending in a less singular way on the pion mass m_π than the leading ones. The scale parameter λ is arbitrary. The isovector electric radius does, however, not depend on it since there is also a counter term contribution which exactly cancels the logarithmic λ-dependence. The expressions in eq.(3) uniquely fix the leading and most singular terms in the pion mass or quark mass expansion of the nucleon isovector electromagnetic radii and they are obtained from evaluating the one loop diagrams shown in Fig.1.

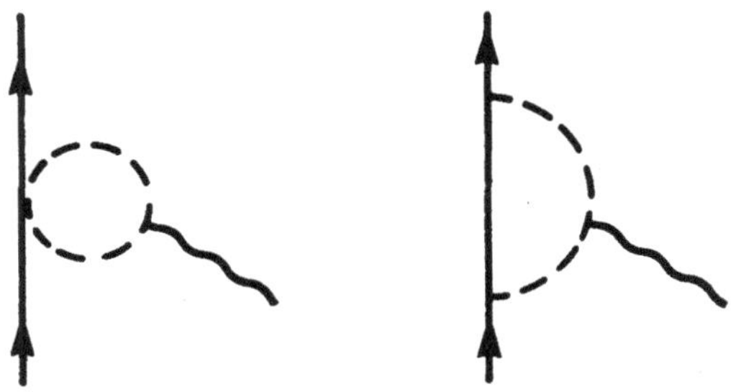

Fig. 1. The one loop diagrams which give the leading terms in the chiral expansion of the isovector electromagnetic radii. Full, dashed and wiggly lines represent nucleons, pions and photons, respectively. The right diagram with the wiggly line representing an external scalar source occurs in the calculation of the nucleon scalar form factor.

Another quantity of interest is the scalar form factor of the nucleon $\sigma(t)$ which characterizes the matrix element of the scalar quark density $\hat{m}(\bar{u}u+\bar{d}d)$ in nucleon states. Its value at zero momentum transfer $\sigma(0)$ is the so-called pion nucleon sigma term measuring the shift of the nucleon mass due to explicit chiral symmetry breaking. The scalar form factor $\sigma(t)$ has been evaluated to one loop in refs.[10,11] and the following expression can be deduced for the scalar radius of the nucleon

$$6\,\sigma'(0) = \frac{15g_A^2}{128\pi f_\pi^2}m_\pi+\ldots \tag{4}$$

Again, the dots stand for higher orders in m_π like $m_\pi^2 \ln m_\pi$.

Recently, the electromagnetic Compton polarizabilities of the proton and neutron have been analyzed in CHPT[11,12]. The evaluation of the one loop Compton graphs shown in Fig.2 (and their crossed partners) leads to the following result for the electric and magnetic polarizabilities

$$\begin{aligned} \alpha_p &= \alpha_n = \frac{5e^2 g_A^2}{384\pi^2 f_\pi^2} \frac{1}{m_\pi} + \ldots \\ \beta_p &= \beta_n = \frac{1}{10} \alpha_p + \ldots \end{aligned} \tag{5}$$

Again, the dots stand for terms less singular than m_π^{-1} such as $\ln m_\pi$. Note that the leading terms in the pion mass expansions eqs.(3,4,5) are completely fixed through g_A and f_π. This is generally the case since the leading terms stem from loops involving only the lowest order effective vertices which are completely determined by g_A and f_π.

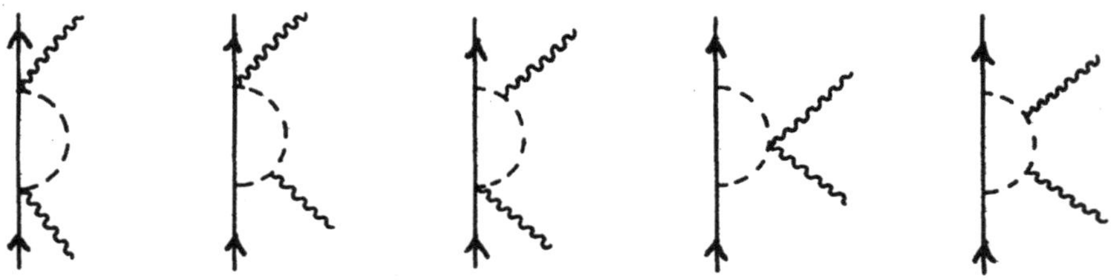

Fig. 2. The one loop Compton diagrams which give rise to electric and magnetic polarizabilities of eq.(5).

One observes that the nucleon electromagnetic polarizabilities and the isovector radii become infinite in the chiral limit $m_\pi = 0$. Intuitively this is not surprising. In the chiral limit the pion cloud around the nucleon becomes long ranged $\sim 1/r^2$ and this leads to infinite mean square radii and polarizabilities. The actual singular dependence of these quantities on the pion mass (logarithmic or inverse or else) with the proper coefficients can only be obtained within the framework of CHPT since this method allows to perform a controlled expansion of the QCD Green's functions in quark masses and external momenta.

With eqs.(3,4,5) we have collected rigorous results from CHPT. Let me again point out that the leading terms in eqs.(3,4,5) are model-independent constraints due to the chiral structure of QCD. Therefore any model which claims to respect chiral symmetry has to fullfil these constraints. Let us now investigate whether the description of the nucleon as a topological soliton in an effective meson theory is consistent with these requirements.

3. Predictions of the soliton model

In the Skyrme soliton model all nucleon matrix elements of interest can be expressed through the chiral angle profile $F(r)$ (and the other profiles for the vector meson fields). The expression for the isovector charge radius takes generically the following form[1,13]

$$6\,F_1^{v\prime}(0) = \frac{8\pi}{3\Theta} f_\pi^2 \int_0^\infty dr r^4 [\sin^2 F(r) + \mathcal{O}(F^4)]\,. \tag{6}$$

Here, Θ is the moment of inertia of the soliton given as

$$\Theta = \frac{8\pi}{3} f_\pi^2 \int_0^\infty dr r^2 [\sin^2 F(r) + \mathcal{O}(F^4)]\,. \tag{7}$$

In the chiral limit $m_\pi = 0$ the chiral angle $F(r)$ falls off only like $1/r^2$ for large distances r and the integrand in eq.(6) approaches a non-zero constant. Obviously, the isovector charge radius is not finite in the chiral limit. The only term responsible for this behaviour is the $F^2(r)$ term in eq.(6). All $\mathcal{O}(F^4)$ and higher terms give a finite result. The moment of inertia eq.(7) is a finite quantity in the chiral limit.

In order to monitor how the isovector charge radius of the nucleon diverges in the chiral limit $m_\pi = 0$ one has to use the asymptotic Yukawa form of the chiral angle

$$F(r) = \frac{3g_A}{8\pi f_\pi^2}\left(\frac{1}{r^2} + \frac{m_\pi}{r}\right)e^{-m_\pi r}\,, \qquad r \to \infty\,. \tag{8}$$

Inserting this asymptotic form into the formula for the isovector electric radius one finds

$$6\,F_1^{v\prime}(0) \sim \frac{1}{m_\pi} \tag{9}$$

as the prediction of the soliton model. The result eq.(9) also holds for those models where the electromagnetic current is given by an approximate vector meson dominance. In that case $\sin^2 F(r)$ in eq.(6) is replaced by a combination of the ρ meson profile functions. The latter have, however, the same asyptotic behaviour for $r \to \infty$ as $F^2(r)$ since the two pion term is the dominant source for the vector meson fields.

In eq.(9) we have now discovered a manifest disagreement with the constraint eq.(3). The soliton model leads to a stronger singularity $(1/m_\pi)$ than allowed by CHPT $(\ln m_\pi)$. This does, however, not mean that the soliton approach to baryons is wrong by itself. The origin of the incorrect chiral singularity eq.(9) must lie in the classical treatment of the nucleon in this model where one first constructs the hedgehog and then performs only a collective quantization of the adiabatic rotation. At such a level the contributions from pionic fluctuations around the soliton are neglected completely. The photon can, however, couple to these fluctuations as well and the electromagnetic properties will get changed correspondingly. Presently, such effects have not been studied in detail and it remains to be seen whether they can solve the problem of the incorrect chiral singularity. (In this context see also the contributions of G. Holzwarth to these proceedings, where the role of the isorotational zero mode is discussed.)

Let me now turn to the other nucleon observables. The isovector magnetic radius takes the following generic form in the soliton model[1,13],

$$\frac{6}{M}F_2^{v\prime}(0) = \frac{16\pi}{15} f_\pi^2 \int_0^\infty drr^4[\sin^2 F(r) + \mathcal{O}(F^4)] = \frac{3g_A^2}{16\pi f_\pi^2}\frac{1}{m_\pi} + \dots \tag{10}$$

One observes that the proper $1/m_\pi$ singularity is indeed reproduced in the soliton model. However, the coefficient of this term is not the correct one. In the soliton model it comes out a factor 3/2 larger than in CHPT (see eq.(3)).

For the electric polarizability of the proton and neutron one has the following expression in soliton models[14] (see also Piet Mulders contribution to these proceedings)

$$\alpha = \frac{2}{9} e^2 f_\pi^2 \int_0^\infty drr^4[\sin^2 F(r) + \mathcal{O}(F^4)] = \frac{5e^2 g_A^2}{128\pi^2 f_\pi^2}\frac{1}{m_\pi} + \dots \tag{11}$$

Here, I used the standard convention $e^2/4\pi = 1/137.036$. Again, one observes that the soliton model gives the proper reciprocal dependence on the pion mass, but in the case of the electric polarizability the coefficient of the $1/m_\pi$ term comes out three times too large. The various calculations of magnetic polarizabilities lead to an expression of the following type[14] (see again Piet Mulders contribution)

$$\beta = -\frac{\alpha}{2} + \frac{e^2 \mu_v^2}{4\pi M^2(M_\Delta - M_N)} + \dots \tag{12}$$

The dots stand for terms which differ in the various calculations of the magnetic polarizability. They are finite quantities in the chiral limit and so do not modify the discussion of chiral singularities. The second term in eq.(12), the paramagnetic contribution which arises from exciting the Δ resonance also stays finite in the chiral limit because the ΔN splitting and the nucleon isovector magnetic moment do so. Therefore we can conclude

$$\lim_{m_\pi \to 0}\left(\frac{\alpha}{\beta}\right)_{p,n} = \begin{cases} 10 & \text{CHPT} \\ -2 & \text{soliton} \end{cases} \tag{13}$$

and find a further disagreement between CHPT and soliton model predictions. Furthermore, in ref.[18] the difference $\alpha_p - \alpha_n$ has been calculated in a soliton model and it was found that this quantity behaves as $1/m_\pi$ in the chiral limit. This prediction is incorrect since CHPT allows at most a $\ln m_\pi$ singularity for $\alpha_p - \alpha_n$[12].

Finally, let me discuss the nucleon scalar form factor. In the soliton model it reads[5],

$$\sigma(t) = 4\pi f_\pi^2 m_\pi^2 \int_0^\infty drr^2\, j_0(\sqrt{-t}r)\,[1 - \cos F(r)] \tag{14}$$

where $j_0(x) = \sin x/x$ is a spherical Bessel function. Exploiting the asymptotic behaviour of the chiral angle $F(r)$ we obtain then for the scalar radius of the nucleon

$$6\,\sigma'(0) = \frac{45 g_A^2}{128\pi f_\pi^2} m_\pi + \dots \tag{15}$$

which comes out a factor three larger than in CHPT.

These investigations have shown that the present treatment of nucleons as chiral solitons is not completely satisfactory since one can find that several nucleon observables are predicted not in agreement with the rigorous constraints following from chiral QCD. I will not be able to repair all these failures of the soliton model, here. This certainly requires a lot of work to be done in the future. I will, however, explain the origin of the factors 3/2 and 3 to which soliton model and CHPT results disagree in some cases.

4. Origin of the factors 3/2 and 3

Consider the loop diagrams shown in Fig.1 and 2 and replace the nucleon line in the pion loop by a Δ isobar. The corresponding diagrams can easily be worked out using the heavy baryon formalism of ref.[15]. I choose $g_{\pi N\Delta} = 3g_{\pi NN}/\sqrt{2}$ for the $\pi N\Delta$ coupling constant which is a prediction of the soliton model[1] and set $\Delta = M_\Delta - M_N$ for the mass splitting. The sum of the corresponding loop graphs with nucleon and Δ intermediate states gives the following results for the various nucleon observables of interest

$$\frac{6}{M}\, F_2^{v\prime}(0) = \frac{g_A^2}{8\pi^2 f_\pi^2}\left[\frac{\pi}{m_\pi} + \frac{1}{\sqrt{\Delta^2 - m_\pi^2}} \ln\left(\frac{\Delta}{m_\pi} + \sqrt{\frac{\Delta^2}{m_\pi^2} - 1}\right)\right] \tag{16a}$$

$$\alpha = \frac{e^2 g_A^2}{192\pi^3 f_\pi^2}\left[\frac{5\pi}{2m_\pi} + \frac{9\Delta}{\Delta^2 - m_\pi^2} + \frac{\Delta^2 - 10m_\pi^2}{(\Delta^2 - m_\pi^2)^{3/2}} \ln\left(\frac{\Delta}{m_\pi} + \sqrt{\frac{\Delta^2}{m_\pi^2} - 1}\right)\right] \tag{16b}$$

$$6\,\sigma'(0) = \frac{15 g_A^2 m_\pi^2}{32\pi^2 f_\pi^2}\left[\frac{\pi}{4m_\pi} + \frac{1}{\sqrt{\Delta^2 - m_\pi^2}} \ln\left(\frac{\Delta}{m_\pi} + \sqrt{\frac{\Delta^2}{m_\pi^2} - 1}\right)\right]. \tag{16c}$$

The first important observation one makes is that the Δ isobar contributions in eq.(16) diverge like $\ln(m_\pi/\Delta)$ in the chiral limit. The leading $1/m_\pi$ term is not affected by the Δ loops. This is in agreement with the general result of Gasser and Zepeda[16], which states that resonances do not modify the leading terms in the chiral expansion.

Things are, however, different in the limit where the nucleon and the Δ isobar are exactly degenerate ($\Delta = 0$). Then one can analytically continue the expressions in eq.(16) via

$$\frac{1}{\sqrt{\Delta^2 - m_\pi^2}} \ln\left(\frac{\Delta}{m_\pi} + \sqrt{\frac{\Delta^2}{m_\pi^2} - 1}\right)\Big|_{\Delta=0} = \frac{1}{im_\pi}\,\frac{i\pi}{2} = \frac{\pi}{2m_\pi} \tag{17}$$

There is no ambiguity in taking the proper branch of the square root and logarithm in eq.(17). The integrals which lead to the expressions with the logarithm in eq.(16) can easily be worked for $\Delta = 0$ and then one convinces oneself that eq.(17) is the proper continuation to $\Delta = 0$. In that limit the $1/m_\pi$ terms get modified, exactly in such a way that one finds the soliton model predictions with the corresponding factors 3/2 and 3 larger than the CHPT results. One observes that the soliton predictions correspond to the unphysical case where the nucleon and the Δ isobar are exactly degenerate. This degeneracy occurs for infinite numbers of colors in QCD, since $\Delta \sim N_c^{-1}$ according to the counting rules of the quark model. The large N_c limit and the chiral limit do, however, not commute. This can easily be seen from the formulae given in eq.(16). For finite N_c or $\Delta \neq 0$ one has a logarithmic dependence on m_π, whereas for $N_c = \infty$ corresponding to $\Delta = 0$ we find a stronger $1/m_\pi$ singularity. The large N_c limit is in some sense inherent in the description of baryons as solitons. Due to the non-commutativity of large N_c and chiral limit one necessarily has to include $1/N_c$ corrections in the soliton model from the start such that the nucleon and Δ isobar will not occur as degenerate states. Only then one will be able to reproduce the proper chiral expansion of all nucleon observables in agreement with the model independent constraints from Chiral Perturbation Theory.

Before finishing this work ref.[19] appeared. There, the authors also find that the incorrect factors of the leading chiral singularities in the soliton model stem from Δ isobar loops in the degeneracy limit $M_N = M_\Delta$. However, I disagree with their explicit form of the Δ isobar loop contribution to the electric polarizability α for $\Delta \neq 0$ (see eq.16b). Furthermore, ref.[18] concludes that the Δ loop effects are quite large for the experimental value of $\Delta/m_\pi = 2.1$ and therefore neither the soliton model nor CPTH would treat them properly. I disagree with this conclusion. In CHPT the effects of the Δ loops are just one part of many next to leading order corrections. At the same order in the chiral expansion *e.g.* $1/M_N$ suppressed one loop effects will contribute and these may partly cancel the Δ pieces. It is only meaningful to compare the sum of all next to leading order corrections to the leading term in the chiral expansion of a certain observable. The observation that the Δ loop contribuiton is not small compared to the leading chiral singularity has no implication for the quality of CHPT.

5. Summary and conclusions

In this short contribution I have investigated the isovector electromagnetic and scalar radii as well as the electromagnetic polarizabilities of the nucleon in the Skyrme soliton model. I studied the behaviour of these quantities in the limit when the pion mass is tending to zero. By comparing to rigorous results of Chiral Perturbation Theory we have seen that the Skyrme soliton model is presently not able to reproduce the model independent constraints of chiral QCD in all cases. For the case of the isovector electric radius of the nucleon a stronger singularity than allowed by CHPT was found. In other cases the coefficients of the leading $1/m_\pi$ singularity were predicted incorrect and differed by factors of 3/2 or 3 from the CHPT results. These factors could be understood to arise from contributions with loops including the Δ isobar in the limit when it is exactly degenerate with the nucleon ($\Delta = M_\Delta - M_N = 0$). This degeneracy occurs in the limit $N_c \to \infty$ which does not commute with the chiral limit. More elaborate treatments of nucleons as Skyrme soliton are necessary in order to resolve the discrepancies discussed here.

6. References

1. Ulf-G. Meißner, *Phys. Reports* **161** (1988) 213 and references therein.
2. G. 't Hooft, *Nucl. Phys.* **B72** (1974) 461; E. Witten, *Nucl. Phys.* **B160** (1979) 57.
3. S. Weinberg, *Phys. Rev. Lett.* **17** (1966) 616.
4. P. Jain, R. Johnson, Ulf-G. Meißner, N.W. Park and J. Schechter, *Phys. Rev.* **D37** (1988) 3252.
5. Ulf-G. Meißner, N. Kaiser, H. Weigel and J. Schechter, *Phys. Rev.* **D39** (1989) 1956.
6. V. Bernard, N. Kaiser and Ulf-G. Meißner, *Phys. Lett.* **B237** (1990) 545.
7. Y. Tomozawa, *Nov. Cim.* **A46** (1966) 707.
8. N.M. Kroll and M.A. Ruderman, *Phys. Rev.* **93** (1954) 233.
9. J. Gasser and H. Leutwyler, *Ann. Phys. (N.Y.).* **158** (1984) 142; *Nucl. Phys.* **B250** (1985) 465.
10. J. Gasser, M.E. Sainio and A. Svarc, *Nucl. Phys.* **B307** (1988) 779.
11. V. Bernard, N. Kaiser, J. Kambor and Ulf-G. Meißner, "Chiral Structure of the Nucleon", *Nucl. Phys.* **B** in print.
12. V. Bernard, N. Kaiser, and Ulf-G. Meißner, *Nucl. Phys.* **B373** (1992) 346.
13. E. Braaten, S.-M. Tse and C. Willcox, *Phys. Rev. Lett.* **56** (1986) 2008.
14. E.M. Nyman, *Phys. Lett.* **B142** (1984) 388; N. Scoccola and W. Weise, *Nucl. Phys.* **A517** (1990) 495.
15. E. Jenkins and A.V. Manohar, *Phys.Lett.* **B255** (1991) 558; *Phys. Lett.* **B259** (1991) 353.
16. J. Gasser and A. Zepeda, *Nucl. Phys.* **B174** (1980) 445.
17. D.G. Caldi and H. Pagels, *Phys. Rev.* **D10** (1974) 3739.
18. W. Broniowski, M.K. Banerjee and T.D. Cohen, *Phys. Lett.* **B283** (1992) 22.
19. T.D. Cohen and W. Broniowski, *Phys. Lett.* **B292** (1992) 5.

ELECTROMAGNETIC POLARIZABILITIES OF THE NUCLEON AND THE DELTA

S. SCHERER
TRIUMF, 4004 Wesbrook Mall, Vancouver B.C., V6T 2A3 Canada

and

P.J. MULDERS
NIKHEF–K, P.O. Box 41882, NL 1009 DB Amsterdam, The Netherlands
and
Department of Physics and Astronomy, Free University, Amsterdam

Abstract

We have calculated the static electromagnetic polarizabilities of the nucleon and the Δ baryon in the Skyrme model [1]. The relation $\alpha = -2\beta_{dia}$ between electric (α) and magnetic (β) polarizabilities, is found to be only approximately correct. Depending on the parameters used, the numerical numbers are roughly a factor 2 - 5 too large in comparison with the available empirical data. We discuss the large N_C limit and illustrate ordering ambiguities by considering a non–adiabatic contribution to the isovector magnetic moment of the nucleon.

1 Introduction

The electromagnetic polarizabilities of the nucleon are quantities of great interest, since they provide information on the structure of the nucleon. The *static* electric (magnetic) polarizability of a particle is a measure for how easily an electric (magnetic) dipole moment may be induced by a constant external field: $\vec{p} = \alpha\vec{E}$ ($\vec{\mu} = \beta\vec{B}$). The *dynamic* or *Compton* polarizabilities $\bar{\alpha}$ and $\bar{\beta}$, as extracted from a low–energy expansion of the Compton scattering cross section, differ from the static polarizabilities due to relativistic and retardation effects caused by the finite extension of the nucleon. A unique interpretation of the Compton polarizabilities in terms of static polarizabilities and effects due to the finite size of the nucleon appears to be difficult. Different prescriptions are available in the literature [2, 3, 4, 5]. In this paper we will only be concerned with the static polarizabilities. The sum of the electric and magnetic Compton polarizabilities is related to the total photoabsorption cross section via a subtracted dispersion relation [6], yielding the following results [7, 8]

$$\begin{aligned}\bar{\alpha}_p + \bar{\beta}_p &= (14.2 \pm 0.5) \times 10^{-4} fm^3, \\ \bar{\alpha}_n + \bar{\beta}_n &= (15.8 \pm 0.5) \times 10^{-4} fm^3. \qquad (1)\end{aligned}$$

All information concerning the polarizabilities of the proton originates from Compton scattering experiments. The most recent numbers are [9] $\bar{\alpha}_p = (10.9 \pm 3.5) \times 10^{-4} fm^3$ and $\bar{\beta}_p = (3.3 \pm 3.5) \times 10^{-4} fm^3$. The electric polarizability of the neutron may also be extracted using Compton scattering on the neutron in the deuteron where one finds

$\bar{\alpha}_n = (11.7\,^{+4.7}_{-11.7}) \times 10^{-4} fm^3$ [10]. Alternatively, low–energetic neutron heavy nucleus scattering is used to extract the static electric polarizability $\alpha_n = (12.0 \pm 3.5) \times 10^{-4} fm^3$ [5]. There is no independent information on the magnetic polarizability of the neutron. A precise experimental determination of the polarizabilities is of great importance, since it may help to discriminate between different models of hadronic structure. Up to now, there is no experimental information on the polarizabilities of the Δ baryon.

Theoretical investigations include calculations in the non–relativistic quark model [11, 12], the MIT bag model [13, 14] and its the chiral extensions [15], different versions of chiral soliton models [16, 17, 18, 19, 20] and chiral perturbation theory [21, 22]. Since the corresponding underlying reaction mechanisms are quite different, it is actually no surprise that the predictions differ substantially. For example, predictions for the electric polarizability of the proton range from $7.1 \times 10^{-4} fm^3$ [13] to $27.3 \times 10^{-4} fm^3$ [20]. The magnetic polarizability of the proton is predicted from $-3.1 \times 10^{-4} fm^3$ [19] to $8.5 \times 10^{-4} fm^3$ [17]. Similar discrepancies are found for the polarizabilities of the neutron.

This paper is organized as follows. In section 2 we discuss the gauging of the Skyrme Lagrangian to include electromagnetism. In section 3 we develop the formalism for calculating the static electric and magnetic polarizabilities of the nucleon and the Δ baryon. We comment on the chiral limit. In section 4 we discuss the numerical predictions and in section 5 we conclude with a short summary.

2 Skyrmions in an external electromagnetic field

The Lagrangian of the Skyrme model [23, 24] including a chiral symmetry breaking pion mass term is in the derivative expansion written as [25]

$$\begin{aligned}
\mathcal{L} &= \mathcal{L}^{(2)} + \mathcal{L}^{(4)} + \mathcal{L}^{(sb)} \\
&= \frac{F_\pi^2}{16} Tr\Big(\partial_\mu U \partial^\mu U^\dagger\Big) + \frac{1}{32\epsilon^2} Tr\Big([U^\dagger\partial_\mu U, U^\dagger\partial_\nu U][U^\dagger\partial^\mu U, U^\dagger\partial^\nu U]\Big) + \frac{m_\pi^2 F_\pi^2}{8} Tr\Big(U - 1\Big),
\end{aligned} \tag{2}$$

where $U(x)$ is an $SU(2)$–valued field, F_π is the pion decay constant (experimental value $F_\pi \approx 186 MeV$) and ϵ determines the strength of the fourth order term which governs the size of the soliton. The superscipts (2), (4), and (sb) refer to the number of derivatives in the corresponding expression and to symmetry breaking, respectively. As we consider only two flavors throughout this paper, the Wess–Zumino term does not contribute to eq. (2) [24, 26]. The Lagrangian of eq. (2) has besides the global $SU(2) \times SU(2)$ symmetry ($m_\pi = 0$) a global $U(1)$ symmetry [26], namely $\mathcal{L}$ is invariant under a global charge transformation over an angle χ

$$U \rightarrow \exp(-i\chi e Q) U \exp(i\chi e Q), \tag{3}$$

where $Q = 1/6 + \tau_3/2$ is the electric charge matrix of up and down quarks, and e is the elementary charge ($e > 0$). In order to promote this symmetry to a local symmetry, one has to replace the ordinary derivatives in eq. (2) by covariant ones

$$\begin{aligned}
\partial_\mu U &\rightarrow D_\mu U = \partial_\mu U + ieA_\mu[Q, U] = \partial_\mu U - eA_\mu \delta U, \\
\partial_\mu U^\dagger &\rightarrow D_\mu U^\dagger = \partial_\mu U^\dagger + ieA_\mu[Q, U^\dagger] = \partial_\mu U^\dagger - eA_\mu \delta U^\dagger,
\end{aligned} \tag{4}$$

and the resulting Lagrangian is invariant under a local gauge transformation ($\chi = \chi(x)$)

$$\begin{aligned} U &\rightarrow \exp(-i\chi eQ)U\exp(i\chi eQ), \\ U^\dagger &\rightarrow \exp(-i\chi eQ)U^\dagger\exp(i\chi eQ), \\ A_\mu &\rightarrow A_\mu + \partial_\mu\chi. \end{aligned} \tag{5}$$

The minimal substitution of eq. (4) into eq. (2) results in interaction terms involving one up to four photon fields. The interaction Lagrangian linear in the photon field is of the form

$$\mathcal{L}_\gamma = -eA_\mu V^{3,\mu}[U], \tag{6}$$

where $V^{3,\mu}[U]$ is the third component of the isovector current density functional

$$\vec{V}^\mu = \frac{F_\pi^2}{8}Tr\Big(\partial^\mu U^\dagger \vec{\delta}^V U\Big) + \frac{1}{8\epsilon^2}Tr\Big([U^\dagger\partial_\nu U, U^\dagger\vec{\delta}^V U][U^\dagger\partial^\nu U, U^\dagger\partial^\mu U]\Big), \tag{7}$$

with $\vec{\delta}^V U = -i[\frac{\vec{\tau}}{2}, U]$. Since we are interested in terms of order e^2 we also have to consider the expressions involving two photon fields (seagull terms), namely

$$\mathcal{L}_{\gamma\gamma} = \mathcal{L}_{\gamma\gamma}^{(2)} + \mathcal{L}_{\gamma\gamma}^{(4)}, \tag{8}$$

where the superscripts indicate from which part of the original Lagrangian they originate. The explicit expressions are given by

$$\begin{aligned} \mathcal{L}_{\gamma\gamma}^{(2)} &= \frac{e^2F_\pi^2}{16}A_\mu A^\mu Tr\Big(\delta U\delta U^\dagger\Big), \\ \mathcal{L}_{\gamma\gamma}^{(4)} &= \frac{e^2}{8\epsilon^2}A_\mu A^\mu Tr\Big(\delta U\partial_\nu U^\dagger\delta U\partial^\nu U^\dagger - \delta U^\dagger\delta U\partial_\nu U^\dagger\partial^\nu U\Big) \\ &\quad -\frac{e^2}{8\epsilon^2}A_\mu A^\nu Tr\Big(\delta U\partial_\nu U^\dagger\delta U\partial^\mu U^\dagger - \delta U^\dagger\delta U\partial_\nu U^\dagger\partial^\mu U\Big)\Big), \end{aligned} \tag{9}$$

where δU and $\delta U^\dagger$ are defined in eq. (4).

However, the minimal substitution only results in the isovector part of the electromagnetic interaction. Clearly, the interaction of a photon with a nucleon also involves an isoscalar piece which is provided by [24, 26]

$$\mathcal{L}_\gamma^{(ano)} = -\frac{1}{2}eA_\mu B^\mu[U], \tag{10}$$

where $B^\mu[U]$ is the functional of the topological baryon current density

$$B_\mu = \frac{1}{24\pi^2}\epsilon_{\mu\nu\rho\sigma}Tr\Big(U^\dagger\partial^\nu UU^\dagger\partial^\rho UU^\dagger\partial^\sigma U\Big), \quad \epsilon_{0123} = 1. \tag{11}$$

This baryon current is the $SU(2)$ restriction of the Noether current which originates from gauging the Wess–Zumino term in $SU(3)$ [26, 24]. It turns out that the interaction term of eq. (10) is not gauge invariant [26] and that one also has to construct an additional piece involving two photon fields,

$$\mathcal{L}_{\gamma\gamma}^{(ano)} = i\frac{e^2}{16\pi^2}\epsilon_{\mu\nu\rho\sigma}(\partial^\mu A^\nu)A^\rho Tr\Big((\partial^\sigma UU^\dagger + U^\dagger\partial^\sigma U)Q\Big) \tag{12}$$

in order to make the combination of $\mathcal{L}_\gamma^{(ano)}$ and $\mathcal{L}_{\gamma\gamma}^{(ano)}$ invariant under the transformations of eq. (5). Note that our additional part differs by a total derivative from the one constructed in ref. [26]. The term $\mathcal{L}_\gamma^{(ano)}$ and $\mathcal{L}_{\gamma\gamma}^{(ano)}$ describes all effects of QCD anomalies in low–energy processes involving photons and Goldstone–pions [26]. The final Lagrange density to second order in the electromagnetic coupling constant then reads

$$\mathcal{L} = \mathcal{L}^{(2)} + \mathcal{L}^{(4)} + \mathcal{L}^{(sb)} + \mathcal{L}_\gamma + \mathcal{L}_\gamma^{(ano)} + \mathcal{L}_{\gamma\gamma}^{(2)} + \mathcal{L}_{\gamma\gamma}^{(4)} + \mathcal{L}_{\gamma\gamma}^{(ano)} + [e^3]. \tag{13}$$

Since in the following we will treat the electromagnetic field as an external classical field, we have deliberately omitted the free field Lagrangian.

3 Static polarizabilities of the nucleon and the delta

The static polarizabilities are defined through the shift of a particle's energy in an external constant electric or likewise magnetic field (using Gaussian units, where $e^2 = 1/137$, in order to be consistent with the convention of eq. 14),

$$\Delta E = -\frac{1}{2}\alpha E^2 - \frac{1}{2}\beta B^2. \tag{14}$$

They are a measure for how easily the charge density or the magnetization of the free particle may be distorted by an external field. For that reason they provide useful information about the internal structure of the particle in question.

We now outline the procedure to calculate the static polarizabilities in the Skyrme model. For that purpose we insert the standard soliton ansatz in the baryon sector ($B = 1$) for a nucleon [24, 27]

$$U(\vec{x}, t) = \exp\left(i\vec{\tau} \cdot R(\alpha(t), \beta(t), \gamma(t))\hat{x}F(r)\right) \tag{15}$$

into the Lagrangian of eq. (13). Here R represents a 3×3 rotation matrix, $R_{ij} = R_{ji}^{-1}$, $det(R) = 1$, inducing an active rotation of the hedgehog's isovector $\hat{x}\sin(F)$. We then define electromagnetic potentials for either a constant electric or magnetic field and determine the classical Lagrange function by integration in ordinary space. The Hamilton function is obtained by treating the Euler angles and the angular velocities as dynamical variables. It is then quantized in the standard way [24, 27] and the energy shift is calculated in first and second order perturbation theory for the electric and magnetic polarizability, respectively.

3.1 Static electric polarizabilities

Let us first consider the electric polarizability. We define $A^\mu = (-zE, 0, 0, 0)$, representing the vector potential for a constant field $\vec{E} = E\hat{e}_z$. The Lagrange function does not contain any terms linear in the electric field. The reason for this is that the spatial integral of $A^0(\vec{x})\rho(\vec{x}, t)$, where $\rho(\vec{x}, t)$ denotes the charge density, involves the integrals of odd functions over symmetric intervals. Furthermore, we note that $\mathcal{L}_{\gamma\gamma}^{(ano)}$ does not contribute in the static situation, since the constant electric field requires the indices ν and ρ in eq. (12) both to be equal to zero, in which case the antisymmetric tensor $\epsilon_{\mu\nu\rho\sigma}$ vanishes. Hence, it is only $\mathcal{L}_{\gamma\gamma}^{(2)}$

and $\mathcal{L}^{(4)}_{\gamma\gamma}$, which give rise to an interaction term in the Lagrange function. Up to quadratic order in the electromagnetic coupling constant it is found to be

$$L(t) = \frac{1}{2}\Theta\vec{\Omega}^2 - M_0 + \frac{1}{2}\gamma^e E^2 \left(1 - \frac{2}{5}D^{(2)}_{0,0}(\alpha,\beta,\gamma)\right) + [e^3], \tag{16}$$

with

$$\gamma^e = \gamma^{e(2)} + \gamma^{e(4)} \tag{17}$$

$$= \frac{2\pi e^2}{9\epsilon^5 F_\pi^3}\int_0^\infty dx x^4 \sin^2(F) + \frac{8\pi e^2}{9\epsilon^5 F_\pi^3}\int_0^\infty dx x^4 \sin^2(F)\left(F'^2 + \frac{\sin^2(F)}{x^2}\right), \tag{18}$$

where the superscripts indicate from which part of eq. (8) the contribution originates. The expressions for the moment of inertia Θ and the static mass M_0 may be found in ref. [24, 27] and we have introduced the dimensionless variable $x = \epsilon F_\pi r$ as in ref. [24]. The convention for the Wigner functions $D^{(J)}_{M,M'}$ is that of ref. [28]:

$$D^{(J)}_{M,M'}(\alpha,\beta,\gamma) = < JM|\mathcal{R}(\alpha,\beta,\gamma)|JM' >^*, \tag{19}$$

where $\mathcal{R}(\alpha,\beta,\gamma) = \exp(-i\alpha J_z)\exp(-i\beta J_y)\exp(-i\gamma J_z)$. When deriving eq. (16) we made use of

$$R_{33}^2 = \frac{1}{3} + \frac{2}{3}D^{(2)}_{0,0}. \tag{20}$$

The vector $\vec{\Omega}$ represents the angular velocity of the global rotation of the hedgehog in isospace and its components are defined through [27]

$$\dot{D}_{ij}D_{kj} = -\epsilon_{ikl}\Omega_l. \tag{21}$$

The corresponding conjugate momentum, the isospin $\vec{T}$, is given by [27]

$$\vec{T} = \frac{\delta L}{\delta\vec{\Omega}} = \Theta\vec{\Omega}. \tag{22}$$

The Hamilton function finally reads

$$H = H_0 + H_I = \frac{\vec{T}^2}{2\Theta} + M_0 - \frac{1}{2}\gamma^e E^2\left(1 - \frac{2}{5}D^{(2)}_{0,0}\right) + [e^3]. \tag{23}$$

Here the first two terms (H_0) represent the Hamilton function without electromagnetic interaction. It is quantized in the usual way [24, 27] treating the generators of rotations in isospace as operators obeying $SU(2)$ commutation relations. The eigenfunctions of H_0 in the Euler angle representation are given by

$$\Psi_{T=J,M_T,M_J}(\alpha,\beta,\gamma) = < \alpha,\beta,\gamma|T = J, M_T, M_J > = (-)^{J+M_J}\sqrt{\frac{2J+1}{8\pi^2}}D^{(J)}_{M_T,-M_J}(\alpha,\beta,\gamma). \tag{24}$$

We note that since H_I commutes with T_3 and J_3, the perturbation is diagonal with respect to the eigenfunctions of eq. (24). The energy shift due to the external electric field turns out to be

$$\Delta E = < J = T, M_T, M_J|H_I|J = T, M_T, M_J >, \tag{25}$$

and we find for the electric polarizability of the nucleon

$$\alpha_N = \gamma^e, \tag{26}$$

with γ^e being defined in eq. (18). The above treatment predicts the same electric polarizability for all nucleon states irrespective of their spin and isospin projection. The situation is different for the Δ baryon. There the polarizability depends on the spin as well as the isospin quantum numbers,

$$\alpha_\Delta^{M_T,M_J} = \begin{cases} \gamma^e\left(1-\frac{2}{25}\right) & \text{if} \quad |M_T| = |M_J|, \\ \gamma^e\left(1+\frac{2}{25}\right) & \text{if} \quad |M_T| \neq |M_J|. \end{cases} \tag{27}$$

When deriving this result we made use of

$$\int d\vec{\alpha}\Psi^*_{T,M_T,M_J} D^{(2)}_{0,0}\Psi_{T,M_T,M_J} = \frac{(3M_T^2 - T(T+1))(3M_J^2 - T(T+1))}{(2T-1)T(T+1)(2T+3)}, \tag{28}$$

where $\int d\vec{\alpha}$ is an abbreviation for $\int_0^{2\pi} d\alpha \int_0^\pi \sin(\beta)d\beta \int_0^{2\pi} d\gamma$. Although the classical hedgehog solution does not correspond to a physical particle we also give the expression for its electric polarizability. It is easily found from eq. (23) by setting all Euler angles equal to zero and by noting that $D^{(2)}_{0,0}(0,0,0) = 1$. The "classical" electric polarizability then is

$$\alpha_{cl} = \frac{3}{5}\gamma^e. \tag{29}$$

With the help of eq. (27) it is straightforward to see that the center of gravity of the electric polarizabilities of the delta is equal to the electric polarizability of the nucleon. It is interesting to note that it is shifted by $2/5\gamma^e$ relative to the "classical" electric polarizability. This is a general result for all the eigenstates of H_0 having $T = J$.

3.2 Static magnetic polarizabilities

The procedure for the magnetic polarizability is more complicated and has partially already been reported in ref. [16]. We write for the vector potential $A^\mu = (0, \frac{1}{2}\vec{B}\times\vec{r})$, where $\vec{B} = B\hat{e}_z$ is a constant magnetic field. Once again, it is easily shown that $\mathcal{L}^{(ano)}_{\gamma\gamma}$ does not contribute to the Lagrange function. In order to see this, we have to realize that either the index μ or the index σ has to be equal to zero in eq. (12). If $\mu = 0$, the expression contains a term $\frac{\partial A^\nu}{\partial t}$, which vanishes for a static field. If on the other hand $\sigma = 0$, the expression is found to be proportional to $\vec{A}\cdot\vec{\nabla}\times\vec{A}$ which vanishes for a *constant* field $\vec{B}$. The Lagrange function now contains linear as well as quadratic terms in the magnetic field,

$$L(t) = \frac{1}{2}\Theta\vec{\Omega}^2 - M_0 + \lambda F_1 + \lambda\vec{\Omega}\cdot\vec{F}_2 + \lambda\vec{\Omega}\cdot(F_3\vec{\Omega}) + \lambda^2 F_4 + \lambda^2\vec{\Omega}\cdot(F_5\vec{\Omega}) + [\lambda^3], \tag{30}$$

where for convenience we have introduced $\lambda = eB$. The functions F_i depend on the Euler angles and contain integrals involving the soliton profile function F. Furthermore, we have used F_3 and F_5 as an obvious shorthand notation for matrices containing 3×3 entries.

In terms of the cartesian components of the rotation matrix R introduced in eq. (15) the explicit expressions for the F_i read

$$\begin{aligned}
F_1 &= I_1 R_{33}, \\
F_{2i} &= I_2 R_{i3}, \\
F_{3ij} &= I_3 (R_{33}\delta_{ij} - R_{i3}\delta_{3j}), \\
F_4 &= I_4^a + R_{33}^2 I_4^b, \\
F_{5ij} &= I_5 \left(\delta_{ij}(1 - R_{33}^2) - \delta_{i3}\delta_{j3} + R_{i3}R_{j3} - 2R_{i3}R_{33}\delta_{j3}\right),
\end{aligned} \tag{31}$$

with the integrals I_i defined as

$$\begin{aligned}
I_1 &= -\frac{\Theta}{2} = -\frac{\pi}{3e^3 F_\pi}\int_0^\infty dx x^2 \sin^2(F)\left(1 + 4(F'^2 + \frac{\sin^2(F)}{x^2})\right), \\
I_2 &= \frac{1}{3\pi e^2 F_\pi^2}\int_0^\infty dx x^2 \sin^2(F) F', \\
I_3 &= \frac{2\pi}{3e^5 F_\pi^3}\int_0^\infty dx x^2 \sin^4(F), \\
I_4^a &= -\frac{\pi}{60 e^5 F_\pi^3}\int_0^\infty dx x^4 \sin^2(F)\left(3 + 12F'^2 + 2\frac{\sin^2(F)}{x^2}\right), \\
I_4^b &= -\frac{\pi}{60 e^5 F_\pi^3}\int_0^\infty dx x^4 \sin^2(F)\left(1 + 4F'^2 + 14\frac{\sin^2(F)}{x^2}\right), \\
I_5 &= \frac{\pi}{15 e^7 F_\pi^5}\int_0^\infty dx x^4 \sin^4(F).
\end{aligned} \tag{32}$$

From eq. (30) the isospin $\vec{T}$ is found to be

$$\vec{T} = \frac{\delta L}{\delta\vec{\Omega}} = \Theta\vec{\Omega} + \lambda\vec{F}_2 + \lambda(F_3 + F_3^T)\vec{\Omega} + \lambda^2(F_5 + F_5^T)\vec{\Omega}, \tag{33}$$

where F_3^T and F_5^T refer to the transposed of F_3 and F_5. In order to obtain the Hamilton function to second order in e^2 we have to express the angular velocity $\vec{\Omega}$ in terms of the isospin $\vec{T}$,

$$\begin{aligned}
\vec{\Omega} &= \left(1 + \lambda\frac{F_3 + F_3^T}{\Theta} + \lambda^2\frac{F_5 + F_5^T}{\Theta}\right)^{-1}\frac{\vec{T} - \lambda\vec{F}_2}{\Theta} \\
&= \frac{\vec{T}}{\Theta} - \lambda\frac{\vec{F}_2}{\Theta} - \lambda\frac{(F_3 + F_3^T)\vec{T}}{\Theta^2} - \lambda^2\frac{(F_5 + F_5^T)\vec{T}}{\Theta^2} + \lambda^2\frac{(F_3 + F_3^T)\vec{F}_2}{\Theta^2} + \lambda^2\frac{(F_3 + F_3^T)^2\vec{T}}{\Theta^3} \\
&\quad +[\lambda^3].
\end{aligned} \tag{34}$$

We are now in the position to write down the Hamilton function to second order in e^2,

$$H = \vec{\Omega}\cdot\vec{T} - L = H_0 + H_1 + H_2, \tag{35}$$

where H_0 is the unperturbed Hamiltonian and the perturbations linear and quadratic in the magnetic field are

$$\begin{aligned} H_1 &= -\lambda F_1 - \lambda\frac{\vec{T}\cdot\vec{F_2}}{\Theta} - \lambda\frac{\vec{T}\cdot(F_3\vec{T})}{\Theta^2}, \\ H_2 &= -\lambda^2 F_4 + \lambda^2\frac{\vec{F_2}^2}{2\Theta} + \lambda^2\frac{\vec{T}\cdot((F_3+F_3^T)\vec{F_2})}{\Theta^2} + \lambda^2\frac{\vec{T}\cdot((F_3+F_3^T)^2\vec{T})}{2\Theta^3} - \lambda^2\frac{\vec{T}\cdot(F_5\vec{T})}{\Theta^2}. \end{aligned} \tag{36}$$

The first and second term of H_1 represent the standard interaction terms of the magnetic field with the isovector and isoscalar magnetic moment [24]. The third term is a non–adiabatic correction to the isovector magnetic moment neglected in ref. [24] because of higher order in the $1/N_C$ expansion. The first as well as the last term of H_2 are derived from $\mathcal{L}_{\gamma\gamma}$ of eq. (8). All other terms in eq. (36) arise due to the fact that the "kinetical isospin" is not the same as the canonical isospin in the presence of the magnetic field. (This is completely analogous to non–relativistic (quantum) mechanics where the mechanical momentum of a particle is no longer equal to the conjugate momentum once an electromagnetic potential $\vec{A}$ is introduced (see e. g. ref. [29])).

The isospin operators will in general not commute with the functions F_i. As a consequence we face ordering ambiguities for those terms of eq. (36) which contain both, a function F_i and an isospin operator. In order to be specific let us discuss the interaction term which would give rise to a non–adiabatic correction of the isovector magnetic moment. The classical expression reads

$$H = -\frac{\lambda I_3}{\Theta^2}\left(\vec{T}^2 R_{33} - \sum_{i=1}^{3} T_i R_{i3} T_3\right). \tag{37}$$

One *possible* way to symmetrize this term is given by

$$H = -\frac{\lambda I_3}{\Theta^2}\left(\frac{1}{2}(\vec{T}^2 R_{33} + R_{33}\vec{T}^2) - \frac{1}{2}\sum_{i=1}^{3}(T_i R_{i3} T_3 + T_3 R_{i3} T_i)\right). \tag{38}$$

We evaluate the symmetrized operators between nucleon states and find

$$\begin{aligned} \vec{T}^2 R_{33} + R_{33}\vec{T}^2 &\rightarrow -\frac{1}{2}\sigma_3\tau_3 \quad \text{between nucleon states,} \\ \sum_{i=1}^{3}(T_i R_{i3} T_3 + T_3 R_{i3} T_i) &\rightarrow -\sigma_3\tau_3 \quad \text{between nucleon states.} \end{aligned} \tag{39}$$

Hence, substituting the result of eq. (39) into eq. (38) we get the following correction to the isovector magnetic moment,

$$\Delta\mu_{I=1} = \frac{eI_3}{2\Theta^2}, \tag{40}$$

with I_3 given in eq. (32). If we use the parameters of ref. [25], this results in a shift of $\mu_V = \mu_p - \mu_n$ in units of the nuclear magneton $e/2M_p$ of

$$\Delta\mu_V = 0.11. \tag{41}$$

Hence, together with the *adiabatic* isovector magnetic moment [25] we get the prediction $\mu_V = 3.21 + 0.11 = 3.32$, which has to be compared with the empirical number of 4.71. The ordering of eq. (38) gives a small correction of about 3% in the right direction. It has, however, to be noted that the ordering of eq. (38) is by no means unique. For the first term of eq. (37) one might as well have chosen

$$\vec{T}^2 R_{33} \to a(\vec{T}^2 R_{33} + R_{33}\vec{T}^2) + b\sum_{i=1}^{3} T_i R_{33} T_i, \tag{42}$$

where a and b are real numbers subject to the the condition $2a + b = 1$. If we evaluate the new term between nucleon states, we find the substitution rule

$$\sum_{i=1}^{3} T_i R_{33} T_i \to \frac{1}{12}\sigma_3\tau_3 \quad \text{between nucleon states.} \tag{43}$$

It is now easily seen, that we may generate any non–adiabatic "correction" we like to the isovector magnetic moment by a corresponding adjustment of the parameters a and b.

In view of this problem we have to restrict our discussion to the presumably dominant terms. For that reason we discuss the Hamilton function of eq. (35) in the large N_C limit [30, 31, 32, 33]. We classify the interaction terms according to their large N_C behaviour employing $F_\pi = O(\sqrt{N_C})$ and $\epsilon = O(1/\sqrt{N_C})$ (although the large N_C behaviour of the Skyrme model is not necessarily equivalent with that of QCD [34]). With the help of eq. (31) and (32) and $\Theta = O(N_C)$, we find that the interaction terms $-\lambda F_1$ and $-\lambda^2 F_4$ are of order N_C, whereas all the other terms are of order N_C^{-1}. For that reason we retain in the large N_C limit

$$H = H_0 + H_1 + H_2, \tag{44}$$

with

$$\begin{aligned} H_1 &= -\lambda F_1 = \frac{1}{2} e\Theta D^{(1)}_{0,0}(\alpha,\beta,\gamma)B, \\ H_2 &= -\lambda^2 F_4 = \frac{1}{2}\gamma^m B^2\left(1 - \frac{2}{5}D^{(2)}_{0,0}(\alpha,\beta,\gamma)\right), \end{aligned} \tag{45}$$

where

$$\begin{aligned} \gamma^m &= \gamma^{m(2)} + \gamma^{m(4)} \qquad (46) \\ &= \frac{\pi e^2}{9\epsilon^5 F_\pi^3}\int_0^\infty dx x^4 \sin^2(F) + \frac{4\pi e^2}{9\epsilon^5 F_\pi^3}\int_0^\infty dx x^4 \sin^2(F)\left(F'^2 + \frac{\sin^2(F)}{2x^2}\right). \qquad (47) \end{aligned}$$

Note that we have kept in H_0 the rotational term which is of order $1/N_C$. This term is required in the semiclassical quantization in order to obtain states of good spin and isospin. We now obtain the energy shift in second order perturbation theory,

$$\Delta E_n = < n|H_2|n > + \sum_{n' \neq n} \frac{|< n|H_1|n' >|^2}{E_n - E_{n'}}, \tag{48}$$

where n is an abbreviation for $J = T, M_J, M_T$ and $n' \neq n$ implies that the sum only contains states with J' different from J. Following the argument of refs. [34, 35] that the states with $J = T$ larger than 3/2 are spurious states for $N_C = 3$, the sum of eq. (48) includes the Δ for the nucleon polarizabilities and viceversa. Note that due to the sign change in the denominator the contribution of the Δ to the nucleon polarizability is paramagnetic whereas the contribution of the nucleon to the Δ polarizability is diamagnetic. We find for the magnetic polarizability of the nucleon

$$\beta_N = \beta_{dia} + \beta_{para} = -\gamma^m + \frac{\mu_V^2}{M_\Delta - M_N}\left(\frac{e}{2M_p}\right)^2, \tag{49}$$

where μ_V is the isovector magnetic moment of the nucleon. The first contribution in eq. (49) is diamagnetic and originates from $\mathcal{L}_{\gamma\gamma}$ of eq. (8). The second term is the standard paramagnetic delta contribution. When deriving this term we made use of $\mu_{I=1} = e\Theta/3$ [24] and $| < \frac{3}{2}, M, M'|D^{(1)}_{0,0}|\frac{1}{2}, M, M' > |^2 = 2/9$ [19]. We note that our expression for β_{dia} disagrees with that of ref. [16] by a factor 1/2 in the last term of $\gamma^{m(4)}$ of eq. (47). Consequently, the relation

$$\beta_{dia} = -\frac{e^2 < r_V^2 >}{4(M_\Delta - M_N)} \tag{50}$$

is not valid in the Skyrme model such as claimed in ref. [16]. In eq. (50) $< r_V^2 >^{\frac{1}{2}}$ is the isovector r. m. s. radius of the nucleon. Once again we find that all nucleon states have the same polarizability whereas the delta polarizabilities depend on spin and isospin quantum numbers,

$$\beta_\Delta^{M_T,M_J} = \begin{cases} -\gamma^m\left(1 - \frac{2}{25}\right) & \text{if} \quad |M_T| = |M_J| = \frac{3}{2}, \\ -\gamma^m\left(1 - \frac{2}{25}\right) - \frac{\mu_V^2}{M_\Delta - M_N}\left(\frac{e}{2M_p}\right)^2 & \text{if} \quad |M_T| = |M_J| = \frac{1}{2}, \\ -\gamma^m\left(1 + \frac{2}{25}\right) & \text{if} \quad |M_T| \neq |M_J|. \end{cases} \tag{51}$$

Had we also taken $J = T = 5/2$ states into account, each delta polarizability would have obtained an additional paramagnetic term similar to the delta contribution to the nucleon polarizability. We finish this section with the result for the polarizability of the classical hedgehog. Here we only have a diamagnetic polarizability

$$\beta_{cl} = -\frac{3}{5}\gamma^m. \tag{52}$$

3.3 Polarizabilities in the chiral limit

Both the electric and the magnetic polarizability diverge in the chiral limit. Taking the large x behaviour of the soliton profile function $F(x)$ for $m_\pi = 0$ into account [24],

$$F(x) \sim \frac{1}{x^2}, x \to \infty, m_\pi = 0, \tag{53}$$

it is easy to see that the contribution derived from $\mathcal{L}^{(2)}_{\gamma\gamma}$ is responsible for the divergence of the polarizabilities. We will find that the polarizabilities diverge as $1/m_\pi$. In order to show this, we split the divergent integral into two pieces,

$$\int_0^\infty dx x^4 \sin^2(F) = \int_0^{x_0} dx \ldots + \int_{x_0}^\infty dx \ldots \quad . \tag{54}$$

For sufficiently large x_0 we make use of the asymptotic behaviour of $F(x)$ [25, 27],

$$\lim_{x\to\infty} F(x) = \frac{B}{x^2}(1+\eta x)\exp(-\eta x) = F_{asym}(x), \tag{55}$$

where $\eta = m_\pi/\epsilon F_\pi$ and B is the coefficient of the asymptotic behaviour. We approximate the second integral of eq. (54) by

$$\int_{x_0}^{\infty} dx x^4 \sin^2(F(x)) \approx \int_{x_0}^{\infty} dx x^4 F^2_{asym}(x) = B^2 \exp(-2\eta x_0)\left(\frac{5}{4\eta} + \frac{3x_0}{2} + \frac{x_0^2\eta}{2}\right). \tag{56}$$

From the first term of this expression it is seen that the polarizabilities diverge as $1/m_\pi$ provided ϵ and F_π approach finite values as $m_\pi \to 0$. This is similar to the behaviour found in a chiral perturbation theory calculation [21, 22]. The difference, however, is that we predict the same leading order coefficient for the neutron and the proton.

4 Results and discussion

We now want to discuss the numerical predictions of the Skyrme model for the polarizabilities of the nucleon and the delta based on eq. (26), (27) and (49), (51). As the numerical results for the static properties of the nucleon and the Δ in the Skyrme model agree with experiment within about 30% [24, 25, 32], we estimate the uncertainties in the predictions for the polarizabilities by including in our discussion two sets of parameters:

- The first set is that of ref. [25], $m_\pi = 138MeV, F_\pi = 108MeV, \epsilon = 4.84$, which is fitted to the empirical values of the pion, nucleon and delta mass.
- The second choice, $m_\pi = 138MeV, F_\pi = 186MeV, \epsilon = 3.55$, reproduces besides the pion mass the weak properties, namely F_π and g_A ($M_N = 2039MeV$) .

The results for the electric polarizabilities are given in table 1. Depending on the parameters used, the predictions for the nucleon are larger by factors 2 - 5 than the experimental numbers $\alpha_p = (7.0 \pm 3.5) \times 10^{-4} fm^3$ [9] and $\alpha_n = (12.0 \pm 3.5) \times 10^{-4} fm^3$ [5]. We explicitly show the contributions derived from $\mathcal{L}^{(2)}_{\gamma\gamma}$ and $\mathcal{L}^{(4)}_{\gamma\gamma}$. As expected from the discussion of the

table 1: *Electric polarizabilities of the classical hedgehog, the nucleon and the Δ in units of $10^{-4} fm^3$. $\alpha^{(2)}$, $\alpha^{(4)}$ and α denote the contributions derived from $\mathcal{L}^{(2)}_{\gamma\gamma}$, $\mathcal{L}^{(4)}_{\gamma\gamma}$ and the sum, respectively. The first three columns refer to the parameter set $F_\pi = 108MeV, \epsilon = 4.84, m_\pi = 138MeV$, the last three columns to $F_\pi = 186MeV, \epsilon = 3.55, m_\pi = 138MeV$. $\Delta^{1)}$, $\Delta^{2)}$ denotes $|M_J| = |M_T|$, $|M_J| \neq |M_T|$, respectively.*

	$\alpha^{(2)}$	$\alpha^{(4)}$	α	$\alpha^{(2)}$	$\alpha^{(4)}$	α
hedgehog	10.3	5.5	15.8	14.5	5.8	20.3
nucleon	17.2	9.2	26.4	24.1	9.6	33.7
$\Delta^{1)}$	15.8	8.5	24.3	22.2	8.8	31.0
$\Delta^{2)}$	18.6	9.9	28.5	26.0	10.4	36.4

table 2: *Magnetic polarizabilities of the classical hedgehog, the nucleon and the Δ in units of $10^{-4}fm^3$. $\beta^{(2)}_{dia}$, $\beta^{(4)}_{dia}$, β_{dis} and β denote the contributions derived from $\mathcal{L}^{(2)}_{\gamma\gamma}$, $\mathcal{L}^{(4)}_{\gamma\gamma}$, the dispersive part and the sum, respectively. The first four columns refer to the parameter set $F_\pi = 108MeV, \epsilon = 4.84, m_\pi = 138MeV$, the last four columns to $F_\pi = 186MeV, \epsilon = 3.55, m_\pi = 138MeV$. $\Delta^{1)}$, $\Delta^{2)}$, $\Delta^{3)}$ denotes $|M_J| = |M_T| = \frac{3}{2}$, $|M_J| = |M_T| = \frac{1}{2}$, $|M_J| \neq |M_T|$, respectively.*

	$\beta^{(2)}$	$\beta^{(4)}$	β_{dis}	β	$\beta^{(2)}$	$\beta^{(4)}$	β_{dis}	β
hedgehog	-5.2	-2.4	0	-7.6	-7.2	-2.5	0	-9.7
nucleon	-8.6	-4.0	5.5	-7.1	-12.0	-4.2	23.3	7.1
$\Delta^{1)}$	-7.9	-3.7	0	-11.6	-11.0	-3.9	0	-14.9
$\Delta^{2)}$	-7.9	-3.7	-5.5	-17.1	-11.0	-3.9	-23.3	-38.2
$\Delta^{3)}$	-9.3	-4.3	0	-13.6	-13.0	-4.5	0	-17.5

chiral limit the larger contribution to α stems from $\alpha^{(2)}$. The transition from parameter set 1 to set 2 results in an increase of the electric polarizabilities of about 30%. It may be seen from table 1 that this increase is mainly due to an increase in $\alpha^{(2)}$ whereas $\alpha^{(4)}$ stays almost constant. From the overall factor $\epsilon^{-5}F_\pi^{-3}$ in eq. (18) one would naively expect a decrease of about 8% using the second set instead of the first. However, this argument does not take the dependence of the integral on $\eta = m_\pi/\epsilon F_\pi$ via the soliton profile F into account (see the discussion of the chiral limit in section 3.3). The second parameter set gives a "softer" pion distribution since η is smaller and hence we expect the integrals to be larger. The slower convergence of the integrand of $\alpha^{(2)}$ as $x \to \infty$ results in a larger increase of $\alpha^{(2)}$ in comparison with $\alpha^{(4)}$. The polarizabilities of the Δ differ from those of the nucleon by a fraction $\pm 2/25$ depending on the spin isospin projection of the Δ.

The magnetic polarizabilities are shown in table 2. The conclusions concerning β_{dia} are completely analogous to the discussion of α. The additional feature here is the presence of a dispersive contribution. If one uses the empirical values to calculate β_{dis}, one obtains $\beta_{para} = 12.0 \times 10^{-4}fm^3$. Of course, as was pointed out in ref. [19], one should consistently use the predictions within the Skyrme model. In that context it is interesting to note that it is impossible to simultaneously reproduce the empirical value for the isovector magnetic moment and the nucleon–delta mass difference entering the determination of β_{dis}. The reason for this is that the Skyrme model gives a parameter free relation for the product of the two quantities, namely $2\mu_{I=1}(M_\Delta - M_N)/e = 1$ [24], whereas with empirical values one finds 1.47 instead of 1. However, qualitatively one can expect a large cancellation between the diamagnetic and the paramagnetic part. In contrast to this we expect for the Δ^0 and the Δ^+ a large diamagnetic polarizability. We conclude that the numerical predictions of tables 1 and 2 may at most be regarded as order of magnitude estimates.

From tables 1 and 2 as well as eq. (18) and (47) it is seen that only the contribution derived from $\mathcal{L}^{(2)}_{\gamma\gamma}$ satisfies the relation $\alpha^{(2)} = -2\beta^{(2)}_{dia}$. This is in agreement with soliton models including vector mesons instead of a quartic stabilizing term [17, 18, 19]. However, it is not correct to generalize this result to the usual Skyrme model as was done in ref. [17, 18, 19]. We have explicitly shown that the above relation is not valid for $\alpha^{(4)}$ and $\beta^{(4)}_{dia}$

derived from $\mathcal{L}^{(4)}_{\gamma\gamma}$, i. e. $\alpha^{(4)} \neq -2\beta^{(4)}_{dia}$. It is, however, interesting to note that α_N in our calculation fulfils

$$\alpha_N = \frac{e^2 < r_V^2 >}{2(M_\Delta - M_N)}, \tag{57}$$

whereas the corresponding relation for $\beta_{N,dia}$ claimed in ref. [16] is not true. However, numerically the difference between $\alpha^{(4)}$ and $-2\beta^{(4)}_{dia}$ is rather small, namely of the order of 15%. Taking the the contributions derived from $\mathcal{L}^{(2)}_{\gamma\gamma}$ into account the relation $\alpha = -2\beta_{dia}$ is satisfied within 5%.

In order to further motivate the use of the large N_C limit, eq. (44), we have also calculated the $1/N_C$ term of eq. (36), which does not suffer from ordering ambiguities, namely the one proportional to $\vec{F_2}^2$. The small values $(-0.43, -0.12) \times 10^{-4} fm^3$ corresponding to the two sets of parameters justify the omission of the higher order terms (see also ref. [19]).

Finally, a more complete treatment would of course have to include dispersive contributions to the electric polarizability of the nucleon. For that purpose one would have to allow for fluctuations around the nucleon solution [36, 37, 38] and consider transitions to states subject to the dipole transition selection rule. An attempt in this direction has been made in ref. [17] in the framework of a Skyrme type model including vector meson degrees of freedom. The dispersive contribution was found to be rather small. However, one may doubt whether the replacement of the full distorted waves for the fluctuations by spherical waves, as was done in ref. [17], leads to a reliable result. In any case, the dispersive contribution will increase the polarizabilities of the nucleon as well as the delta. A final caveat concernes the fact that the ansatz of eq. (15) is only an approximate solution to the Euler Lagrange equations of the Skyrme model. Of course it would be desirable to perform a perturbative calculation where the exact solution of H_0 is known.

5 Conclusions

We have calculated the electromagnetic polarizabilities of the nucleon and the Δ baryon in the Skyrme model. We have shown that the contributions derived from the quadratic part of the Skyrme Lagrangian fulfil $\alpha^{(2)} = -2\beta^{(2)}_{dia}$ [17, 18, 19] whereas the general relation $\alpha = -2\beta_{dia}$ is satisfied within 5%. The numerical predictions may at most be regarded as order of magnitude estimates. We have argued that the $1/N_C$ contributions may not unambiguously be determined. The present framework does not allow for different polarizabilities of the proton and the neutron. However it seems to indicate that due to the cancellation between the diamagnectic pion contribution and the paramagnetic dispersive part the absolute value of the magnetic polarizability of the nucleon may be much smaller than the electric polarizability. In this case one would expect for the corresponding Δ baryon states a large enhancement of the magnetic polarizabilities.

This work was made possible by financial support from the Foundation for Fundamental Research on Matter (FOM) and the Netherlands Organization for Scientific Research (NWO).

References

[1] S. Scherer and P.J. Mulders, report NIKHEF-92-P2, to be publ. Nucl. Phys. **A**

[2] T. E. O. Ericson and J. Hüfner, Nucl. Phys. **B57** (1973) 604

[3] J. L. Friar, Ann. Phys. **95** (1975) 170

[4] V. A. Petrun'kin, Sov. J. Part. Nucl. **12** (1981) 278

[5] J. Schmiedmayer et al., Phys. Rev. Lett. **66** (1991) 1015

[6] A. M. Baldin, Nucl. Phys. **18** (1960) 310

[7] P. S. Baranov et al., Sov. J. Nucl. Phys. **21** (1975) 355

[8] A. I. L'vov, Sov. J. Nucl. Phys. **29** (1979) 651

[9] F. J. Federspiel et al., Phys. Rev. Lett. **67** (1991) 1511

[10] K. W. Rose et al., Phys. Lett. **B234** (1990) 460

[11] D. Drechsel and A. Russo, Phys. Lett. **B137** (1984) 294

[12] F. Schöberl and H. Leeb, Phys. Lett. **B166** (1986) 355

[13] P. C. Hecking and G. F. Bertsch, Phys. Lett. **B99** (1981) 237

[14] A. Schäfer et al., Phys. Lett. **B143** (1984) 323

[15] R. Weiner and W. Weise, Phys. Lett. **B159** (1985) 85

[16] E. M. Nyman, Phys. Lett. **B142** (1984) 388

[17] M. Chemtob, Nucl. Phys. **A473** (1987) 613

[18] N. N. Scoccola and W. Weise, Phys. Lett. **B232** (1989) 287

[19] N. N. Scoccola and W. Weise, Nucl. Phys. **A517** (1990) 495

[20] W. Broniowski et al., Phys. Lett. **B283** (1992) 22

[21] V. Bernard et al., Phys. Rev. Lett. **67** (1991) 1515

[22] V. Bernard et al., Nucl. Phys. **B373** (1992) 346

[23] T. H. R. Skyrme, Proc. Roy. Soc. **A260** (1961) 127

[24] G. S. Adkins et al., Nucl. Phys. **B228** (1983) 552

[25] G. S. Adkins and C. R. Nappi, Nucl. Phys. **B233** (1984) 109

[26] E. Witten, Nucl. Phys. **B223** (1983) 422

[27] G. Holzwarth und B. Schwesinger, Rep. Prog. Phys. **49** (1986) 825

[28] A. Bohr and B. Mottelson, Nuclear Structure, vol. I (Benjamin, New York, 1969)

[29] J. J. Sakurai, Modern Quantum Mechanics (Addison–Wesley, Redwood City, 1985)

[30] G. 't Hooft, Nucl. Phys. **B72** (1974) 461

[31] E. Witten, Nucl. Phys. **B160** (1979) 57

[32] U.-G. Meissner and I. Zahed, Adv. Nucl. Phys. **17** (1986) 143

[33] I. Zahed and G. E. Brown, Phys. Rep. **142** (1986) 1

[34] G. Karl and J. E. Paton, Phys. Rev. **D30** (1984) 238

[35] A. Jackson et al., Nucl. Phys. **A 432** (1985) 567

[36] H. Walliser and G. Eckart, Nucl. Phys. **A429** (1984) 514

[37] M. P. Mattis and M. Karliner, Phys. Rev. **D31** (1985) 2833

[38] T. Ikehashi and K. Ohta, Nucl. Phys. **A536** (1992) 521

U-SPIN SYMMETRY BREAKING AND ELECTROMAGNETIC PROPERTIES OF OCTET AND DECUPLET BARYONS

BERND SCHWESINGER
FB Physik, University Siegen, Postfach 101240
D-5900 Siegen, Germany

ABSTRACT
The SU(3) Skyrme model is used to investigate the effects of symmetry breaking imposed by different pseudoscalar meson masses and decay constants. Since the soliton is allowed to deform under the pressure of the symmetry breaking we find significant variations in magnetic moments, charge radii and polarizabilities with respect to strangeness.

1. Introduction

In the mesonic sector of the chiral lagrangian for SU(3)[1]

$$
\begin{aligned}
L &= L^{(2)} + L^{(4A)} + L^{(WZ)} + L^{(CSB)} + L^{(FSB)} \quad , \\
L^{(2)} &= -\frac{f_\pi^2}{4}\int d^3r \quad tr\ell_\mu\ell^\mu \quad , \qquad \ell_\mu = U^\dagger\partial_\mu U \quad , \\
L^{(4A)} &= \frac{1}{128g^2}\int d^3r \quad tr[\ell_\mu,\ell_\nu][\ell^\mu,\ell^\nu] \quad , \qquad U\in SU(3) \quad , \\
\int L^{(WZ)}dt &= \frac{iN_c}{240\pi^2}\int_Q d\Sigma^{\mu\nu\lambda\rho\sigma} tr\ell_\mu\ell_\nu\ell_\lambda\ell_\rho\ell_\sigma \quad , \\
L^{(CSB)} &= \frac{f_\pi^2 m_\pi^2}{4}\int d^3r \quad tr(U+U^\dagger-2) \\
L^{(FSB)} &= -\frac{f_K^2-f_\pi^2}{12}\int d^3r \quad tr(1-\sqrt{3}\lambda_8)(U\ell_\mu\ell^\mu+\ell_\mu\ell^\mu U^\dagger) \\
&+ \frac{f_K^2m_K^2-f_\pi^2m_\pi^2}{6}\int d^3r \quad tr(1-\sqrt{3}\lambda_8)(U+U^\dagger-2)
\end{aligned}
\tag{1}
$$

the size of SU(3) flavor symmetry breaking is entirely fixed by the masses and decay constants of the pseudoscalar meson octet. In Skyrme type models for baryons the same form of the action is also used in the winding number one sector. As far as the parameters of the action are concerned different attitudes have been adopted. From considerations in the meson-baryon scattering system, also located in the winding number one sector, evidence has, however, increased that the same parameters as in the mesonic sector must be used. Otherwise, wrong meson masses would lead to the wrong thresholds for different reaction channels[2] and wrong decay constants would lead to the wrong low energy behaviour of the meson baryon phase shifts. The latter point only has been understood after the Tomozawa-Weinberg relation was established[3] in Skyrme type models, too.

Given the mesonic SU(3) symmetry breaking it is a non-trivial matter to show, that it is of the right magnitude and form to also reproduce the pattern of symmetry breaking for different baryonic properties. The demonstration of this remarkable fact which is part of this report has been obtained in collaboration with H. Weigel[4,5]. In this contribution I will concentrate on the mass splittings, magnetic moments and charge radii of octet and decuplet baryons as well as their electric polarizabilities. Although the effective action is not complete in the sense that it provides a perfect fit to the mesonic masses and decay constants - for instance the η mass follows from a slightly modified Gell-Mann - Okubo relation

$$m_\eta^2 = \frac{4f_K^2 m_K^2 - f_\pi^2 m_\pi^2}{4f_K^2 - f_\pi^2}, \tag{2}$$

the missing terms to be found in the publications of Gasser and Leutwyler[1] are small and numerically unimportant for our purposes.

2. Mass splittings in the slow rotator aproach

When the SU(2) hedgehog is embedded into the SU(2) subgroup of SU(3)

$$U_H = A(t) \quad e^{i\sum_{p=1}^{3}\lambda_p \hat{r}_p \chi(r)} \quad A^\dagger(t) \tag{3}$$

and performs slow rotations $A(t) \in SU(3)$ it has time to adjust its chiral angle χ to its momentary orientation in flavor space according to the forces exerted by the SU(3) symmetry breaking term $L^{(FSB)}$. Thus, at each moment the minimization of a static orientation dependent mass

$$\begin{aligned} M(\nu) = \int d^3r \Bigg[&\frac{f_\pi^2}{2}(\chi'^2 + \frac{2s^2}{r^2}) + \frac{1}{4g^2}\frac{s^2}{r^2}(\chi'^2 + \frac{s^2}{r^2}) + m_\pi^2 f_\pi^2 (1-c) \\ &+ \sin^2\nu \left(\frac{1}{2}(f_K^2 - f_\pi^2)(\chi'^2 + \frac{2s^2}{r^2})c + (m_K^2 f_K^2 - m_\pi^2 f_\pi^2)(1-c)\right)\Bigg], \\ &s = \sin\chi,\ c = \cos\chi, \end{aligned} \tag{4}$$

determines the shape $\chi(\nu; r)$ of the soliton's chiral angle[4,5]. The strangeness changing Euler angle ν is defined by our parametrization of the rotations

$$A = R(\alpha,\beta,\gamma) \quad e^{-i\nu\lambda_4} \quad R(\alpha',\beta',\gamma') \quad e^{-i(\rho/\sqrt{3})\lambda_8} \tag{5}$$

where the R's are SU(2) rotations. With respect to these collective angular degrees of freedom

$$A^\dagger \dot{A} = -\frac{i}{2}\sum_{a=1}^{8}\lambda_a \Omega_a. \tag{6}$$

one finds the hamiltonian

$$H = M(\nu) + \left(\frac{1}{2\Theta_\pi(\nu)} - \frac{1}{2\Theta_K(\nu)}\right)\mathbf{J}^2 + \frac{1}{2}\left\{\frac{1}{2\Theta_K(\nu)}, C_2\left[SU(3)\right]\right\} + \frac{1}{16}\left\{\frac{1}{\Theta_K(\nu) + \frac{1}{2}\Delta(\nu)} - \frac{1}{\Theta_K(\nu)}, p_\nu^2\right\}. \quad (7)$$

expressed in terms of $C_2[SU(3)]$, the quadratic Casimir operator of SU(3), $\mathbf{J}$ the total angular momentum and p_ν, the conjugate momentum to the strangeness changing angle ν. The soliton mass M, the pionic and kaonic moments of inertia Θ_π and

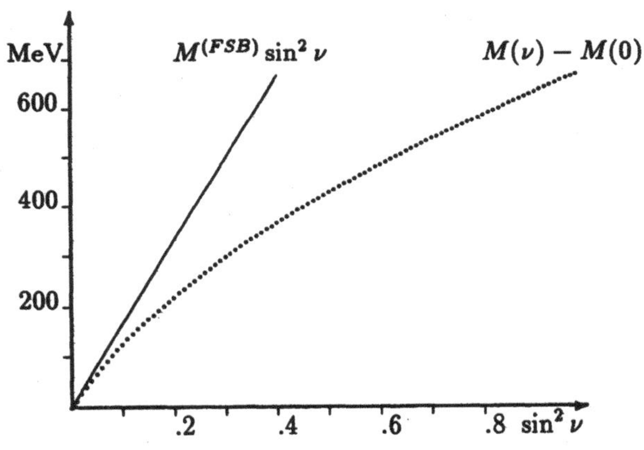

Figure 1: The mass of the soliton as a function of its orientation in flavor space.

Θ_K all are integrals involving the chiral angle $\chi(\nu;r)$ and thus depend implicitly and explicitly on the strangeness changing angle ν. Figure 1. shows that the variations of these integrals with changing ν are non-negligible relative to a symmetry breaking $M^{(FSB)}\sin^2\nu$ (second line in eq.(4) evaluated at $\nu = 0$). This difference will alter SU(3) baryon properties qualitatively as compared to either perturbative approaches[6], diagonalizations[7] as initiated by Yabu and Ando or the strong symmetry breaking approximation[8] of Callan-Klebanov; latter approaches are all based on a hedgehog minimizing the mass functional for a fixed orientation pointing into the isospin subgroup of SU(3), $\nu = 0$. The first two types of approaches will be termed "rigid rotator" approaches, here.

The hamiltonian is symmetrized by taking anti-commutators which is a reasonable approximation to more evolved and complicated quantization procedures. Due to the implicit ν-dependence of χ there is an additional kinetic energy

$$\dot{\nu}^2\Delta(\nu) = \dot{\nu}^2 \int d^3r (\frac{\partial\chi}{\partial\nu})^2 \left[\frac{f_\pi^2}{2} + \frac{1}{4g^2}\frac{s^2}{r^2} + \frac{1}{2}(f_K^2 - f_\pi^2) c \sin^2\nu\right] \tag{8}$$

which after Legendre transformation to the hamiltonian results in the additional term in (7) proportional to the squared canonical momentum p_ν.

As shown by Yabu and Ando[7] first, hamiltonian operators of the form (7) may be diagonalized exactly by expressing the quadratic Casimir operator as a second order differential operator in the Euler angles (5). Splitting the corresponding wave-function into its various SU(2) parts and an isoscalar function $f_{M_L M_R}(\nu)$

$$\Psi_{YII_3,Y_RJJ_3}(A) = (-)^{J-J_3} \sum_{M_L M_R} D^{(I)*}_{I_3 M_L}(\alpha,\beta,\gamma) f_{M_L M_R}(\nu) D^{(J)*}_{M_R -J_3}(\alpha',\beta',\gamma') e^{iY_R\rho} \tag{9}$$

the eigenvalue equation $H\Psi = E\Psi$ reduces to a set of coupled second order differential equations for $f_{M_L M_R}(\nu)$ which may be integrated for each isospin multiplet separately[7]. Since the angular velocity Ω_8 appears only linearly in the collective lagrangian the right hypercharge, Y_R is constrained to $Y_R = B = 1$. This fixes the spin to half integer values, i.e. fermions. Due to the additional implicit ν dependence of the classical mass and the moments of inertia the diagonalization of the hamiltonian (7) is much more involved than in the approach of Yabu and Ando. Therefore we would like to simplify our numerical procedure by setting $\Delta = 0$. We have checked that its action on the eigenvalues of H may be mocked up by an appropriate readjustment of the free parameters of the model. Thus we will put $\Delta = 0$ in what follows and choose g to yield a best fit for the baryon mass differences. Figure 2. shows that a good fit to the mass splittings may be obtained by a considerable reduction of the Skyrme parameter to $e = 2g = 3.46$ indicating a large size of the soliton when pointing towards $\nu = 0$; the other parameters of the lagrangian were kept at their experimental mesonic values: $f_\pi = 93MeV, m_\pi = 138MeV, m_K = 495MeV$ apart from a slight increase to $f_K = 118MeV$ in order to fine tune the splittings. The effects of the contribution proportional to p_ν^2 mentioned above leads to very similar results after a slight refit of $f_K = 110MeV$ and $e = 2g = 3.57$.

As an unavoidable consequence of employing physical values for the decay constants we have to face a very large nucleon mass ($M_N = 2769.$MeV), which might now have found an explanation in recent work by Moussalam and Kalafatis[9] and by Holzwarth[10], both presented in these proceedings, too. Their evaluation of the zero point energies of fluctuating fields relative to those of the vacuum of order one in N_C counting yields a negative contribution which is roughly proportional to the number of normalizable zero modes and inversely proportional to the size of the soliton. For SU(2) they quote numbers of the order of one GeV[9] depending slightly on the parameters used accompanied by uncertainties which probably must be estimated to be of the order of 30%. In unbroken SU(3) the number of normalizable zero modes is augmented from six to ten pointing to sizable corrections there, too. Being of order one these Casimir energies will not affect the mass splittings of baryons. Because of these observations we only present mass differences and ignore the problem of absolute masses completely.

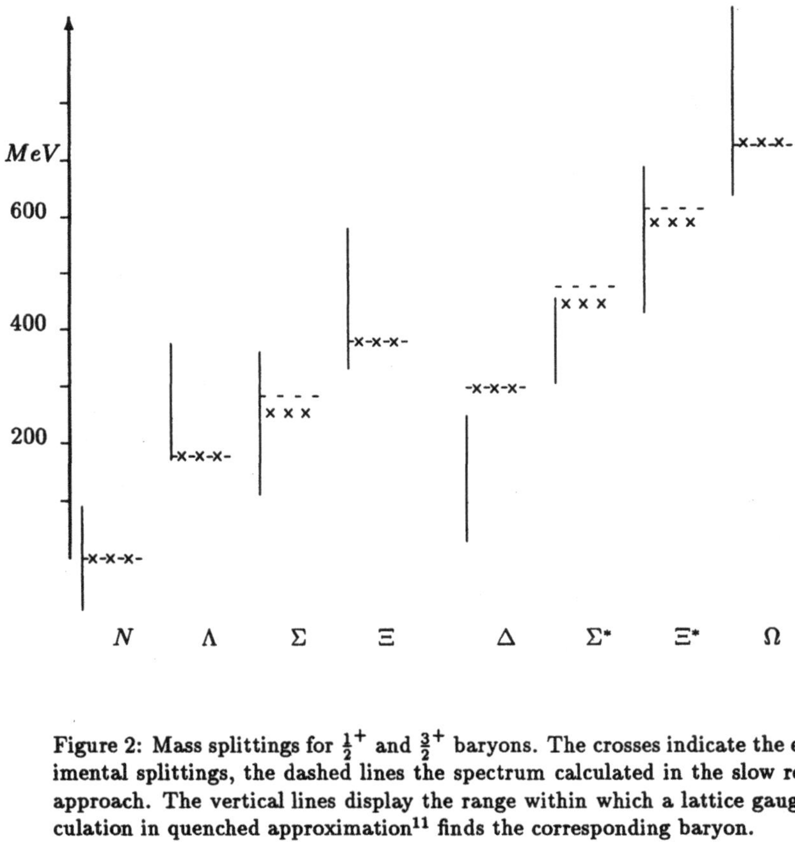

Figure 2: Mass splittings for $\frac{1}{2}^+$ and $\frac{3}{2}^+$ baryons. The crosses indicate the experimental splittings, the dashed lines the spectrum calculated in the slow rotator approach. The vertical lines display the range within which a lattice gauge calculation in quenched approximation[11] finds the corresponding baryon.

Apart from the Ω^- all $\frac{3}{2}^+$ resonances are unstable with respect to strong decays, therefore one must expect shifts in the resonance positions of the calculated $\frac{3}{2}^+$ masses which could be as large as our 30MeV deviation from the equal spacing rule. The quantity which doesn't suffer from the coupling to the decay channels, $(M_\Omega - M_N)$, is reproduced. Thus we think that although the calculated results are far away from those obtainable by first order perturbation theory the calculation has managed to keep the masses within reasonable deviations from the Gell-Mann - Okubo relation.

3. Electromagnetic properties and U-spin symmetry violations

3.1. *Magnetic moments*

The photon couples to the derivative terms of the effective action. Minimally the photon is introduced by gauge covariant derivatives

$$\partial_\mu U \rightarrow \nabla_\mu U = \partial_\mu U + i a_\mu [Q, U]. \tag{10}$$

Generally, terms in the lagrangian containing derivatives are flavor symmetric, i.e. symmetry breaking is exclusively due to non-derivative mass terms apart from the

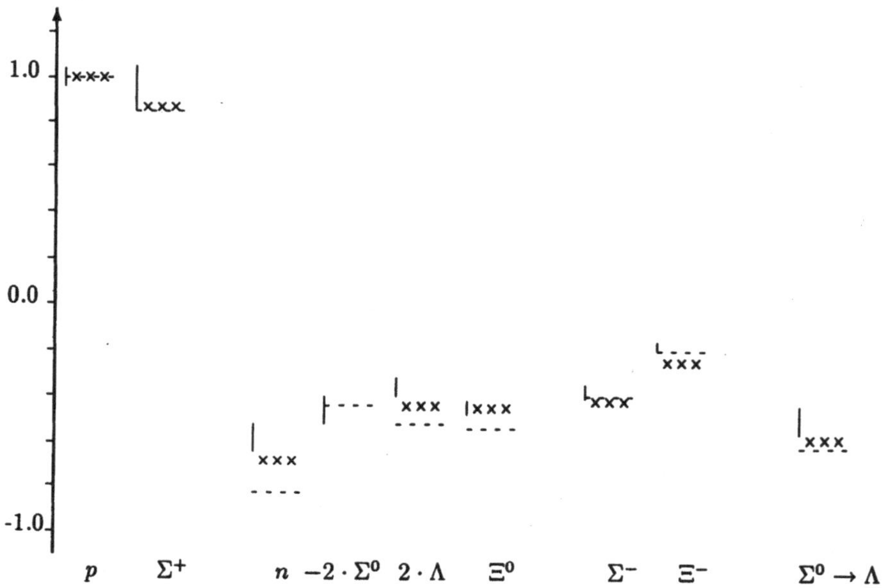

Figure 3: Magnetic moment ratios of octet baryons relative to the proton. Experimentally (crosses) $\mu_p = 2.79$, slow rotator approach (dashes) $\mu_p = 1.8$, lattice gauge calculation (vertical lines) $\mu_p \sim 2.3$. The last column gives the transition moment indicated.

small term in (1) responsible for a difference in the weak decay constants $f_K \neq f_\pi$. The charge matrix in SU(3)

$$Q = \frac{e}{2}(\lambda_3 + \frac{1}{\sqrt{3}}\lambda_8) = e\begin{pmatrix} \frac{2}{3} & 0 & 0 \\ 0 & -\frac{1}{3} & 0 \\ 0 & 0 & -\frac{1}{3} \end{pmatrix}. \tag{11}$$

doesn't distinguish between the second (d-quark) and third (s-quark) position. Thus the photon coupling is explicitly U-spin symmetric:

$$L_\gamma(U) = L_\gamma(CUC^\dagger) \tag{12}$$

for any constant, unitary matrix

$$C = \begin{pmatrix} 1 & 0 & 0 \\ 0 & c_{22} & c_{23} \\ 0 & c_{32} & c_{33} \end{pmatrix}. \tag{13}$$

Deviations of electromagnetic operators away from U-spin symmetry can only arise through an implicit hypercharge dependence of the photocouplings, e.g. the ν-dependence of the chiral angle χ in the case of the slow rotator approximation, where there is a feedback of the SU(3)-symmetry breaking terms on the soliton's shape. Additional deviations from U-spin symmetry arise due to admixtures of higher flavor representations in the baryonic wave-functions, but these contributions turn out to be small[7] because the admixtures induced by the flavor symmetry breaking terms do not suffice.

Thus it is the implicit dependence on strangeness which is responsible for the apparent deviations of the $\frac{1}{2}^+$ baryonic moments, figure 3.,and for $\frac{3}{2}^+$ moments, figure 4. from U-spin symmetry. All magnetic moments in figure 3. suffer from the fact that they come out too small by a factor of roughly 1.5 which seems to be a problem for all models of the nucleon which derive the photon coupling from the same action that also determines mass and size of the hadron: the SU(2) Skyrme model with mesonic parameters fitting the $N\Delta$ -split gives $\mu_p = 1.75$, the MIT-bag model[12] $\mu_p = 1.9$, the SU(2) Skyrme model with vector mesons: $\mu_p = \frac{939}{1575}2.77 = 1.65$ where we scaled the result of ref.[13] back to magnetons defined with the empirical nucleon mass as opposed to natural units. Our numbers have additional shortcomings with respect to the ratio of neutron to proton moments which comes out too high in magnitude because the isoscalar contribution is so small (see the contribution of F. Meier in these proceedings).

The point we would like to emphasize here, however, are the variations of the moments as a function of increasing strangeness. The ratios in all figures presented are scaled such that in case of U-spin symmetry baryons with same charges would lie on a horizontal line.

There is an interplay of three effects:
(i) variations from U-spin symmetry due to configuration mixing,
(ii) variations due to $f_K \neq f_\pi$,
(iii) variations due to the ν-dependent compression of the soliton.
For the $p\,\Sigma^+$ case configuration mixing goes in the right direction but the size is far too small, the different decay constants (ii) go in the opposite direction and these changes must then be overcompensated by the smaller soliton size in case of the Σ^+. For the $n\,\Xi^0$ we have the same situation apart from the fact that configuration mixing is not negligible any more. For the $\Sigma^-\,\Xi^-$ case configuration mixing is just of right

size for the split, the different decay constants go in the opposite direction and these changes are now compensated by the smaller soliton size of the Ξ^-. The latter effect is small now because the compression effect levels off with increasing angle ν,(c.f. fig.1.).

A very similar U-spin symmetry breaking pattern arises for the moments of the $\frac{3}{2}^+$

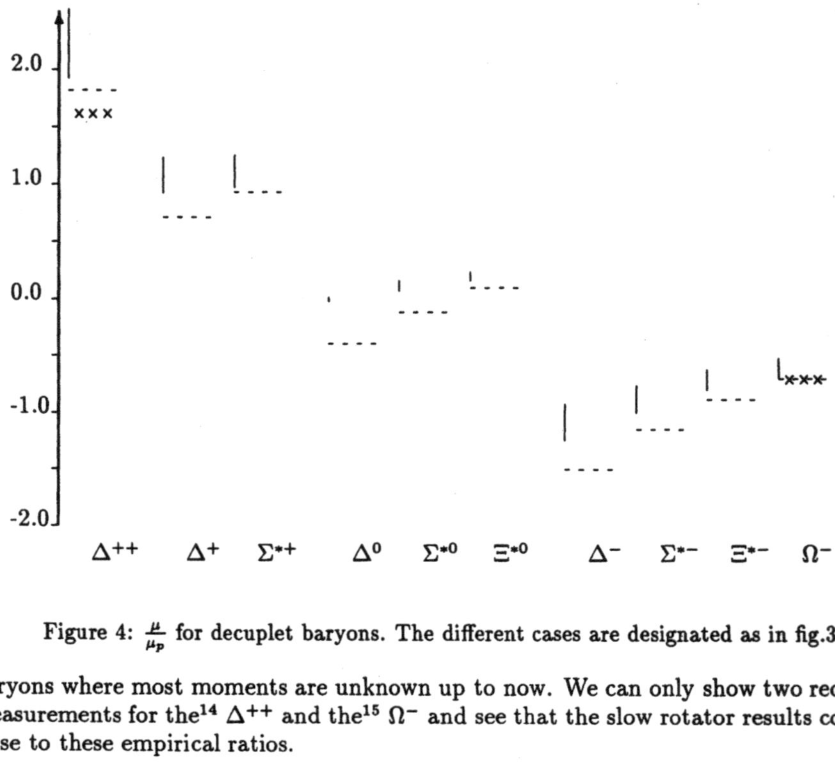

Figure 4: $\frac{\mu}{\mu_p}$ for decuplet baryons. The different cases are designated as in fig.3.

baryons where most moments are unknown up to now. We can only show two recent measurements for the[14] Δ^{++} and the[15] Ω^- and see that the slow rotator results come close to these empirical ratios.

3.2. *Electric properties and non-minimal couplings*

Within models using the pseudoscalars only the expressions for isovector charge formfactors like the isovector radii $\langle r^2 \rangle_V$, isovector magnetic moments, Δ-nucleon mass split and the electric polarizability α all may immediatly be obtained from the density of the soliton's moments of inertia $\Theta_{ab}(\mathbf{r})$. The inertia are introduced through a rotation of the hedgehog configuration by time dependent matrices $A(t)$

$$\dot{U} = [\dot{A}A^\dagger, U] = A[-\frac{i}{2}\Omega_a\lambda_a, U_H]A^\dagger \tag{14}$$

giving rise to a kinetic energy $\frac{1}{2}\int d^3r\Theta_{ab}(\mathbf{r})\Omega_a\Omega_b$. The gauge covariant coupling of the time component of the electromagnetic field introduces the modifications

$$\dot{U} \to \nabla^0 U = \dot{U} + ia^0[Q,U] = \dot{U} + A\left(ia^0\frac{e}{2}D_{ea}(A)[\lambda_a, U_H]\right)A^\dagger \tag{15}$$

such that the expressions for the isovector radii or quadrupole moments and the electric polarizability may simply be obtained from the rotational energy density by appropriate replacements $\Omega_a \to -ea^0 D_{ea}(A)$. We use the usual definitions $D_{ab} = \frac{1}{2} tr \lambda_a A \lambda_b A^\dagger$ and the abbreviations that the index e stands for the linear combination of flavors entering in the charge operator, ($D_{ea} = D_{3a} + \frac{1}{\sqrt{3}} D_{8a}$ for SU(3) e.g.).

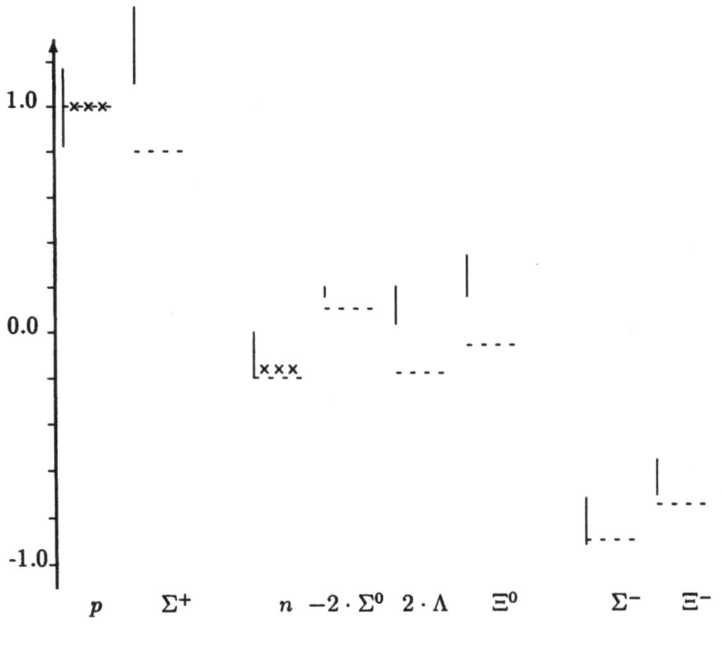

Figure 5: Ratios of squared charge radii of octet baryons relative to the proton. Experimentally (crosses) $\langle r^2 \rangle_p = .74 fm^2$, slow rotator approach (dashes) $\langle r^2 \rangle_p = .41 fm^2$, lattice gauge calculation[11] (vertical lines[a]) $\langle r^2 \rangle_p \sim .42 fm^2$.

It therefore follows that the isovector radii are given by ($a_0 = -r^2$)

$$\frac{e}{2}\langle r^2 \rangle_V = e\langle \Omega_a D_{eb} \int d^3 r r^2 \Theta_{ab}(\mathbf{r}) \rangle \tag{16}$$

[a]The uncharged radii have been deduced from the formfactors $G_E(q^2)$ of ref.[11] by assuming a straight line in q^2.

and the seagull contributions, i.e. two photon contact terms, to the electric polarizability ($a_0 = -zE$) by

$$L_{\gamma\gamma}^{seagull} = \frac{1}{2}\alpha^{seagull}\mathbf{E}^2 = \frac{1}{2}e^2E^2\langle D_{ea}D_{eb}\int d^3rz^2\Theta_{ab}(\mathbf{r})\rangle \quad . \tag{17}$$

In chiral perturbation theory up to fourth order there are two non-minimal terms coupling the photon to the chiral fields[1]:

$$\begin{aligned}\Delta\mathcal{L} = \quad & il_9\, a_{\mu\nu}\, trQ(\nabla^\mu U\nabla^\nu U^\dagger + \nabla^\mu U^\dagger\nabla^\nu U) \\ & +l_{10}\, a_{\mu\nu}a^{\mu\nu}\, tr(QUQU^\dagger - Q^2), \\ & l_9 = (7.4 \pm .7)\cdot 10^{-3}, \quad l_{10} = (-6.0 \pm .7)\cdot 10^{-3}\end{aligned} \tag{18}$$

which loosen the tight connection between e.g. the isovector radius and the polarizability of the soliton. The term proportional to l_9 fixes the radius of the pion and

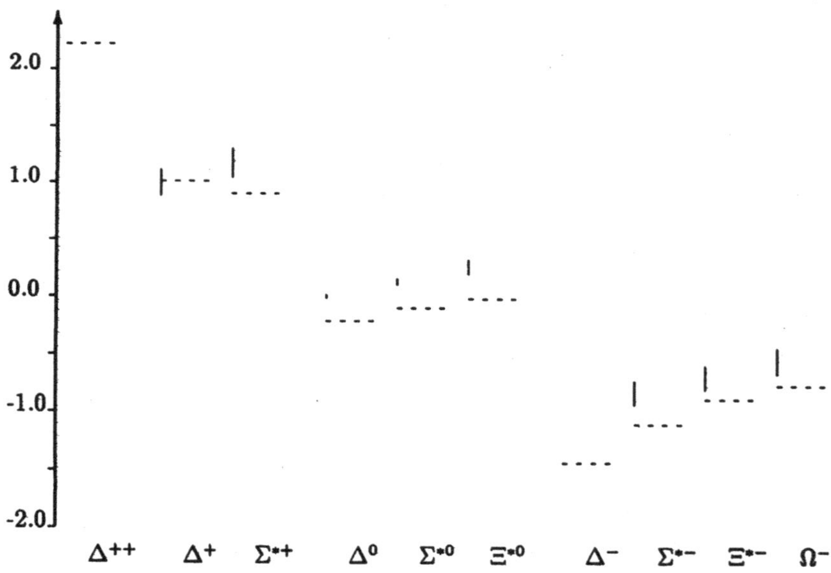

Figure 6: Ratios of squared charge radii of decuplet baryons relative to the proton. Different cases are designated as in fig.3.

also contributes to the isovector radius of solitons by adding finite pion size effects[5]; this term, however, gives no contribution to the polarizability. On the other hand, the term proportional to l_{10}, fixed by $\pi \to e\nu\gamma$, does not change the isovector radius but reduces the polarizability by

$$\Delta\alpha^{seagull} = 4e^2 l_{10} \langle \frac{2}{3} D_{ep} D_{ep} \int s^2 d^3r + D_{ek} D_{ek} \frac{1}{2} \int (1-c)) d^3r \rangle \qquad (19)$$
$$p \in \{1,2,3\}, k \in \{4,5,6,7\}.$$

The charge radii which we present in figure 5. for spin $\frac{1}{2}^+$ baryons and in figure 6. for $\frac{3}{2}^+$ behave very much the same way as the magnetic moments. Firstly, their overall size is too small by a similar factor of 1.5 despite of the finite pion size correction (18) which contributes about 20%. However, the ratio between proton and neutron comes out fairly well. As far as the dependence on strangeness is concerned we once again may see large violations of U-spin symmetry where increasing strangeness means smaller charge radii, very much in the direction discussed in ref.[16]. The Skyrme model predicts roughly a 15 − 20% decrease of the charge radii per unit decrease of hypercharge in baryons. By comparison to the results of the rigid rotator[7] we deduce that only minor contributions originate from the representation mixing in the baryonic wave-functions.

A similar pattern for electromagnetic observables is found when the size of the soliton is treated as a collective, breathing mode, variable[17]: its quantization in the presence of symmetry breaking also yields the analogous feature that baryonic radii decrease with hypercharge and this alternative approach demonstrates again that a dependence on symmetry breaking of the soliton's shape is the main ingredient to account for deviations of electromagnetic properties from symmetry relations.

For completeness we also present the charge radii for $\frac{3}{2}^+$ baryons in figure 6. although there is even less information for these quantities as compared to the already sparse data for the $\frac{1}{2}^+$ baryons.

3.3. *Electric polarizabilities in SU(2)*

In hedgehog models the second order change of the baryon's energy in static electromagnetic fields

$$\Delta E = -\frac{1}{2}\alpha \mathbf{E}^2 - \frac{1}{2}\beta \mathbf{B}^2 \qquad (20)$$

is, as already has been stated, tightly correlated with other hadronic quantities. As such it therefore offers another clue in probing the hedgehog structure of real baryons. Let us concentrate for a moment on the seagull contributions which are just the leading $\mathcal{O}(N_C)$ terms in a $\frac{1}{N_C}$ classification. The result of corresponding calculations in Skyrme type models[18,19], all carried out in SU(2), have met with mixed success. Within SU(2) and with minimal couplings the pure pseudoscalar soliton model faces the difficulty that if the only free parameter of the model - which here is the strength of the fourth order stabilizer g - is chosen to reproduce the Δ-nucleon split or the electromagnetic radii then the electric polarizability is too large by roughly a factor of two[18]

$$\alpha_p = \alpha_n = \frac{e^2}{2M_{\Delta N}} \langle r^2 \rangle_V . \qquad (21)$$

Also, the contributions for proton and neutron turn out to be the same which is in contrast to the experimental determination of these quantities[20]:
$\alpha_p = 7 \cdot 10^{-4} fm^2, \quad \alpha_n = 12 \cdot 10^{-4} fm^2$.

Now we examine the non-minimal photocouplings. Both non-minimal terms (18), with very similar coefficients, also appear once a local approximation on lagrangians including vector mesons is performed. One is therefore lead to expect that the pure pseudoscalar model with these non-minimal terms will yield similar results as the case with explicit vector mesons[19]. Fortunately, for a soliton stabilized by vector meson fields the situation appeared to be more favorable, because at least the magnitude of the polarizability seemed to agree[19] with the data leaving only the proton-neutron difference as an open problem. This improvement persists once all two photon terms emerging after a diagonalization to physical photons and vector mesons are taken into account. In the resulting expression

$$\alpha = \frac{e^2}{9} \int [2 f_\pi^2 r^2 (s^2 + 2(1-c)^2) + \frac{1}{g^2} G^2] \, d^3 r \tag{22}$$

the last term $\sim G^2$, not quoted in ref.[19], brings the polarizabilities in the vector meson model back to the size obtained in purely pseudoscalar versions of the model as shown in table 1.

	PS min	PS non-min	VM	exp
$\langle r^2 \rangle_p [fm^2]$	.46	.55(.75)	.94	.74
$\langle r^2 \rangle_n [fm^2]$	-.23	-.32(-.12)	-.063	-.12
$\langle r^2 \rangle_V [fm^2]$	.69	.87	1.00	.86
$\langle r^2 \rangle_S [fm^2]$	.23	.23(.63)	.88	.62
$M_{\Delta N} [MeV]$	300	300	442	300
$\alpha_n \, [10^{-4} \cdot fm^3]$	16.8	13.4	14.4	12.
$\frac{\alpha_n}{\langle r^2 \rangle_V} \, [10^{-4} \cdot fm]$	24.4	15.4	14.4	13.9

Table 1. SU(2) polarizabilities α correlated with different mean square radii and the Δ-nucleon mass split for various models and experiment: first and second column for the Skyrme model with pseudoscalar mesons only ($f_\pi = 93.MeV, m_\pi = 138.MeV, g = 2.14$) obtained without and with the non-minimal photon contributions, respectively. The third column shows the results with explicit vector mesons using $g = 2.92$ and $m_V^2 = 8 g^2 f_\pi^2$. Numbers in round brackets include the non-minimal isoscalar contribution, eq. (23).

Evidently, the non-minimal photon couplings lead to a considerable improvement of the electromagnetic quantities examined here leaving only shortcomings due to a too small isoscalar radius $\langle r^2 \rangle_S$. We are missing a further isoscalar non-minimal coupling

$$\begin{aligned} \Delta \mathcal{L}_\gamma &= -\frac{e}{2 m_V^2} a_{\mu\nu} \partial^\mu B^\nu(U) \quad , \\ B^\nu(U) &= \frac{\epsilon^{\nu\alpha\beta\gamma}}{24\pi^2} \, tr l_\alpha l_\beta l_\gamma \end{aligned} \tag{23}$$

which is not present in the list of Gasser and Leutwyler because it is already of sixth order in chiral perturbation theory. However, it emerges immediately, once a local approximation on the isoscalar vector mesons is performed. Such a term adds $\frac{3}{m_V^2} = .20 fm^2$ to the neutron and proton radii and twice this amount to $\langle r^2 \rangle_S$ but it doesn't contribute to the polarizabilities. Note that the relation $\alpha = -2\beta_{dia}$ found in earlier publications[18,19] will no longer hold in pure pseudoscalar models in the presence of non-minimal terms nor in the vector meson model due to the presence of the last contribution $\sim G^2$ in eq.(22).

The results obtained so far urge us to conclude that once mesonic parameters are used there is no principle difference whether one restricts oneself to purely pseudoscalar type soliton models in SU(2) or attempts a description within the technically much more cumbersome models augmented by vector mesons. In both cases the leading seagull contributions give the correct magnitude of the polarizability and in both cases the split between neutron and proton polarizability is missed.

3.4. *Electric polarizabilities in SU(3)*

We now turn to the modifications due to an extension of the flavor group to SU(3) in order to examine the possibility of nondegenerate polarizabilities for $\frac{1}{2}^+$ baryons. The complete expressions for the moments of inertia appearing in the minimal photon polarizabilities (17) may be found in ref.[5] where not only the contributions of the flavor symmetric terms but also those of the important, i.e. numerically large, kinetic symmetry breakers $L^{(FSB)}$ have been spelled out explicitly. The contributions from the non-minimal photocouplings have been given in eqn.(19).

The perturbative result for the polarizabilities in table 2. clearly displays the explicit U-spin symmetry of the photocouplings, e.q. $\alpha_{\Sigma^+} = \alpha_p$.

	30.4		29.0			17.9		17.7	
		28.3					11.1		
30.4		26.9		21.9	10.7		9.3		8.0
	29.0		21.9			8.0		5.5	
		PT					SRA		

Table 2. Ratios of calculated electric polarizabilities for $\frac{1}{2}^+$ baryons relative to $\langle r^2 \rangle_V = \langle r^2 \rangle_p - \langle r^2 \rangle_n$ in $[10^{-4} \cdot fm]$. The entries are positioned by isospin decreasing from left to right and hypercharge decreasing from top to bottom - the leftmost highest position thus corresponds to the proton. The two cases: perturbation theory (PT, $\langle r^2 \rangle_V = .97 fm^2$) and slow rotator approximation (SRA, $\langle r^2 \rangle_V = .49 fm^2$) are calculated with the parameters $f_\pi = 93.MeV, m_\pi = 138.MeV, f_K = 118.MeV, m_K = 495.MeV, g = 1.73$.

From the observation that the slow rotator approximation gives a reasonable U-spin symmetry breaking in the case of the magnetic moments we believe that the decrease of the polarizability with decreasing hypercharge as displayed in table 2. is a meaningful prediction. It is a direct consequence of the compression of the soliton with decreasing hypercharge.

Within SU(3) the degeneracy of the proton-neutron polarizabilities is lifted. The resulting splitting is, however, far too small, of opposite sign and signalizes the importance of the dispersive contributions of higher order in the $\frac{1}{N_C}$ classification.

4. Conclusions

This contribution has presented a comprehensive study of electromagnetic single baryon properties within a dynamically complete model. The version considered here contains essentially only one free parameter and enabled us to calculate mass splittings of $\frac{1}{2}^+$ and $\frac{3}{2}^+$ baryons, their magnetic moments and charge radii. The main emphasis of our investigations was focussed on SU(3) symmetry breaking effects which seem to display contradictory aspects in the data: perturbative mass splittings, SU(3) symmetric weak hyperon decays as opposed to strong U-spin symmetry violation in magnetic moments and charge radii as well as strongly suppressed strange quark contributions in the proton.

The principal mechanism imposing the symmetry breaking within the model is provided by the symmetry breaking mass term the size of which is fixed by the pseudoscalar meson masses and decay constants. The soliton reacts to the symmetry breaking forces (i) by increased compression at decreasing values of its hypercharge and (ii) decreased strangeness content in the baryon wave-functions, i.e. admixtures of higher representations beyond octet or decuplet. Our results reveal that the size changes of the baryon for different hypercharges is the dominant cause moving the magnetic moments and the charge radii away from their U-spin symmetric values. This is understood from the fact that the electromagnetic current operator is explicitly U-spin invariant since the large symmetry breaking terms involve no derivatives. Thus U-spin symmetry violation in the current operator can only proceed by its implicit dependence on the shape of the soliton. Our results also show that the symmetry breaking manifest in the mesonic sector is just of the right size to reproduce the mass splittings of the baryons leading to distortions of baryon size and strangeness content sufficient to reproduce the seemingly contradictory aspects of the data.

Quantitativly the model produces a fair agreement to the known data which is nontrivial in view of the few free parameters entering into the calculation. The missing clue which would promote the fair agreement to a good agreement is the overall size of electromagnetic and axial quantities, which, all together, are too small. As far as the parameters are important in this respect, we have adhered to two principles: (i) fit of the mesonic sector which is mandatory as soon as one requires that the model must also describe meson baryon scattering correctly and (ii) fit of the mass splittings. Fortunately, it appears that ratios among the calculated electromagnetic or the axial matrix elements are more reliable quantities.

5. Acknowledgements

Major parts of this contribution have been obtained in collaboration with Herbert Weigel. I thank Frank Meier for supplying me with the numbers for the polarizability in the vector meson model.

6. References

1. J. Gasser and H. Leutwyler, *Nucl. Phys.* **B250** (1985) 465
2. B. Schwesinger, *Nucl. Phys.* **A537** (1992) 253
3. H. Walliser, *Nucl. Phys.* **A524** (1991) 706
   M. Uehara and H. Kondo, *Prog. Theor. Phys.* 75 (1986) 981
4. B. Schwesinger and H. Weigel, *Phys. Lett.* **B267** (1991) 438
5. B. Schwesinger and H. Weigel, *Nucl. Phys.* **A540** (1992) 461
6. E. Guadagnini, *Nucl.Phys.* **B236** (1984) 35
   M. Praszalowicz,*Phys.Lett.* **158B** (1985) 264
   G. Adkins and C. Nappi, *Nucl.Phys.* **B249** (1985) 507
   M. Chemtob, *Nucl.Phys.* **B256** (1985) 600
   A. Kanazawa, *Prog.Theor.Phys.* **77** (1987) 1240
   D. Masak, *Phys.Rev.* **D39** (1989) 305
7. H. Yabu and K. Ando, *Nucl.Phys.* **B301** (1988) 601
   H. Weigel, J. Schechter, N. W. Park and Ulf-G. Meißner,
   Phys. Rev. **D42** (1990) 3177.
   G. Pari, B. Schwesinger and H. Walliser, *Phys. Lett.* **255B** (1991) 1
   N. W. Park, J. Schechter and H. Weigel, *Phys. Rev.* **D43** (1991) 869
8. C. G. Callan and I. Klebanov, *Nucl.Phys.* **B262** (1985) 365
   C. G. Callan, K. Hornbostel, I. Klebanov, *Phys.Lett.* **B202** (1988) 269
   E. Nyman and D. O. Riska, *Nucl. Phys.* **B325** (1989) 593
   N. N. Scoccola, D. P. Min, H. Nadeau and M. Rho, *Nucl. Phys.* **A505** (1989) 497.
   J. Kunz and P. J. Mulders, *Phys. Rev.* **D41** (1990) 1578
9. B. Moussalam and D. Kalafatis, *Phys. Lett.* **B272** (1992) 196
10. G. Holzwarth, *Phys. Lett.* **B291** (1992) 218
11. D. B. Leinweber, R. M. Woloshyn and T. Draper, *Phys. Rev.* **D43** (1991)1659,
    D. B. Leinweber, T. Draper and R. M. Woloshyn, *Phys. Rev.* **D46** (1992) 3067
12. T. deGrand, R. L. Jaffe, K. Johnson and J. Kiskis, *Phys. Rev.* **D12** (1975) 2060
13. U.-G. Meißner, N. Kaiser and W. Weise, *Nucl. Phys.* **A446** (1987) 685.
14. A. Bosshard et al.,*Phys. Rev.* **D44** (1991) 1962
15. H. T. Diehl et al., *Phys. Rev. Lett.* **67** (1991) 804
16. B. Povh and J. Hüfner, *Phys. Lett.* **B245** (1990) 653
17. J. Schechter and H. Weigel,*Phys. Rev.* **D44** (1991) 2916
    J. Schechter and H. Weigel, *Phys. Lett.* **B261** (1991) 235
18. E. Nyman, *Phys. Lett.* **B142** (1984) 388
    M. Chemtob, *Nucl. Phys.* **A473** (1987) 613
    S. Scherer and P. J. Mulders, *Preprint NIKHEF-92-P2* (1992)

19. N. N. Scoccola and W. Weise, *Phys. Lett.* **B232** (1989) 287,
Nucl. Phys. **A517** (1990) 495
20. F. J. Federspiel, R. A. Eisenstein, M. A. Lucas, B. E. MacGibbon,
K. Mellendorf, A. M. Nathan, A. O'Neill and D. P. Wells,
Phys. Rev. Lett. **67** (1991) 1511
J. Schmiedmayer, P. Riehs, J. Harvey and N. Hill, *Phys. Rev. Lett.* **66** (1991) 1015
P. C. Hecking and G. F. Bertsch, *Phys. Lett.* **B99** (1981) 237

ELECTROMAGNETIC FORMFACTORS IN THE SKYRME MODEL

FRANK MEIER
Fachbereich Physik, Universität Siegen
Adolf-Reichwein-Str., W-5900 Siegen, Germany

ABSTRACT

Structural difficulties of the Skyrme model for electromagnetic properties of baryons and their improvement through vector mesons are discussed. Elimination of the vector mesons in an extended local approximation creates nonminimal photon-couplings for the pseudoscalars which solve some, but not all of the problems.

1. Introduction

The revival of Skyrme's idea of baryons being topological solitons in the mesonic nonlinear σ-model plus stabilising term[1], which is nowadays considered a low-energy 'remnant' of QCD, started after Witten et al.[2] had shown

- that the model, using a semiclassical quantization of the hedgehog, reproduces static nucleon properties within 30% of experimental data, and
- that the WZ-anomalous action constrains the quantized object to half integer spin provided the theory is formulated in groups $SU(N)$ with $N > 2$.

These results were achieved by simultaneously varying the Skyrme-model's free parameter e and letting the pion decay constant f_π deviate from its experimental value. However, the nucleon formfactors as calculated first by Braaten, Tse and Wilcox[3] were shown to be unsatisfactory at momentum transfers markedly different from zero.

To improve on that situation and to account for additional mesonic degrees of freedom which should simulate nucleon core effects, vector- and axialvector mesons were introduced into the model.

But so far, there is no unique procedure for that. Consequently there are ambiguities, mainly from different ways of treating the anomaly.

The aim of the present note now is threefold: After reviewing e.m. properties in the Skyrme-model, this time keeping f_π fixed on its experimental value, we will first reconsider the improvements which are possible by the inclusion of vector mesons. To avoid the ambiguities related with the treatment of the anomaly, we will argue that, for these purposes, it seems sufficient to use the minimal model[4] as a generic vector-meson (VM) lagrangian, because it is obtained as the 'local' approximation to the general form suggested by Kaymakcalan and Schechter[5], where each independent term of the anomalous part appeared with its own free coupling constant.

Secondly, we will demonstrate how to construct, by extending the local approximation on the vector mesons, a lagrangian for the pseudoscalars only. In comparison with the original Skyrme-model, this will lead to partially improved e.m. properties for the nucleons.

So, most of the material that will be presented here is not new; but it is reconsidered under the critical aspect to bring forth the structural deficiencies with respect to electromagnetism which prevail in the models considered. We will argue, however, that they are not necessarily inherent in the models themselves, but that they are rather due to the procedures by which we extract information from these models. From this we have to conclude that remaining difficulties only can be cured by going in some way beyond the semiclassical quantization prescription.

2. Starting Points: The Models

2.1. *The Skyrme model*

Starting point of all the considerations is the Skyrme-lagrangian[1]

$$L = \int d^3x \ (\mathcal{L}^{(2)} + \mathcal{L}^{(4)} + \mathcal{L}^{(mass)}) \tag{1}$$

with action

$$\Gamma = \int dt \ L + \Gamma^{WZ} \tag{2}$$

including the WZ-term.

In all calculations we will keep f_π and m_π fixed at their experimental values of 93 and 138 MeV, respectively.

Quantization proceeds along the well known lines; a parametrisation $U(\boldsymbol{x},t) = A(t)U_h A(t)^\dagger$ where U_h is the familiar hedgehog-ansatz and $A \in SU(2)$ obeys $A^\dagger \dot{A} = \frac{i}{2}\boldsymbol{\tau} \cdot \boldsymbol{\Omega}$ leads to the rotational energy of the skyrmion $E_{rot} = \frac{1}{2}\Theta \boldsymbol{\Omega}^2$ with ($s = \sin(F), c = \cos(F)$)

$$\begin{aligned} \Theta &= \int d^3x [t_2(r) + t_4(r)] \\ t_2 &= \frac{2}{3} s^2 f_\pi^2 \\ t_4 &= \frac{2}{3} s^2 \frac{1}{e^2} (F'^2 + \frac{s^2}{r^2}) \end{aligned} \tag{3}$$

Coordinates are then replaced by operators

$$\boldsymbol{\Omega} \rightarrow \boldsymbol{J}/\Theta, \quad \boldsymbol{J} : \text{spinoperator} \quad \boldsymbol{J} = \boldsymbol{\sigma}/2 \quad tr(\tau^i A^\dagger \tau^j A) \rightarrow -(2/3)\sigma^i \tau^j$$

where in the latter case matrix elements are understood to be taken between nucleon states. To reproduce the NΔ-split of 293MeV, Θ has to be equal $1.008 fm$.

2.2. *Vector Mesons*

2.2.1. *Gauging of* $\mathcal{L}^{(2)}$

Vector and axialvector mesons have been introduced into the 'normal' part of the Lagrangian through differently looking, yet equivalent procedures, namely the hidden gauge[6] and the massive Yang-Mills[7] scheme. In the latter left and right handed gauge fields are introduced through covariant derivatives

$$D^{\mu}U = \partial^{\mu}U - igA_L{}^{\mu}U + igUA_R{}^{\mu}. \tag{4}$$

Then, the axial vector fields are eliminated by the chirally invariant constraint

$$A_R{}^{\mu} = U^{\dagger}(A_L{}^{\mu} + \frac{i}{g}\partial^{\mu})U. \tag{5}$$

Mass terms for the gauge bosons are

$$\mathcal{L}_{mass} = \frac{m_0^2}{2}tr(A_{L\mu}A_L{}^{\mu} + A_{R\mu}A_R{}^{\mu}) - \frac{B}{2}tr(A_{L\mu}UA_R{}^{\mu}U^{\dagger}) \tag{6}$$

where the constants m_0 and B are tied to pion decay constant and vector meson masses by

$$\frac{B}{2} = \frac{f_\pi^2 g^2}{2} - \frac{m_V^2}{8} \qquad m_0^2 = \frac{f_\pi^2 g^2}{2} + \frac{m_V^2}{8}$$

The only free parameter g is related to the $\rho\pi\pi$-decay constant $g_{\rho\pi\pi}$ via

$$g = \frac{m_\rho{}^2}{2\sqrt{2}g_{\rho\pi\pi}f_\pi^2}. \tag{7}$$

With $g_{\rho\pi\pi} = 8.66$ it follows that $g = 2.8$. This deviates only very little from $g = 2.925$ as given by the KSRF-relation $m_V^2 = 8g^2f_\pi^2$ (using $m_V = 770\text{MeV}$). For simplicity we shall therefore equate the vector-meson masses with m_V and relate them to g and f_π by KSRF. Although there is no compelling reason from the point of fundamental QCD to prefer a specific scheme to introduce the vector mesons, the phenomenological concepts of vector-meson dominance together with chiral symmetry serve as guidelines which give some confidence at least for the normal part of the lagrangian.

2.2.2. *The Anomaly*

For Γ^{WZ}, things are more complicated. Witten's original idea[8] to trial-and-error-gauge Γ^{WZ} requires a Bardeen subtraction to get agreement of theory and experimental data on anomalous processes in the mesonic sector. This destroys global chiral invariance of the gauged Γ^{WZ} and therefore violates the construction principles for our lagrangians .

An alternate way to treat the anomaly, suggested by Kaymakcalan and Schechter[5], is to write down all globally chiral invariant parts of the object resulting from gauging Γ^{WZ} with independent, a priori unknown coefficients. After elimination of the axial vectors, there are three such constants. This line of thought is equivalent, at least for unitary unimodular groups, to construct out of A_L and $\alpha = dUU^\dagger$ the most general, P and C-invariant 4-form with minimum number of derivatives.

The structure then reads

$$\begin{aligned}\Gamma^a &= \Gamma^{WZ} + \int [ic_1 tr(A_L\alpha^3) \\ &+c_2 tr(dA_L\alpha A_L - A_L\alpha dA_L + A_L\alpha A_L\alpha) \\ &+c_3 tr(-2iA_L{}^3\alpha + \frac{1}{g}A_L\alpha A_L\alpha)] \qquad (8)\end{aligned}$$

It is interesting to note that this form collapses to one single and simple expression by adopting the (first) local approximation for the ρ fields: with the definitions

$$l^\mu = \xi^\dagger\partial^\mu\xi \qquad r^\mu = \partial^\mu\xi\xi^\dagger$$

we have

$$\rho_\mu = \frac{i}{2g}(l_\mu - r_\mu). \qquad (9)$$

A_L then reduces to

$$A_L = \omega - \frac{i}{2g}\alpha, \qquad (10)$$

and insertion into Γ^a yields

$$\Gamma^a = \int i(c_1 - \frac{c_3}{2g^2})\omega tr(\alpha^3). \qquad (11)$$

This is of course the minimal model[4]. It supplies the important ω-stabilisation, yet no additional ω-ρ-correlations. But by our derivation, we see that

- such correlations are contained implicitly in this simplified action and that
- all experimental information used so far[9] to determine the coefficients in Γ^a is still contained in the factor in front of $\omega tr(\alpha^3)$.

Moreover, it is immediately clear that a better approximation to ρ creates terms with higher number of derivatives of pseudoscalar fields. Therefore in the spirit of chiral perturbation theory (which has, however, to be taken with a grain of salt, because constructing the soliton takes the theory beyond its original justification) we may expect the additional terms to be less important, and work with the computationally far more convenient form (11) for the anomalous part which contains only one additional parameter.

From ref.10 we can get the coefficient as

$$i(c_1 - \frac{c_3}{2g^2}) = \frac{i}{8}(-\frac{\tilde{h}}{\sqrt{2}} + \frac{\tilde{g}_{VV\Phi}}{g}) \equiv \frac{h}{24\pi^2}$$

(note that our definition of c_i and g is different from the one used in ref.10). With $\bar{h} = 0.4$, $\bar{g}_{VV\Phi} = 1.9$, we have $h \cong 12$. But late experimental data[11] suggest $\bar{h} = 0.5$, $\bar{g}_{VV\Phi} = 1.94$, and this means $h \cong 10$. Since the values of $\bar{h}$, $\bar{g}_{VV\Phi}$ depend quadratically on the deviation from ideal mixing in $SU(3)$, and since h involves the difference of both, even deviations from this value may be justified.

2.2.3. *VM-lagrangian*

After all the previous remarks we will use for the following considerations the VM-lagrangian

$$
\begin{aligned}
\mathcal{L} &= \frac{f_\pi^2}{4}\cdot tr(\partial_\mu U \partial^\mu U^\dagger) + \frac{f_\pi^2}{4}\cdot m_\pi^2 tr(U + U^\dagger - 2) \\
&\quad -\frac{1}{8} tr(\rho_{\mu\nu}\rho^{\mu\nu}) + \frac{m_V^2}{4} tr(\rho_\mu - \frac{i}{2g}[l_\mu - r_\mu])^2 \\
&\quad -\frac{1}{4}\omega_{\mu\nu}\omega^{\mu\nu} + \frac{1}{2} m_V^2 \omega_\mu \omega^\mu + h\omega_\mu B^\mu
\end{aligned} \tag{12}
$$

This is evaluated in the soliton sector by means of ansätze

$$\omega^\mu = \omega(r)\delta^{\mu 0} \tag{13}$$

$$\rho^{ia} = \epsilon^{ika}\hat{r}^k \frac{G(r)}{2gr} \tag{14}$$

$$\rho^{0a} = 0 \tag{15}$$

for static and

$$\boldsymbol{\omega} = -\frac{\Phi(r)}{2r}\hat{\boldsymbol{r}} \times \boldsymbol{\Omega} \tag{16}$$

$$\tau^a \rho^{0a} = \frac{1}{g} A\tau^a[\Omega^a \xi_1(r) + \hat{r}^a(\boldsymbol{\Omega}\cdot\hat{\boldsymbol{r}})\xi_2(r)]A^\dagger \tag{17}$$

for induced VM-components and of course the hedgehog ansatz for U. Ω and A are defined as in the purely pionic case. The moment of inertia after employing equations of motion for the meson profiles is given by

$$
\begin{aligned}
\Theta &= 4\pi \int dr \left\{ \frac{2f_\pi^2 r^2}{3}[(1-c)^2 + (1+2c)\xi_1 + \xi_2] - \frac{h}{3}\Phi B_0 r^2 \right\} && (18) \\
&\equiv \int d^3r\, t_{VM} && (19)
\end{aligned}
$$

3. Form Factors

3.1. *Definitions*

Electric and magnetic form factors are defined through

$$G_E = \int d^3x \; e^{(iqx)} j^0(x) \tag{20}$$

$$G_M = M_N \int d^3x \; e^{(iqx)} (x \times j(x)\,) \tag{21}$$

where M_N is understood to be 938MeV. The respective current densities are obtained by evaluating the current operators determined from the model under consideration between appropriate nucleon states.

Electromagnetic radii are defined through

$$< r_i^2 >^j = -6 \frac{\partial_{q^2} G_i^j}{G_i^j(0)} |_0 \tag{22}$$

where i, j stand for (n,p) , (E,M) respectively.

3.2. *Purely Pionic SM: Formfactors*

Calculation of the currents in the pionic case merely means performing a local symmetry transformation $U \rightarrow e^{(i\epsilon Q)} U e^{(-i\epsilon Q)}$ with charge operator $Q = \frac{1}{2}(\frac{1}{3} + \tau^3)$, and then reading off the coefficient of $-\partial_\mu \epsilon$. The results are then inserted in the relations (20). This yields

$$\begin{aligned} G_E^S &= \frac{1}{2} \int d^3r \; j_0(qr) \; B_0 && (23) \\ G_E^V &= \frac{1}{2\Theta} \int d^3r \; j_0(qr) \; [t_2 + t_4] && (24) \\ G_M^S &= \frac{M_N}{2\Theta} \int d^3r \; \frac{j_1(qr)}{qr} \; r^2 B_0 && (25) \\ G_M^V &= M_N \int d^3r \; \frac{j_1(qr)}{qr} \; [t_2 + t_4] && (26) \end{aligned}$$

where $B_0 = -\frac{F'_s{}^2}{2\pi^2 r^2}$ is the baryon charge density. These form factors imply several stringent relations:

$$\begin{aligned} G_E^p(0) &= G_E^S(0) + G_E^V(0) = 1 && (27) \\ G_E^n(0) &= G_E^S(0) - G_E^V(0) = 0 && (28) \\ G_M^p(0) + G_M^n(0) &= \mu_p + \mu_n = 2\mu^S = \frac{M_N}{3\Theta} <r_B^2> && (29) \\ G_M^p(0) - G_M^n(0) &= \mu_p - \mu_n = 2\mu^V = \frac{2M_N}{3}\Theta && (30) \\ <r_E^2>_S &= <r_E^2>_p + <r_E^2>_n = <r_B^2> && (31) \\ <r_E^2>_p &= \frac{1}{2}(<r_B^2> + \frac{1}{\Theta} \int d^3r \; r^2 \; [t_2 + t_4]) && (32) \end{aligned}$$

$$
\begin{aligned}
<r_E^2>_n &= \frac{1}{2}(<r_B^2> - \frac{1}{\Theta}\int d^3r\ r^2\ [t_2+t_4]) && (33)\\
<r_M^2>_p &= \frac{\mu^S}{\mu^S+\mu^V}<r_M^2>_S + \frac{\mu^V}{\mu^S+\mu^V}<r_M^2>_V && (34)\\
<r_M^2>_n &= \frac{\mu^S}{\mu^S-\mu^V}<r_M^2>_S - \frac{\mu^V}{\mu^S-\mu^V}<r_M^2>_V && (35)\\
<r_M^2>_S &= \frac{3}{5}\frac{<r_B^4>}{<r_B^2>} && (36)\\
<r_M^2>_V &= \frac{3}{5}\frac{1}{\Theta}\int d^3r\ r^2\ [t_2+t_4] && (37)
\end{aligned}
$$

With experimental values for the magnetic moments and Θ fixed by the observed NΔ mass split Eq.29 implies $<r_B^2>= 0.56fm^2$, while Eq.31 implies $<r_B^2>= 0.62fm^2$ in fair mutual agreement, whereas (30) implies 4.7(=)3.2, which is quite bad.

But in actual calculations, Eq.29 causes even greater difficulties than (30). This is evident from table 1, where the numbers obtained for different values of e are displayed: For the experimental NΔ-split $<r_B^2>$ comes out too low by a factor of two (so does the isoscalar magnetic moment). Consequently, the electric proton- and neutron-radii are much too small. For the magnetic ones, the situation is slightly different, because in this case the isoscalar- and isovector contributions to $<r_M^2>^{p,n}$ are weighted by factors of $\mu^S/\mu^{p,n}$, $\mu^V/\mu^{p,n}$ respectively, which for experimental values differ by a factor of five. If we use theoretical predictions for the moments, the difference gets even bigger. Thus, as $<r_M^2>^S$ ($0.29fm^2$ as opposed to experimental $0.675fm^2$) and $<r_M^2>^V$ ($0.42fm^2$ as opposed to experimental $0.75fm^2$) miss the measurements by an equal margin, here the isovector relation is responsible for the bad fit.

If on the other hand $<r_B^2>$ is brought to its 'experimental' value (defined by Eq.29)(the remaining radii then also come out close to measurements), Θ is four times too large and μ^S is thus even worse than in the previous case.

Furthermore, figure 1 shows that the q^2 dependence of the formfactors is quite unsatisfactory.

Altogether, this is discourageing: keeping f_π fixed at its physical value, the results of the simple Skyrme model are far from the 30% correspondence with experiments. This fact should not be hidden by using smaller values for f_π, because all the Low-Energy-Theorems in meson-baryon dynamics are sensitive on that value. To sum up, in a better model

- relations (29), (30) should be changed
- the electric radii should be tied to $<r_B^2>$ in a way other than given by (31) and the magnetic ones should be connected to the moment of inertia in a way other than given by (37)
- the q^2-dependence of the formfactors should be changed.

In the following we now discuss if and to which extent inclusion of vector mesons can cure these shortcomings.

Table 1. Values of the pionic Skyrme model. Masses are given in MeV, radii in fm^2.

e	2.35	2.5	3.0	3.525	3.85	4.25	exp.
$\Theta[fm]$	4.767	4.063	2.529	1.656	1.331	1.008	1.008
M_H	3119.1	2913.3	2388.2	2008.6	1829.0	1647.8	
M_N	3134.6	2931.5	2417.4	2053.2	1884.5	1721.1	939.0
μ_p	7.676	6.565	4.150	2.786	2.250	1.780	2.79
μ_n	−7.470	−6.344	−3.887	−2.478	−1.917	−1.418	−1.91
$<r_B^2>$	0.619	0.561	0.420	0.320	0.275	0.232	
$<r_E^2>_p$	0.955	0.890	0.71	0.583	0.524	0.465	0.74
$<r_E^2>_n$	−0.338	−0.330	−0.292	−0.264	−0.249	−0.234	−0.119
$<r_M^2>_p$	0.776	0.727	0.599	0.502	0.456	0.407	0.74
$<r_M^2>_n$	0.778	0.730	0.606	0.517	0.476	0.437	0.77

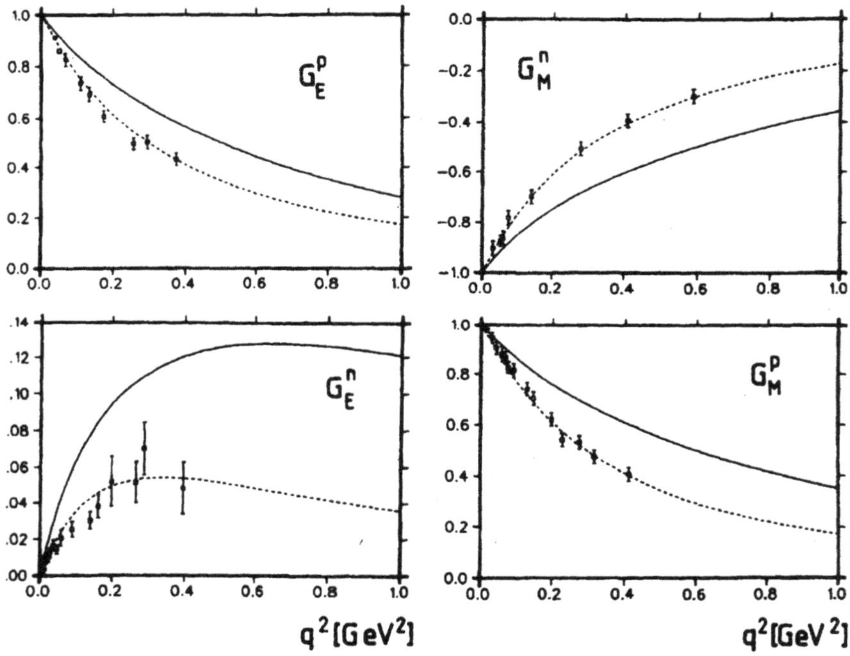

Fig.1: Formfactors of the pionic SM for $e = 4.25$. Dashed line: Dipole fit. Data taken from ref.13. $G_M(0)$ is normalised to one.

3.3. *Vector mesons: Formfactors*

The vector meson analogue of our procedure for computation of currents in the purely pionic case is to perform the same symmetry transformation on the pseudoscalars as before and a gauge transformation on the vector fields:

$$A^{\mu}_{L,R} \rightarrow e^{(i\epsilon_0 Q_0 + i\epsilon_V Q_V)}(A^{\mu}_{L,R} + \frac{Q_V}{g}\partial^{\mu}\epsilon_V + \frac{Q_0}{g_0}\partial^{\mu}\epsilon_0)e^{(-i\epsilon_0 Q_0 - i\epsilon_V Q_V)} \tag{38}$$

with $Q_0 = \frac{1}{6}$, $Q_V = \frac{\tau^3}{2}$. Here we come upon two additional types of freedom:

Firstly, the gauge parameters g and g_0 for isovector- and isoscalar vector mesons, respectively, need not be the same. Secondly, g_0 is not necessarily equal to $\frac{h}{3}$, because the neutral VMs can couple to the photon through a constant different from (the elementary charge) e (we use e=1 throughout).

Thus, we arrive at currents and, consequently, formfactors which are slightly different from those calculated earlier[4]. They are

$$G_E^S = \int d^3r\, j_0(qr)\left\{-\frac{m_V^2}{6g_0}\omega + \frac{1}{2}(1-\frac{h}{3g_0})B_0\right\} \tag{39}$$

$$G_E^V = \frac{1}{\Theta}\int d^3r\, j_0(qr)\left\{\frac{f_\pi^2}{3}\left[(1-c)^2 + (1+2c)\xi_1 + \xi_2\right] - \frac{h}{6}\Phi B_0\right\} \tag{40}$$

$$G_M^S = -\frac{M_N}{2\Theta}\int d^3r\, \frac{j_1(qr)}{qr}\left\{\frac{m_V^2}{6g_0}\Phi - (1-\frac{h}{3g_0})r^2 B_0\right\} \tag{41}$$

$$G_M^V = 2M_N\int d^3r\, \frac{j_1(qr)}{qr}\left\{\frac{f_\pi^2}{3}\left[(1-c)^2 - 2Gc\right] - \frac{h}{3}\omega r^2 B_0\right\} \tag{42}$$

The eqs. of motion for ω and Φ can be directly inserted into (39) and (41) which leads to

$$G_E^S = \int d^3r\, j_0(qr)\left\{\frac{m_V^2 h}{6g_0(q^2+m_V^2)} + \frac{1}{2}(1-\frac{h}{3g_0})\right\}B_0 \tag{43}$$

$$G_M^S = \frac{M_N}{2\Theta}\int d^3r\, \frac{j_1(qr)}{qr}\left\{\frac{m_V^2 h}{3g_0(q^2+m_V^2)} + 1 - \frac{h}{3g_0}\right\}r^2 B_0 \tag{44}$$

From the limit $q^2 \rightarrow 0$ it becomes evident, that the functional form of (29) for the isoscalar magnetic moment is unchanged, whereas (30) now reads

$$\mu^V = \frac{2M_N}{3}\int d^3r\, \left\{\frac{f_\pi^2}{3}\left[(1-c)^2 - 2Gc\right] - \frac{h}{3}\omega r^2 B_0\right\}. \tag{45}$$

Therefore, if the VM-model leads to baryon radii similar to the ones obtained from the SM (which is in fact the case), the problem with μ^S will remain.

Comparison with (18) shows that also the functional relation of G_E^V with the moment of inertia density remains unchanged. The propagators in the isoscalar formfactors give rise to corrections in the VM-counterparts of relations (31), (36). The changed relations are now given by

$$<r_E^2>_S = <r_B^2> + \frac{2h}{g_0 m_V^2} \tag{46}$$

$$<r_M^2>_S = \frac{3}{5}(\frac{<r_B^4>}{<r_B^2>} + \frac{5}{3}\frac{2h}{g_0 m_V^2}). \tag{47}$$

In the same manner as μ^V in (45), $<r_M^2>_V$ has lost its connection to the moment of inertia, which was desirable from our study of the pionic SM. By inspection of table 2, one sees that indeed all changes are improvements. Eq.45 now means 4.7(=)4.3, provided Θ is fixed at the experimental value, and in the same case, $<r_E^2>^S = 0.754$. For $g = g_0 = \frac{h}{3}$, this radius is even better ($0.636 fm^2$), however, then Θ and (30) are worse. The remaining discrepancies with regard to electric radii stem from the fact that $<r_E^2>_V$ is still connected to Θ (although t_{VM} is of course different from the pionic $t_2 + t_4$). For the magnetic radii, the situation is even better, with numbers being within 5% of experimental data for most parameter values.

From the same table it can also be seen that the radii with the exception of $<r_E^2>_n$ are not very sensitive to variations of g_0. Thus it seems reasonable to fix g_0 at g.

Summing up again, we recognize that

- the radii and the NΔ-split now simultaneously can be brought close to experimental values
- the q^2-dependence of formfactors is much improved (figure 2)
- the problem with μ^S is apparently independent of model and choice of parameters.

Table 2. Values of the VM-model. Radii are given in fm^2.

g	2.925	"	"	"	"	"	exp.
h	8.775	10.5	"	"	11.4	12.375	exp.
g_0	2.925	"	2.0	5.0	2.925	"	exp.
$\Theta[fm]$	0.823	1.019	"	"	1.128	1.252	1.008
μ_p	2.018	2.370	"	"	2.566	2.787	2.79
μ_n	−1.544	−1.918	"	"	−2.126	−2.359	−1.91
$<r_B^2>$	0.245	0.289	"	"	0.312	0.337	
$<r_E^2>_p$	0.853	0.941	1.050	0.844	0.987	1.036	0.74
$<r_E^2>_n$	−0.217	−0.1897	−0.076	−0.282	−0.167	−0.148	−0.119
$<r_M^2>_p$	0.703	0.740	0.761	0.720	0.758	0.776	0.74
$<r_M^2>_n$	0.715	0.730	0.704	0.750	0.740	0.752	0.77

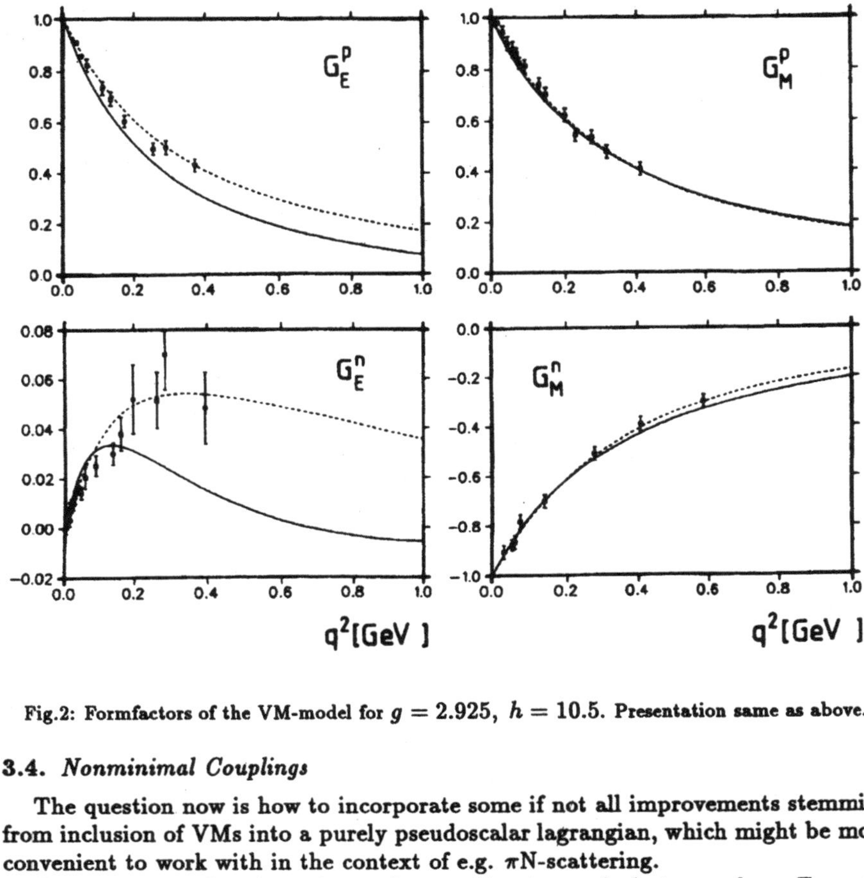

Fig.2: Formfactors of the VM-model for $g = 2.925$, $h = 10.5$. Presentation same as above.

3.4. *Nonminimal Couplings*

The question now is how to incorporate some if not all improvements stemming from inclusion of VMs into a purely pseudoscalar lagrangian, which might be more convenient to work with in the context of e.g. πN-scattering.

To answer, we first investigate the coupling term of photon and ω. Formally solving the ω's equation of motion, we find

$$\frac{m_V^2}{6g_0}\int d^4x A_\mu \omega^\mu = \frac{1}{(2\pi)^4}\frac{m_V^2 h}{6g_0}\int d^4x d^4x' d^4q \; A_\mu(x)\frac{e^{iq(x-x')}}{q^2 - m_V^2}B^\mu(x'). \quad (48)$$

Expansion of the propagator yields a lowest correction to the baryon current already present in the SM, which we can rewrite as a nonminimal coupling between photon and B^μ:

$$-\frac{1}{(2\pi)^4}\frac{h}{6g_0}\int d^4x d^4x' d^4q \; A_\mu(x)e^{iq(x-x')}\frac{q^2}{m_V^2}B^\mu(x') = -\frac{1}{2}\frac{h}{6g_0 m_V^2}\int d^4x \; A_{\mu\nu}B^{\mu\nu}. \quad (49)$$

From the definitions (22) and the structure of (49) it is clear that such an addition will change only the radii, but not the magnetic moments.

Regarding the isovectors, the situation is more complicated because of their nonabelian nature. This prevents us from expanding the propagator in a similar fashion as with the isoscalars. To solve this problem, we first show for the ω that expanding the propagator is equivalent to adopting a next-to-leading-order local approximation on the vectors.

For the proof, let

$$\omega_\mu = -h/(m_V)^2 B_\mu + 1/(m_V)^4 W_\mu \tag{50}$$

Inserting into the eqn. of motion and neglecting lowest-order-terms in m_V then gives

$$W_\mu = h\Box B_\mu \tag{51}$$

which in turn when put into the coupling term yields the same result as the expansion.

Therefore, we proceed and write

$$\rho_\mu = (i/2g)(l_\mu - r_\mu) + (1/m_V{}^2)\bar{R}_\mu \equiv v_\mu + (1/m_V{}^2)\bar{R}_\mu \tag{52}$$

The ρ's eqn. of motion

$$-\frac{1}{2}(\partial_\nu \rho^{\mu\nu} - ig[\rho_\nu, \rho^{\mu\nu}]) + \frac{m_V^2}{2}(\rho^\mu - v^\mu) = 0 \tag{53}$$

then implies for $\bar{R}$, omitting $\mathcal{O}(\frac{1}{m_V^2})$-terms,

$$\bar{R}^\mu = -\frac{i}{8g}\xi^\dagger(2\partial^\nu[\alpha^\mu, \alpha_\nu] - [\alpha^\nu, [\alpha^\mu, \alpha_\nu]])\xi. \tag{54}$$

Now the photon-ρ coupling has to be evaluated in this approximation. In order to do so we note that the isovector part of the minimal model's e.m. current, expressed in terms of A_L, reads

$$j_V^\mu = -\frac{m_0{}^2}{g} tr\left(Q_V[A_L^\mu + U^\dagger A_L^\mu U + \frac{i}{g}\beta^\mu]\right) + \frac{B}{2g} tr\left(Q_V[U^\dagger A_L^\mu U + A_L^\mu + \frac{i}{g}\alpha^\mu]\right). \tag{55}$$

With the definition $R^\mu = \xi\bar{R}^\mu\xi^\dagger$, A_L itself collapses to (neglecting for the moment the irrelevant ω)

$$A_L^\mu = -\frac{i}{2g}\alpha^\mu + \frac{1}{m_V^2}R^\mu \tag{56}$$

Using this, the resulting object then contains three contributions, namely

- the current j_2^μ of $\mathcal{L}^{(2)}$

$$-\frac{if_\pi^2}{2} tr\,(Q_V[-\alpha^\mu + \beta^\mu]), \tag{57}$$

- the Skyrme-term's current j_4^μ

$$-\frac{i}{32g^2} tr\,\{Q_V\,([\alpha^\nu,[\alpha^\mu,\alpha_\nu]] - [\beta^\nu,[\beta^\mu,\beta_\nu]])\}\,, \tag{58}$$

and

- (given as part of the action)

$$-\frac{i}{32g^2}\int d^4x\ A^{\mu\nu} tr\,\{Q_V\,([\alpha_\mu,\alpha_\nu] + [\beta_\mu,\beta_\nu])\} \tag{59}$$

or equivalently

$$-\frac{i}{18g^2}\int d^4x\ A^{\mu\nu}\partial_\nu tr\,(Q_V[-\alpha_\mu+\beta_\mu]) \tag{60}$$

which is a nonminimal coupling of photon and j_2^μ.

By inspection of the chiral perturbation series[12], one sees that the term

$$-\,il_9 tr\left(F^L_{\mu\nu}D^\mu U D^\nu U^\dagger + F^R_{\mu\nu}D^\mu U^\dagger D^\nu U\right) \tag{61}$$

where $F_{\mu\nu}$ and D_μ are respectively field tensor and covariant derivative containing vector and axialvectorfields, upon specialising to electromagnetism and keeping only parts linear in the photon field and its derivatives becomes

$$-\,i\frac{l_9}{2}A^{\mu\nu} tr\,\{Q_V\,([\alpha_\mu,\alpha_\nu] + [\beta_\mu,\beta_\nu])\}\,. \tag{62}$$

From this we have the identification $l_9 = \frac{1}{18g^2} = 0.0065$ for $g = 2.925$, which is in good agreement with $l_9 = 0.007(4 \pm 7)$ as given in ref.12 .

How then should the improved pseudoscalar lagrangian look like? If we were to write down all pseudoscalar terms coming from the local approximation on some VM-model up to a given order in chiral perturbation series, then from that on ω the resulting object had to contain at least terms of order ten. This would imply that our approximation on ρ as performed so far would not be sufficient and had to be extended again.

This is beyond the scope of the work presented here, and our objective is more limited: We shall collect all the necessary ingredients for a most simple lagrangian from which an improvement over the conventional SM can be expected.

This reads

$$\begin{aligned} L \;=\; \int d^3x\ \{&\mathcal{L}^{(2)} + \mathcal{L}^{(4)} + \mathcal{L}^{(mass)} \\ &-A_\mu(j_2^\mu + j_4^\mu + \frac{1}{2}B^\mu) \\ &+A_\mu(\frac{1}{9f_\pi^2 g^2}\Box j_2^\mu + \frac{h}{6g_0 m_V^2}\Box B^\mu)\} \end{aligned} \tag{63}$$

and its formfactors are

$$G_E^S = \frac{1}{2}(1 - \frac{h}{3gm_V^2}q^2)\int d^3r\ j_0(qr)\ B_0 \tag{64}$$

$$G_E^V = \frac{1}{2\Theta}(1 - \frac{1}{9f_\pi^2 g^2}q^2)\int d^3r\ j_0(qr)\ t_2 + \frac{1}{2\Theta}\int d^3r\ j_0(qr)\ t_4 \tag{65}$$

$$G_M^S = \frac{M_N}{2\Theta}(1 - \frac{h}{3gm_V^2}q^2)\int d^3r\ \frac{j_1(qr)}{qr}\ r^2 B_0 \tag{66}$$

$$G_M^V = M_N(1 - \frac{1}{9f_\pi^2 g^2}q^2)\int d^3r\ \frac{j_1(qr)}{qr}\ t_2 + M_N\int d^3r\ \frac{j_1(qr)}{qr}\ t_4 \tag{67}$$

Magnetic moments and charges remain, of course, unchanged; for the radii we have

$$<r_E^2>_S = <r_B^2> + \frac{2h}{gm_V^2} \tag{68}$$

$$<r_E^2>_V = \frac{1}{\Theta}\int d^3r\ r^2\ (t_2 + t_4) + \frac{2}{3\Theta f_\pi^2 g^2}\int d^3r\ t_2 \tag{69}$$

$$<r_M^2>_S = \frac{3}{5}(\frac{< r_B^4 >}{<r_B^2>} + \frac{10h}{3gm_V^2}) \tag{70}$$

$$<r_M^2>_V = \frac{3}{5}\left(\frac{1}{\Theta}\int d^3r\ r^2\ (t_2 + t_4) + \frac{10}{9f_\pi^2 g^2}\frac{1}{\Theta}\int d^3r\ t_2\right) \tag{71}$$

Because the variation of g_0 in the VM-model has been shown not to improve results, we have set $g = g_0$ in the above formulae. To reproduce the NΔ-split, we use $e = 4.25$ in our numerical evaluations.

Radii are then computed (and listed below in table 3) by

- taking the VM-values for the remaining parameters (using $h = 8.775$ and $h = 10.5$ as well)
- varying g and h so as to fit electric radii
- varying g and h so as to fit magnetic radii.

Table 3 shows that already for the first choice, electric radii are very well reproduced, and that magnetic radii are clearly improved over the conventional SM. Apparently the numerical differences to the VM-model in either of the first two cases listed above come from the isovector electric formfactor.

This is not surprising, because the form of the isoscalar relations now is the same as for the VM-lagrangian, and differences could only come from different chiral angles used in $<r_B^2>$, which effect is apparently small. Put in another way, the structure of the local approximation on ω shows that higher corrections will enter only through higher moments of the formfactors.

By the same token, in the isovector case, there are corrections to even the lowest moments from all orders of the approximation. Moreover, our isovector formfactor neglects effects from the $\omega_\mu B^\mu$-term. The magnetic radii are not reproduced as well, because they are determined, apart from $<r_M^2>^S$ and $<r_M^2>^V$ by the magnetic moments which are not the same in our and the VM-model, respectively.

If we fit magnetic radii, then the situation more closely resembles the VM-model; whereas the proton electric radius stays close to its experimental value, the neutron radius is clearly off.

A look at the formfactors in figures 3, 4 shows that their q^2-dependence is not much improved, as was to be expected.

Table 3. Nonminimal couplings. Radii are given in fm^2

h	8.775	10.5	8.607	4.758	VM ($h = 8.775$)	exp.
g	2.925	2.925	2.886	1.971	VM ($g = 2.925$)	exp.
$<r_E^2>_p$	0.736	0.775	0.74	0.792	0.85	0.74
$<r_E^2>_n$	−0.113	−0.074	−0.118	−0.244	−0.22	−0.119
$<r_M^2>_p$	0.581	0.589	0.587	0.739	0.706	0.74
$<r_M^2>_n$	0.557	0.547	0.564	0.774	0.723	0.77

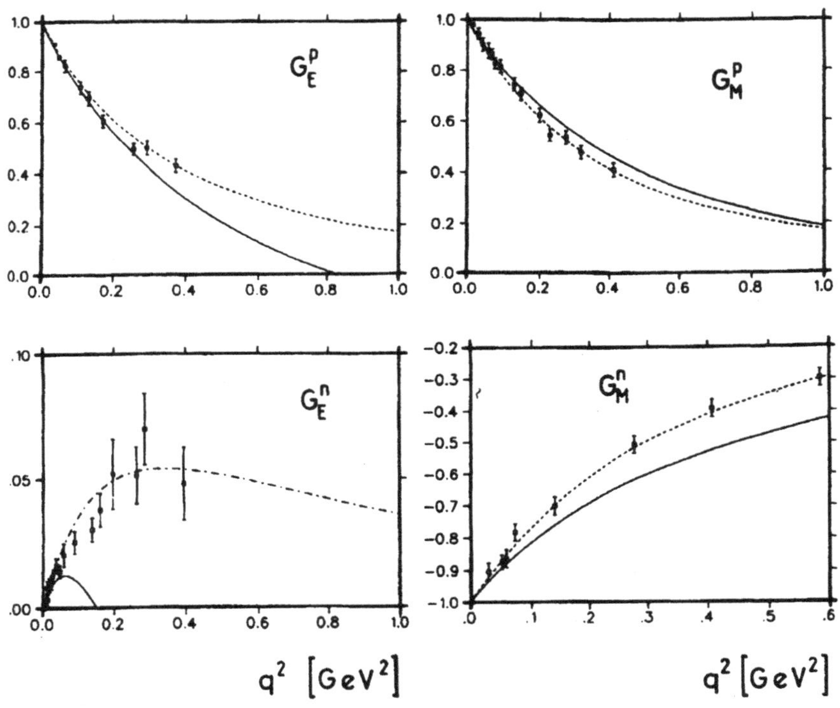

Fig.3: Formfactors of the pionic SM with nonminimal couplings.
$e = 4.25$, $g = 2.886$, $h = 8.607$. Presentation same as above.

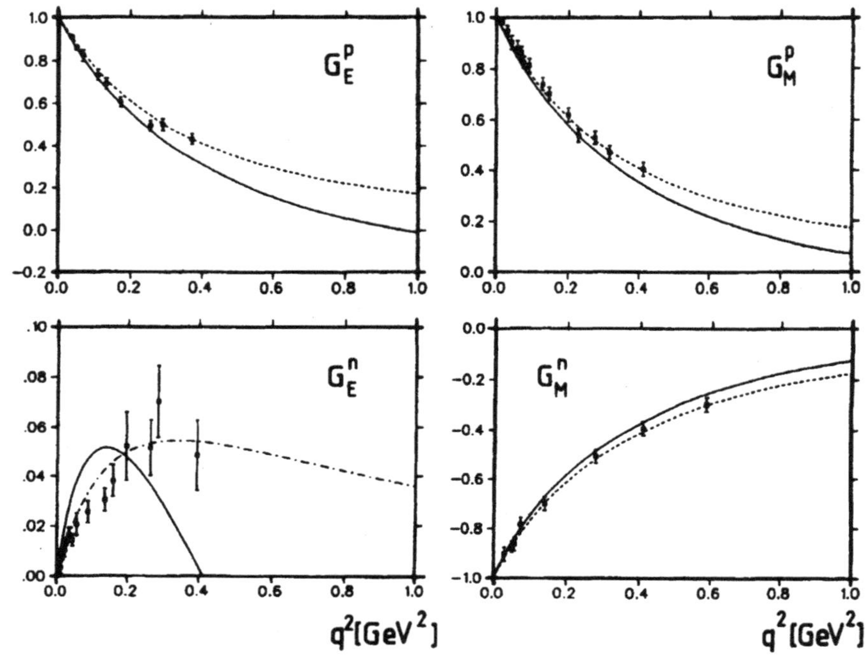

Fig.4: Same as above, but $g = 1.971$ $h = 4.758$.

Altogether, our quite simple model with essentially two free parameters is able to simultaneously reproduce Θ and e.m. radii within a small error margin, yet there is no solution to the problem of the isoscalar magnetic moment.

3.5. *The Failure: Reasons*

We now want to detail the reasons for the failure of the models discussed above to reproduce the isoscalar magnetic moment.

It is clear by construction of our lagrangien that the isoscalar formfactor is supplied only by the anomaly and, given the VM-case, the ω mass-term. Now note that in $SU(2)$ any conceivable 'gauged' anomaly with minimum number of derivatives and preserving global chiral invariance can contain terms at most linear in ω (this time thought of as a one-form). This is the case, because

- powers of ω greater than one vanish and

- a term $\omega d\omega tr(M)$ has to contain a traceless one-form M and therefore also vanishes.

To see this, one notes that because Γ^a as in Eq.8 is a four-form and VMs are introduced as gauge fields, M has to be purely pseudoscalar. Then from global chiral invariance it has to be of the (Maurer-Cartan-) form $PdP^\dagger$, and thus is traceless.

Hence we can write

$$atr(\omega_\mu J^\mu) \tag{72}$$

for the coupling of ω to pseudoscalars and other vector mesons. The isoscalar current then is

$$-\frac{m_V^2}{6g_0}\omega^\mu - \frac{a}{6g_0}J^\mu + \frac{1}{2}B^\mu. \tag{73}$$

Now writing down the formfactor, we can integrate out the ω. Since it is driven by J^μ, the first two contributions in (73) cancel at $q^2 = 0$. Therefore, the isoscalar magnetic moment reads

$$\mu^S = \frac{M_N}{6\Theta} <r_B^2>, \tag{74}$$

as previously. Thus, no conceivable vector meson model in $SU(2)$ can go beyond the Skyrme model with respect to this quantity as long as it is not to produce, at $\Theta = 1.008 fm$, a chiral angle vastly different from the ones obtained in the purely pionic case.

Consequently we argue that the problem has to be solved if it can be solved at all in the purely pionic model.

But then it is clear that changes can only be introduced through the use of some different kind of ansatz for $U(x, t)$ within the topological current, and this means going beyond the semiclassical quantization procedure used so far to compute formfactors.

4. Conclusions

We have reviewed e.m properties of baryons in a purely pionic Skyrme model and shown them to be worse as in previous work[2] because we kept f_π fixed on its physical value. We then discussed the improvements that can be achieved through vector mesons, argueing that the minimal model is a suitable generic model for that purpose. A purely pseudoscalar lagrangian was then constructed, which contained nonminimal couplings of the photon to pseudoscalars and showed improvements over the conventional SM with respect to electromagnetism. This lagrangian was shown to be consistent with chiral perturbation theory up to order four, yet neglected terms which appear in order six and these therefore would have to be incorporated in a more elaborate calculation.

In all calculations, the isoscalar magnetic moment could not be brought to its experimental value, which implies that μ^p and μ^n cannot be simultaneously reproduced by the models.

Eventually, we showed that, at least in $SU(2)$, this is a structural difficulty which one cannot get rid of through incorporation of vector mesons, and which has to be cured in the purely pionic SM if it is to be cured at all.

A natural and physically appealing way would be to incorporate time-dependent pionic fluctuations in $U(\mathbf{x},t)$, which would allow the photon to couple not only to the classical soliton field of the baryon, but also to the quantized pion cloud.

5. Acknowledgements

The author wishes to thank Prof. G. Holzwarth for his support in completing this work. Thanks also to B. Schwesinger, B. Stötzel and H. Walliser for their patience in listening and responding to my questions.

6. References

1. T.H.R. Skyrme *Nuc. Phys.* **31** (1961) 550
2. G.S. Adkins, C. Nappi and E. Witten *Nuc. Phys.* **B228** (1983) 552
3. E. Braaten, S. Tse and C. Wilcox *Phys. Rev.* **D34** (1986) 1482
4. U.G. Meissner *Phys. Rep.* **161** (1988) 213
5. Ö. Kaymakcalan and J. Schechter *Phys. Rev.* **D31** (1985) 1109
6. M. Bando, T. Kugo and K. Yamawaki *Nuc. Phys.* **B259** (1985) 493
7. Ö. Kaymakcalan, S. Rajeev and J. Schechter *Phys. Rev.* **D30** (1984) 594
8. E. Witten *Nuc. Phys.* **B223** (1983) 552
9. P. Jain, R. Johnson, U.G. Meissner, N.W. Park and J. Schechter *Phys. Rev.* **D37** (1988) 3252
10. U.G. Meissner, N. Kaiser, H. Weigel and J. Schechter *Phys. Rev.* **D39** (1989) 1956
11. Particle Data Properties Booklet (1990)
12. J. Gasser and H. Leutwyler *Nuc. Phys.* **B250** (1985) 465
13. G. Höhler, E. Pietarinen, I. Sabba-Stefanescu, F. Borkowski, G.G. Simon, V.H. Walther and R.D. Wendling *Nuc. Phys.* **B114** (1976) 505

III. Hyperons

MASSIVE-FLAVOR BARYONS IN THE BOUND STATE APPROACH

NORBERTO N. SCOCCOLA
Physics Department, Comisión Nacional de Energía Atómica
Av. Libertador 8250, (1429) Buenos Aires, Argentina

ABSTRACT
The structure of massive-flavor baryons is studied within the context of the bound state soliton model. In this approach hyperons are described as bound states of an $SU(2)$ topological soliton and flavored mesons. Masses, electromagnetic properties and coupling constants of these baryons are calculated and compared with available empirical data and quark model calculations. The possibility of extending this scheme to the description of exotic baryons like the H-dibaryon, pentaquark hyperons and quarkonium baryons is also discussed.

1. Introduction

During the last few years it has become clear that the applicability of the Skyrme's topological soliton model for baryon structure[1] goes beyond all the original expectations. Indeed, as I will discuss in this contribution, this model provides a very useful tool not only for the description of light flavored baryons and nuclei, but of the massive flavor hyperons as well. The main difficulty in describing the structure of such baryons within the soliton approach is the existence of explicit chiral symmetry breaking terms associated with the massive flavor degrees of freedom. One way to deal with this problem is based on the assumption that one can enlarge the original $SU(2)$ isospin (zero mode) space to $SU(3)$ (or more in general to $SU(N_f)$) to encompass the flavor degrees of freedom in such a way that both isospin and heavy flavors are treated as collective coordinates[2] . The symmetry breaking terms, however, are to be diagonalized exactly within this truncated basis[3] . This method is discussed in other contributions in this volume. An alternative approach (the so-called "bound state approach") was suggested by Callan and Klebanov[4] . In such an approach, which is clearly more suited for heavier flavors, this problem is treated by considering isospin and flavored excitations around the soliton in a different way. Hyperons are then composed by a flavored pseudoscalar doublet (i.e. kaons) bound to an $SU(2)$ soliton. Although the model was originally applied to the study of strange hyperons[5] it has recently been shown[6] that the properties of hyperons with heavier flavors (i.e. charm) are equally well described by it. Here, I will present some of the bound state model predictions for such hyperon properties. Moreover, I will show that this scheme can be extended to the study of exotic baryons like i.e. H-dibaryon, pentaquarks hyperons and quarkonium baryons.

The plan of this contribution is as follows. In Sec.2 I will give a general description of the bound state approach. In Sec.3 the predictions for the low-lying hyperon spectra will be given. In Sec.4 I will discuss some hyperon electromagnetic properties and coupling constants. In Sec.5 the model will be extended to the study of exotic baryons. Finally, in Sec.6 conclusions will be given.

2. The Bound State Model

I start by considering the effective action for the simple Skyrme model with an appropriate chiral symmetry breaking, expressed in terms of the $SU(3)$ valued field $U(x)$ as

$$\Gamma = \int d^4x \left\{ \frac{F_\pi^2}{16} \mathrm{Tr}\left[\partial_\mu U \partial^\mu U^\dagger\right] + \frac{1}{32e^2} \mathrm{Tr}\left[[U^\dagger \partial_\mu U, U^\dagger \partial_\nu U]^2\right] \right\} + \Gamma_{WZ} + \Gamma_{SB}, \quad (1)$$

where F_π is the pion decay constant ($= 186\ MeV$ empirically) and e is the so-called Skyrme parameter. In Eq.(1), Γ_{WZ} is the Wess-Zumino action given by

$$\Gamma_{WZ} = -\frac{iN_c}{240\pi^2} \int_{M_4 \times [0,1]} \mathrm{Tr}[(U^\dagger dU)^5] \quad (2)$$

where M_4 denotes $3+1$ dimensional space. Finally, Γ_{SB} is responsible for the explicit symmetry breaking of chiral symmetry. Its explicit form is

$$\begin{aligned} \Gamma_{SB} = \int d^4x \Bigg\{ & \frac{F_\pi^2 m_\pi^2 + 2F_M^2 m_M^2}{48} \mathrm{Tr}\left[U + U^\dagger - 2\right] \\ & + \frac{F_\pi^2 m_\pi^2 - F_M^2 m_M^2}{24} \mathrm{Tr}\left[\sqrt{3}\lambda^8 \left(U + U^\dagger\right)\right] \\ & + \frac{F_M^2 - F_\pi^2}{48} \mathrm{Tr}\left[\left(1 - \sqrt{3}\lambda^8\right)\left(U\partial_\mu U^\dagger \partial^\mu U + U^\dagger \partial_\mu U \partial^\mu U^\dagger\right)\right] \Bigg\} \end{aligned} \quad (3)$$

where λ^8 is the eighth Gell-Mann matrix and m_π and m_M represent the pion and flavored meson masses respectively. Eq.(3) accounts not only for the finite mass of the pseudoscalar mesons but also for the empirical difference between their decay constants. In previous calculations[5] the flavored meson was found to be overbound to the soliton. It was recently shown[7] that this defect can be mostly eliminated if the difference in the decay constants is properly taken into account as done in Eq.(3).

An ansatz for the U-field is then introduced in such a way that the flavor carrying meson component U_M is separated from the $SU(2)$ soliton component U_π. It reads[4]

$$U = \sqrt{U_\pi}\, U_M \sqrt{U_\pi}. \quad (4)$$

Here the soliton field component is defined as

$$U_\pi = \begin{pmatrix} u & 0 \\ 0 & 1 \end{pmatrix}, \quad (5)$$

with u being the usual $SU(2)$ hedgehog field

$$u = \exp\left(i\vec{\tau} \cdot \hat{r}\theta(r)\right). \quad (6)$$

The flavored meson field has the form

$$U_M = \exp\left(i\frac{2\sqrt{2}}{F_M}\begin{bmatrix} 0 & M \\ M^\dagger & 0 \end{bmatrix}\right). \tag{7}$$

where M is one of the isodoublets

$$K = \begin{pmatrix} K^+ \\ K^0 \end{pmatrix}, \; D = \begin{pmatrix} \bar{D}^0 \\ D^- \end{pmatrix}, \; B = \begin{pmatrix} B^+ \\ B^0 \end{pmatrix}, \tag{8}$$

respectively. In general the soliton field U should be taken as an $SU(N_f)$ field in order to encompass all the flavor components, but as long as the flavor carrying component is treated perturbatively to second order the different flavor sectors can be treated as separate $SU(3)$ flavor groups[6] .

The soliton-meson action is obtained by inserting the ansatz (4) into Eq.(1) and expanding to second order in M. The corresponding lagrangian density can be separated in two classes of terms. The first one reduces to the lagrangian of the interacting pion fields $\mathcal{L}_{SU(2)}$. The minimization of the static soliton energy obtained from such a lagrangian provides the soliton profile $\theta(r)$ [8] . The rest of the terms describe the system of quantum M fields moving in the presence of the soliton background. The corresponding hamiltonian is quadratic in the dynamical variables and therefore can be diagonalized exactly. Because of the particular spin-isospin structure of the soliton background it is convenient, however, to partial-wave-decompose the meson field M in the basis (Λ, l, Λ_z)

$$M = \sum_{\Lambda,l,\Lambda_z} \varphi_{\Lambda,l}(r)\, \mathcal{Y}_{\Lambda,l,\Lambda_z}. \tag{9}$$

Here, $\mathcal{Y}_{\Lambda,l,\Lambda_z}$ is the spinor spherical harmonics and $\vec{\Lambda}$ is the sum of the meson isospin $\vec{I}_M$ and its angular momentum $\vec{L}$. The corresponding eigenvalue equation in each (Λ, l, F) channel reads

$$\left[-\frac{1}{r^2}\partial_r\left(r^2 h\, \partial_r\right) + m_M^2 + V^{eff}_{\Lambda,l} - f\omega^2 \pm 2\, F\, \lambda\, \omega\right] \varphi_{\Lambda,l}(r) = 0, \tag{10}$$

where F denotes the flavor quantum number ($F = \pm 1$). The positive sign in the term linear in ω corresponds to the cases $M = K, B$ while the negative is for $M = D$. The functions λ, f, h, and $V^{eff}_{\Lambda,l}$ can be written in terms of the background profile. They depend on the radial coordinate $r = |\vec{r}|$, only. Their explicit form can be found in Ref.[4] . The function λ is primarily related to the anomalous terms of the effective action (i.e. the Wess-Zumino term). In the present formalism this term plays the essential role of breaking the degeneracy between states with different sign of the flavor quantum number.

As it will be shown in Sec.3, using soliton profiles that provide a good description of the $SU(2)$ sector, the meson eigenvalue equation Eq.(10) has only two solutions with $\omega < m_M$ when kaons are considered. The lowest one is in the $(\Lambda, l, S) =$

$(1/2, 1, -1)$ channel and gives rise to the low-lying positive parity strange hyperons. The other solution is in the $(1/2, 0, -1)$ channel and describes the $\Lambda(1405)$ negative parity resonance which experimentally lies below the KN threshold. For heavier flavors bound states in other channels are also predicted.

To describe the splittings between baryons with the same value of F but different spin and/or isospin, we have to go beyond the static ansatz Eq.(6). We need to take into account the isospin zero-modes of the soliton. This can be done by introducing the $SU(2)$ collective coordinates $A(t)$ [4]

$$\begin{aligned} u(\vec{r}, t) &= A(t)\ u(\vec{r})\ A^{\dagger}(t), \\ M(\vec{r}, t) &= A(t)\ \tilde{M}(\vec{r}, t) \end{aligned} \tag{11}$$

Here, $\tilde{M}$ is the flavored meson field observed from the rest frame of the soliton.

Substitution of Eq.(11) into the effective action leads to a $\mathcal{O}\ (N_c^{-1})$ contribution to the hamiltonian of the system,

$$\frac{\left(\vec{J}_{sol} + c\vec{J}_M\right)^2}{2\Theta} \tag{12}$$

where $\vec{J}_{sol}$ is the angular momentum of the rotating soliton, $\vec{J}_M$ the one of the meson field and Θ is the $SU(2)$ moment of inertia. The quantity c in Eq.(12) measures the hyperfine splitting. Its lengthy expression in terms of the background fields can be found in Ref.[6] and it is different for mesons bound in different (Λ, l) channels. When more than one type of mesons are bound to the soliton the term proportional to $\vec{J}_M$ in Eq.(12) has to be replaced by the sum of the corresponding hyperfine splitting constants multiplied by the spin of each meson field.

The hyperon mass formula can be obtained by calculating the mean value of the hamiltonian in the different baryon states. In the case where K and D mesons are present one obtains

$$M(I, J, n_1, n_2, J_1, J_2, J_m) = M_{\rm sol} + n_1\omega_1 + n_2\omega_2 + M_{\rm rot}, \tag{13}$$

$$\begin{aligned} M_{\rm rot} = \frac{1}{2\Theta}\Big\{ & I(I+1) + (c_1 - c_2)[c_1 J_1(J_1+1) - c_2 J_2(J_2+1)] + c_1 c_2 J_m(J_m+1) \\ & + \Big[J(J+1) - J_m(J_m+1) - I(I+1)\Big]\left[\frac{c_1+c_2}{2} + \frac{c_1-c_2}{2}\frac{J_1(J_1+1) - J_2(J_2+1)}{J_m(J_m+1)}\right]\Big\}, \end{aligned} \tag{14}$$

where $M_{\rm sol}$ is the soliton mass. Here, n_1 is the absolute value of strangeness, n_2 the charm quantum number and ω_1 and ω_2 are, respectively, the bound-state energies of the K and D mesons. In addition, c_1 is the hyperfine splitting constant corresponding to K and c_2 the one corresponding to D. The angular momenta J_1 and J_2 are defined as $J_i = n_i j_i$ with $j_i \equiv \Lambda_i$ being the angular momentum of the bound state orbital and J_m is given by $J_m = J_1 + J_2, \cdots, |\ J_1 - J_2\ |$. The total angular momentum $\vec{J}$ is the sum of the rotor spin and the total meson spin $\vec{J}_m$. The quantum numbers of the physical hyperons can be obtained by using the quantization rules described in Ref.[9]

3. The hyperon spectrum

The masses of the low-lying positive parity hyperons predicted by the present model in comparison with the available experimental data[10] and with the quark model predictions of Refs.[11,12] are shown in Table 1. In the soliton model these states are obtained by populating the $K-$ and/or $D-$meson ground states (which have quantum numbers $l=1$, $j=\frac{1}{2}$).

Table 1: Masses (in MeV) of the low-lying positive parity strange and charmed hyperons.

Particle	Exp.	SET I	SET II	Ref.[11]	Ref.[12]
Λ	1116	1106	1086		
Σ	1193	1203	1205		
Σ^*	1385	1350	1320		
Ξ	1318	1332	1311		
Ξ^*	1530	1480	1425		
Ω	1672	1621	1549		
Λ_c	2285	2172	2209	2200	2260
Σ_c	2453	2327	2379	2360	2440
Σ_c^*	?	2387	2417	2420	2510
Ξ_{cc}	?	3513	3601	3550	
Ξ_{cc}^*	?	3574	3639	3610	
Ω_{ccc}	?	4791	4898	4810	
Ξ_c	2470	2381	2426	2420	2480
Ξ_c'	?	2509	2514	2523	2575
Ξ_c^*	?	2524	2539	2531	2645
Ω_c	2740	2643	2647	2680	2730
Ω_c^*	?	2674	2662	2720	2790
Ω_{cc}	?	3700	3764	3730	
Ω_{cc}^*	?	3730	3778	3770	

Two sets of parameters are considered in the soliton model. SET I corresponds to

$$m_\pi = 0, \quad F_\pi = 129\ MeV, \quad e = 5.45, \quad \chi_1 = 1.22, \quad \chi_2 = 1.8,$$

while SET II corresponds to

$$m_\pi = 138\ MeV, \quad F_\pi = 108\ MeV, \quad e = 4.84, \quad \chi_1 = 1.22, \quad \chi_2 = 2.0.$$

Here, χ_i is the ratio of the corresponding flavored meson decay constant to the pion one. In SET I the chiral limit in the $SU(2)$ sector is taken while SET II corresponds to the case of massive pions. In both cases F_π and e are fitted to reproduce the empirical nucleon and $\Delta_{3,3}$ masses and χ_1 is set to its empirical value. On the other hand the empirical value of χ_2 is not very well established. In SET I χ_2 is set to $\chi_2 = 1.8$ that falls well within the range given in Ref.[13] , while in SET II a slightly larger value is used in order to get a better D-meson binding energy.

We observe that the soliton model, with or without pion mass, works very well for both strange and charmed baryons. Moreover the similarity between its predictions and those of the quark model is striking. They predict the same number of low-lying states as well as the same energy ordering. As mentioned in Ref.[6] , the soliton model even predicts a certain number of mass formulae that agree with those of quark model.

Hyperon resonances can also be described in the present model[14] . As already mentioned, in the strange sector there is only one excited bound state which has $l = 0$, $j = \frac{1}{2}$. It provides a natural description of the odd parity resonance $\Lambda(1405)$. The predictions for the mass of this state are around 100 MeV too low (1325 MeV for SET I and 1297 MeV for SET II). This situation contrasts with typical quark model calculations where this state appears too high in energy. On the other hand the model predicts a far richer spectrum for the charmed hyperons. The D-meson has bound states in all partial waves with $l \leq 3$ and additional excited states in the P and S-waves. As discussed in Ref.[6] the spectra of single D-meson (Λ_c, Σ_c) states are in good qualitative agreement with quark-model predictions[12] . Like in the strange sector however negative-parity states are too low as compared with the quark model calculations.

4. Electromagnetic properties and coupling constants of the hyperons

In this section I will concentrate on the predictions of the model for some strong and electromagnetic hyperon properties. This provides a further check of the validity of the bound state approach. First I consider the hyperon magnetic moments. Given the electromagnetic current derived from the effective lagrangian by means of Noether's theorem, one can readily obtain the magnetic moment operator[15] . To calculate the matrix elements of this operator the hyperon wave functions written in terms of the rotor and bound meson wave functions are used. In this way one obtains the magnetic moment of the different hyperons. Before presenting the results it is worthwhile noting that soliton models with Skyrme term stabilization are known to predict a somewhat small value for the proton magnetic[8] ($\mu_p = 1.88$ using SET I and $\mu_p = 1.97$ using SET II to be compared with the empirical value $\mu_p^{emp} = 2.79$). On the other hand, the ratio μ_n/μ_p is predicted quite accurately. As we shall see, a similar situation occurs for the hyperon magnetic moments.

In Table 2 the values of the magnetic moments of the spin 1/2 strange and charmed baryons as obtained in the bound state approach with the two sets of parameters given in the previous section are shown. Also listed in Table 2 are the available empirical results[10] and the quark model and bag model predictions of Ref.[16] and Ref.[17] respectively. All the values are given relative to the proton magnetic moment. Similar results are obtained for the spin 3/2 hyperons magnetic moment[18] . As in the case of the hyperon masses the agreement of the predicted magnetic moments with experiments and quark model predictions is quite remarkable. The discrepancies appearing in the magnetic moments of the non-zero isospin hyperons are due to the rather small value of the meson contribution to the isovector meson magnetic moment. The explicit incorporation of vector mesons in the effective action is expected to increase such a contribution.

Table 2: Magnetic moments of the spin 1/2 strange and charmed hyperons.

Particle	Exp.	SET I	SET II	Quark Model	Bag Model
Λ^0	−0.22	−0.21	−0.27	−0.21	−0.25
Σ^+	0.87	1.07	1.10	0.96	0.97
Σ^0	–	0.27	0.34	0.29	0.31
Σ^-	−0.41	−0.54	−0.42	−0.38	−0.36
Ξ^0	−0.45	−0.58	−0.66	−0.50	−0.56
Ξ^-	−0.24	−0.07	−0.19	−0.16	−0.23
Λ_c^+	–	0.11	0.10	0.13	0.18
Σ_c^{++}	–	0.98	0.99	0.85	0.70
Σ_c^+	–	0.16	0.21	0.18	0.13
Σ_c^0	–	−0.65	−0.56	−0.49	−0.44
Ξ_{cc}^{++}	–	−0.17	−0.17	−0.04	0.06
Ξ_{cc}^+	–	0.35	0.32	0.29	0.31
Ξ_c^+	–	0.11	0.10	0.13	0.18
Ξ_c^0	–	0.11	0.10	0.13	0.18
$\Xi_c'^+$	–	0.44	0.39	0.26	0.17
$\Xi_c'^0$	–	−0.59	−0.57	−0.41	−0.39
Ω_c^0	–	−0.31	−0.40	−0.32	−0.35
Ω_{cc}^+	–	0.21	0.23	0.25	0.30

The strong meson-hyperon couplings can be evaluated from the explicit expression of the meson-soliton lagrangian[19] . In Table 3 the calculated values of the ΛKN and ΣKN coupling constants are given for the sets of parameter values used above. Also shown are the values obtained when the chiral symmetry breaking terms Γ_{SB} are left out. These values are useful for the comparison with the predictions of the phenomenological pseudovector coupling model. The coupling constant values in Table 3 fall well within the range of values obtained by different approaches and phenomenological analyses that are given in Ref.[20] . We note that when both the kaon and pion masses take their empirical values (SET II) the effect of the chiral symmetry breaking terms is small. This shows that the bound state model leads to results that are similar to those obtained with the phenomenological pseudovector coupling model. When the pion mass is neglected the longer range of the soliton profile leads to a larger relative correction from the chiral symmetry breaking terms.

The $\Lambda_c ND$ and $\Sigma_c ND$ coupling constants calculated with and without the inclusion of Γ_{SB} are also listed in Table 3. Because of the much larger D-meson mass the effect of the chiral symmetry breaking terms on the calculated coupling constants are somewhat greater than in the case of the kaons.

Table 3: The meson-hyperon coupling constants.

		SET I	SET II
$S=-1$	$g_{KN\Lambda}$	-1.82	-2.79
	$g_{KN\Lambda}$ (with c.s.b.)	-1.06	-2.17
	$g_{KN\Sigma}$	0.63	0.97
	$g_{KN\Sigma}$ (with c.s.b.)	0.37	0.76
$C=+1$	$g_{DN\Lambda_c}$	-1.80	-2.44
	$g_{DN\Lambda_c}$ (with c.s.b.)	-2.73	-5.20
	$g_{DN\Sigma_c}$	0.63	0.86
	$g_{DN\Sigma_c}$ (with c.s.b.)	0.96	1.82

Finally, I present some results for the threshold photoproduction amplitudes of hyperons. They can be obtained from the explicit expression of the electromagnetic current. Their general form is [19]

$$\langle H|T|N\rangle \;=\; i\,e\,g_A^{HN}\,\vec{\sigma}\cdot\vec{\varepsilon}+\frac{i\,e}{4}\Big(\frac{1}{m_N}+\frac{1}{m_H}\Big)\,f_A^{HN}\,\vec{\sigma}\cdot\vec{q}\,\vec{q}\cdot\vec{\varepsilon} \tag{15}$$

where g_A^{HN} is the axial coupling constant for the particular hyperon–nucleon vertex, given in terms of its isoscalar and isovector components as

$$g_A^{HN}=g_A^S(H,0)+g_A^V(H,0)\,\tau_3\,,$$

and f_A^{HN} is the corresponding "induced" pseudoscalar transition form factor (this contribution vanishes at threshold) given in the same way as

$$f_A^{HN}\;=\;f_A^S(H,0)+f_A^V(H,0)\,\tau_3\,.$$

Here τ_3 should be understood as a matrix element between the isospin wave functions of the outgoing kaon and the incoming nucleon. Finally $\vec{\varepsilon}$ is the polarization unit vector of the electromagnetic field $\vec{A}^{e.m.}$. The explicit expressions of the coefficients $g_A^{S,V}(H,0)$ and $f_A^{S,V}(H,0)$ in terms of the soliton profile θ and the bound meson wave function can be found in Ref.[19] . The numerical values of the axial transition form factors $g_A^{S,V}(H,0)$ for both the Λ and Σ hyperons as well as for the corresponding charmed Λ_c and Σ_c hyperons are given in Table 4. The numerical values are calculated with the same parameter sets as in the previous calculations. The corresponding pseudoscalar transition form factors $f_A^{S,V}(H,0)$ are given in Table 5.

Table 4: The axial transition coupling constants.

		g_A^S		g_A^V	
		SET I	SET II	SET I	SET II
$S=-1$	$KN\Lambda$	−0.28	−0.44	0.18	0.17
	$Kn\Sigma^0$	−0.09	−0.15	0.23	0.29
	$Kp\Sigma^+$	−0.13	−0.21	−0.04	−0.08
$C=+1$	$DN\Lambda_c$	−0.19	−0.26	0.05	0.05
	$Dn\Sigma_c^0$	−0.06	−0.09	0.10	0.12
	$Dp\Sigma_c^+$	−0.09	−0.12	−0.03	−0.05

Table 5: The induced pseudoscalar coupling constants (in MeV).

		f_A^S		f_A^V	
		SET I	SET II	SET I	SET II
$S=-1$	$KN\Lambda$	48.63	69.02	62.58	92.57
	$Kn\Sigma^0$	16.82	23.87	11.61	10.85
	$Kp\Sigma^+$	23.81	33.80	37.75	60.32
$C=+1$	$DN\Lambda_c$	4.24	6.89	10.08	16.41
	$Dn\Sigma_c^0$	1.44	2.34	1.57	1.56
	$Dp\Sigma_c^+$	2.04	3.31	6.16	10.74

By comparing the numerical values for the axial transition couplings in Table 4 to the meson–hyperon–nucleon coupling constants in Table 3 using the phenomenological pseudovector relations

$$g_{\Lambda NK} = -\frac{m_\Lambda + m_N}{F_K}\, g_A^{\Lambda N}(0) \tag{16}$$

$$g_{\Sigma NK} = -\frac{m_\Sigma + m_N}{F_K}\, g_A^{\Sigma N}(0) \tag{17}$$

one notes that they are approximately satisfied. If the photoproduction amplitudes obtained with the g_A–values in Table 4 are multiplied by the elementary charge $e \simeq 0.3$ their magnitudes agree well with those obtained for the corresponding contact term amplitude with the "cloudy" bag model[21] .

5. Exotic baryons

In the previous sections I have shown that the bound state soliton model provides a satisfactory description of the stable hyperon properties. In this situation it is natural to use such a model to obtain some insight into the possibility of having stable exotic baryons that cannot be formed of 3 quarks.

As a first example of such baryons I consider the strange dibaryons. Of special interest is the $S = -2$ flavor singlet H-dibaryon which has $(I, J^P) = (0, 0^+)$ quantum numbers. In the quark models this state is appreciably lower than other states in the spectrum[22] . In previous calculations[23] the H-dibaryon was found to be unbound when described as a two kaon state interacting with an $B = 2$ soliton background. Moreover, a certain number of states which are forbidden in the quark model were present in the bound state soliton model. However, a more recent analysis[24] shows that in fact the mass of the H particle can be below the $\Lambda\Lambda$ threshold when an improved axially symmetric soliton configuration is used. In addition, all the spurious states are eliminated when the constrains imposed by the symmetries of the soliton-meson system are satisfied. As a result of this the Skyrme model dibaryon spectrum looks quite similar to the quark model one.

Another type of exotic baryons that have received some attention are the negative charm and positive bottom hyperons. In the quark model the simplest such hyperons would have the configuration $\bar{Q}qqqq$, where q denotes u, d and s quarks and $\bar{Q}$ a $\bar{c}$ or $\bar{b}$ quark. In the constituent quark model it has been found that the colour magnetic hyperfine interaction may be strong enough to stabilize the strange antiquark hyperons ($C = -1, S = -1$), in which one of the q quarks is strange, but not the non-strange $C = -1$ hyperons[25] . The bound state soliton model however predicts both the strange and non-strange $C = -1$ hyperons to be stable[26] . Within this model anticharm hyperons are formed by a $\bar{D}$ isodoublet bound to the soliton. The effective lagrangian density for this meson has the same form as that for the D isodoublet, except for a change in sign of the contribution from the Wess-Zumino. This leads to the sign change indicated by the presence of $F = \pm 1$ in Eq.(10). It should be noted that the appearance of opposite flavor bound states is a consequence of the decrease of the amplitude of the Wess-Zumino term as the meson mass increases[27] and of the existence of a remaining attraction supplied by the quartic Skyrme term. The predicted masses of the $C = -1$ ground state hyperon are given in Table 6. In fact the hyperfine correction is large enough to push the energies of the predicted $I = 1, J = 3/2$ and those of both $I = 2$ states above the ND threshold at 2808 MeV. Those states are therefore very likely to be unbound.

Table 6: Predicted non-strange $C = -1$ ground state hyperon masses (in MeV).

I	J	SET I	SET II
0	1/2	2697	2617
1	1/2	2861	2781
1	3/2	2908	2828
2	3/2	3236	3156
2	5/2	3314	3234

The stable pentaquark states predicted by the quark model (i.e. $P_{\bar{c}s}$) are also obtained in the soliton model. They are two-meson states formed by a soliton and a

$S = -1$ kaon and a $C = -1$ D-meson. The predictions of the energy levels of these two-meson states obtained in the present version of the model, which does not take into account the interaction between the mesons, cannot however be assumed to be as good as those for the single meson non–strange pentaquarks. The small empirical mass difference of only 100 MeV between the D and the D_s mesons indicates that there is an effective attractive interaction between the $\bar{c}$ and s quarks. The $P_{\bar{c}s}$ baryons have the isospin values 1/2 and 3/2. Using the parameter values denoted SET II above one obtains a bound $I = J = 1/2$ $P_{\bar{c}s}$ pentaquark at 2860 MeV in the symmetric meson spin state ($J_M = 1$). With the parameter values denoted SET I this state would be predicted to lie at 2949 MeV, which is 40 MeV above the ND_s threshold.

Finally I consider the heavy quarkonium baryons (i.e. baryons containing a $c\bar{c}$ pair). In the present formalism such baryons would be described as η_c—soliton bound systems. As first noticed in Ref.[28] in order to have some η-soliton interaction the Skyrme quartic term in Eq.(1) has to be replaced by the linear combination

$$\int d^4x \left\{ \frac{x}{32e^2} Tr\left[[U^\dagger\partial_\mu U, U^\dagger\partial_\nu U]^2 \right] \right. \\ \left. + \frac{1-x}{16e^2} \left[\left(Tr\left[U^\dagger\partial_\mu U U^\dagger\partial_\nu U \right] \right)^2 - \left(Tr\left[U^\dagger\partial_\mu U U^\dagger\partial^\mu U \right] \right)^2 \right] \right\} \tag{18}$$

where the parameter x takes values between 0 and 1. The case $x = 1$ corresponds to the Skyrme model case which has no coupling between the η-like states and the soliton. It should be noted that in the pure $SU(2)$ limit Eq.(18) reduces to the standard Skyrme term independently of x. As shown in Ref.[28] the η-soliton attraction increases with decreasing x. In the strange sector the η-soliton system describes the $N(1535)$ resonance that lies above the ηN threshold. Therefore one should choose a value of x which is close enough to 1 to ensure that η is not bound to the soliton. For this reason x is set to $x = 0.85$. It has been shown[29] that with this x-value one also obtains a satisfactory description of the strange and charmed hyperons. Since in general the meson-soliton attraction increases with the meson mass when we turn to the η_c case there is a bound S state with energy $\omega = 2817$ MeV [30]. This result is obtained with $m_{\eta_c} = 2980$ MeV and the parameter values SET I. Similar results can be found with SET II. This bound state is the $c\bar{c}$ analog of the $N(1535)$. A similar bound analog to the $\Lambda(1670)$ is also predicted.

6. Conclusions

I have shown that the static and dynamical properties of the massive-flavor baryons are well described in the bound meson-soliton model. It is particularly noteworthy that with only two parameters needed for light-quark (up and down) systems and masses and decay constants taken from empirical sources, the model is able to fit not only the masses but also the magnetic moments and coupling constants of strange and charmed baryons. The agreement with experiments and quark-model predictions is quite remarkable and suggests strongly that the model is close to nature in its physics content. At first sight, this is surprising since the skyrmion model looks so different from the quark description. The crucial feature of the model is that a massive scalar doublet carrying the flavor quantum number of the massive quark gets bound to and wrapped by the $SU(2)$ soliton, the quantum numbers arising through topologically induced transmutation. As discussed in Ref.[31] , the essential dynamics can be understood in terms of a hierarchy of induced gauge (Berry) connections generated in integrating out layers of length scales.

The results presented in this contribution correspond mainly to the strange and charm sector. Similar calculations can be done in the bottom sector[6] . In that case however the predictions are at most of qualitative nature since, for example, Λ_b is predicted to be overbound by 600 MeV. To overcome these difficulties the vector partners of the flavored mesons (i.e. B^*) should be explicitly included. In addition such degrees of freedom are essential to make the model fully consistent with the heavy-quark Isgur-Wise symmetry of QCD [32] . A first step in this direction has been taken in Ref.[33] . A more detailed calculation is under way.

Finally, I would like to mention that several exotic states like i.e. the H-dibaryon and the anticharm pentaquarks are predicted in the model. The experimental search for the H-dibaryon is a topic of current interest[34] . The search for the strange anticharm pentaquark is difficult because of the small cross sections for simultaneous production of charm and strangeness. Empirical verification or disproof of the non-strange pentaquark should be much easier since it requires production of only one heavy quark. Its possible existence as a bound state or a near threshold resonance can in principle be decided by present experiments designed to search of the strange pentaquark.

Acknowledgments

I would like to thank the members of the Organizing Committee for their invitation to participate in this workshop. I am also grateful to my collaborators C.Gobbi (Pavia), G.Pari (Toronto), M.Rho (Saclay), D.O.Riska (Helsinki), G.Thomas (Porto Alegre) and A.Wirzba (Copenhagen) together with whom different parts of the work reported in this contribution was done.

References

1. T.H.R. Skyrme, *Proc.R.Soc.London* **A260** (1961) 127; *Proc.R.Soc.London* **A262** (1961) 237; *Nucl. Phys.* **31** (1962) 556.
2. M.A. Nowak, P.O. Mazur and M. Praszalowicz, *Phys. Lett.* **B147** (1984) 137; E. Guadagnini, *Nucl. Phys.* **B236** (1984) 35.
3. H. Yabu and K. Ando, *Nucl. Phys.* **B301** (1988) 601.
4. C.G. Callan and I. Klebanov, *Nucl. Phys.* **B262** (1985) 365.
5. N.N. Scoccola, H. Nadeau, M. Nowak and M. Rho, *Phys. Lett.* **B201** (1988) 425; C.G. Callan, K. Hornbostel and I. Klebanov, *Phys. Lett.* **B202** (1988) 269; U. Blom, K. Dannbom and D.O. Riska, *Nucl. Phys.* **A493** (1989) 384; N.N. Scoccola, D.P. Min, H. Nadeau and M. Rho, *Nucl. Phys.* **A505** (1989) 497.

6. M. Rho, D.O. Riska and N.N. Scoccola, *Phys. Lett.* **B251** (1990) 597; *Z. Phys.* **A341** (1992) 343.
7. D.O. Riska and N.N. Scoccola, *Phys. Lett.* **B265** (1991) 188.
8. G.S. Adkins, C.R. Nappi and E. Witten, *Nucl. Phys.* **B228** (1983) 552; G.S. Adkins and C.R. Nappi, *Nucl. Phys.* **B233** (1984) 109.
9. N.N. Scoccola and A. Wirzba, *Phys. Lett.* **B258** (1991) 451.
10. Particle Data Group, *Phys. Lett.* **B239** (1990) 1.
11. A. De Rújula, H. Georgi and S.L. Glashow, *Phys. Rev.* **D12** (1975) 147.
12. L.A. Copley, N. Isgur and G. Karl, *Phys. Rev.* **D20** (1979) 768; K. Maltman and N. Isgur, *Phys. Rev.* **D22** (1980) 1701; S. Capstick and N. Isgur, *Phys. Rev.* **D34** (1986) 2809.
13. C. Dominguez and N. Paver, *Phys. Lett.* **B197** (1987) 423.
14. N.N. Scoccola, *Phys. Lett.* **B236** (1990) 245.
15. E. Nyman and D.O. Riska, *Nucl. Phys.* **B325** (1989) 593; J. Kunz and P.J. Mulders, *Phys. Lett.* **B231** (1989) 335; D.P. Min, Y.S. Koh, Y. Oh and H.K. Lee, *Nucl. Phys.* **A530** (1991) 698.
16. D.B. Lichtenberg, *Phys. Rev.* **D15** (1977) 345.
17. T. DeGrand, R.L. Jaffe, K. Johnson and J. Kiskis, *Phys. Rev.* **D12** (1975) 2060; S.K. Bose and L.P. Singh, *Phys. Rev.* **D22** (1980) 773.
18. Y. Oh, D.P. Min, M. Rho and N.N. Scoccola, *Nucl. Phys.* **A534** (1991) 493.
19. C. Gobbi, D.O. Riska and N.N. Scoccola, *Nucl. Phys.* **A544** (1992) 671.
20. R.A. Adelseck and B.Saghai, *Phys. Rev.* **C42** (1990) 108.
21. Y.S. Zhong, A.W. Thomas, B.K. Jennings and R.C. Barrett, *Phys. Rev.* **D38** (1988) 837.
22. R.L. Jaffe, *Phys. Rev. Lett.* **38** (1977) 195.
23. J. Kunz and P.J. Mulders, *Phys. Lett.* **B215** (1988) 449.
24. G. Thomas, N.N. Scoccola and A. Wirzba, in preparation.
25. H.J. Lipkin, *Phys. Lett.* **B195** (1987) 484; C. Gignoux, B. Silvestre-Brac and J.M. Richard, *Phys. Lett.* **B193** (1987) 323.
26. D.O. Riska and N.N. Scoccola, "Anticharm and antibottom hyperons", Helsinki Univ. preprint HU-TFT-92-46 (1992).
27. H.K. Lee, M.A. Nowak, I. Zahed and M. Rho, "Induced gauge potentials in baryons", Stony Brook Univ. preprint SUNY-NTR-92-50 (1992).
28. G. Pari, *Phys. Lett.* **B261** (1991) 347.
29. G. Pari and N.N. Scoccola, *Phys. Lett.* **B** in press.
30. C. Gobbi, D.O. Riska and N.N. Scoccola, *Phys. Lett.* **B** in press.
31. M. Rho, contribution to this conference.
32. N. Isgur and M.B. Wise, *Phys. Rev. Lett.* **66** (1991) 1130.
33. D.P. Min, Y.S. Oh, B.Y. Park and M. Rho, "Soliton structure of heavy baryons", Seoul National Univ. preprint SNUTP-92/78 (1992).
34. B. Quinn, Lecture given at *Baryons 92*, 1-5 June 1992, Yale Univ.

HEAVY-QUARK BARYONS AS SKYRMIONS

Mannque RHO
Service de Physique Théorique, C.E. Saclay
91191 Gif-sur-Yvette, France

ABSTRACT

I discuss recent development on the description of heavy-quark (such as charmed and bottom) baryons as one or more heavy mesons "wrapped" by a skyrmion. Amazingly enough, such a description naturally arises when light-quark chiral symmetry and heavy-quark spin symmetry are incorporated in an effective Lagrangian. I interpret the resulting spectrum in terms of nonabelian induced (Berry) potentials in analogy to diatomic molecular systems.

1. Introduction

There is a wide-spread feeling that the skyrmion description of the baryons is a highly predictive and coherent one, ranging from the structure of the nucleon and Δ to nuclear forces. Indeed many of the no-go arguments raised against the skyrmion picture are disappearing rapidly. As we heard in this meeting, we no longer seem to have any problem with S- and P-wave πN scatterings[1,2]; yesterday's too large energy of the ground state baryon seems no longer too large once Casimir contributions of $O(N_c^0)$ are taken into account[1,3]; the missing attraction at medium range in NN potentials that binds nuclei is no longer missing[4]; the soliton instability problem is just a red herring; whether explicit quark degrees of freedom should be present or not is becoming a non-issue. At this moment, there seem to exist no really serious arguments against the notion that the nucleon is a skyrmion in and outside of nuclear medium. Whatever deviation from nature there may be, may simply be a matter of poorly understanding the intricacy of the skyrmion, not the deficiency of the model itself. As far as I know, it is the only model available which can address simultaneously one-nucleon *and* many-nucleon problems with equal ease. It is of course not derived from QCD but it is consistent with it.

In this talk, I will argue that even when one has very heavy quarks in the baryons, the skyrmion picture still makes sense and in fact, it comes out strikingly close to the quark-model description which we expect gets better as the quark mass becomes heavier. This may sound amazing or, to some, unbelievable. In fact, when Riska, Scoccola and I submitted a paper a few years ago showing that the skyrmion model worked just as well in charmed and bottom baryon sectors, a referee promptly rejected it. It appeared in a different journal[5] only after a long delay because of the referee's repeated refusal to accept our arguments. As I shall describe below, our claim is vindicated by the recent development.

2. Diatomic Molecules

To better clarify the basic idea involved in the workings of the soliton model in complex strong interactions, let me start with a quantum mechanical problem of a diatomic molecule. This quantum mechanical problem can be put in a context that closely mimics the baryon excitation we are interested in. It shows a generic setting in which nonabelian gauge potentials are *induced* by interactions.

Consider the dynamics of a diatomic molecule where two atoms are separated by the internuclear separation denoted by the vector $\vec{R}$. We shall restrict ourselves to the relative motion only. Because of the symmetry of the diatom, rotational symmetry of the electrons is broken. The electronic state is characterized by the eigenvalue Ω of the operator $\hat{R}\cdot\mathbf{J}_{el}$. In my discussion, I will confine myself to the triplet of the states $\Omega=0,\pm1$. This system was recently studied by Zygelman[6] and reanalyzed by Lee and myself[7] to gain useful insight into its generic structure.

This system can be roughly categorized into two according to the size of R. For small R, the potential curve for the $\Omega=0$ state (called Σ) which is higher lying does not overlap with that of the degenerate doublet $\Omega=\pm1$ (called π). Thus one can focus on the doublet, ignoring the singlet. For large R, however, the potential curves start overlapping, the complete overlap occurring at $R=\infty$. Then the triplets become degenerate, with the *restoration of the electronic rotational symmetry.* We will exploit this feature later on in discussing heavy-quark symmetry.

Let me now describe how the above structure can be understood.[7] We will consider the molecular excitation described by the dynamical variables $\mathbf{R}(t)$ which could be vibration or rotation as *slow* compared with the electronic excitation which we consider to be *fast.* We wish to integrate out the fast degree of freedom and express the whole system on the coordinate $\mathbf{R}(t)$ of the slow degree of freedom. The resulting system can be described by the following Lagrangian

$$\mathcal{L}=\frac{1}{2}\mu\dot{\vec{R}}^2+i\theta_a^\dagger\left(\frac{\partial}{\partial t}-i\vec{A}^\alpha T^\alpha_{ab}\cdot\dot{\vec{R}}\right)\theta_b \tag{1}$$

where μ is the reduced mass, θ_a is a Grassmann variable labeled by a, $\vec{A}^\alpha$ is the Berry potential inherited from integrating out the fast degree of freedom and T^α is the matrix representation of the vector space in which the Berry potential lives. In our case, we have a triplet of states and $\mathbf{A}\cdot\mathbf{T}\in SU(2)$. I have kept only the essential pieces in the Lagrangian (1), leaving out some "junks", to make the argument as simple as possible. The "junks" do not change the main thrust of my argument. Let me also mention that the Grassmannian variables for each a are used as a trick to avoid writing the Lagrangian in matrix form. There is nothing deep in it at least for our purpose. Quantizing (1) in a standard way, one gets the Hamiltonian

$$H=\frac{1}{2\mu}(\vec{p}-\vec{\mathbf{A}})^2. \tag{2}$$

Since there is gauge invariance in the theory as one can see from (1), we are allowed to make a gauge transformation on (2) and obtain the following Hamiltonian [7]

$$H = -\frac{1}{2\mu R^2}\frac{\partial}{\partial R}R^2\frac{\partial}{\partial R} + \frac{1}{2\mu R^2}(\vec{L}_o + (1-\kappa)\vec{I})^2 - \frac{1}{2\mu R^2}(1-\kappa)^2(\vec{I}\cdot\hat{R})^2 \tag{3}$$

and the corresponding gauge field [a]

$$\vec{\mathbf{A}}' = (1-\kappa)\frac{\hat{R}\times\vec{I}}{R^2},$$

and the magnetic field

$$\vec{\mathbf{B}}' = -(1-\kappa^2)\frac{\hat{R}(\hat{R}\cdot\mathbf{I})}{R^2}$$

where κ indicates the extent to which electronic rotational symmetry is present, with $\kappa = 1$ indicating full symmetry and $\kappa = 0$ a complete absence of symmetry. In (3), $\vec{L}_o$ is the angular momentum lodged in the dumb-bell and $\vec{I}$ is the angular momentum contributed by the Berry potential. Neither $\vec{L}_o$ nor $\vec{I}$ separately commutes with the Hamiltonian. What commutes is the the total angular momentum $\vec{L} = \vec{L}_o + \vec{I}$.

We are now ready to analyze what happens in the two extreme cases of $\kappa = 0$ which results when $R \to 0$ and $\kappa = 1$ which results when $R \to \infty$. For $\kappa = 1$, the degenerate Σ and π states form a representation of the rotation group and hence the Berry potential (and its field tensor) vanishes or becomes a pure gauge. The spectrum is then independent of the angular momentum carried by the electronic system. *This just means that rotational symmetry is restored in the electronic sector.* For $\kappa = 0$, the Σ and π states are completely decoupled and only the $U(1)$ monopole field can be developed on the π states. κ goes to zero and the Hamiltonian can be written as

$$H = -\frac{1}{2\mu R^2}\frac{\partial}{\partial R}R^2\frac{\partial}{\partial R} + \frac{1}{2\mu R^2}(\vec{L}\cdot\vec{L} - 1) \tag{4}$$

which is a generic form for a system coupled to an $U(1)$ monopole field. In this case, the "magnetic charge" is quantized to ± 1 or twice the basic Dirac monopole charge $\pm\frac{1}{2}$. A truly nonabelian Berry potential with non-quantized charge can be obtained only for κ which is not equal to zero or one.

3. Spectrum of Heavy Baryons

I will now turn to the structure of heavy baryons described as skyrmions and summarize the recent development. What I will present below is based on work done recently in collaboration with Min and his coworkers in Seoul.[8] Instead of making detailed analysis to compare with experiments or with quark models, let me start with a simplified Lagrangian consistent with hidden gauge symmetry (HGS).[9] I will

---

[a] We should note that the magnetic field is of the nonabelian monopole type of 't Hooft and Polyakov with however a charge which is *not* quantized. This is a generic feature of induced gauge fields we encounter in various systems.

consider two light flavors $q = u, d$ and a third flavor Q which will be taken to be heavy later on. For the moment I will consider u, d, Q on the same footing and write a Lagrangian built from chiral symmetry. Obviously when the mass of Q, m_Q, becomes large compared with the chiral scale $\Lambda_\chi \sim 1$ Gev, it would make no sense to talk about chiral symmetry associated with the quark Q but let me blindly start with an $SU(3)$ chiral Lagrangian anyway and take m_Q become large. I will write the Lagrangian as the sum of the $SU(2)$ Skyrme Lagrangian of the (u, d) sector, $\mathcal{L}_{SU(2)}$, the HGS Lagrangian without (with) the ω meson coupling, $\mathcal{L}_\Phi^{HGS}$ ($\mathcal{L}_\omega^{HGS}$), and the "anomalous parity" Lagrangian, $\mathcal{L}_{an}$;

$$
\begin{aligned}
\mathcal{L}^{HGS} &= \mathcal{L}_{SU(2)} + \mathcal{L}_\Phi^{HGS} + \mathcal{L}_\omega^{HGS} + \mathcal{L}_{an}, \\
\mathcal{L}_{SU(2)} &= \frac{F_\pi^2}{16}\mathrm{Tr}\left[\partial_\mu\Sigma\partial^\mu\Sigma^\dagger\right] + \frac{1}{32e^2}\mathrm{Tr}\left[\Sigma^\dagger\partial_\mu\Sigma, \Sigma^\dagger\partial_\nu\Sigma\right]^2, \\
\mathcal{L}_\Phi^{HGS} &= (D_\mu\Phi)^\dagger(D^\mu\Phi) - m_\Phi^2\Phi^\dagger\Phi \\
&\quad -\frac{1}{2}\Phi_{\mu\nu}^{*\dagger}\Phi^{*\mu\nu} + m_{\Phi^*}^2\left[\Phi_\mu^{*\dagger} + \frac{2i}{F_\pi g_{\Phi^*}}\Phi^\dagger A_\mu\right]\left[\Phi^{*\mu} + \frac{2i}{F_\pi g_{\Phi^*}}A^\mu\Phi\right] \\
\mathcal{L}_\omega^{HGS} &= -\frac{iN_c}{2F_\pi^2}B_\mu\left[\left(\Phi^\dagger D^\mu\Phi - (D^\mu\Phi)^\dagger\Phi\right) - \left(\Phi_\nu^{*\dagger}D^\mu\Phi^{*\nu} - (D^\mu\Phi_\nu^*)^\dagger\Phi^{*\nu}\right)\right] \\
\mathcal{L}_{an} &= -\frac{iN_c}{F_\pi^2}B_\mu\left(\Phi^\dagger D^\mu\Phi - (D^\mu\Phi)^\dagger\Phi\right) + \delta\mathcal{L}_{an}
\end{aligned} \tag{5}
$$

where

$$
\begin{aligned}
D_\mu &= \partial_\mu + V_\mu, \qquad \Sigma = \xi\cdot\xi, \\
\begin{pmatrix} V_\mu \\ A_\mu \end{pmatrix} &= \frac{1}{2}\left(\xi^\dagger\partial_\mu\xi \pm \xi\partial_\mu\xi^\dagger\right), \\
B_\mu &= \frac{1}{24\pi^2}\epsilon_{\mu\nu\alpha\beta}\mathrm{Tr}\left\{\Sigma^\dagger\partial^\nu\Sigma\Sigma^\dagger\partial^\alpha\Sigma\Sigma^\dagger\partial^\beta\Sigma\right\}, \\
\Phi_{\mu\nu}^* &= \partial_\mu\Phi_\nu^* - \partial_\nu\Phi_\mu^* + V_\mu\Phi_\nu^* - V_\nu\Phi_\mu^*,
\end{aligned} \tag{6}
$$

with $\epsilon_{0123} = +1$. Here, Σ is the $SU(2)$ chiral field, Φ and Φ_μ^* are, respectively, the pseudoscalar and vector meson doublets of the form $Q\bar{q}$, F_π represents the pion decay constant and g_{Φ^*} is the Φ^* "gauge" coupling to matter fields. For instance, if we take the kaons to be heavy mesons, $\Phi^\dagger = (K^-, \overline{K}^0)$, $\Phi_\mu^{*\dagger} = (K_\mu^{*-}, \overline{K}_\mu^{*0})$. This Lagrangian is obtained from a hidden gauge symmetric Lagrangian [10] by integrating out the ω and ρ meson fields and then taking the limit $m_\Phi = m_{\Phi^*} \to \infty$, neglecting the terms that vanish as m_Φ^{-1} and $m_{\Phi^*}^{-1}$ or faster. For the purpose of comparing with the heavy-quark limit [11] which we will refer to as Isgur-Wise (IW) symmetric limit, it is necessary to keep the vector mesons explicitly instead of integrating them out as done in Scoccola et al.[10] The reason for this will become clear later on.

We need to explain a bit what $\mathcal{L}_{an}$ is in the context of the heavy-meson limit that we are interested in. The first term is what one obtains from the topological Wess-Zumino term written down by Witten [12] when expanded à la Callan-Klebanov

(CK).[13] This is intrinsically tied to anomalies in effective theory. Later, as the heavy quark mass increases, this term will disappear. However the second term, which is intrinsic-parity odd as the Wess-Zumino term is and involves the vectors P^*'s, need not vanish in the heavy-quark limit. We expect it to modify the constants of the main term responsible for the binding of the mesons Φ and Φ^* to a soliton. They are fixed at low energy by low-energy theorems and in the heavy-quark limit by heavy-quark symmetry. *A priori*, there is no reason why they should be related.

To see what remains of the Lagrangian (5) in the heavy-quark limit, we make the meson-field redefinition,[14] (taking $m = m_\Phi = m_{\Phi^*} \to \infty$)

$$\begin{aligned} \Phi_\mu^{*\dagger} &= e^{-imv\cdot x} P_\mu^* / \sqrt{m}, \\ \Phi^\dagger &= e^{-imv\cdot x} P / \sqrt{m}, \end{aligned} \tag{7}$$

so that the fields P_μ^* and P are independent of the meson mass and obtain

$$\mathcal{L}_\Phi^{\rm HGS} = -iPv\cdot \overleftrightarrow{D}\, P^\dagger + iP_\mu^* v\cdot \overleftrightarrow{D}\, P^{*\mu\dagger} + i\sqrt{2}\left(PA^\mu P_\mu^{*\dagger} + P_\mu^* A^\mu P^\dagger\right), \tag{8}$$

$$\mathcal{L}_\omega^{\rm HGS} = \frac{N_c}{F_\pi^2} B_\mu \left(Pv^\mu P^\dagger - P_\nu^* v^\mu P^{*\nu\dagger}\right) \tag{9}$$

$$\mathcal{L}_{an} = \delta\mathcal{L}_{an} \tag{10}$$

where

$$\begin{aligned} (D_\mu P)^\dagger &= (\partial_\mu + V_\mu) P^\dagger, \\ A\, \overleftrightarrow{D}\, B^\dagger &= A(DB)^\dagger - (DA)B^\dagger. \end{aligned} \tag{11}$$

We have not written out the term $\delta\mathcal{L}_{an}$ since while the coefficients are known phenomenologically in the light-quark sector, they are not known in the regime we are concerned with. We expect that it will include terms of the form

$$\frac{iN_c}{F_\pi \pi^2} \epsilon^{\mu\nu\alpha\beta} v_\mu \left(aPA_\nu A_\alpha P_\beta^{*\dagger} - bP_\beta^* A_\alpha A_\nu P^\dagger\right) \tag{12}$$

with a and b unknown constants. Note that since in [9] we started with an apparently $SU(3)$ symmetric Lagrangian (apart from the meson mass term) with the flavor Q put on the same footing as the light quarks, Eq. (9) results from the ω-meson coupling terms and hence the constant N_c/F_π^2 is fixed. In the heavy-quark limit, the "primordial" Wess-Zumino term is absent. However, the rest will remain to modify effectively the coefficient of Eq.(9) which came from the ω-meson coupling with the heavy mesons P and P^*. That such a term must be present can be seen by bosonizing light and heavy quarks from QCD.[15]

If the P and P^* are degenerate, the intrinsic-parity odd Lagrangian in which the Wess-Zumino term figures in the HGS Lagrangian[8] contains a term that survives in the heavy-meson mass limit

$$c_4\mathcal{L}_{(4)} = -c_4 2ig_{\Phi^*}^2 \epsilon^{\mu\nu\alpha\beta} v_\mu P_\nu^* A_\alpha P_\beta^{*\dagger}, \tag{13}$$

with c_4 the coefficient of $\mathcal{L}_{(4)}$.[8] This is effectively a four-derivative term that belongs to the same intrinsic-parity class as $\delta\mathcal{L}_{an}$ discussed above. The coefficient c_4 is fixed in the light-quark sector to $c_4 = iN_c/16\pi^2$ from the decay $\omega \to \rho\pi$. We will see what the coefficient is in the heavy-quark sector.

Our Lagrangian (8)–(13) can now be compared with the one implied by IW symmetry [16,13]

$$\begin{aligned} \mathcal{L}_\Phi^{\rm JMW} &= -i{\rm Tr}\overline{H}_a v^\mu \partial_\mu H_a + i{\rm Tr}\overline{H}_a H_b v^\mu (V_\mu)_{ba} \\ &\quad + ig{\rm Tr}\overline{H}_a H_b \gamma^\mu \gamma_5 (A_\mu)_{ba} + \cdots \quad , \end{aligned} \tag{14}$$

with the heavy meson field H_a (where a labels the light-quark flavor) defined as

$$H = \frac{(1+ \not{v})}{2}\left[P^*{}_\mu \gamma^\mu - P\gamma^5\right]. \tag{15}$$

In terms of P and P^*, (14) reads

$$\begin{aligned} \mathcal{L}_\Phi^{\rm JMW} &= -iPv\cdot \overleftrightarrow{D}\, P^\dagger + iP^*_\mu v\cdot \overleftrightarrow{D}\, P^{*\mu\dagger} \\ &\quad + 2ig\left\{P^*_\mu A^\mu P^\dagger + PA^\mu P^{*\dagger}_\mu\right\} + 2g\epsilon^{\lambda\mu\nu\kappa} v_\lambda P^*_\mu A_\nu P^{*\dagger}_\kappa. \end{aligned} \tag{16}$$

We see that the HGS Lagrangian (8) with (13) is identical – except for the term (9) – to the IW symmetric Lagrangian (16) if we identify $g = 1/\sqrt{2}$ and $c_4 g^2_{\Phi^*} = ig$. Surprisingly $g = 1/\sqrt{2}$ is rather close to the quark-model value $g \sim 0.75$ and also to the recent CLEO collaboration data [17] $g \approx 0.6$. Furthermore low-energy theorem [8] gives $c_4 g^2_{\Phi^*} \approx i0.65 \approx ig$. In (16), the term of the form (9) is missing for the simple reason that B_μ involves three derivatives, so higher order in derivative expansion. Ignoring it in pion dynamics may be justified but it is not in soliton dynamics. In fact we will see later that it is the most important contribution in our way of describing heavy baryons as it is in the Callan-Klebanov scheme.[13] Since we do not know its coefficient in the IW limit, we will take it in the form

$$\begin{aligned} \mathcal{L}_\omega^{\rm HGS} &= \alpha B_\mu j^\mu, & (17) \\ j^\mu &= {\rm Tr}\left(\overline{H} v^\mu H\right). & (18) \end{aligned}$$

with α an unknown constant. The Lagrangian (17) obviously satisfies both chiral symmetry and IW symmetry. Such a term arises in an approximate bosonization of QCD, through the coupling of H to the ω meson.[15] In (17), j^μ is the $U(1)$ current of the Lagrangian $\mathcal{L}_\Phi^{\rm JMW}$ corresponding to the heavy-quark flavor which is conserved in our case. Although as mentioned above, the coupling constant α cannot be determined by chiral and Isgur-Wise symmetries alone, we will analyze the structure of heavy baryons in units of $-N_c/2F_\pi^2$, *i.e.*, the coefficient of $\mathcal{L}_\omega^{\rm HGS}$ in Eq.(9). We will normalize the meson field as

$$\int d^3r j^0 = -2\int d^3r \left(PP^\dagger + P^*_i P^{*\dagger}_i\right) = -1 \tag{19}$$

and work in the rest frame of the heavy meson, $v_\mu = (1,0,0,0)$. Note that $P_0^* = 0$ since $v \cdot P^* = 0$.

The Lagrangian correct to order $O(m_\Phi^0 \cdot N_c^0)$ is given by

$$
\begin{aligned}
L_B &= -\mathcal{M}_{sol} - m_\Phi + \int d^3r (\mathcal{L}_P + \mathcal{L}_W), \\
-\mathcal{L}_P &= 2gi\left\{P^{*i}A^iP^\dagger + PA^iP^{*i\dagger} - i\epsilon^{0ijk}P^{*i}A^jP^{*k\dagger}\right\} \\
-\mathcal{L}_W &= 2\alpha B_0\left(PP^\dagger + P_i^*P_i^{*\dagger}\right).
\end{aligned}
\tag{20}
$$

One can readily see that $\mathcal{L}_P$ and $\mathcal{L}_W$ are invariant with respect to the global rotation $S \in SU(2)_V$ in the light flavor space (*i.e.*, the isospin space) provided that the fields transform

$$
\begin{aligned}
P(x) &= \phi(x)S^\dagger, \\
P_i^*(x) &= \phi_i^*(x)S^\dagger, \\
\xi(x) &= S\xi_0(\vec{x})S^\dagger,
\end{aligned}
\tag{21}
$$

with $x = (t, \vec{x})$ and $\xi_0(\vec{x}) = \exp(i\vec{\tau} \cdot \hat{r}F(r)/2)$ in the hedgehog configuration. Following the standard procedure for collective quantization, we elevate S to a dynamical variable by endowing it with the time dependence $S(t) = a_0(t) + \vec{a}(t) \cdot \vec{\tau}$. As defined, the fields $\phi(x)$ and $\phi_i^*(x)$ are fields living in the rotating frame

The equations of motion for $\phi(x)$ and $\phi^*(x)$ gotten from the Lagrangian valid at $O(m_\Phi^0 \cdot N_c^0)$ imply that

$$
\begin{aligned}
|\phi(x)|^2 &\propto \delta^3(\vec{x}), \\
|\phi_i^*(x)|^2 &\propto \delta^3(\vec{x}).
\end{aligned}
$$

Given these solutions, the energy shift coming from $-(\mathcal{L}_P + \mathcal{L}_W)$ of (20) is

$$
\begin{aligned}
E_I &= -\int (\mathcal{L}_P + \mathcal{L}_W) d^3r \\
&= -\frac{1}{2\pi^2}\alpha\left\{F'(0)\right\}^3
\end{aligned}
\tag{22}
$$

with the contribution of $\mathcal{L}_P$ vanishing identically. This differs from the result of the recent work by Manohar and his collaborators[18] who get the entire action from $\mathcal{L}_P$ whose contribution is non-vanishing since the meson H is not rotated in their scheme in contrast to our scheme (21). I will discuss the difference of these two approaches later.

3.1. *Spectrum in IW limit*

We take $g = 1/\sqrt{2} \simeq 0.7$, $F'(0) \approx -0.89$ GeV from the literature and the experimental value of $F_\pi = 186$ MeV and $e = 4.75$, with which we find the α value in the b-quark sector should be

$$
\alpha \approx -\frac{1}{2.8}\left(\frac{N_c}{2F_\pi^2}\right)
\tag{23}
$$

to reproduce $M_{\Lambda_b} - M_N = 4.65$ GeV, the predicted value of the quark model. This corresponds to the binding energy of

$$E_I \approx -0.55 \text{ GeV}. \tag{24}$$

Next we consider the effects of $O(m_\Phi^0 \cdot N_c^{-1})$ term in the Lagrangian. For this we define

$$S^\dagger \dot{S} = i\vec{\tau} \cdot \vec{\Omega} \tag{25}$$

and write the corresponding Lagrangian to $O(m_\Phi^0 \cdot N_c^{-1})$

$$L_{(-1)} = \int d^3r \mathcal{L}_{(-1)} = 2\mathcal{I}\Omega^2 - 2\vec{\Omega} \cdot \vec{Q}, \tag{26}$$

where

$$\begin{aligned}
\vec{Q} &= -\int d^3r \left(\phi \vec{n}(\vec{r}) \phi^\dagger + \phi_i^* \vec{n}(\vec{r}) \phi_i^{*\dagger}\right), \\
\vec{n} &= \frac{1}{2}\left(\xi_0^\dagger \vec{\tau} \xi_0 + \xi_0 \vec{\tau} \xi_0^\dagger\right) \\
&= \cos F(r) \vec{\tau} - (\cos F(r) - 1)\hat{r}(\vec{\tau} \cdot \hat{r}),
\end{aligned} \tag{27}$$

and $\mathcal{I}$ is the moment of inertia of the $SU(2)$ soliton determined from the properties of the N and Δ. As suggested by Manohar et al,[18] because of the δ-function structure of the meson wavefunctions and a parity-flip at the origin, it is more convenient to transform the heavy-meson fields to

$$\begin{aligned}
\phi &\to \phi' = \phi\, \xi_0, \\
\phi_\mu^* &\to \phi_\mu^{*'} = \phi_\mu^* \, \xi_0, \\
\vec{n} &\to \xi_0 \, \vec{n} \, \xi_0^\dagger.
\end{aligned} \tag{28}$$

The binding energy is not affected by this transformation. With the primed fields, $\vec{Q}$ is of the form

$$\vec{Q} = -\frac{1}{2}\int d^3r \left\{\phi' \left(\Sigma^\dagger \vec{\tau} \Sigma + \vec{\tau}\right) \phi'^\dagger + \phi_i^{*'} \left(\Sigma^\dagger \vec{\tau} \Sigma + \vec{\tau}\right) \phi_i^{*'\dagger}\right\}. \tag{29}$$

Now since in the soliton rotating frame, the "isospin" of the meson is transmuted to spin, we can identify

$$\vec{Q} = c\vec{J}_Q, \tag{30}$$

namely, proportional to the angular momentum lodged in the meson which is 1/2. Canonical quantization of (26) leads to an $O(m_\Phi^0 \cdot N_c^{-1})$ splitting[b] in energy given by the Hamiltonian

$$\Delta H = 2\mathcal{I}\Omega^2 = \frac{1}{2\mathcal{I}}\left(\vec{J}_l + c\vec{J}_Q\right)^2 \tag{31}$$

---

[b] Modulo a hidden m_Φ^{-1} dependence in c explained below.

with the spectrum

$$\Delta E_{hf} = \frac{1}{2\mathcal{I}} \left\{ cJ(J+1) + (1-c)J_\ell(J_\ell+1) + c(c-1)J_Q(J_Q+1) \right\}, \tag{32}$$

where $\vec{J}_\ell$ is the spin lodged in the rotor. The total spin $\vec{J}$ of the system is

$$\vec{J} = \vec{J}_\ell + \vec{J}_Q. \tag{33}$$

The Hamiltonian (31) is a heavy-baryon analog to the diatomic molecular spectrum (3) and c an analog to $(1-\kappa)$. One can show by an explicit calculation that with (29)

$$c = 0. \tag{34}$$

This is the analog of the vanishing $(1-\kappa)$ in diatomic molecules, a consequence of restored rotational symmetry. What happens here is that the first term of (29) coming from the P mesons gets cancelled exactly by the second coming from the P^* mesons. If the P and P^* were not degenerate the cancellation would not occur. This suggests the following: For not too large m_Φ, say, m_K, c can be substantial, of $O(1)$, since the K^* is rather high-lying compared with the K. As m_Φ becomes large, the P^* comes near the P, thus decreasing c such that in the heavy-quark limit, we get $c = 0$.

Given that $c = 0$ in the Isgur-Wise limit, we have the splitting

$$\Delta E_{hf} = \frac{1}{2\mathcal{I}} J_\ell(J_\ell+1). \tag{35}$$

This ΔE_{hf} predicts that there is an effective "fine" splitting of right sign and magnitude between Λ and the degenerate Σ and Σ^*. The predicted mass spectrum (denoting the mass by the particle symbol) for b-quark baryons, with $\Lambda - N = 4.65$ GeV to fix α, is

$$\Sigma_b - N = \Sigma_b^* - N = 4.84 \text{ GeV}. \tag{36}$$

These are comparable to the predictions of quark potential models

$$\begin{aligned}(\Lambda_b - N)^{QM} &= 4.65 \text{ GeV},\\ (\Sigma_b - N)^{QM} &= 4.86 \text{ GeV},\end{aligned}$$

and to those of bag models

$$\begin{aligned}(\Lambda_b - N)^{BM} &= 4.62 \text{ GeV},\\ (\Sigma_b - N)^{BM} &= 4.80 \text{ GeV}.\end{aligned}$$

3.2. *Hyperfine spectrum*

It is possible, within the scheme described so far, to discuss hyperfine splitting with a nonzero c. For a finite heavy-quark mass for which $m_\Phi < m_\Phi^*$, the CK model indicates that $c \sim 1/m_\Phi$. This is the hidden m_Φ^{-1} dependence buried in the hyperfine coefficient c alluded above which we conjecture may have an intricate connection to a Berry potential. For a sufficiently large m_Φ, we may therefore assume $c = a/m_\Phi$. Now using (32), we can write for baryons with one heavy quark Q

$$\Sigma_Q - \Lambda_Q = \frac{1}{\mathcal{I}}(1 - c_Q) \simeq 195\text{MeV}(1 - c_Q). \tag{37}$$

With the experimental value $\Sigma_c - \Lambda_c \approx 168$MeV for the charmed baryons, we get $c_c \simeq 0.14$. This means that with $m_D = 1869$MeV, the constant $a \simeq 262$ MeV. So [c]

$$c_\Phi \simeq 262\text{MeV}/m_\Phi. \tag{38}$$

Now for b-quark baryons, using $m_B = 5279$ MeV, we find $c_b \simeq 0.05$ which with (37) predicts

$$\Sigma_b - \Lambda_b \approx 185\text{MeV}. \tag{39}$$

This agrees well with the quark-model prediction. Furthermore the $\Sigma^* - \Sigma$ splitting comes out correctly also. For instance, it is predicted that

$$\frac{\Sigma_b^* - \Sigma_b}{\Sigma_c^* - \Sigma_c} \simeq \frac{m_D}{m_B} \approx 0.35 \tag{40}$$

to be compared with the quark-model prediction ~ 0.32. If one assumes that the heavy mesons Φ are weakly interacting, then we can put more than one Φ's in the soliton [10] and obtain the spectra for Ξ's and Ω's in good agreement with quark-model results.[19]

4. Discussions

I have discussed how one can continuously "dial" from chiral symmetry to IW symmetry in the spectrum of baryons. This is a surprising outcome. In doing so, a Wess-Zumino like term plays an essential role. For $m_Q < \Lambda_\chi$, the Wess-Zumino term plays a key role in binding a pseudoscalar Φ to an $SU(2)$ soliton. For $m_Q >> \Lambda_\chi$, the Wess-Zumino term vanishes but a term of the form $\text{Tr}\bar{H}v_\mu H B^\mu$ contributes through coupling with the light vector meson ω which given a reasonable strength again dominates the binding. The baryon structure is exactly the same as the one given by the CK model with the meson shrunk to the center of the soliton. It is not clear whether all this is just a coincidence or something profound but it is certainly intriguing.

The description of Manohar and his collaborators [18] differs from the above scenario in that in their scheme, the heavy meson H gets bound to a *rotating* skyrmion through

[c]It is amusing to note that this formula works satisfactorily even for the kaon for which one predicts $c_s \simeq 0.53$ to be compared with the empirical value 0.62.

a residual interaction given by $\mathcal{L}_P$ in (20) with no contribution from a Wess-Zumino like term. Since one starts here with a Lagrangian in the IW symmetry limit, the hyperfine splitting comes from an IW symmetry breaking term of the form

$$\frac{c}{m_H}\mathrm{Tr}(\bar{H}\sigma_{\mu\nu}H\sigma^{\mu\nu}) \tag{41}$$

which splits the degeneracy of P and P^*. Surprisingly the results of both approaches seem rather close. The connection between the two is not yet understood.

In terms of the "Berry charge," the limit $c = 0$ clearly corresponds to the restoration of the IW symmetry, namely the symmetry that makes P and P^* degenerate. It remains to be seen how this result can be obtained in the general setting of Berry potentials in the strong interactions developed recently by Lee, Nowak, Zahed and myself.[20] More work is needed in this area.

A matter of potential interest on which I have no result to discuss is the possibility of having the chiral partner S of H which arises naturally in approximate bosonization of QCD [15] bound to a soliton. A rich variety of spectra associated with this excitation is predicted.

Acknowledgments

I would like to acknowledge fruitful discussions with Hyun Kyu Lee, Dong-Pil Min, Maciej Nowak, Yongseok Oh, Dan Olof Riska, Norberto Scoccola and Ismail Zahed.

References

1. G. Holzwarth, talk at this conference.
2. A. Hayashi, talk at this conference.
3. B. Moussallam, talk at this conference.
4. T. Walhout, talk at this conference.
5. M. Rho, D.O. Riska and N.N. Scoccola, *Z. Phys.* **A341**, 343 (1992).
6. B. Zygelman, *Phys.Rev.Lett.* **64**, 256 (1990).
7. H.K. Lee and M. Rho, "Rotational Symmetry and Nonabelian Berry Potential," Hanyang University preprint HYUPT-92-07.
8. D.-P. Min, Y. Oh, B.-Y. Park and M. Rho, "Soliton Structure of Heavy Baryons," SNUTP-92/78
9. M. Bando, T. Kugo and K. Yamawaki, *Phys. Rep.* **164**, 317 (1988); T. Fujiwara, T. Kugo, H. Terao, S. Uehara and K. Yamawaki, *Prog. Theor. Phys.* **73**, 926 (1985).
10. N.N. Scoccola, D.-P. Min, H. Nadeau and M. Rho, *Nucl. Phys.* **A505**, 497 (1989).
11. N. Isgur and M.B. Wise, *Phys. Rev. Lett.* **66**, 1130 (1991) and references given therein.
12. E. Witten, *Nucl. Phys.* **B223**, 422, 433 (1983).
13. C.G. Callan and I. Klebanov, *Nucl. Phys.* **B262** (1985) 365; C.G. Callan, K. Hornbostel and I. Klebanov, *Phys. Lett.* **B202**, (1988) 269

14. T.-M. Yan, H.-Y. Cheng, C.-Y. Cheung, G.-L. Lin, Y.C. Lin and H.-L. Yu, *Phys. Rev.* **D46**, 1148 (1992).
15. M.A. Nowak, M. Rho and I. Zahed, "Chiral Effective Action with Heavy Quark Symmetry," Stony Brook preprint SUNY-NTG-92/27
16. M.B. Wise, *Phys. Rev.* **D45**, 2118 (1992).
17. See *e.g.* P. Cho and H. Georgi, "Electromagnetic Interactions in heavy-hadron chiral theory," Harvard preprint HUTP-92/A043.
18. E. Jenkins, A.V. Manohar and M.B. Wise, "Baryons Containing a Heavy Quark as Solitons", preprint CALT-68-1783; Z. Guralnik, M. Luke and A.V. Manohar, "Properties of Baryons Containing a Heavy Quark in the Skyrme Model," preprint UCSD/PTH 92-24; E. Jenkins and A.V. Manohar, "Hyperfine Splittings of Baryons Containing a Heavy Quark in the Skyrme Model," UCSD/PTH 92-26.
19. Y. Oh, D.-P. Min, M. Rho and N.N. Scoccola, *Nucl. Phys.* **A534**, 493 (1991).
20. H.K. Lee, M.A. Nowak, M. Rho and I. Zahed, "Induced gauge potentials in baryons," Stony Brook preprint SUNY-NTR-92-50 (1992).

SYMMETRY BREAKING IN A GENERALIZED SKYRME MODEL

J. SCHECHTER

Department of Physics, Syracuse University,
Syracuse, NY 13244-1130, USA

ABSTRACT

We first outline the calculations of the neutron-proton mass difference and of the axial singlet matrix element (relevant to the "proton spin" puzzle) in a generalized Skyrme model of pseudoscalars and vectors. These two calculations are, perhaps surprisingly, linked to each other and furthermore are sensitive to some fine details of symmetry breaking in the effective meson Lagrangian. This provides a motivation for us to examine these symmetry breaking terms more closely. We find a qualitatively new feature in the symmetry breaking pattern of the vector meson system and discuss its significance.

1. Introduction

This material is based on work with Anand Subbaraman and Herbert Weigel which will be described in more detail elsewhere[1]. The original motivation was to update some older papers[2,3,4], which should be consulted for adequate references to background work.

There are three reasons for taking the Skyrme model approach seriously. a) It *works* fairly well for baryon mass *differences*[5], static properties , scattering amplitudes etc. b) the picture is suggested by the consistent large N_c approximation scheme for QCD. c) It makes use of a nice feature of "old physics"– the "pion cloud" of the Yukawa theory – and builds upon it in a constructive way. Roughly speaking, the original Skyrme model replaces the "core" of the nucleon by a boundary condition. We shall consider here that the vector mesons describe more of the core physics. Other approaches utilize explicit quarks for this purpose.

The calculation of the non-electromagnetic part of the neutron-proton mass difference (as given in ref. 2) is reviewed in section 2. This has the interesting aspect that it forces one to go beyond the Skyrme model of pions to include at least the η field and also some information (via the introduction of vector mesons, for example) about "short distance" physics in the model. The extension to the full SU(3) model of pseudoscalars and vectors is, in fact, indicated.

In section 3, the calculation [3,6] of the protons's axial singlet matrix element in the pseudoscalar-vector model is briefly reviewed and the connection with the so-called "proton-spin puzzle[7]" is given. It is shown how a conjectured decomposition of the matrix element into "matter" and "glue" parts is related[4] to the calculation of the $n-p$ mass difference in this model.

In order to improve the accuracy of the calculations discussed in sections 2 and 3, or at the least, to test their sensitivity to reasonable parameter changes it is necessary to employ the *full three flavor* pseudoscalar-vector model. The first step of carefully discussing symmetry breaking in the underlying meson Lagrangian is given in section 4. It is noted that the introduction of derivative type symmetry breaking terms for the vectors dramatically improves the agreement with experiment. This encourages us to suggest that the three-flavor chiral perturbation theory program[8] be extended to include the vectors as well as pseudoscalars so that it has a chance of describing low energy physics up to about 1 GeV. The relevant soliton calculations, using the full three flavor meson Lagrangian, are discussed in the following talk of Herbert Weigel[9].

2. Neutron-proton mass difference

We write

$$m(n) - m(p) = \Delta_{EM} + \Delta \tag{2.1}$$

where Δ_{EM} is due to photon exchange and Δ is due to the different u and d quark masses in the fundamental QCD Lagrangian. Subtracting the conventional estimate[10] for Δ_{EM} gives $\Delta = (2.05 \pm 0.30)$ MeV as the number to be understood in the present calculation. Let us first try to explain this at the two flavor level, which seems very reasonable. But there is a problem: In the original Skyrme model containing only pions, the mesonic term mocking up the fundamental iso-spin breaker has the form $\mathrm{Tr}[\tau_3(U+U^\dagger)]$ which vanishes identically when we plug in $U = \cos\omega + i\underline{n}\cdot\underline{\tau}\sin\omega$ as the general parameterization of a 2×2 unitary unimodular matrix. A way out at the two flavor level would appear to be to include also an isoscalar meson, $\eta \sim (\overline{u}u + \overline{d}d)/\sqrt{2}$. In this case U is no longer unimodular and the mesonic symmetry breaker does not vanish when we plug in the more general form $U = e^{i\chi}(\cos\omega + i\underline{n}\cdot\underline{\tau}\sin\omega)$. In particle language we get a symmetry breaker like $\epsilon\eta\pi^0$. We would expect an $n-p$ splitting in the Yukawa theory from graphs in which the nucleon emits a virtual π^0 which converts, due to the $\epsilon\eta\pi^0$ interaction, into an η which is then reabsorbed by the nucleon. Unfortunately the nucleon matrix elements of the operator $\epsilon\eta\pi^0$ vanish for the original Skyrme Lagrangian; the η field does not get excited. This is because the η field only appears quadratically in the Lagrangian and, for example, the static Hamiltonian terms $\frac{1}{2}(\nabla\eta)^2 + \frac{1}{2}m_\eta^2\eta^2$ are minimized for $\eta = 0$. Clearly a term linear in η is required to act as a source (the term $\epsilon\eta\pi^0$ does so to a small extent but leads to $\Delta = \mathcal{O}(\epsilon^2)$ which is negligible). It is interesting that exciting an η forces us to extend our original model. We may think of this effect as arising from the "core" of the nucleon, described, for example, by vector mesons. It was shown that in a model of pseudoscalars *and* vectors there are terms[11] linear in η proportional to the Levi-Civita tensor,

$$\epsilon_{\mu\nu\alpha\beta}\partial_\mu\eta(\omega_\nu\partial_\alpha\omega_\beta + \text{ many others})\ .$$

These terms do excite the η field to the required strong order when account is taken of the need to "crank" the η field. This amounts to allowing the η to get excited by centrifugal effects; the ansatz

$$\eta(\underline{r}) = \eta(r)\hat{\underline{r}} \cdot \underline{\Omega}, \tag{2.2}$$

where $\frac{i}{2}\tau_a \Omega_a = A^\dagger \dot{A}$ is the angular velocity, (A is defined from $U(t) = A(t)U_c(\underline{x})A^\dagger(t)$, $U_c(x)$ being the classical soliton solution) is substituted in and the profile $\eta(r)$ is chosen so as to maximize the moment of inertia. Using (2.2) in the symmetry breaking terms finally gives an iso-spin breaking piece in the collective Hamiltonian of the form

$$H_{SB} = -\Delta I_3, \tag{2.3}$$

where I_3 is the iso-spin *operator*. Note that there will now be a number of different types of symmetry breaking terms in our effective meson Lagrangian. For example there will also be an $\omega\rho^0$ transition piece and others involving more derivatives. Several years ago, Δ was estimated[2] on this basis to be about 1.3 MeV, just a little low compared to (2.1). It was also shown that the extra amount was likely due to inclusion of the strange degree of freedom. This is plausible in the Yukawa theory since K meson exchange with unequal K^+ and K^0 meson masses is also expected to contribute to Δ.

However there are a number of interesting complications when we consider the three flavor case in the generalized Skyrme model approach. First, of course, we have both η and η' fields appearing. Structurally, the η' appears in the mesonic Lagrangian as the phase of the 3×3 matrix U in analogy to the η in the two flavor case. But there is now a qualitatively new feature in that one has a contribution to Δ even without cranking the η. This is proportional to

$$< n|D_{38}(A)|n > - < p|D_{38}(A)|p >, \tag{2.4}$$

where $D_{ab}(A) = \frac{1}{2}$ Tr $(\lambda_a A \lambda_b A^\dagger)$ is the SU(3) adjoint representation matrix element. The λ_a are the Gell-Mann matrices. Evaluating (2.4) using for $|n>$ and $|p>$ the eigenstates of the symmetric (no symmetry breaking terms) collective Hamiltonian of the pseudoscalar only Skyrme model actually gives the fairly large value of about 1.3 MeV as the contribution to Δ. This might suggest that perhaps there was no need to excite the η field by extending the model to include vectors. However that is wrong. Yabu and Ando[12] showed that the baryon eigenstates are very sensitive to symmetry breaking. When the collective Hamiltonian is diagonalized *exactly* the contribution to Δ is reduced by a factor of $\frac{1}{2}$. Furthermore, improving the Yabu-Ando calculation by cranking the K mesons leads[13] to another factor of $\frac{1}{2}$ reduction (in the model of pseudoscalars only). Altogether we thus *estimate*

$$\Delta \approx 1.3 + (1.3)/4 \approx 1.6 \text{ MeV}, \tag{2.5}$$

where the first piece is due to cranking the η at the two flavor level and the second piece is due to (2.4). A similar estimate was presented in ref. 2. We stress that this is just an estimate since the first part was calculated in a two flavor model of vectors and pseudoscalars while the second was gotten from a three flavor model of pseudoscalars only. It is thus desirable to see what the result would be in a treatment using the full three flavor model with both pseudoscalars and vectors throughout. This is part of our motivation for a more thorough analysis of symmetry breaking in the full model.

3. Axial Singlet Matrix Element

The current interest in this "hot topic" will turn out to perhaps justify the complicated meanderings to which we have exposed the reader in the last section. The form factors of the ath flavor axial current in the proton are defined by

$$\sqrt{\frac{p_0 p_0' V^2}{M_N^2}} < p'|i\overline{q}_a \gamma_\mu \gamma_5 q_a|p >= i\overline{u}(\underline{p}')[\gamma_\mu \gamma_5 H_a(q^2) + \frac{iq_\mu}{2M_N}\gamma_5 \widetilde{H}_a(q^2)]u(\underline{p}). \qquad (3.1)$$

The *singlet* form factors are defined by

$$H(q^2) = \sum_{a=1}^{3} H_a(q^2); \ \widetilde{H}(q^2) = \sum_{a=1}^{3} \widetilde{H}_a(q^2), \qquad (3.2)$$

where $(1,2,3) = (u,d,s)$.

A great deal of excitement was caused by the EMC (European Muon Collaboration) experiment[14] on deep inelastic polarized μ - polarized p scattering. The result can be interpreted as a measurement of

$$\frac{4}{9}H_1(0) + \frac{1}{9}H_2(0) + \frac{1}{9}H_3(0). \qquad (3.3)$$

Another linear combination of these form factors is known from neutron beta decay:

$$H_1(0) - H_2(0) \simeq 1.25, \qquad (3.4)$$

while a third can be estimated from hyperon beta decay plus SU(3) flavor covariance:

$$R \equiv H_1(0) + H_2(0) - 2H_3(0) \approx 0.68. \qquad (3.5)$$

For the first time it was possible to get information on the matrix element of the axial *singlet* current, J_μ^5. This is especially interesting since J_μ^5 obeys the *anomalous* divergence equation

$$\partial_\mu J_\mu^5 = 2i \sum_{a=1}^{3} m_a \overline{q}_a \gamma_5 q_a + \partial_\mu K_\mu, \qquad (3.6)$$

where K_μ is the Chern-Simons current of the Yang-Mills theory. Note that $\partial_\mu J_\mu^5$ does not vanish when the m_a go to zero. From the EMC result for (3.3) together with (3.4) and (3.5) one gets:

$$H(0) = 0.03 \pm 0.18, \qquad (3.7a)$$

$$H_3(0) = -0.22 \pm 0.06. \tag{3.7b}$$

Eq. (3.7a) is surprising since it is expected to be 1.0 in the naive non-relativistic quark model wherein $\frac{1}{2}H(0)$ is identified as the proton expectation value of the *quark spin* part of the total angular momentum operator. Also, (3.7b) is surprising since the "strangeness content" of the proton is expected to be small. It seems to us that, of these two surprising results, only the first is a reliable deduction from the measurement of (3.3). That is because the use of flavor SU(3) to obtain R in (3.5) is debatable. The SU(3) Skyrme model, in fact, suggests[15] much smaller values of R (while not giving big SU(3) deviations for the flavor *changing* currents which enter into the Cabibbo theory). Taking, somewhat arbitrarily, $R = 0.3$ would lead to the deduction from experiment

$$H(0) = 0.12 \pm 0.18, \tag{3.8a}$$

$$H_3(0) = -0.06 \pm 0.06. \tag{3.8b}$$

Eq. (3.8a) is *still* small and surprising while (3.8b) is no longer surprising!

Now in the Skyrme model containing just pseudoscalar fields, there is a nice result[16]:

$$H(0) = 0. \tag{3.9}$$

So, what is *mysterious* in the quark model is the *expected* feature in the Skyrme model. Let us try to understand this from a different[17] point of view. In the Skyrme model of psudoscalars the η' field enters as $U = \exp\left(\frac{2i\eta'}{\sqrt{3}F_\pi}\right)\widetilde{U}$, where $\det \widetilde{U} = 1$. By Noether's theorem the axial singlet current is simply

$$J_\mu^5 = -\sqrt{3}F_\pi \frac{\partial \mathcal{L}}{\partial(\partial_\mu \eta')} = \sqrt{3}F_\pi \partial_\mu \eta', \tag{3.10}$$

where the second step holds because η' enters only quadratically in $\mathcal{L}$ (we just pick up its kinetic term for (3.10)). Now the result (3.9) is evident since J_μ^5 is a pure gradient and hence proportional to q_μ in momentum space. From the form factor decomposition in (3.1) we see that (3.10) only contributes to the "induced" form factor $\widetilde{H}(q^2)$ and hence $H(q^2)$ must vanish in this simple model.

But this is troublesome from the point of view of the $n-p$ calculation described in the last section. There we *needed* a term linear in η' in order to excite the pseudoscalar iso-singlet. We saw that there *was* such a term when vectors were added; it can be written as

$$\mathcal{L} = \ldots - \frac{1}{\sqrt{3}F_\pi}\partial_\mu \eta' \widetilde{J}_\mu^5. \tag{3.11}$$

Here $\widetilde{J}_\mu^5$ is proportional to $\epsilon_{\mu\nu\alpha\beta}$ and is due to the vector fields. It can be identified, using (3.10), as a new (short distance) contribution to J_μ^5:

$$J_\mu^5 = \sqrt{3}F_\pi \partial_\mu \eta' + \widetilde{J}_\mu^5. \tag{3.12}$$

Since $\widetilde{J}_\mu^5$ is not a pure gradient[18] it does contribute to $H(0)$. Does this destroy the nice prediction of the Skyrme model? A couple of years ago it was found that $H(0) \approx 0.30$

in this model.[3,6] This is still qualitatively small (and in agreement with experiment if (3.8a) is taken rather than (3.7a)).

Another aspect of this problem which the Skyrme model approach might illuminate concerns the so-called "2-component decomposition".[19] This is an attempt to make the small value of $H(0)$ intuitively plausible from the QCD parton model point of view. It is certainly legal to write:

$$< p|J_\mu^5|p > = < p|(J_\mu^5 - K_\mu)|p > + < p|K_\mu|p > . \tag{3.13}$$

Since the first term on the right hand side is conserved when the light quark masses are neglected it is tempting to consider it, in some sense, the usual "matter" contribution and to be around the naive quark model value of 1. It is hoped that this would be largely cancelled by the second "glue" term. There has been considerable discussion of this possibility in the literature.[20] One criticism is that the decomposition (3.13) is not gauge invariant. It has been suggested that a gauge invariant and hence better way is to give an analogous decomposition based on a generalized Goldberger-Treiman relation. This reads[21]:

$$H(0) = \frac{\sqrt{3}F_\pi}{2M_N}\left(g_{\eta' NN} - \sqrt{3}F_\pi m_{\eta'}^2 g_{GNN}\right), \tag{3.14}$$

where $g_{\eta' NN}$ is the Yukawa coupling constant of the (pure SU(3) singlet) η' field with the nucleons while g_{GNN} is the Yukawa coupling constant for the composite glue field $G \equiv \partial_\mu K_\mu$. The first term is supposed to be the "matter" contribution, the second the "glue" piece. Now it is very difficult[22] to obtain a reliable value for $g_{\eta' NN}$ from experiment and even harder to get an experimental handle on g_{GNN}. However it is possible to make a theoretical estimate in the generalized Skyrme model of pseudovectors and vectors. We showed[4] that the existence of the second term in (3.14) corresponds to the presence of a new term,

$$\frac{t}{3F_\pi^2 m_{\eta'}^2}\partial_\mu G \widetilde{J}_\mu^5, \tag{3.15}$$

in the effective Lagrangian. t is a new dimensionless constant. The effect of this term is to provide an extra contribution to the η' excitation (the "auxiliary" field G must be eliminated by its equation of motion in terms of η' to see this) but *not* to the current, J_μ^5. This means that the η contribution to Δ in (2.3) should be multiplied by a factor of $(1-t)$. Thus we can try to improve the predicted value of Δ by adjusting t. This predicts the coupling constants in the two component decomposition as[4]:

$$g_{\eta' NN} = (1-t)\frac{2M_N H(0)}{\sqrt{3}F_\pi},$$

$$g_{GNN} = \frac{t}{t-1}\frac{g_{\eta' NN}}{\sqrt{3}F_\pi m_{\eta'}}. \tag{3.16}$$

Here $H(0)$ should still be taken to be 0.30 as discussed after (3.12). Finally, "improving" the prediction for Δ (assuming the central value, 2.05 MeV) gives[4] for (3.14):

$$H(0) = \text{“matter”} + \text{“glue”}$$
$$0.3 = 0.4 - 0.1. \tag{3.17}$$

This indicates a rather small gluonic as well as a small matter contribution. We should stress, that since this estimate is related to perhaps arcane details of symmetry breaking and η' excitation, (see the discussion at the end of section 2) it is desirable to investigate further the sensitivity of the result to these factors.

4. Back to the Meson Lagrangian

The first step in a more detailed investigation of symmetry breaking in the generalized Skyrme model is clearly to examine the underlying meson Lagrangian itself. Since the soliton energy is in the neighborhood of 1 GeV it would seem prudent for the meson Lagrangian to adequately describe mesonic physics up to this energy. This would imply that we include in addition to the pseudoscalars, the vectors (because, among other things, they are "there"). We will also include the composite glue field G in order to simplify the implementation of the $U(1)$ axial anomaly. (A composite scalar glue field, $H = \Theta_{\mu\mu}$ can also, if desired, be introduced in order to conveniently implement the trace anomaly).

The dynamical variables we use in the Lagrangian are, in addition to the chiral field U, $\xi \equiv U^{1/2}$ and a vector meson nonet, ρ_μ related to linearly transforming objects A^L_μ and A^R_μ by

$$A^L_\mu = \xi\rho_\mu\xi^\dagger + \frac{i}{\tilde{g}}\xi\partial_\mu\xi^\dagger; \; A^R_\mu = \xi^\dagger\rho_\mu\xi + \frac{i}{\tilde{g}}\xi^\dagger\partial_\mu\xi. \tag{4.1}$$

The chiral symmetric and OZI rule conserving terms of the Lagrangian have been given in many places.[23] Here we shall concentrate our discussion on the symmetry breaking terms. For this purpose we want to introduce some notation associated with the mass terms of the fundamental QCD Lagrangian:

$$\mathcal{L}_{\text{mass}} = -m_u\bar{u}u - m_d\bar{d}d - m_s\bar{s}s \equiv -\hat{m}\bar{q}\mathcal{M}q, \tag{4.2}$$

where $\hat{m} = (m_u + m_d)/2$ and the dimensionless matrix $\mathcal{M}$ is written as

$$\mathcal{M} = y\lambda_3 + T + xS,$$
$$\lambda_3 = \text{ diag } (1,-1,0), \; T = \text{ diag } (1,1,0), \; S = \text{ diag } (0,0,1),$$
$$x = m_s/\hat{m}, \; y = \frac{m_u - m_d}{m_u + m_d}. \tag{4.3}$$

In addition we note the redundant, but convenient, quantity[10],

$$R = \frac{m_s - \hat{m}}{m_d - m_u} = \frac{\text{“strange” splitting}}{\text{isospin splitting}}. \tag{4.4}$$

The interest of R lies in the fact that it may, in principle, be straightforwardly extracted from SU(3) multiplets other than 0^- and 1^-. We have introduced so much notation for $\mathcal{M}$ since all symmetry breaking terms will be taken to be proportional to it. Rather than writing down all symmetry breaking terms as in the CPT (chiral perturbation theory) program[8] of pseudoscalars only, we shall only write the ones we believe are most important. These are the ones which (except for those needed to solve the η' problem) obey the OZI rule. Incidentally, we may note that the OZI violating symmetry breaking terms turn out to have small coefficients in the CPT approach too. We implement the OZI rule in Okubo's original form[24]: all mesons are described by nonets and only terms which are a single trace in flavor space are included. This eliminates the "hairpin" diagrams. Without further ado we list the OZI rule conserving, symmetry breaking terms to be included in the effective Lagrangian:

$$\alpha' \text{ Tr } [\mathcal{M}(A^L_\mu U A^R_\mu + A^R_\mu U^\dagger A^L_\mu)] + \beta' \text{ Tr } [\mathcal{M}(\partial_\mu U \partial_\mu U^\dagger U + U^\dagger \partial_\mu U \partial_\mu U^\dagger)]$$
$$+\gamma' \text{ Tr } [\mathcal{M}(F^L_{\mu\nu} U F^R_{\mu\nu} + F^R_{\mu\nu} U^\dagger F^L_{\mu\nu})] + \delta' \text{ Tr } [\mathcal{M}(U + U^\dagger - 2)]$$
$$+\lambda'^2 \text{ Tr } [\mathcal{M}U^\dagger \mathcal{M}U^\dagger + \mathcal{M}U\mathcal{M}U - 2\mathcal{M}^2], \tag{4.5}$$

where

$$F^{L,R}_{\mu\nu} = \partial_\mu A^{L,R}_\nu - \partial_\nu A^{L,R}_\mu - i\tilde{g}[A^{L,R}_\mu, A^{L,R}_\nu].$$

The explanation for these terms is as follows. First, the δ' term is the usual, main symmetry breaker for the pseudoscalars. The α' term plays the same role for the vectors. However, noting (4.1), the α' term also supplies an undesirably large amount of derivative-type symmetry breaking for the pseudoscalars. Most of this is cancelled by the β' term. Together, the α' and β' terms enable us to fit both F_K and F_π. The γ' term is a derivative type symmetry breaker for the vectors. It was set to zero for simplicity in ref. 2 but we shall see here that it may be rather important. Finally, the $(\lambda')^2$ term, though of second order in $\mathcal{M}$, is argued to be of the same order as the β' term in the CPT program[8] ($\mathcal{M}$ counts as two derivatives). This is reasonable because the symmetry breaking for the pseudoscalars is relatively large. Such a term may be less important for the vectors but, in any event, it would have the form μ' Tr $(\mathcal{M}A^L_\mu \mathcal{M}A^R_\mu)$.

Now our task is to fit the quantities

$$\alpha', \beta', \gamma', \delta', \lambda'^2; \ \tilde{g}, m_v; \ x = \frac{m_s}{\hat{m}}, \ y = \frac{m_u - m_d}{m_u + m_d},$$

where $\tilde{g}$ is the "gauge coupling constant" in (4.1) and m_v is a "bare" vector meson mass, from the experimental mass spectrum of 0^- and 1^- mesons, the 0^- decay constants and the $V \to \phi\phi$ decay widths. Since there are many unknowns one might think that the work could be simplified if we had reliable information on the quark mass ratios x and y from outside the 0^- and 1^- multiplets. Let us make a small ***detour*** to see if this is possible.

Certainly the best source of information is the baryon octet. One may make a perturbation around zero quark masses, $m_u = m_d = m_s = 0$ (chiral perturbation theory) or around, say, the symmetric point

$$m_u = m_d = m_s = \frac{1}{3}(m_u + m_d + m_s), \tag{4.6}$$

which we may denote as "flavor" perturbation theory. We can then use group theory (generalized Wigner-Eckart theorem) to derive mass relations in a well known way. To first order in perturbation it doesn't matter which scheme is used and we find the Gell-Mann Okubo formula

$$\Lambda - N = \frac{1}{3}[2(\Xi - N) - (\Sigma - N)]$$
$$176.9 \text{ MeV} = 168.1 \text{ MeV}\ , \tag{4.7}$$

(particle symbols stand for their masses) and also

$$R = \frac{m_s - \hat{m}}{m_d - m_u} = \frac{\Xi - \Sigma}{n - p} = \frac{\Sigma - N}{\Xi^- - \Xi^0}$$
$$60.7 \pm 9.1 = 46.2 \pm 5.9. \tag{4.8}$$

In (4.8) the photon exchange contributions have been subtracted out of the iso-spin violating mass differences. The large deviation between the central values in (4.8) shows that the values of R indicated there may not be reliable. Together with the observed deviation in (4.7) this suggests that we go on to second order. There is a known difficulty[8,10] in going beyond first order in the CPT scheme since non-analytic terms like $m^{\frac{3}{2}}$ and $m \ln m$ arise. Since these are due to massless Goldstone boson exchange it would seem that this effect is already taken into account at zeroth order if we expand around the point (4.6). Recently, a new relation at second order was found[25]:

$$R = \frac{3\Lambda + \Sigma - 2N - 2\Xi}{2\sqrt{3}m_T + (n - p) + (\Xi^0 - \Xi^-)}, \tag{4.9}$$

where m_T is the "corrected" value of the $\Lambda - \Sigma^0$ transition mass. In principle, m_T may be determined from precision measurements of the difference between, say, $pK^- \to \Lambda\eta$ and $n\overline{K}^0 \to \Lambda\eta$. Thus (4.9) may be predictive in the future. For the moment the best we can do is to use the structure of (4.9) and the assumption that second order quantities don't deviate by more than 20% from their first order values to derive a lower bound,

$$R > 38 \pm 10. \tag{4.10}$$

However this isn't very precise and we are forced to end our detour by concluding that the quark mass ratios are not known well enough from "outside" to be useful for our purpose. The fundamental quark mass ratios should, in fact, thus be obtained as results from our fit.

The formulas for the meson one, two and three point functions used for fitting are fairly standard and are explicitly given in ref. 1. It is useful to keep $x = m_s/\hat{m}$ as a parameter during the fit and to evaluate everything for each value of x. Then once the other decay widths and masses are used as inputs we have four remaining predictions[1] for each x. These are displayed below with the experimental values indicated in parentheses:

x	$(K^{*0} - K^{*+})_{\text{non-EM}}$	$\frac{\Gamma(\rho\to 2\pi)}{\Gamma(K^*\to K\pi)}$	$\frac{\Gamma(\rho\to 2\pi)}{\Gamma(\phi\to K\bar{K})}$	$m(\phi)$
20	2.44 MeV	4.97	163	1.02 GeV (1.02)
25	3.05	4.45	81	1.04
30	3.97	3.90	57	1.09
32	4.47	3.67	47	1.13
36	5.84 $\begin{pmatrix} 5.2 \\ 7.4 \end{pmatrix}$	3.21 (3.0)	28 (40.3)	1.26

(Note that the experimental value for the non-electromagnetic part of the $K^{*0} - K^{*+}$ mass difference is 5.2 MeV if the individual masses given in the Particle Data Tables are simply subtracted but is 7.4 MeV if attention is restricted to the "dedicated" experiments.) We see that three quantities are well fit for x in the range $34 - 38$ but that $m(\phi)$ is fit for $x = 20$. A compromise "best fit" is characterized by the fundamental quark mass ratios:

$$x = 31.5, \qquad y = -0.42, \qquad R = 36. \tag{4.11}$$

This may be compared with the values obtained by Gasser and Leutwyler in their big Physics Report[10]:

$$x = 25.0 \pm 2.5, \qquad y = -0.28 \pm 0.03, \qquad R = 43.5 \pm 2.2. \tag{4.12}$$

As a kind of check we note that we would also get y= -0.27 when we set $x = 25$. Note that a recent determination[26] using ψ decays finds $y = -0.54 \pm 0.09$ which is much closer to our result than to (4.12).

To see what physics lies behind this "numerology" let us attempt to understand why our result (4.11) differs from (4.12). We remark that the Gasser-Leutwyler values are essentially what one gets by applying first order perturbation theory to the vector multiplet. However, there are some problems with this. For one thing the SU(3) relation between $(M_{\rho\omega})_{\text{non-EM}}$, the quark mass induced part of the $\rho\omega$ transition mass, and $(K^{*0} - K^{*+})_{\text{non-EM}}$ appears to be very badly broken. In fact, they used[10] $(M_{\rho\omega})_{\text{non-EM}}$ as an input and assumed something was wrong with the extraction of $(K^{*0} - K^{*+})_{\text{non-EM}}$ from experiment. Furthermore even though the width relation between $\Gamma(K^* \to K\pi)$ and $\Gamma(\rho \to \pi\pi)$ was one of the historical successes of SU(3) invariance, the measured width of the ρ has increased from 100 MeV to 150 MeV since those days and the simple SU(3) relation no longer holds. Also the ordinary SU(3) width relation between $\Gamma(\phi \to K\overline{K})$ and $\Gamma(\rho \to 2\pi)$ does not hold. In the present approach $(K^{*0} - K^{*+})$, $\Gamma(K^* \to K\pi)/\Gamma(\rho \to 2\pi)$ and

$\Gamma(\phi \to K\overline{K})/\Gamma(\rho \to 2\pi)$ can be explained with the help of a relatively important derivative symmetry breaker for the vectors (the γ' term in (4.5)). This modifies the SU(3) relations via non-trivial wave-function renormalizations for vector mesons containing strange quarks; explicitly[1],

$$Z_\rho = 0.99, \qquad Z_{K^*} = 0.84, \qquad Z_\phi = 0.65 \tag{4.13}$$

for the best fit parameters (Note, the physical ρ field, ρ_p, is given by $\rho_p = Z_\rho \rho$ etc.). To fine tune our prediction for $m(\phi)$, more exotic terms and loop corrections should be included. In any case it is clear that the above feature represents a new qualitative effect for describing the vector nonet system.

For our purposes the η' -OZI rule violating-interactions are also very important. To mock up the axial anomaly and produce an η' mass we add the terms:

$$\frac{1}{\kappa}G^2 + \frac{i}{12}G \ \ln(\det U/\det U^\dagger) + n \ \mathrm{Tr}\ (\alpha_\mu)\ \mathrm{Tr}\ (\alpha_\mu) + i\epsilon' G \ \mathrm{Tr}\ [\mathcal{M}(U - U^\dagger)], \tag{4.14}$$

where $\alpha_\mu \equiv \partial_\mu U U^\dagger$ and $G = \partial_\mu K_\mu$. No kinetic energy terms are included for G so it acts as an auxiliary field and gets eliminated by its equation of motion:

$$G = \frac{\sqrt{3}\kappa}{8} F_\pi \eta' - \frac{i\kappa\epsilon'}{2}\ \mathrm{Tr}\ [\mathcal{M}(U - U^\dagger)], \tag{4.15}$$

which is to be substituted back into (4.14). The Lagrangian which results[27] when one keeps just the first two terms in (4.14) (i.e. setting $n = \epsilon' = 0$) has been known for quite a while to be able to explain the η' mass and the $\eta - \eta'$ mixing angle but not [28] the η mass. With the two additional terms[29] we are now able to also explain the η mass as well as $\pi^0 \to \gamma\gamma$, $\eta \to \gamma\gamma$ and $\eta' \to \gamma\gamma$. Note that substituting (4.15) into (4.14) yields, among other things, a symmetry breaker of "type 7" but not one of "type 6" in the second order CPT program[8]. Thus the assumption that the OZI rule violating interactions are dominated by the pseudoscalar iso-singlet channel (and given by (4.14)) evidently selects a unique "frame" from the one parameter family allowed by the Kaplan-Manohar ambiguity[30].

With the present Lagrangian explaining more features of low energy meson physics, we may feel more confident about investigating symmetry breaking in the soliton sector. This aspect is discussed in refs. 1 and 9. Qualitatively, the effect of the important γ' term in (4.5) is to decrease the relative contribution of the η excitation ($\propto \delta'$) to the $n - p$ mass difference while leaving the net prediction about the same. This has the consequence[1] that, while the "glue" contribution to $H(0)$ remains relatively small, it could be somewhat larger than the value in (3.17).

The question of symmetry breaking in the pseudoscalar-vector system is certainly of interest outside the soliton sector too. It seems interesting to also calculate loop diagrams and include other "second-order" interaction terms (although we believe we have the dominant symmetry breakers) to construct an analog of the CPT scheme which would be useful up to about 1 GeV. Of course, this covers a lot of physics so we expect that progress in this direction will be incremental or evolutionary in character.

Acknowledgements

I would like to thank the organizers for holding an exceptionally pleasant and worthwhile workshop.

This work was supported in part by the U.S. Department of Energy under contract number DE-FG-02-85ER40231.

References

1. J. Schechter, A. Subbaraman and H. Weigel, Syracuse-Tübingen report, in preparation.
2. P. Jain, R. Johnson, N.W. Park, J. Schechter and H. Weigel, *Phys. Rev.* **D40** 855 (1989).
3. R. Johnson, N.W.Park, J. Schechter, V. Soni and H. Weigel, *Phys. Rev.* **D42**, 2998 (1990).
4. J. Schechter, V. Soni, A. Subbaraman and H. Weigel, *Phys. Rev. Lett.* **65**, 2955 (1990).
5. Progress, discussed in the talks of G. Holzwarth and B. Moussallam at this workshop, has been made on the problem of too large absolute baryon masses.
6. V. Bernard, N. Kaiser and U.-G. Meissner, *Phys. Lett.* **B237**, 545 (1990).
7. Reviews are provided by R. Jaffe and A. Manohar, *Nucl. Phys.* **B337**, 509 (1990); G. Altarelli and W.J. Stirling, CERN report Th 5249/88; H.-Y. Cheng, Taipei report IP-ASTP-01-91. A "brief review" of the material in sec. 3 is given in J. Schechter, V. Soni, A. Subbaraman and H. Weigel, *Mod. Phys. Lett. A* **7**, 1 (1992); E **7**, 1199 (1992).
8. References may be traced from J. Gasser and H. Leutwyler, *Nucl. Phys.* **B250**, 465 (1985).
9. Talk of H. Weigel at this workshop.
10. This is reviewed in J. Gasser and H. Leutwyler, *Phys. Rep.* **87**, 77 (1982).
11. See (5.6) of P. Jain, R. Johnson, Ulf-G. Meissner, N.W. Park and J. Schechter, *Phys. Rev.* **D 37**, 3252 (1988) and also Ulf-G. Meissner, N. Kaiser, H. Weigel and J. Schechter, *Phys. Rev.* **D39**, 1956 (1989).

12. H. Yabu and K. Ando, *Nucl. Phys.* **B301**, 601 (1988).

13. H. Weigel, J. Schechter, N.W. Park and Ulf-G. Meissner, *Phys. Rev.* **D42**, 3177 (1990).

14. E.M. Collaboration, J. Ashman et al., *Phys. Lett.* **B206**, 364 (1988); *Nucl. Phys.* **B328**, 1 (1989).

15. N.W. Park, J. Schechter and H. Weigel, *Phys. Lett.* **B228**, 420 (1989); *Phys. Rev.* **D41**, 2836 (1990); *Phys. Rev.* **D43**, 869 (1991); N.W. Park and H. Weigel, *Nucl. Phys.* **A541**, 453 (1992). The possibility of such a situation in the quark model approach is discussed by H. Lipkin, *Phys. Lett.* **B256**, 284 (1991).

16. S. Brodsky, J. Ellis and M. Karliner, *Phys. Lett.* **B206**, 309 (1988).

17. See the first of ref. 15 and T. Cohen and M. Banerjee, *Phys. Lett.* **B230**, 129 (1989).

18. It is explicitly given in (4.5) of ref. 3.

19. R. Carlitz, J. Collins and A. Mueller, *Phys. Lett.* **B214**, 229 (1988); A.V. Efremov and D.J. Teryaev, Dubna report JINR E2-88-297; G. Altarelli and G.G. Ross, *Phys. Lett.* **B212**, 391 (1989).

20. In addition to the reviews in ref. 7, see J. Mandula, *Phys. Rev. Lett.* **65**, 1403 (1990).

21. G.M. Shore and G. Veneziano, *Phys. Lett.* **B244**, 75 (1990). See also J. Schechter, V. Soni, A. Subbaraman and H. Weigel, *Mod. Phys. Lett.* **A5**, 2543 (1990); T. Hatsuda, Univ. of Washington report INT-5-90.

22. See A. Efremov, J. Soffer and N. Tornqvist, *Phys. Rev. Lett.* **64**, 1495 (1990).

23. See ref. 11 and references therein.

24. S. Okubo, *Phys. Lett.* **5**, 165 (1963).

25. J. Schechter and A. Subbaraman, *Int. Jour. Mod. Phys. A* to be published.

26. J. Donoghue and D. Wyler, *Phys. Rev.* **D45**, 892 (1992).

27. C. Rosenzweig, J. Schechter and G. Trahern, *Phys. Rev.* **D21**, 3388 (1980); P. Di Vecchia and G. Veneziano, *Nucl. Phys.* **253** (1980); E. Witten, *Ann. Phys.* **128**, 1789 (1981); P. Nath and R. Arnowitt, *Phys. Rev.* **D23**, 473 (1981).

28. V. Mirelli and J. Schechter, *Phys. Rev.* **D15**, 1361 (1977).

29. See also P. Di Vecchia, F. Nicodemi, R. Pettorino and G. Veneziano, *Nucl. Phys.* **B181**, 318 (1981).

30. D. Kaplan and A. Manohar, *Phys. Rev. Lett.* **56**, 2004 (1986).

VECTOR MESONS IN THE COLLECTIVE APPROACH TO SU(3)

HERBERT WEIGEL
Institut for Theoretical Physics, Tübingen University
D-7400 Tübingen, FR Germany

ABSTRACT

We study the SU(3) extension of the Skyrme model with vector mesons in the collective quantization scheme. The parameters of the model are fixed in its mesonic sector. Fields which are excited by the collective rotation of the classical soliton are determined by a variational method. The resulting collective Hamiltonian which includes various symmetry breaking terms is diagonalized exactly. This yields reasonable predictions for the mass differences of the low-lying baryons. We furthermore investigate static properties of these baryons in this model by evaluating all relevant matrix elements using the exact eigenstates of the collective Hamiltonian. Electromagnetic observables are found to approximately obey the predictions obtained from SU(3) symmetry. Strong deviations from SU(3) symmetry are, however, found for the matrix elements of strange vector and scalar operators indicating a small amount of strangeness in the proton. The exploration of axial currents shows that the results of the EMC experiment may be explained with a small polarization of the strange sea of the proton without violating the successful Cabibbo model for semileptonic hyperon decays.

1. Introduction and Motivation

Assuming the point of view that Quantum Chromo Dynamics (QCD) is well approximated by an effective meson Lagrangian[1,2] with baryons emerging as soliton solutions[3,4] we shall investigate such a Lagrangian for the case of three flavors. The exploration of SU(3) models is very interesting since it provides extensive information about the low lying $\frac{1}{2}^+$ and $\frac{3}{2}^+$ baryons. Of course, the first results extracted from such a model concern the baryon mass spectrum, however, it is possible to go one step further and obtain predictions for static properties like electromagnetic form factors or matrix elements of the axial current which are measured in semileptonic hyperon decays. In view of recent or upcoming experiments it is very challenging to consider the structure of the proton in a three flavor model. At this point the famous EMC measurement[5] of scattering polarized muons off polarized protons which led to the "proton spin puzzle" and experiments on electron-hadron scattering via neutral currents[6] are to be mentioned. SU(3) models furthermore may shed some light on the strangeness content of the proton which appears to be an important ingredient for the determination[7] of the pion-nucleon Σ term, $\Sigma_{\pi N}$ as well as for parity violation processes[8].

The first step in constructing an effective meson action naturally consists of employing the non-linear $\sigma-$model which allows for solitonic solutions once stabilizing terms are added. Most commonly Skyrme's[3] antisymmetric fourth order term is employed for this purpose but also a six-order term has frequently been used.[9] Since these additional terms are of higher orders in the derivative operator and are only restricted by demanding them to be chirally symmetric they introduce arbitrary parameters which cannot be determined from mesonic observables. However, stability of the soliton may as well be gained by the incorporation of vector meson fields like ρ and ω.[10,11] The stability is then provided by the "anomalous" terms in the action which contain the Levi-Cevita tensor $\epsilon_{\mu\nu\rho\sigma}$. In the meson sector these terms describe vertices like $\omega \rightarrow \pi\pi\pi$ or $\omega \rightarrow \rho\pi$. The experimentally determined decay widths for these processes fix the associated parameters. An additional coupling constant appearing in the non-anomalous sector may be determined from the decay $\rho \rightarrow \pi\pi$. Furthermore it is important to note that for the description of baryons in the SU(2) model vector mesons have been proven to be necessary ingredients to improve the predictions for static properties of the nucleon[12] (*e.g.* electromagnetic form factors), the neutron proton mass difference[13] and the "high" energy behavior of the phase shifts in πN scattering.[14,15]

Unfortunately the extension to SU(3) cannot be done straightforwardly due to the presence of flavor symmetry breaking. In models consisting of pseudoscalar mesons only, two (on first sight quite different) treatments of SU(3) symmetry breaking in the baryon sector have been relatively successful. In the bound state (BS) approach collective coordinates are only introduced for the rotations into "non-strange" directions, the real zero modes.[16] Canonical quantization of these coordinates yields states with good quantum numbers for spin and isospin. A kaonic bound state in the background field of the SU(2) soliton is constructed and each occupation of this bound state lowers the hypercharge by one unit. On the other hand, in the Yabu and Ando (YA) approach to symmetry breaking SU(3) is considered approximative and collective coordinates are therefore introduced for the complete SU(3) rotation.[17] Employing an "Euler angle" parametrization of these rotations the collective Hamiltonian which contains all relevant information about flavor symmetry breaking may be diagonalized exactly. The YA treatment furthermore allows to easily take the zero symmetry breaking limit and thus it enables one to compare with ordinary SU(3) baryon phenomenology.[18] This comparison shows that the $\frac{1}{2}^+$ ($\frac{3}{2}^+$) baryons are no longer pure octet (decouplet) states but contain significant admixtures of higher dimensional representations like $\overline{10}$ or 27 ($27, 35, \overline{35}$). A detailed study of the YA approach in models with pseudoscalar mesons only[19,20] has shown that symmetry breaking effects are large for mass differences and matrix elements of strangeness conserving operators but negligible for those of strangeness changing operators, needed e.g. in the Cabibbo model of hyperon decays. In view of the success of the YA treatment in the pseudoscalar model we shall generalize this approach to models containing vector mesons as well.

Gauging the vector meson Lagrangian with external fields allows to read off the corresponding currents. For the coupling of vector mesons to photons from the non-anomalous terms this approximately yields the phenomenologically successful concept of vector meson dominance (VMD).[21] It is therefore very interesting to investigate electromagnetic properties of strange and non-strange baryons since the predictions for magnetic moments and electromagnetic radii are expected to improve due to VMD.[12,22] In pure pseudoscalar models the magnitudes of these observables commonly come out too small.[23] Furthermore the vector meson model appears to give more reliable information on the axial singlet current matrix element[24,25] as well as the two flavor contribution to the neutron proton mass difference[13] which both vanish identically in pseudoscalar models.[13,26]

This talk is organized as follows. In section 2) we shall construct the mesonic action as well as the soliton solution in the baryon sector. The baryon mass spectrum is explored in section 3). Section 4) contains the the discussion of the static properties of the low-lying $\frac{1}{2}^+$ and $\frac{3}{2}^+$ baryons. A few conclusive remarks are found in section 5).

2. The Model

2.1. *The mesonic part*

The mesonic action, Γ, which we shall adopt here, was first presented in ref.[27] in its manifest chirally invariant form. The authors employed the non-linear realization $U = \exp(i\sqrt{2}\phi/f_\pi)$ of the pseudoscalar nonet field ϕ as well as left and right transforming vector mesons fields A_μ^L and A_μ^R, respectively. The axial vector meson was eliminated by imposing the non-linear constraint:

$$A_\mu^R = U^\dagger A_\mu^L U + \frac{i}{g} U^\dagger \partial_\mu U, \tag{1}$$

g being a coupling constant to be determined later. The constraint (1) may be regarded as the defining equation for the vector meson nonet field ρ_μ by writing:

$$\begin{aligned} A_\mu^L &= \xi \rho_\mu \xi^\dagger + \tfrac{i}{g} \xi \partial_\mu \xi^\dagger \\ A_\mu^R &= \xi^\dagger \rho_\mu \xi + \tfrac{i}{g} \xi^\dagger \partial_\mu \xi, \end{aligned} \tag{2}$$

wherein $\xi = U^{\frac{1}{2}}$. Defining furthermore, for convenience:

$$\begin{aligned} p_\mu &= \partial_\mu \xi \xi^\dagger + \xi^\dagger \partial_\mu \xi, \quad v_\mu = \xi^\dagger \partial_\mu \xi - \partial_\mu \xi \xi^\dagger \\ R_\mu &= \rho_\mu - \frac{i}{2g} v_\mu, \quad F_{\mu\nu}(\rho) = \partial_\mu \rho_\nu - \partial_\nu \rho_\mu - ig[\rho_\mu, \rho_\nu], \end{aligned} \tag{3}$$

the quantities v_μ and p_μ obviously transform as vector and pseudovector fields.

We may now list the mesonic action Γ by splitting it up into four pieces:

$$\Gamma = \int d^4x (\mathcal{L}_S + \mathcal{L}_{SB} + \mathcal{L}_G) + \Gamma_{an}. \tag{4}$$

The symmetric, non-anomalous part of the Lagrangian is given by:

$$\mathcal{L}_S = -\frac{f_\pi^2}{4} Tr(p_\mu p^\mu) - \frac{1}{2} Tr[F_{\mu\nu}(\rho) F^{\mu\nu}(\rho)] + m_V^2 Tr(R_\mu R^\mu), \tag{5}$$

while the anomalous part of the action is most clearly arranged by using the notation of one-forms ($p = p_\mu dx^\mu$, etc.):

$$\begin{aligned}\Gamma_{an} &= \frac{iN_c}{240\pi^2}\int Tr(p^5)\\ &+\int Tr[\frac{1}{6}(\gamma_1+\frac{3}{2}\gamma_2)Rp^3 - \frac{ig}{4}\gamma_2 F(\rho)(pR-Rp) - g^2(\gamma_2+2\gamma_3)R^3 p], \quad (6)\end{aligned}$$

wherein the first expression on the RHS represents the Wess-Zumino term, an integral over a five dimensional manifold whose boundary is Minkowski space.

To complete the SU(3) action we include the symmetry breaking terms as discussed in ref.[13] which, by using the parametrization (3), acquire the form:

$$\begin{aligned}\mathcal{L}_{SB} = Tr\Big[&-2R_\mu R^\mu(\alpha' T^+ + \alpha'' S^+) - \tfrac{i}{g}[R_\mu, p^\mu](\alpha' T^- + \alpha'' S^-)\\ &+p_\mu p^\mu[(\beta' - \tfrac{\alpha'}{2g^2})T^+ + (\beta'' - \tfrac{\alpha''}{2g^2})S^+] + \delta'(T^+ - 2T) + \delta''(S^+ - 2S)\Big], \quad (7)\end{aligned}$$

where we defined, for convenience, $T^\pm = \xi T\xi \pm \xi^\dagger T\xi^\dagger$, $S^\pm = \xi S\xi \pm \xi^\dagger S\xi^\dagger$ and $T = diag(1,1,0), S = diag(0,0,1)$. Under chiral $SU(3)_L \times SU(3)_R$, $\mathcal{L}_{SB}$ transforms according to the representation $(3,3^*) + (3^*,3)$.

The parameters appearing in $\mathcal{L}_S$ and $\mathcal{L}_{SB}$ may be determined by noting, *e.g.*, that the physical pion decay constant $f_{\pi p} = 0.093GeV$ and the vector meson masses $m_\rho = m_\omega = 0.773GeV$ are related to the parameters in (5) and (7) via:

$$f_{\pi p}^2 = f_\pi^2 + 4(\frac{\alpha'}{g^2} - 2\beta'), \quad m_\rho^2 = m_V^2 - 4\alpha'. \quad (8)$$

Together with the condition that the symmetry breaking matrices are proportional to the current quark mass matrix $\mathcal{M} = diag(m_u, m_d, m_s)$, imposing:

$$\alpha' : \alpha'' = \beta' : \beta'' = \delta' : \delta'' \stackrel{!}{=} \hat{m} : m_s. \quad (9)$$

with $\hat{m} = \frac{1}{2}(m_u + m_d)$. At this point it is important to mention that the quark masses do *not* serve as input quantities, on the contrary, the ratio $\hat{m} : m_s$ will be a prediction. This ratio is identical to the quantity x in ref.[28]. Noting furthermore that the third term in (5) contains the $\rho\pi\pi$ vertex, which may be used to determine g, one finds:[13]

$$\begin{aligned} f_\pi &\approx 0.092GeV, \quad m_V \approx 0.766GeV, \quad g \approx 5.57,\\ \alpha' &\approx -0.0028GeV^2, \quad \alpha'' \approx -0.102GeV^2,\\ \beta' &\approx -7.14\times 10^{-5}GeV^2, \quad \beta'' \approx -0.00266GeV^2,\\ \delta' &\approx 4.15\times 10^{-5}GeV^4, \quad \delta'' \approx 1.55\times 10^{-3}GeV^4, \end{aligned} \quad (10)$$

wherein errors due to uncertainties in the decay constants and masses are not denoted.

In ref.[11] two of the three unknown constants $\gamma_{1,2,3}$ were determined from purely strong interaction processes. Defining $\tilde{h} = -\frac{2\sqrt{2}}{3}\gamma_1$, $\tilde{g}_{VV\phi} = g\gamma_2$ and $\kappa = \frac{\gamma_3}{\gamma_2}$ the central values $\tilde{h} = 0.4$ and $\tilde{g}_{VV\phi} = 1.9$ were found. Within experimental uncertainties (stemming from the uncertainty in the $\omega-\phi$ mixing angle) these may vary in the range $\tilde{h} = -0.15, .., 0.7$ and $\tilde{g}_{VV\phi} = 1.3, .., 2.2$ subject to the constraint $|\tilde{g}_{VV\phi}-\tilde{h}| \approx 1.5$. The third parameter, κ could not be fixed in the meson sector, however, from the study[12] of nucleon properties in the U(2) reduction of the model it was argued that $\kappa \approx 1$. Furthermore note that the overall signs of γ_i cannot be fixed from the knowledge of mesonic properties.

Finally we may add a term to mock up the $U(1)$ anomaly:[29]

$$\mathcal{L}_G = \frac{G^2}{12f_\pi^2 m_{\eta'}^2} + \frac{i}{12} G \ln\Big(\frac{\det U}{\det U^\dagger}\Big). \tag{11}$$

The pseudoscalar glueball G is related to the QCD field strength tensor $F_{\mu\nu}$ and coupling constant g_{QCD} by:

$$G = \frac{3ig_{QCD}^2}{16\pi^2}\epsilon_{\mu\nu\rho\sigma}\mathrm{tr}(F^{\mu\nu}F^{\rho\sigma}) \tag{12}$$

obviously there is no kinetic term for G and thus it may be eliminated via the equation of motion:

$$G = 2\sqrt{3}m_{\eta'}^2\eta' \tag{13}$$

providing the mass term for pseudoscalar singlet η'.

2.2. *The baryon sector*

We assume the well established point of view that baryons emerge as soliton solutions of the mesonic theory[2,3,4]. In the model under consideration this solution is found[10] by invoking the hedgehog *ansätze* in the isospin subgroup of SU(3):

$$\xi_\pi(\mathbf{r}) = \exp(i\hat{\mathbf{r}}\cdot\boldsymbol{\tau}F(r)/2), \quad \omega_0 = \frac{\omega(r)}{2g}, \quad \rho_{i,a} = \frac{G(r)}{gr}\epsilon_{ija}\hat{r}_j. \tag{14}$$

Substitution of these *ansätze* yields the classical mass M_{cl}. An analytic expression for M_{cl} may be found in appendix A of ref.[30]. Minimization of M_{cl} yields Euler-Lagrange equations which are integrated in conjunction with boundary conditions appropriate to the baryon number one sector. The resulting profiles F, ω and G are shown in figure 1.

The field configuration (14) obviously does not lead to states with either good spin or isospin. In order to project on spin and SU(3) quantum numbers the zero modes are quantized semiclassically via the introduction of suitable time dependent collective coordinates. As motivated in the introductory section, we consider SU(3) as an approximative symmetry. Therefore we introduce collective coordinates for the rotations in SU(3) and not only $SU_I(2) \times U_Y(1)$ which are the real zero modes of the Lagrangian.

Special care is required concerning fields which vanish as long as only static configurations are considered. These fields may, however, be excited by the time-dependent

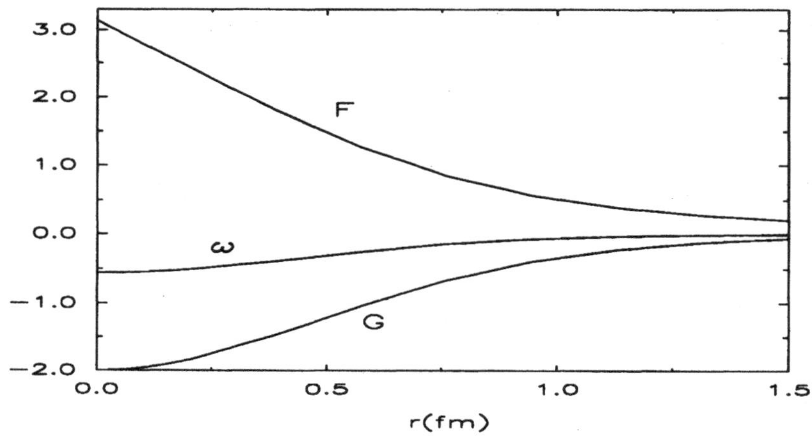

Fig. 1. The profiles of the classical soliton. ω is measured in GeV. The input parameters are as in eq. 28.

rotation of the classical fields F, ω and G. Besides introducing collective coordinates $A(t) \in SU(3)$ we generalize the *ansätze* for the fields in the body-fixed frame via:[19]

$$\xi(\mathbf{r},t) = A(t)\xi_k\xi_\pi(\mathbf{r})\xi_k A^\dagger(t) \tag{15}$$

The unitary (but not unimodular) matrix ξ_k is most conveniently parametrized by

$$\xi_k = e^{iz}, \quad z = \begin{pmatrix} \eta_T & K \\ K^\dagger & \eta_S \end{pmatrix}. \tag{16}$$

For the vector fields we write:[30,31]

$$\rho_\mu(\mathbf{r},t) = A(t)\begin{pmatrix} \rho_\mu^\pi + \omega_\mu & K_\mu^* \\ K_\mu^{*\dagger} & 0 \end{pmatrix} A^\dagger(t). \tag{17}$$

The time dependence of the collective rotations is exhibited most clearly by defining angular velocities Ω_a (a=1,...,8):

$$A^\dagger \dot{A} = \frac{i}{2}\sum_{a=1}^{8}\lambda_a\Omega_a = i\begin{pmatrix} \Omega_\pi + \Omega_\eta & \Omega_K \\ \Omega_K^\dagger & -2\Omega_\eta \end{pmatrix}. \tag{18}$$

The pseudoscalar nonet contains both, components which are excited by the isospin rotation

$$\eta_T = \frac{1}{4}(\chi + \chi_8)\hat{\mathbf{r}}\cdot\boldsymbol{\Omega}, \quad \eta_S = \frac{1}{4}(\chi - 2\chi_8)\hat{\mathbf{r}}\cdot\boldsymbol{\Omega}, \tag{19}$$

as well as components which are excited by the rotations into strange direction

$$K = W(r)\hat{\mathbf{r}} \cdot \boldsymbol{\tau}\Omega_K. \tag{20}$$

In view of the constraint (13) the decomposition of the η fields into singlet (χ) and octet (χ_8) seems to be most appropriate. From the parity and isospin properties we get the most general *ansatz* for the vector mesons which are induced by the isospin rotation:[22]

$$\rho_0^\pi = \frac{1}{2g}[\xi_1(r)\boldsymbol{\Omega} + \xi_2(r)(\hat{\mathbf{r}} \cdot \boldsymbol{\Omega})\hat{\mathbf{r}}] \cdot \boldsymbol{\tau} \quad \omega_i = \frac{\Phi(r)}{2g}\epsilon_{ijk}\Omega_j\hat{r}_k. \tag{21}$$

Suitable *ansätze* for the vector meson fields which are excited by the kaonic angular velocity Ω_K are given by[31]

$$K_0^* = \frac{S(r)}{\tilde{g}}\Omega_K, \quad K_i^* = \frac{1}{2\tilde{g}}[iE(r)\hat{r}_i + \frac{D(r)}{r}\epsilon_{ijk}\hat{r}_j\tau_k]\Omega_K. \tag{22}$$

Substituting these field configurations into the action and expanding up to second order in the angular velocities yields the collective Lagrangian:

$$\begin{aligned} L = & - M_{cl} + \frac{1}{2}\alpha^2\sum_{i=1}^{3}\Omega_i^2 + \frac{1}{2}\beta^2\sum_{\alpha=4}^{7}\Omega_\alpha^2 - \frac{\sqrt{3}}{2}\Omega_8 - \frac{\gamma}{2}[1 - D_{88}(A)] \\ & + \alpha_1\sum_{i=1}^{3} D_{8i}\Omega_i + \beta_1\sum_{\alpha=4}^{7} D_{8\alpha}\Omega_\alpha \end{aligned} \tag{23}$$

D_{ij} denote the rotation matrices $\frac{1}{2}Tr(\lambda_i A \lambda_j A^\dagger)$. Terms of the order (symmetry breaking)$\times$(angular velocity)^2 have been neglected.[a]

The non-strange moment of inertia, α^2 is a functional of the induced fields ξ_1, ξ_2, Φ, η_1 and η_2 with the classical profiles (14) as source fields: $\alpha^2[\xi_1, \xi_2, \Phi, \eta_1, \eta_2; F, \omega, G]$. The excitations are obtained by extremizing α^2 together with boundary conditions which render α^2 finite. The radial dependence of the excitations is displayed in figure 2 for a typical set of parameters (28). In the same manner extremization of the strange moment of inertia, $\beta^2 = \beta^2[W, S, E, D; F, \omega, G]$ provides the radial functions W, S, E and D which are shown in figure 3. Details of these calculations as well as explicit expression for α^2 and β^2 may be found in ref.[30].

3. The Baryon Mass Spectrum

In order to obtain the baryon mass spectrum we need to derive the collective Hamiltonian from the collective Lagrangian (23). Canonical quantization is most conveniently carried out by introducing SU(3) right generators, R_a ($a = 1, .., 8$):

$$R_a = -\frac{\partial L}{\partial \Omega_a} = \begin{cases} -(\alpha^2\Omega_a + \alpha_1^0 D_{8a}) = -J_a, & \text{a=1,2,3} \\ -(\beta^2\Omega_a + \beta_1^0 D_{8a}), & \text{a=4,..,7} \\ \frac{1}{2}\sqrt{3}, & \text{a=8,} \end{cases} \tag{24}$$

[a]Note that only the classical fields are contained in α_1 and β_1 in order to preserve isospin invariance(cf. appendix C of ref.[19]).

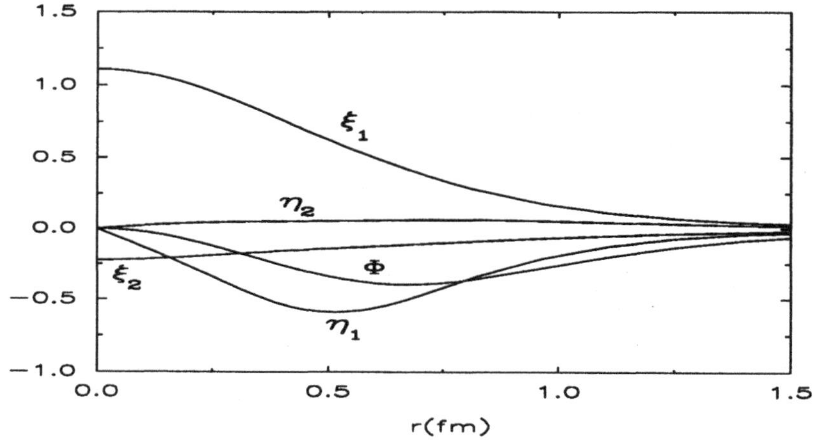

Fig. 2. The non-strange excitations $\xi_1, \xi_2, \Phi, \eta_1$ and η_2 as functions of the radial distance r. The radial functions ξ_1 and ξ_2 are dimensionless while Φ, η_1 and η_2 are measured in GeV^{-1}.

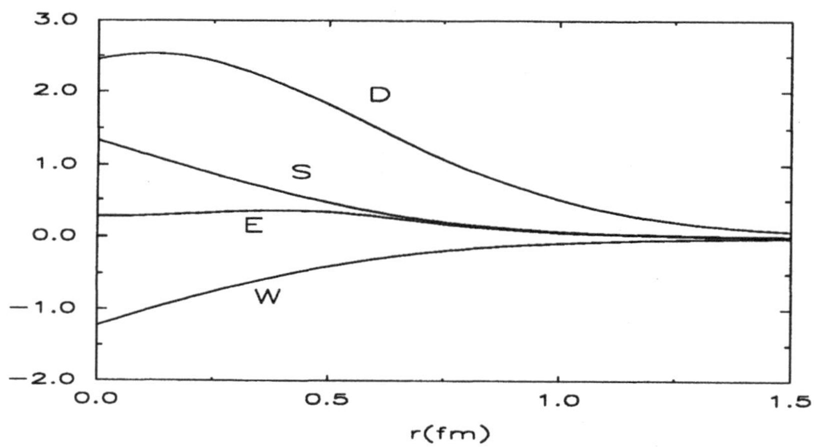

Fig. 3. The strange excitations W, S, E and D as functions of the radial distance r. The radial functions S and E are dimensionless while W and D are measured in GeV^{-1}.

wherein J_i $(i = 1,2,3)$, are the spin operators. It is helpful to define the eigenvalues of

$$C_2 + \beta^2\gamma(1 - D_{88}) + \beta^2\frac{\alpha_1^0}{\alpha^2}\sum_{i=1}^{3} D_{8i}(2R_i + \alpha_1^0 D_{8i}) + \beta_1^0 \sum_{\alpha=4}^{7} D_{8\alpha}(2R_\alpha + \beta_1^0 D_{8\alpha}) \quad (25)$$

as ϵ_{SB}. $C_2 = \sum_{a=1}^{8} R_a^2$ denotes the quadratic Casimir operator of SU(3). ϵ_{SB} may be obtained by introducing an "Euler angle" parametrization of the collective rotation matrices A.[17] Then the generators R_a are expressed in terms of partial differential operators[b] Furthermore the operator (25) is a second order differential operator whose explicit form depends on the isospin multiplet under consideration. Integrating these differential equations finally gives ϵ_{SB}. The eigenfunctions are "distored" SU(3) D-functions. Of course, this procedure represents just a generalization of the original YA treatment who only took the symmetry breaker $\beta^2\gamma(1 - D_{88})$ into account.[17] The collective Hamiltonian:

$$H = -\sum_{a=1}^{8} R_a\Omega_a - L \quad (26)$$

is now easily diagonalized exactly, resulting in the energy formula:

$$E = M_{cl} + \frac{1}{2}\left(\frac{1}{\alpha^2} - \frac{1}{\beta^2}\right)J(J+1) - \frac{3}{8\beta^2} + \frac{1}{2\beta^2}\epsilon_{SB}. \quad (27)$$

It is well known that this procedure gives too large absolute values for the masses of the baryons. However, it has been shown in the simple SU(2) model of pseudoscalars only[32] that there are significant subtractions originating from quantum corrections due to pion loops. Moreover, these subtructions are dominated by the $O(N_C^0)$ contribution and are therefore identical for nucleon and Δ.[33] Thus it is suggestive and justified to only consider mass differences.

In table 1 we list our numerical results for the mass differences of the low-lying $\frac{1}{2}^+$ and $\frac{3}{2}^+$ baryons with respect to the nucleon. The second column in table 1 displays the best fit to the mass spectrum which is obtained for the input parameters:

$$\begin{aligned}
\tilde{h} &= 0.3, \quad \tilde{g}_{VV\phi} = 1.8, \quad \kappa = 1.2, \\
\alpha' &= -0.0028GeV^2, \quad \alpha'' = -0.100GeV^2, \\
\beta' &= -7.14 \times 10^{-5}GeV^2, \quad \beta'' = -0.00270GeV^2, \\
\delta' &= 4.15 \times 10^{-5}GeV^4, \quad \delta'' = 1.55 \times 10^{-3}GeV^4,
\end{aligned} \quad (28)$$

$f_{\pi p}, m_V$ and g remaining unchanged. It should be noted that this set is compatible with the parameters needed to describe the relevant observables in the meson sector. For the central values of input parameters (10) the splitting between multiplets of different spin is predicted too large. This is due to the small non-strange moment of inertia α^2. Changing to parameter set (28) leads to a larger soliton. This increases α^2

[b]Explicit expressions are given in appendix B of ref.[20].

Table 1. The mass differences of various baryons with respect to the nucleon in MeV compared to experimental data. The equation numbers refer to the input parameters and PS denotes the results in the pseudoscalar model of ref.[19].

	eq. 10	eq. 28	Expt.	PS
Λ	136	159	177	157
Σ	252	270	254	238
Ξ	349	398	379	371
Δ	351	311	293	246
Σ^*	467	448	446	381
Ξ^*	586	592	591	520
Ω	681	718	733	661

and gives therefore a better fit to the $N-\Delta$ splitting. The larger soliton also provides an increased symmetry breaking parameter γ leading to a better description of the SU(3) mass differences. Obviously the problem is reversed in the pseudoscalar model (denoted PS in table 1): For a suitable γ the non-strange moment of inertia is too large. Amusingly we find in the vector meson model that the predicted overall mass difference in the spin $\frac{1}{2}$ multiplet, $M_\Xi - M_N = 398MeV$ is larger than the experimental value ($379MeV$) while for the spin $\frac{3}{2}$ multiplet our result for the overall splitting, $M_\Omega - M_\Delta = 407MeV$ is smaller than the experimental number ($440MeV$). The fact that the model describes the (relative) mass spectrum reasonably well provides a strong motivation to also explore static properties of $\frac{1}{2}^+$ baryons.

4. Static Properties of Baryons

Most of the static properties of baryons are related to matrix elements of Noether currents. It is therefore essential to extract these currents from the Lagrangian. This is conveniently performed by adding external gauge fields for local vector (V) and axial vector (A) transformations: $B_\mu^{(V,A)} = \sum_{a=0}^{8} B_\mu^{(V,A)a} Q_a$ ($Q_0 = diag(\frac{1}{3}, \frac{1}{3}, \frac{1}{3})$ and $Q_a = \frac{\lambda_a}{2}$ ($a = 1,..,8$), with λ_a being the Gell-Mann matrices) such that the total action $\Gamma[U, A^L, B^{(V)}, B^{(A)}]$ is gauge invariant.[11] The covariant expressions for the vector (V_μ^a) and axial vector (A_μ^a) currents are then read off as the coefficient of the terms linear in the gauge fields:[12]

$$V_\mu^a = \frac{\delta\Gamma(U, A^L, B^{(V)}, B^{(A)})}{\delta B^{(V)a\mu}}\Big|_{B^{(V,A)}=0}, \qquad A_\mu^a = \frac{\delta\Gamma(U, A^L, B^{(V)}, B^{(A)})}{\delta B^{(A)a\mu}}\Big|_{B^{(V,A)}=0}. \quad (29)$$

We should mention that gauging the anomalous part of the action (6) allows for a non-minimal term which vanishes as long as only strong fields are present (cf. eq. 3.11 of ref.[11].). This new term gives rise for a new coupling constant d_1 which, however, may be determined from the decay width of $\omega \rightarrow \pi^0\gamma$.[12]

4.1. *Electromagnetic properties of $\frac{1}{2}^+$ baryons*

Since the generator for the $U_{\text{e.m.}}(1)$ gauge transformation in SU(3) is given by $Q_{\text{e.m.}} = diag(2/3, -1/3, -1/3)$ the electromagnetic current $J_\mu^{\text{e.m.}}$ is related to a special linear combination of vector currents:

$$J_\mu^{\text{e.m.}} = V_\mu^3 + \frac{1}{\sqrt{3}} V_\mu^8. \tag{30}$$

The electromagnetic form factors of baryon B are defined as the matrix elements of $J_\mu^{\text{e.m.}}$:

$$\frac{\sqrt{p_0 p_0'}}{M_B} \langle B(\mathbf{p}') | J_\mu^{em}(0) | B(\mathbf{p}) \rangle = i\bar{u}(\mathbf{p}')[\gamma_\mu F_1^B(Q^2) + \frac{\sigma_{\mu\nu} Q^\nu}{2M_B} F_2^B(Q^2)] u(\mathbf{p}) \tag{31}$$

where $Q_\mu = p_\mu - p'_\mu$ is the momentum transfer and M_B is the physical mass of baryon B. For the discussion of electromagnetic properties the introduction of electric, $G_E^B(Q^2)$, and magnetic, $G_M^B(Q^2)$, form factors is more suitable:

$$\begin{aligned} G_E^B(Q^2) &= F_1^B(Q^2) - \frac{Q^2}{4M_B^2} F_2^B(Q^2) \\ G_M^B(Q^2) &= F_1^B(Q^2) + F_2^B(Q^2). \end{aligned} \tag{32}$$

At zero momentum transfer the electric form factor yields the charge of B, while the magnetic moment is obtained from:

$$\mu_B = G_M^B(0). \tag{33}$$

To evaluate $G_M^B(Q^2)$ we require the Fourier transformation of the space components of the electromagnetic current, $J_i^{\text{e.m.}}$ in the Breit frame. We thus substitute the *ansätze* for the pseudoscalar and vector meson fields into the expression for the current gained from eq. (29). The angular velocities are eliminated via the quantization rule (24) and matrix elements of the operators are evaluated using the *exact* eigenstates of the collective Hamiltonian (26). Details of this calculation as well as analytic expressions for the currents are given in ref.[30] .

Table 2 contains our numerical results for the magnetic moments using the parameters (28). As expected, the incorporation of vector mesons supplies the desired effect of increasing the contribution from the isovector current, V_i^3 to the magnetic moments while leaving the isoscalar part, V_i^8 almost unaltered. This leads to improved predictions for the magnetic moments of $p, n, \Lambda, \Sigma^+, \Sigma^-$ and also the $\Sigma^0 \to \Lambda$ transition moment. However, we observe that most of our predictions do not deviate significantly from the SU(3) relations:[34]

$$\begin{aligned} \mu_{\Sigma^+} &= \mu_p \quad \mu_{\Sigma^0} = \frac{1}{2}(\mu_{\Sigma^+} + \mu_{\Sigma^-}), \quad \mu_{\Sigma^-} = \mu_{\Xi^-}, \\ 2\mu_\Lambda &= -(\mu_{\Sigma^+} + \mu_{\Sigma^-}) = -2\mu_{\Sigma^0} = \mu_n = \mu_{\Xi^0} = \frac{2}{\sqrt{3}} \mu_{\Sigma^0 \to \Lambda} \end{aligned} \tag{34}$$

Table 2. The magnetic moments for the spin $\frac{1}{2}$ baryons predicted by our model compared to experimental numbers. Also the results obtained in a model of pseudoscalars(PS) only are listed (ref.[20]).

	This model	Expt.	PS
p	2.36	2.79	2.03
n	-1.87	-1.91	-1.58
Λ	-0.60	-0.61	-0.71
Σ^+	2.41	2.42±0.05	1.99
Σ^0	0.66	—	0.60
Σ^-	-1.10	-1.16±0.03	-0.79
Ξ^0	-1.96	-1.25±0.01	-1.55
Ξ^-	-0.84	-0.68±0.03	-0.64
$\Sigma^0 \to \Lambda$	-1.74	-1.61±0.08	-1.39

although the baryon wave functions are strongly distorted SU(3) D-functions. A major disagreement from the relations (34) is only found for the relation involving the neutron whose magnetic moment, in agreement with experiment, is much smaller than $-(\mu_{\Sigma^+} + \mu_{\Sigma^-})$. Unfortunately $\mu_n = \mu_{\Xi^0}$ is still satisfied approximately, predicting a too large value for μ_{Ξ^0}. Also the deviation from $\mu_p = \mu_{\Sigma^+}$ is only small and opposite to the direction required by experiment, resulting in a somewhat too small μ_p. Recent investigations have shown that a deviation from (34) in the desired direction and thus a fine tuning of the magnetic moments may be achieved when the shape of the profile functions is sensitive to symmetry breaking.[35,36] Nevertheless, we may state that our model provides a reasonable description of the magnetic moments of $\frac{1}{2}^+$ baryons.

The vector meson fields play an even more pronounced role for the momentum dependence of the form factors. This is, for example, indicated by the magnetic radii:

$$\langle r^2 \rangle_M^B = -\frac{6}{\mu_B}\frac{\partial G_M^B(Q^2)}{\partial Q^2} \tag{35}$$

the results for different baryons are listed in table 3. As shown in ref.[22] vector mesons provide an additional contribution to the squared radii (of the order $6/m_V^2$) due to a mechanism comparable to VMD. This effect also shows up in the SU(3) model thereby slightly overestimating the experimental values for the magnetic radii of the proton and the neutron. Also the fact that we are dealing with a large soliton is recognizable from the predicted values of the magnetic radii. In any event the vector meson model gives significantly better agreement with experiment than the model containing only pseudoscalar fields.

Next we consider the electric form factors $G_E^B(Q^2)$ which are obtained as the Fourier transform of $J_0^{\text{e.m.}}$. Again we employ the exact eigenstates of the collective Hamiltonian to evaluate the matrix elements of the relevant SU(3) operators. In table 4 we

Table 3. Same as table 2) for the magnetic radii defined in eq. 35.

	This model	Expt.	PS
p	0.94	0.74	0.43
n	0.94	0.77	0.43
Λ	0.78	—	0.36
Σ^+	0.96	—	0.45
Σ^0	0.86	—	0.36
Σ^-	1.07	—	0.58
Ξ^0	0.90	—	0.43
Ξ^-	0.84	—	0.38
$\Sigma^0 \to \Lambda$	0.97	—	0.48

Table 4. Same as table 2) for the electric radii defined in eq. 36.

	This model	Expt.	PS
p	1.20	0.74	0.59
n	-0.15	-0.12	-0.22
Λ	-0.06	—	-0.08
Σ^+	1.20	—	0.59
Σ^0	-0.01	—	-0.02
Σ^-	-1.21	—	-0.63
Ξ^0	-0.10	—	-0.15
Ξ^-	-1.21	—	-0.49

display the results for the electric radii defined by:

$$\langle r^2 \rangle_E^B = -6 \frac{\partial G_E^B(Q^2)}{\partial Q^2}. \tag{36}$$

The absolute values of the radii of charged particles are drastically increased by the inclusion of vector mesons demonstrating once again that the soliton is quite large. On the other hand the experimental value for the neutron electric radius is nicely reproduced. This may also be observed from figure 4 where the momentum dependence of the proton and neutron electic form factors are shown. While there is some deviation of the predicted proton electric form factor from the dipole fit: $1/(1 + Q^2/0.71\text{GeV}^2)^2$, the neutron electric form factor follows closely the phenomenological parametrization of Platchkov et al.[37] who extracted the neutron electric form factor from the deuteron form factor using Paris potential wave functions:

$$G_E^n(Q^2) = -1.25\mu_n \frac{\tau}{1 + 18.3\tau} \frac{1}{(1 + Q^2/0.71GeV^2)^2}, \quad \tau = \frac{Q^2}{4M_N^2} \tag{37}$$

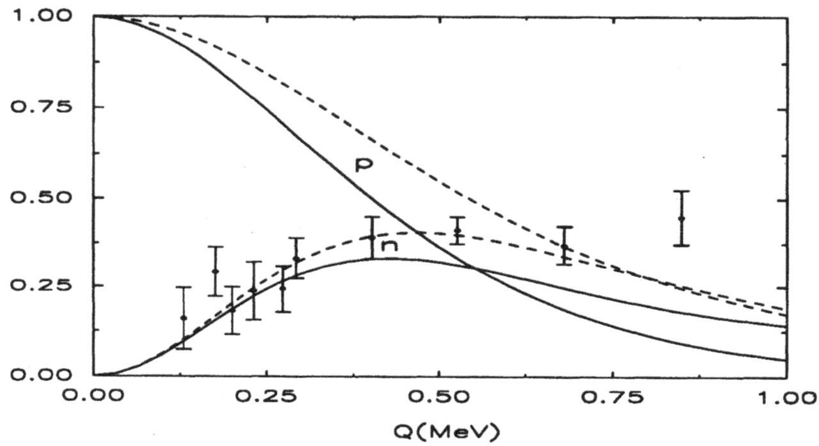

Fig. 4. The electric form factors of the neutron (n) and the proton (p) as functions of the momentum transfer Q. The solid lines correspond to the model's predictions. The dashed lines are the dipole fit $1/(1+Q^2/0.71\text{GeV}^2)^2$ for the proton and the fit of Platchkov et al.[37] for the neutron. Also shown are some selected expermental data for the neutron electric form factor. Note that all numbers for the neutron electric form factor have been scaled by a factor 10.

up to momentum transfers of about $0.3GeV$ possessing, however, a somewhat smaller maximum than the Platchkov et al.[37] fit. Summarizing this subsection on electromagnetic properties we may state that the model provides reasonable predictions for the magnetic moments and the corresponding radii. However, we suspect from the discussion of the proton electric form factor that for charged baryons the absolute values for the electric radii are overestimated.

4.2. *Strangeness in the proton*

Having seen that the vector meson model extended to SU(3) gives a fair description of the static electromagnetic properties of $\frac{1}{2}^+$ baryons we shall now proceed and employ this model for predictions on the relevance of strange quarks in the nucleon. *I.e.* we will evaluate matrix elements of operators involving strangeness between nucleon states. Let us firstly consider the strange vector current:

$$V_\mu^s = V_\mu^0 - \frac{2}{\sqrt{3}} V_\mu^8, \qquad (38)$$

which in the quark language corresponds to $\bar{s}\gamma_\mu s$. The matrix elements of V_μ^s are of special interest for precision analyses of eN and νN scattering experiments[6] in the standard model. Again it proves useful to define Dirac ($F_3(Q^2)$) and Pauli ($\tilde{F}_3(Q^2)$) type form factors for the strangeness current V_μ^s which are the analogues of $F_1^p(Q^2)$ and $F_2^p(Q^2)$ in eq. 31, respectively. Obviously we have $F_3(0) = 0$ since the net strangeness in the proton is zero. For the corresponding radius which is defined as the slope of the Dirac form factor:

$$\langle r_s^2 \rangle = -6 \frac{\partial F_3(Q^2)}{\partial Q^2} |_{Q^2=0} \tag{39}$$

we find numerically $\langle r_s^2 \rangle = 0.05 fm^2$. Amusingly this result has the opposite sign of the prediction in the pseudoscalar model[20]:$-0.10 fm^2$. The positiveness of $\langle r_s^2 \rangle$ may be linked to VMD since a one pole VMD approach[20] together with $\omega - \phi$ mixing leads to $\langle r_s^2 \rangle = 0.01 fm^2$; which, however, is much smaller in magnitude. Also Jaffe[38] obtained (by using a three pole vector meson fit to dispersion relations suggested by Höhler et al.[39]) a positive value $\langle r_s^2 \rangle = 0.16 \pm 0.06 fm^2$, although this is larger than our result. Actually the fit of ref.[38] has the unusual feature that the ϕ meson has a fairly substantial coupling to the nucleon.

Our prediction for the strange magnetic moment of the proton:

$$\mu_s = \tilde{F}_3(0) = -0.05 \tag{40}$$

has the same sign as in the model of pseudoscalars only[20]: -0.13 with a smaller absolute value. Our result is even much smaller (in magnitude) than Jaffe's estimate[38]: -0.31 ± 0.09.

We should stress once again that all these results are obtained by taking matrix elements between exact eigenstates of the Hamiltonian (26) including symmetry breaking. It is, of course, interesting to ask how symmetry breaking influences the matrix elements of strange operators. Using the SU(3) symmetric wave function of the proton yields $\mu_s = -0.54$, *i.e.* the distorted wave function reduces the strange magnetic moment by about 90%! On the other hand the already small result for $\langle r_s^2 \rangle$ is only mildly changed to $0.04 fm^2$ when the symmetric wave-function is employed.

Turning to the scalar strangeness content of the nucleon we are immediately led to the puzzle of the nucleon Σ term. The nucleon Σ term is defined by:

$$\Sigma_\pi(Q^2) = \frac{1}{3} \sum_{i=1}^{3} \langle P(\mathbf{p}') | [Q_i^5, [Q_i^5, H]] | P(\mathbf{p}) \rangle \tag{41}$$

wherein Q_i^5 are the generators of an infinitesimal axial transformation and $|P(\mathbf{p})\rangle$ denots a proton state of momentum $\mathbf{p}$. Using SU(3) symmetric wave functions for the baryons it has been observed long ago[41] that

$$\Sigma_\pi(0) = \frac{3\hat{m}}{\hat{m} - m_s}(M_\Lambda - M_\Xi)\frac{1}{1-y}, \quad \text{with} \quad y = \frac{2\langle \bar{s}s \rangle_P}{\langle \bar{u}u + \bar{d}d \rangle_P}. \tag{42}$$

y obviously measures the scalar strange content fraction in the proton. Analyses of the S-wave amplitude in $\pi - N$ scattering leads to[42] $\Sigma_\pi(0) \approx 45 MeV$[c] The puzzle

[c]The authors of ref.[42] pointed out that further low-energy measurements would be needed to reduce the uncertainty in $\Sigma_\pi(0)$.

now is obvious: Using the experimental mass difference $M_\Xi - M_\Lambda = 202MeV$ as well as established values for the ratio $\hat{m}/m_s = 25...30$ one finds $y \approx 0.4$ which, of course, is unexpectedly large. Skyrme type models provide a solution to this puzzle by incorporating SU(3) symmetry breaking effects. Working out the double commutator (41) gives the generic form of the nucleon Σ-term:

$$\Sigma_\pi(Q^2) = \frac{4}{3} I(Q^2)\big(2 + \langle D_{88}\rangle_P\big) \tag{43}$$

wherein $I(Q^2)$ is the Fourier transform the integrand of the symmetry breaker γ, especially we have $I(0) = \frac{3\delta'}{8(\delta''-\delta')}\gamma$. Using eq. 9 and performing a perturbation expansion[18] of the matrix element $\langle D_{88}\rangle_P$ in terms of the effective symmetry breaking parameter $\beta^2\gamma$ we find:

$$\Sigma_\pi(0) = \frac{3\hat{m}}{\hat{m} - m_s}(M_\Lambda - M_\Xi - 0.031\gamma(\beta^2\gamma + \cdots)) \tag{44}$$

the dots refer to higher order terms in $\beta^2\gamma$. Values for β^2 and γ which fit the baryon mass spectrum reasonable well are of the order $\beta^2 \approx 4GeV^{-1}$ and $\gamma \approx 1.5GeV$. This demonstrates that SU(3) symmetry breaking effects may easily double the "naïve" estimate (42) for the nucleon Σ-term.

Of course, we shall carry out the exact evaluation of $\Sigma_\pi(0)$ which also takes into account the dots in eq. 44. Actually, our prediction for $M_\Xi - M_\Lambda$ is somewhat larger than the experimental value (cf. table 1). Also, due to the large soliton, our numerical results for β^2 and γ are slightly larger than the above mentioned estimates. This results in an even increased value for $\Sigma_\pi(0) = 62MeV$. Recently Gasser et al.[43] have converted the difference between the value at the Cheng-Dashen point, $\Sigma_\pi(t^2 = -\mathbf{Q}^2 = 2m_\pi^2)$[d] and zero momentum transfer into a radius (employing constraints from chiral perturbation theory). They obtained the surprisingly large value $\langle r_\Sigma^2\rangle = 1.6fm^2$ which is about twice as large as the experimental value for r_E^2 (cf. section 4). However, our result for $\langle r_\Sigma^2\rangle = 1.03fm^2$ is actually smaller than the predicted electric radius of the proton (cf. table 3). Here it is, of course, appropriate to compare the results of ref.[43] with our predicted differences $\Sigma_\pi(t^2 = -\mathbf{Q}^2 = 2m_\pi^2) - \Sigma_\pi(0) = 13.3MeV$ ($\approx 15MeV$) and $\Sigma_\pi(0) - \Sigma_\pi(t^2 = -\mathbf{Q}^2 = 2m_\pi^2) = 9.2MeV$ ($\approx 10MeV$); the numbers in parentheses are extracted from fig.3 of ref.[43]. This seems to indicate that our smaller prediction for $\langle r_\Sigma^2\rangle$ is linked to the too large value for $\Sigma_\pi(0)$ which enters in the denominator in the definition of $\langle r_\Sigma^2\rangle$. We may speculate about the origin of the large $\Sigma_\pi(0)$. In the calculation of Σ_π we have restricted ourselves to leading order in $\frac{1}{N_C}$. On the other hand the Σ_π-term is extracted from the s-wave amplitude in πN scattering which can only be described reasonably well in soliton models when higher order corrections in $\frac{1}{N_C}$ are taken into account.[46] These higher order corrections may thus also have a significant influence on the Σ-term. Also, the inclusion of explicit

[d] $\Sigma_\pi(-\mathbf{Q}^2 = 2m_\pi)$ has been extracted from πN scattering data by Koch[44] to be $64 \pm 8MeV$; interpolating the πN phase shifts from Weinberg's[45] result for the scattering length $a_0^0 = 0.16m_\pi^{-1}$ reduces this result to $56 \pm 8MeV$.

scalar degrees of freedom (like *e.g.* via the trace anomaly[47]) might alter the results since the Σ-term is associated to the matrix element $< p|\overline{u}u+\overline{d}d|p >$. Technically this is realized by scaling the symmetry breaking terms in eq. 7 with powers of $\exp(\sigma) < 1$ wherein σ is a gluonium scalar order parameter which is introduced to mock up the QCD trace anomaly. This procedure reduces the integrand of $I(Q^2)$ and thus also Σ_π. Surely, this is only speculative since the incorporation of scalar degrees of freedom is somewhat arbitrary and might change other baryon observables as well.

Since we have seen that Σ_π is related to the scalar strange content fraction[48,49] X_S of the proton we will also quote the predictions of our model for X_S. Using a σ - model interpretation of the quark bilinears we define:

$$X_S = \frac{< p|\overline{s}s|p > - < 0|\overline{s}s|0 >}{< p|\overline{u}u + \overline{d}d + \overline{s}s|p > - < 0|\overline{u}u + \overline{d}d + \overline{s}s|0 >} = \frac{1}{3}\langle p|1 - D_{88}|p\rangle. \qquad (45)$$

Evaluating this matrix element with the exact eigenstates of the collective Hamiltonian yields $X_S = 0.134$. This is still fairly large although it already represents an almost 50% reduction compared to the value $X_S = \frac{7}{30} \approx 0.233$ which is obtained by using the SU(3) symmetric proton wave function.[48] In the language of eq. 42 the obtained value $X_S = 0.134$ corresponds to $y = 0.31$.

4.3. *Axial current matrix elements*

To discuss matrix elements of the axial current, A_μ^a (corresponding to $\overline{q}\gamma_\mu\gamma_5\frac{\lambda_a}{2}q$ in the quark language) between proton states we define the proton axial form factors of the lth axial quark current

$$\frac{\sqrt{p_0 p_0'}}{M_P}\langle P(\mathbf{p}')|\overline{q}_l\gamma_\mu\gamma_5 q_l|P(\mathbf{p})\rangle = \overline{u}(\mathbf{p}')[\gamma_\mu\gamma_5 H_l(Q^2) + \frac{Q_\mu}{2M_P}\gamma_5\tilde{H}_l(Q^2)]u(\mathbf{p}), \qquad (46)$$

with $l = 1,2,3 = u,d,s$ and $|P(\mathbf{p})\rangle$ again denotes a proton state with momentum $\mathbf{p}$. Here we will only be concerned with the main form factor $H_l(Q^2)$ and therefore omit total derivatives in the calculation of the axial current from eq. 29 since they only contribute to the induced form factor $\tilde{H}_l(Q^2)$.[26] For the evaluation of $H_l(Q^2)$ it is actually sufficient to only consider the spatial components A_i^a. As in the case of the electromagnetic properties we insert the *ansätze* made for the meson fields eqs. 14 and 19-22 into the expression for axial current gained from (29). Finally the Fourier transform of A_i^a is interpreted as the corresponding form factor in the Breit frame.

Of course, the final goal is to disentangle the three form factors $H_l(Q^2)$. However, considering the linear combinations

$$\begin{aligned} G_a(Q^2) &= H_1(Q^2) - H_2(Q^2) \\ R(Q^2) &= H_1(Q^2) + H_2(Q^2) - 2H_3(Q^2) \\ H(Q^2) &= H_1(Q^2) + H_2(Q^2) + H_3(Q^2) \end{aligned} \qquad (47)$$

turns out to be more instructive. The momentum dependence of these form factors is displayed in figure 5. The form factor for neutron β decay, $G_a(Q^2)$ is obtained

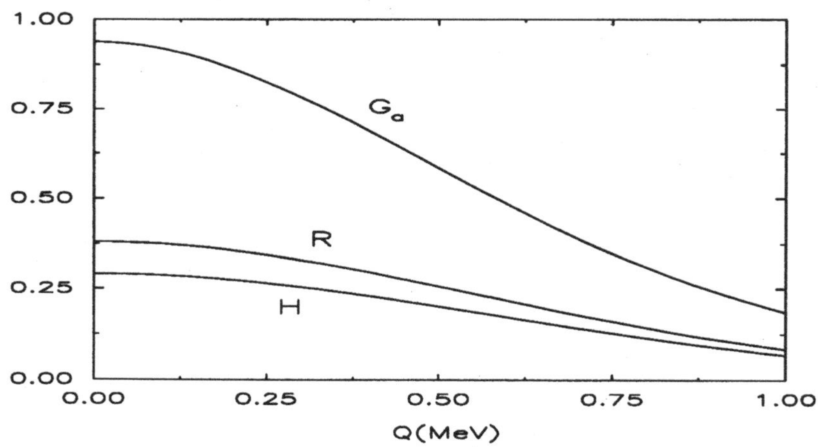

Fig. 5. The momentum dependence of the axial form factors of the nucleon (cf. eq. 47). The input parameters are as in eq. 28.

by Fourier transforming $2A_3^3$. Its value at zero momentum transfer $g_a = G_a(0)$ is very well known experimentally to be 1.26. There are no restrictive informations on $R(Q^2)$ and $H(Q^2)$ from experiment. The only additional measurement on axial from factors is the EMC experiment of scattering polarized muons off polarized protons.[5] There the linear combination $4H_1(Q^2) + H_2(Q^2) + H_3(Q^2)$ was investigated. Only by employing SU(3) symmetry relations and informations on semileptonic hyperon decays further inside into $R(Q^2)$, the Fourier transform of $2\sqrt{3}A_3^8$, and $H(Q^2)$, the Fourier transform of the axial singlet current $3A_3^0$, may be gained. These investigations have led to the famous "proton spin crisis" stating the quarks' contribution to the proton spin, $2H(0)$, might vanish identically.[50] However, we will abandon the use of SU(3) symmetry relations and explore the effects of SU(3) symmetry breaking in the baryon wave functions on these form factors as well as for the Cabibbo model[51] which provides various relations between the form factors of *strangeness changing* axial currents.

Using the parameters listed in eq. 28 we find for the form factors at zero momentum transfer:

$$g_a = 0.93, \;\; R(0) = 0.38, \;\; H(0) = 0.29. \tag{48}$$

Our result for g_a is still about 30% lower than the experimental value although it represents a noticeable improvement against pseudoscalar models.[20] A similar result for g_a is already obtained in the U(2) reduction of the present model indicating that strange degrees of freedom only play a minor role for axial matrix elements of the proton.[11] This gets more transparent by disentangling the individual quark form factors at zero momentum transfer

$$H_1(0) = 0.63, \ H_2(0) = -0.31, \ H_3(0) = -0.03, \tag{49}$$

showing that strange quarks contribute as few as 10% of the down quarks to the matrix elements of the axial current.

Amusingly g_A and $R(0)$ approximately satisfy the SU(3) relation $R = \frac{3}{7}g_A(= 0.40)$, which is obtained by comparing the matrix elements of D_{83} and D_{33} assuming the wave functions to be SU(3) symmetric. However, we already observe from figure 5 that the momentum dependence of R follows more closely that of H than of G_a and we will immediately show that this coincidence is completely accidental for the parameter set (28). One striking evidence for this behavior is observed form the radii of the axial form factors:

$$\langle r_3^2 \rangle = 0.51 fm^2, \ \langle r_8^2 \rangle = 0.34 fm^2, \ \langle r_0^2 \rangle = 0.35 fm^2, \tag{50}$$

the lower index denotes the flavor component under consideration. The fact that $\langle r_8^2 \rangle$ is almost identical to $\langle r_0^2 \rangle$ indicates that the eighth and singlet components of the axial current are both dominated by $H_1 + H_2$. Note, that on the contrary the above quoted SU(3) relation would enforce $\langle r_3^2 \rangle \approx \langle r_8^2 \rangle$. The numerical results (48) and (50) seem to justify the assumption $R(Q^2) = H(Q^2)$ made in ref.[25] although there is no *a priori* reason for this identification. As in the case of the electromagnetic radii the model gives a somewhat larger value for the axial isovector radius $\langle r_3^2 \rangle$ than the empirical data[52]: $0.40 \pm 0.04 fm^2$.

We may now ask for the relevance of these predictions for the analysis of the EMC experiment. In ref.[24] it was shown that $H(0)$ being as large as 0.30 is not in contradiction to the experimental data as long as $R(0)$ is smaller than 0.5; employing, however, the experimental value for g_a. Let us therefore, for the time being, consider a different set of parameters which gives the experimental value for g_a

$$\tilde{h} = -0.1, \ \tilde{g}_{VV\phi} = 1.4, \ \kappa = 1.0 \tag{51}$$

all other parameters remain as in (28). This set gives too large mass differences, since we find $\gamma = 2.545 GeV$ yielding, *e.g.* $M_\Lambda - M_N = 236 MeV$. Nevertheless (51) may shed some light on the analysis of the EMC experiment. Although g_A has increased to its experimental number, the values for $R(0)$ and $H(0)$ have decreased to 0.29 and 0.25, respectively. This shows once more that $R(0)$ is governed by the terms it has in common with $H(0)$. The set (51) also leads to a very small strange contribution $H_3(0) = -.015$ and a 50% deviation from the SU(3) relation between g_a and $R(0)$. This impressively demonstrates that employing SU(3) relations to relate matrix elements of strangeness conserving currents to those of strangeness changing currents in unreliable. We finally have to question whether our treatment which obviously disagrees with one of the basic assumptions underlying the interpretation of the EMC measurement is compatible with the results of the EMC experiments. In our model we find

$$\frac{1}{9}[4H_1(0) + H_2(0) + H_3(0)] = \begin{cases} 0.24 \text{ for set eq. 28} \\ 0.28 \text{ for set eq. 51} \end{cases} \tag{52}$$

which actually is not very sensitive to the change of parameters and compares reasonably well with the experimental value[5]:0.252 ± 0.036.

Finally let us also mention that additional information on $H_3(0)$ may be extracted from elastic νp scattering. Ahrens et al.[53] obtain $H_3(0) = -0.15 \pm 0.09$, however, this result is claimed[54] to be subject to change due to some theoretical uncertainties of analysis.

It remains to check whether or not this strong deviation from the SU(3) relations contradicts the successful Cabibbo model for semileptonic hyperon decays since in this model various strangeness changing currents are related by assuming SU(3) symmetry. In such an approach only two combinations (antisymmetric and symmetric) exist to construct flavor singlets which are bilinear in the baryon octet and linear in the octet currents. Establishing the electromagnetic charge normalization completely fixes the structure of the vector current while for the axial current two constants of proportionality, F and D for the antisymmetric and symmetric combinations, respectively are left undetermined. The relevant matrix elements at zero momentum transfer are parametrized[55] as linear combinations of F and D (ignoring the spinor bilinears $i\overline{u}(\mathbf{p}')\gamma_\mu\gamma_5 u(\mathbf{p})$):

$$\begin{aligned} \langle p|A^{k-}|\Lambda\rangle &= -\frac{3F+D}{\sqrt{6}}, \ \langle\Sigma^-|A^{k-}|n\rangle = D - F \\ \langle\Xi^-|A^{k-}|\Lambda\rangle &= \frac{3F-D}{\sqrt{6}}, \ \langle\Xi^-|A^{k-}|\Sigma^0\rangle = \frac{F+D}{\sqrt{2}}, \end{aligned} \tag{53}$$

furthermore relations for the diagonal matrix elements of the proton are obtained: $g_a = F + D$ and $R(0) = 3F - D$. The SU(3) symmetric Skyrme model predicts[34] $D/F = 9/5$ which has been considered to be in nice agreement with a typical fit[56,57] of (6.1) to experimental data: $D/F = 1.63$. This agreement has been viewed as one of the successes of the SU(3) symmetric Skyrme model. However, it has been shown in several treatments[20,35,36,55] of the Skyrme model of pseudoscalars that a reasonable description of hyperon decay processes may even be achieved when the baryon wave functions deviate strongly from SU(3).

Specifically we evaluate the matrix elements of $\int V_0^{k-} d^3r$ and $\int A_i^{k-} d^3r$ between baryons, B and B' of different strangeness which yield the vector and axial vector charges g_v^- and g_a^-, respectively. Since $\int V_0^{k-} d^3r$ corresponds to the Noether charge, g_v^- is just the matrix element of the left SU(3) generator[e] $L^- = L_4 + iL_5$:

$$g_v^- = \langle B'|L^-|B\rangle. \tag{54}$$

[e]The left SU(3) generators are defined as $L_a = \sum_{b=1}^{8} D_{ab} R_b$.

Table 5. The g_a^-/g_v^- ratios as predicted by our model. The equation numbers refer to the input parameters. The SU(3) symmetry predictions are obtained using $D/F = 1.8$ as well as $F + D = g_a = 1.26$. The star indicates input quantities.

	g_a^-/g_v^- eq. 28	eq. 51	Expt.	SU(3)
$\Lambda \to p$	0.69	0.74	0.70±0.03	0.74
$\Sigma^- \to n$	0.24	0.22	0.35±0.05	0.30
$\Xi^- \to \Lambda$	0.19	0.22	0.25±0.05	0.22
$\Xi^- \to \Sigma^0$	1.22	1.31	—	1.26
$n \to p$	0.93	1.26*	1.26	1.26*
R	0.38	0.29	—	0.66

The deviation of g_v^- from its SU(3) symmetric value is at least of second order in SU(3) symmetric breaking since the matrix elements of L_a vanish between states of different SU(3) representations. This, of course, agrees with the Ademollo-Gatto theorem.[58] Nevertheless, the deviations are not necessarily small. Actually, experimental data do only exist for the ratios g_a^-/g_v^- and we present our results for the two sets of parameters (28) and (51) in table 5 and compare the predictions to those gained from the SU(3) relations. For the realistic set (28) the model describes the processes $\Lambda \to p$ and $\Xi^- \to \Lambda$ fairly well as the comparison with the data extracted from experiment reveals. Our prediction of g_A^-/g_V^- for the decay $\Sigma^- \to n$ is on the low side which is mostly due to the strong enhancement of g_V^- for which the SU(3) symmetric wave functions give unity (see also figure 1 of ref.[55]). Changing to set (51) which reproduces the experimental value of g_a the g_a^-/g_v^- ratios change only mildly. For the processes $\Lambda \to p$ and $\Xi^- \to \Lambda$ we find again a fair agreement with SU(3) predictions and experimental data. Unfortunately g_A^-/g_V^- deviates significantly from experimental data as well as from SU(3) for $\Sigma^- \to n$ because $g_v^- = 1.48$ is almost 50% larger than the SU(3) result while $g_a^- = 0.327$ is close to the SU(3) prediction $\frac{2}{7}g_a = 0.357$. However, we should stress that this deviation from SU(3) is considerably smaller than that found for $R(0)$.

Concluding this sub-section on axial form factors we may state that a small value for the singlet matrix element of the proton may be explained by Skyrme type models with a small polarization of the strange sea of the proton once the condition of SU(3) symmetry is relaxed. The deviations from SU(3) symmetry are sizable for strangeness conserving matrix elements but small for those components of the current which carry strangeness changing quantum numbers.[f] And therefore the extraction of $H_3(0)$ from the EMC data and SU(3) symmetry relations is questionable.

[f]Similar results have also been found in ref.[59].

4.4. *The neutron proton mass difference*

Finally let us consider isospin violation. In this context the neutron proton mass difference is of special interest since it may be used to get a deeper understanding of the problem discussed in the preceding section, the "proton spin puzzle". It has been demonstrated that this mass difference may be employed to disentangle the "matter" and "glue" contributions to the proton's axial singlet current matrix element.[60] Here we shall not deal with this subject since it is covered by another contribution to these proceedings.[61] Nevertheless, as a first step it is appropriate to discuss the various contributions to this quantities since their relative strengh is crucial for the "matter -glue" decomposition of $H(0)$.

Experimentally the neutron proton mass difference is obtained to be $1.29 MeV$. The strong as well as the electromagnetic interaction contribute to this quantity:

$$(M_{\text{neutron}} - M_{\text{proton}}) = \Delta_{\text{st.}} + \Delta_{\text{e.m.}} \tag{55}$$

Using experimental electromagnetic form factors $\Delta_{\text{e.m.}}$ is estimated as[62]: $(-0.76 \pm 0.30) MeV$. The negative sign may be understood by noting that $\Delta_{\text{e.m.}}$ is dominated by the Coulomb repulsion. Thus we may extract:

$$\Delta_{\text{st.}} = (2.05 \pm 0.30) MeV. \tag{56}$$

which is the quantity we wish to explore in the vector meson model. Actually it was found some time ago that vector mesons are of utmost importance in order to predict a non-vanishing $\Delta_{\text{st.}}$ in a two flavor Skyrme type model.[13] Of course, a two flavor model should give the main share to $\Delta_{\text{st.}}$ since one naïvely expects that $\Delta_{\text{st.}} \sim m_d - m_u$.

We firstly have to incorporate isospin breaking in the meson sector before discussing $\Delta_{\text{st.}}$. This is done by generalizing the symmetry breaking matrices in eq. 7:

$$T + \frac{m_s}{\hat{m}} S \longrightarrow Y\lambda_3 + T + \frac{m_s}{\hat{m}} S, \tag{57}$$

i.e. we have to introduce only one additional parameter $Y = (m_u - m_d)/\hat{m}$ which may for example be fixed from the $\omega - \rho^0$ mixing.[13,28]

Having set up the isopin breaking part of the effective mesonic action we return to the baryon sector. After substitution of the *ansätze* (14), (19-22) into this action and canonical quantization we obtain the collective Hamiltonian including isospin violating terms:

$$H = H_{I=0} + H_{I=1}. \tag{58}$$

$H_{I=0}$ is identical to the Hamiltonian derived in section 3. The isospin violating terms read:

$$\begin{aligned} H_{I=1} &= \Gamma_3 D_{38} + \Delta_3 \sum_{i=1}^{3} (D_{3i} D_{8i} + D_{38} D_{88}) \\ &\quad + \frac{\alpha_3}{\alpha^2} \sum_{i=1}^{3} D_{3i}(R_i + \alpha_1 D_{8i}) + \frac{\beta_3}{\beta^2} \sum_{\alpha=4}^{7} D_{3\alpha}(R_\alpha + \beta_1 D_{8\alpha}). \end{aligned} \tag{59}$$

Analytic expressions for the parameters measuring isospin breaking in the baryon sector, $\Gamma_3, \Delta_3, \alpha_3$ and β_3 may *e.g.* be traced in the appendix of ref.[28]. Γ_3 and Δ_3 are functionals of the classical fields F, ω and G while α_3 and β_3 also contain the non-strange and strange excitations, respectively. Note, however, that only the linear combination $\eta_T \sim (\chi + \chi_8)$ appears in α_3. In the evaluation of the Hamiltonian only terms linear in the isospin breaking have been retained. Finally we have the non-electromagnetic contribution to the neutron proton mass difference:

$$\begin{aligned}
\Delta_{\text{st.}} &= \langle n|H_{I=1}|n\rangle - \langle p|H_{I=1}|p\rangle \\
&= -2\Gamma_3\langle p|D_{38}|p\rangle - 2\Delta_3\langle p|\sum_{i=1}^{3}(D_{3i}D_{8i} + D_{38}D_{88})|p\rangle \\
&\quad -2\frac{\alpha_3}{\alpha^2}\sum_{i=1}^{3}\langle p|D_{3i}(R_i + \alpha_1 D_{8i})|p\rangle - 2\frac{\beta_3}{\beta^2}\sum_{\alpha=4}^{7}\langle p|D_{3\alpha}(R_\alpha + \beta_1 D_{8\alpha})|p\rangle, \qquad (60)
\end{aligned}$$

Let us now discuss the numerical results for $\Delta_{\text{st.}}$. For the best fit parameters (28) we obtain:

$$\Delta_{\text{st.}} = 1.65 MeV \qquad (61)$$

which is somewhat smaller than the value extracted from experimental data (56). The contribution due to the classical fields in Γ_3 is only $0.20MeV$ since the matrix element $\langle p|D_{38}|p\rangle$ turns out to be reduced to about a quarter of its SU(3) symmetric value $\sqrt{3}/30$. Of course, we use the exact proton eigenstate of $H_{I=0}$ to evaluate the matrix element of $H_{I=1}$. The dominant contribution is due to the α_3 term $(0.95MeV)$[g] This is the only term which survives in the two flavor limit. In this limit we have $\sum_{i=1}^{3}\langle p|D_{3i}(R_i + \alpha_1 D_{8i})|p\rangle = \langle p|T_3|p\rangle = \frac{1}{2}$. Its crucial to note that the η mesons give the main share to this part. Setting $\chi \equiv \chi_8 \equiv 0$ in eq. 19 and extremizing the non-strange moment of inertia with this *ansatz* only gives $0.38MeV$! This value is now solely due to vector mesons. It is actually the fact that η_T needs to be increased by about 60% in order to reproduce the experimental value for $\Delta_{\text{st.}}$ which sheds some light on the possible decomposition of the singlet axial form factor of the proton into "matter" and "glue" portions.[60,61] Finally let us mention that the strange excitations (through β_3) contribute about $0.50MeV$ to the strong interaction part of the neutron proton mass difference.

5. Conclusions

We have presented the treatment of the SU(3) Skyrme model including vector mesons in the context of the collective approach which considers SU(3) symmetry as approximative. This assumption motivates the introduction of collective coordinates for the whole SU(3) group although these coordinates no longer correspond to real

[g] In ref.[13] the neutron proton mass difference in the U(2) reduction of the current model was investigated. For the η meson the *ad hoc* mass of $549MeV$ was assumed.[12] This gives an even larger value: $1.10MeV$.

zero modes. Special care has been taken of fields which vanish at the classical level but get excited by the collective rotation of the static soliton. SU(3) symmetry breaking has been treated exactly on the level of the collective Hamiltonian by employing a generalized Yabu-Ando method for diagonalization. The resulting eigenstates are no longer pure octet (decouplet) states but contain significant admixtures of higher dimensional representations like $\overline{10}$ or 27 $(27, 35, \overline{35})$. The experimental numbers for the mass differences of the low lying $\frac{1}{2}^+$ and $\frac{3}{2}^+$ baryons have been reproduced within 10% although a quite large soliton was needed to achieve that goal.

Vector and axial vector currents have been constructed by gauging the mesonic action with external fields. Matrix elements of the currents have yielded predictions for the static properties of the baryons. Especially good agreement with experiment has been obtained for the magnetic moments. Amusingly the results do not deviate strongly from the predictions obtained by assuming SU(3) symmetry. Nevertheless the strange magnetic moment of the proton has come out very small, indicating, as naïvely expected, that strange effects in the nucleon are strongly suppressed. Also reasonable results for the magnetic radii were obtained while the the electric radii of charged spin 1/2 baryons seemed to be somewhat overestimated. This effect is most likely due to the large soliton needed to fit the mass difference. Exposing the effects of symmetry breaking in the context of matrix elements of the axial currents was somewhat tricky since for the standard set of parameters (28) the SU(3) relation between g_a and $R(0)$ has accidentally been recovered. Choosing, however, to work with a set of parameters which gives the experimental value for g_a reveals that the form factor $R(Q^2)$ is closely related to the form factor of the axial singlet current, $H(Q^2)$. In this way the model predicts a very small contribution of strange quarks to the matrix element of the axial current between nucleon states contradicting neither the EMC experiment nor the successful Cabibbo model of semileptonic hyperon decays.

Let us finally comment on new developments concerning symmetry breaking in the meson sector. In the current model the predicted ratio $2m_s/\hat{m} \approx 36$ is rather large compared to *e.g.* the chiral perturbation program of Gasser and Leutwyler[62] who find 25.0 ± 2.5. The small value of ref.[62] is mostly due to the additional term

$$Tr\Big[(\lambda'T + \lambda''S)U(\lambda'T + \lambda''S)U + (\lambda'T + \lambda''S)U^\dagger(\lambda'T + \lambda''S)U^\dagger - 2(\lambda'T + \lambda''S)^2\Big] \tag{62}$$

which is of second order in SU(3) symmetry breaking. However this term has only a very minor effect on the baryon mass spectrum in the soliton picture. More crucial is the incorporation of a symmetry breaking term for the wave-functions of the vector mesons:

$$Tr\Big[(\gamma'T^+ + \gamma''S^+)F_{\mu\nu}(\rho)F^{\mu\nu}(\rho))\Big] \tag{63}$$

which is needed to give the correct ratio $\Gamma(K^*)/\Gamma(\rho)$. We have performed first calculations with this term in the soliton sector.[28] In table 6 we present the corresponding

Table 6. Best fit to the experimental baryon mass differences with the symmetry breaking term (63) included.

	Λ	Σ	Ξ	Δ	Σ^*	Ξ^*	Ω
Fit	168	263	404	327	470	617	766
Expt.	177	254	379	293	446	591	733

best fit to the mass spectrum which is gained for parameters

$$\tilde{h} = 0.36, \quad \tilde{g}_{VV\phi} = 1.88, \quad \kappa = 1.0 \tag{64}$$

which are closer to the central values $\tilde{h} = 0.4$ and $\tilde{g}_{VV\phi} = 1.9$ than eq. 28. Changing from set (28) to (64) also decreases the size of the soliton somewhat. This is seen most easily by noting that the baryonic radius is reduced from $0.30fm^2$ to $0.28fm^2$. Thus we may speculate that the inclusion of (63) improves the predictions for the baryon radii. However, g_a might come out even smaller than 0.93.

The generalization of (63) to also include isospin breaking increases the effects of vector mesons on the neutron proton mass difference by about 30% yielding $\Delta_{st.} = 1.77MeV$ which is within the experimental error bars (56).

Acknowledgements

The author would like to thank Prof. G. Holzwarth for the warm hospitality extended at the workshop. Also the nice atmosphere which stimulated a lot of fruitful discussions is appreciated.

The author furthermore wants to thank his colleagues J. Schechter and N. W. Park who participated in parts of the work underlying this talk and R. Alkofer for valuable remarks on the manuscript.

This work is supported by the Deutsche Forschungsgemeinschaft (DFG) under contract Re 856/2-1.

References

1. G. 't Hooft, *Nucl. Phys.* **B72** 461 (1974).
2. E. Witten *Nucl. Phys.* **B160** 57 (1979).
3. T.H.R. Skyrme, *Proc. R. Soc.* **A260**, 127 (1961).
4. G.S. Adkins, C.R. Nappi and E. Witten, *Nucl. Phys.* **B228**, 552 (1983).
5. EM Collaboration, J. Ashman et al., *Phys. Lett.* **B206**, 364 (1988); *Nucl. Phys.* **B328**, 1 (1989).
6. W. Heil et al., *Nucl. Phys.* **B327**, 1 (1989); P. Souder et al. in *Intersection between Particle and Nuclear Physics*, proceedings of the Second Conference, Lake Louise, Canada 1986, edited by Donald F. Geesaman, AIP Conf. Proc. No. 150 (AIP, New York, 1986).
7. J. Gasser, *Ann. Phys.* (NY), **136**, 62 (1981).
8. Ulf-G. Meißner, these proceedings.

9. A. Jackson, A. D. Jackson, A. S. Goldhaber, G. E. Brown, and L. C. Castillejo, *Phys. Lett.* **154B** 101 (1985); M. Lacombe, B. Loiseau, R. Vinh Mau and W. N. Cottingham, *Phys. Rev.* **D40**, 3012 (1989).
10. Ulf-G. Meißner, *Phys. Rep.* **161**, 213 (1988).
11. P. Jain, R. Johnson, Ulf-G. Meißner, N.W. Park and J. Schechter, *Phys. Rev.* **D37**, 3252 (1988).
12. Ulf-G. Meißner, N. Kaiser, H. Weigel and J. Schechter, *Phys. Rev.* **D39**, 1956 (1989).
13. P. Jain, R. Johnson, N.W. Park, J. Schechter and H. Weigel, *Phys. Rev.* **D40**, 855 (1989).
14. B. Schwesinger, H. Weigel, G. Holzwarth and A. Hayashi, *Phys. Rep.* **173**, 173 (1989).
15. B. Schwesinger and H. Weigel, *Nucl. Phys.* **A465**, 733 (1987); H. Weigel, B. Schwesinger and A. Hayashi, *Phys. Lett.* **B197**, 11 (1987); H. Weigel, *Phys. Lett.* **B215**, 24 (1988)
16. C. Callan and I. Klebanov, *Nucl. Phys.* **B262**, 365 (1985); C. Callan, K. Hornbostel and I. Klebanov, *Phys. Lett.* **B202**, 296 (1988); I.Klebanov in *Hadrons and Hadronic Matter*, page 223, proceedings of the NATO Advanced Study Institute, Cargese, 1989, edited by D. Vautherin, J. Negele and F. Lenz (Plenum Press 1989).
17. H. Yabu and K. Ando, *Nucl. Phys.* **B301**, 601 (1988).
18. N.W. Park, J. Schechter and H. Weigel, *Phys. Lett.* **B224**, 171 (1989).
19. H. Weigel, J. Schechter, N.W. Park and Ulf-G. Meißner, *Phys. Rev.* **D42**, 3177 (1990)
20. N.W. Park, J. Schechter and H. Weigel, *Phys. Rev.* **D43**, 869 (1991).
21. J. Schechter, *Phys. Rev.* **D34**, 868 (1986).
22. Ulf-G. Meißner, N. Kaiser and W. Weise, *Nucl. Phys.* **A466**, 685 (1987).
23. G. Holzwarth and B. Schwesinger, *Rep. Prog. Phys.* **49**, 825 (1986); G. E. Brown and I. Zahed, *Phys. Rep.* **142**, 1 (1986).
24. R. Johnson, N.W. Park, J. Schechter, V. Soni and H. Weigel, *Phys. Rev.* **D42**, 2998 (1990).
25. V. Bernard, N. Kaiser and Ulf-G. Meißner, *Phys. Lett.* **B237**, 545 (1990); *Phys. Lett.* **B243**, 439 (1990).
26. N.W. Park, J. Schechter and H. Weigel, *Phys. Lett.* **B228**, 420 (1989).
27. Ö. Kaymakcalan, S. Rajeev and J. Schechter, *Phys. Rev.* **D31**, 1109 (1985).
28. J. Schechter, A. Subbaraman and H. Weigel, in preparation.
29. C. Rosenzweig, J. Schechter and G. Trahern, *Phys. Rev.* **D21**, 3388 (1980); P. DiVecchia and G. Veneziano, *Nucl. Phys.* **B171**, 153 (1980); E. Witten, *Ann. Phys.* **128**, 1789 (1981); P. Nath and R. Arnowitt *Phys. Rev.* **D23**, 473 (1981).
30. N. W. Park and H. Weigel, *Nucl. Phys.* **A541**, 453 (1992).
31. N. W. Park and H. Weigel, *Phys. Lett.* **B268**, 155 (1991).
32. B. Moussallam and D. Kalafatis, *Phys. Lett.* **B272**, 196 (1991).
33. G. Holzwarth, "Quantum corrections to Nucleon and Delta mass in the Skyrme model", Siegen University preprint 1992; see also these proceedings.

34. G.S. Adkins and C.R. Nappi, *Nucl. Phys.* **B249**, 507 (1985); G.S. Adkins in *Solitons in Nuclear and Particle Physics*, page 47, proceedings of the Lewes workshop, 1984, edited by A. Chodos, E. Hadjimichael and C. Tze.
35. J. Schechter and H. Weigel, *Phys. Rev.* **D44**, 2916 (1991); *Phys. Lett.* **B261**, 235 (1991).
36. B. Schwesinger and H. Weigel, *Phys. Lett.* **B267**, 438 (1991); *Nucl. Phys.* **A540**, 461 (1992).
37. S. Platchkov et al., *Nucl. Phys.* **A510**, 740 (1990).
38. R. Jaffe, *Phys. Lett.* **B229**, 275 (1989).
39. G. Höhler et al., *Nucl. Phys.* **B114**, 505 (1974).
40. H. Pagels, *Phys. Rep.* **16**, 219 (1975).
41. See *e.g.* T.-P. Cheng and L.-F. Li, *Gauge theory of elementary particle physics*, Oxford University press, chapter 5.
42. J. Gasser, H. Leutwyler and M.E. Sainio, *Phys. Lett.* **B253**, 252 (1991).
43. J. Gasser, H. Leutwyler and M.E. Sainio, *Phys. Lett.* **B253**, 260 (1991).
44. R. Koch, *Z. Phys.* **C15**, 161 (1982).
45. S. Weinberg, *Phys. Rev. Lett.* **17**, 616 (1966).
46. H. Walliser, *Nucl. Phys.* **A524**, 706 (1991).
47. H. Gomm, P. Jain, R. Johnson and J. Schechter, *Phys. Rev.* **D33**, 801 (1986); *ibid.*, 3476 (1986).
48. J. Donoghue and C.R. Nappi, *Phys. Lett.* **B168**, 105 (1986).
49. H. Yabu, *Phys. Lett.* **B218**, 124 (1989).
50. S. Brodsky, J. Ellis and M. Karliner, *Phys. Lett.* **B206**, 309 (1988); J. Ellis and M. Karliner, *Phys. Lett.* **B213**, 73 (1988).
51. N. Cabibbo, *Phys. Rev. Lett.* **10**, 531 (1963).
52. L.A. Ahrens et al., *Phys. Lett.* **B202**, 284 (1988).
53. L.A. Ahrens et al., *Phys. Rev.* **D35**, 785 (1987).
54. D. Kaplan and A. Manohar, *Nucl. Phys.* **B310**, 527 (1988).
55. N.W. Park, J. Schechter and H. Weigel, *Phys. Rev.* **D41**, 2836 (1990).
56. M. Bourquin et al., Z. Phys. **C21**, 27 (1983).
57. R.L. Jaffe and A. Manohar, *Nucl. Phys.* **B337**, 509 (1990).
58. M. Ademollo and R. Gatto, *Phys. Rev. Lett* **13**, 264 (1964).
59. H.J. Lipkin, *Phys. Lett.* **B256**, 284 (1991).
60. J. Schechter, V. Soni, A. Subbaraman and H. Weigel, *Phys. Rev. Lett.* **65**, 2955 (1990); *Mod. Phys. Lett.* **A5**, 2543 (1990).
61. J. Schechter, these proceedings.
62. J. Gasser and H. Leutwyler, *Phys. Rep.* **87**, 77 (1982).

AN EXTENSION OF THE STANDARD SKYRME MODEL

H. WALLISER
Fachbereich 7, Universität Siegen
Adolf-Reichwein-Str., W-5900 Siegen, Germany

ABSTRACT
The spacing of the {8} baryon spectrum, the np split and related quantities require a type of symmetry breaker not contained in the standard Skyrme model. Additional terms obtained from the Gasser-Leutwyler lagrangian prove unimportant in the soliton sector. A local approximation on vector mesons cut off at six derivatives, seems to supply all relevant terms to be used in an extended pseudoscalar lagrangian for a satisfactory description of low-energy baryon properties.

1. Introduction

The developments of this contribution are guided by the prejudice that the low-energy baryon properties may be described by solitons of a purely pseudoscalar effective meson action. There are, however, several quantities which are not satisfactory reproduced in the standard Skyrme model. For some of these quantities this may be attributed to an inadequate calculation. An example is the too large soliton mass, where we have learned that quantum corrections are important and must be taken into account[1]. Others, as the {8} baryon splittings, the np split, and related quantities discussed here, indicate that there are terms missing in the standard lagrangian.

Because the standard symmetry breakers naturally supply only less than half of the np split, the missing rest was suspected to arise from more complicated terms not considered in the $B = 1$ sector so far. A persistent search led to the question whether the Gasser-Leutwyler lagrangian[2] could possibly provide the desired terms. The answer was essentially negative (section 3). Consequently, the operator structure of the missing term was analysed more carefully (section 2), and it was noticed, that it must originate in an antisymmetric contribution to the lagrangian. This almost inevitably leads to vector meson models which contain such contributions automatically through the anomaly[3,4]. In fact, this seems to be a strong argument in favor of these models. To further persue the goal of describing the baryonic low-energy properties in a purely pseudoscalar model, a local approximation on the vector meson fields was considered (section 5) and a six-derivative term with the desired structure was finally isolated.

This investigation may be useful in several respects. First of all it shows how a missing term in the standard Skyrme model may be pinned down and systematically traced to the vector meson model. By the way, it is noticed that many other terms in question are quite unimportant in the soliton sector. Because many rotationally induced components[3,4] make the vector meson model rather unwieldy, the quite simple results obtained in local approximation may possibly help to get more insight into the mechanisms, which among other things are responsible for the np split.

2. Standard SU(3) Skyrme model

The absolute minimum of terms, needed in the pseudoscalar lagrangian, for a fair description of the $B = 1$ sector, is given by

$$
\begin{aligned}
L &= L^0 + L^{SB} + L^{WZ} \\
L^0 &= -\frac{f_\pi^2}{4}\int d^3r\, tr\alpha_\mu\alpha^\mu + \frac{1}{32e^2}\int d^3r\, tr[\alpha_\mu,\alpha_\nu][\alpha^\mu,\alpha^\nu]\,, \qquad \alpha_\mu = \partial_\mu U U^\dagger \\
L^{SB} &= \frac{f_\pi^2 m_\pi^2}{4}\int d^3r\, tr(U + U^\dagger - 2) \\
&\quad + \frac{f_K^2 m_K^2 - f_\pi^2 m_\pi^2}{6}\int d^3r\, tr(1-\sqrt{3}\lambda_8)(U + U^\dagger - 2) \\
&\quad - \frac{f_K^2 - f_\pi^2}{12}\int d^3r\, tr(1-\sqrt{3}\lambda_8)(\alpha_\mu\alpha^\mu U + U^\dagger\alpha_\mu\alpha^\mu) \qquad (1) \\
\int dt L^{WZ} &= \frac{iN_C}{240\pi^2}\oint d\Sigma^{\mu\nu\lambda\sigma\tau}\, tr\alpha_\mu\alpha_\nu\alpha_\lambda\alpha_\sigma\alpha_\tau .
\end{aligned}
$$

The chiral invariant part L^0 is constructed from the nonlinear sigma model and the Skyrme term, and the chiral symmetry breaking part L^{SB} contains mass terms and the kinetic symmetry breaker. The corresponding isospin symmetry breakers must be added if such effects are considered. The Wess-Zumino term L^{WZ} fixes the correct quantum numbers for the baryons in the $B = 1$ sector. Everywhere mesonic parameters are used, except for the Skyrme parameter $e = 3.96$, which is as usually fitted to the $N\Delta$ split. This is refered to as standard Skyrme model in the following.

In the soliton sector, the static hedgehog configuration U_0, embedded into the isospin subgroup of $SU(3)$, is collectively rotated with the $SU(3)$ matrix A

$$
\begin{aligned}
&U = AU_0A^\dagger, \qquad U_0 \in SU(3)\,, \qquad A \in SU(3) \\
&D_{ab} = \frac{1}{2}tr\lambda_a A\lambda_b A^\dagger\,, \qquad A^\dagger\dot{A} = \frac{i}{2}\sum_{a=1}^{8}\lambda_a\Omega_a^R\,. \qquad (2)
\end{aligned}
$$

The resulting collective lagrangian

$$
\begin{aligned}
L^0 &= -M + \frac{1}{2}\Theta_\pi\sum_{a=1}^{3}(\Omega_a^R)^2 + \frac{1}{2}\Theta_K\sum_{a=4}^{7}(\Omega_a^R)^2 - \frac{N_C B_0}{2\sqrt{3}}\Omega_8^R \\
L^{SB} &= -\frac{1}{2}\Gamma_3(1 - D_{88}) + \frac{\sqrt{3}}{4}\Gamma_2 D_{38}, \qquad \Gamma_3/\Gamma_2 = 40 \qquad (3)
\end{aligned}
$$

correspondingly is divided into a symmetric and a symmetry breaking part. The soliton mass M, the pionic and kaonic moments of inertia Θ_π and Θ_K and the symmetry breakers Γ_2 and Γ_3 are functionals of the chiral angle which follow from the lagrangian

(1). The ratio of $SU(3)$ and $SU(2)$ symmetry breaking $\Gamma_3/\Gamma_2 \simeq 40$ is fixed by the meson masses and decay constants.

With the right and left angular momenta

$$R_a = -\frac{\partial L^0}{\partial \Omega_a^R}, \qquad L_a = \sum_{b=1}^{8} D_{ab} R_b , \tag{4}$$

the hamiltonian is obtained by a Legendre transfomation

$$H^0 = M + \frac{1}{2\Theta_\pi} C(SU(2)_R) + \frac{1}{2\Theta_K}\left(C(SU(3))_R - C(SU(2)_R) - \frac{3}{4}\right) \tag{5}$$

$$H^{SB} = -L^{SB} = \frac{1}{2}\Gamma_3(1 - D_{88}) - \frac{\sqrt{3}}{4}\Gamma_2 D_{38} \tag{6}$$

together with the Wess-Zumino-Witten constraint

$$R_8 = \frac{N_C B_0}{2\sqrt{3}}, \tag{7}$$

which fixes the right hypercharge $Y_R = 1$ and the lowest multiplets to be the $\{8\}$ with spin $J = \frac{1}{2}$ and the $\{10\}$ with spin $J = \frac{3}{2}$. Physical baryon states are obtained by diagonalising the symmetry breakers (6) in the basis of the unperturbed eigenstates $|B\rangle$ of H_0, this procedure is equivalent to the Yabu-Ando approach[5].

In fig.1 the calculated spectrum for the octet and decuplet baryons marked by (a) is compared with the experimental masses. Let me repeat that there is no further manipulation: everything is deduced from the standard Skyrme lagrangian (1), with $e = 3.96$ fitted to the $N\Delta$ split and otherwise mesonic parameters. The decuplet splitting turns out somewhat too small. This may be repaired by the slow rotator approach of Schwesinger and Weigel[6]. Also centrifugal effects, which should pull the solitons with $J = \frac{3}{2}$ more apart could explain this effect. The overall octet splitting is alright in magnitude, but the spacings are not well reproduced, in particular the $\Sigma\Lambda$ split is too large. This shortcoming is more serious and will require another type of symmetry breaker as will be dicussed in the next section.

There are also exotic $B = 1$ states in the Skyrme model with hypercharge $Y = 2$ and isospin $T = 0, 1$. These states, which cannot be explained as qqq states in a simple quark model, possess main components in the $\{\overline{10}\}$ and $\{27\}$ multiplets. Unfortunately, there is only very vage experimental information about the two possible candidates $Z_0(1780)P_{01}$ and $Z_0(1900)P_{13}$. The situation is formulated in ref.7: "The skepticism about baryons not made of three quarks, and the lack of any experimental activity in this area, make it likely that another 20 years will pass before the issue is decided." Because in the Skyrme model these states react very sensitively to the explicit form of the symmetry breakers, more concrete experimental information would help to fix this part of the lagrangian.

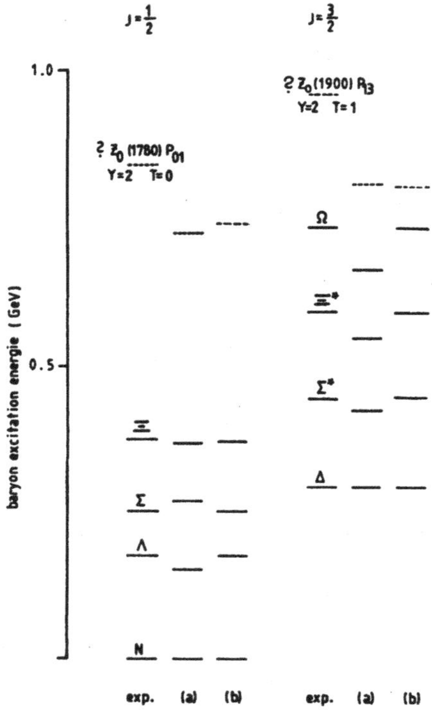

Fig. 1. The experimental baryon spectrum is compared to the spectrum calculated in the Skyrme model, (a) with standard symmetry breakers and (b) with the additional symmetry breaker (10) included.

3. The missing symmetry breaker

In this section we will argue that the {8} baryon spectrum, i.e. the $\Sigma\Lambda$ split requires a different symmetry breaker in addition to the standard ones in eq.(1).

From the quark mass expansion of the {8} baryons[8], which is also valid in the Skyrme model at least linearly in the symmetry breakers, we find

$$\begin{aligned} M_\Lambda - M_N &= (m_s - \hat{m})[\tfrac{2}{3}(B_u - B_d) + \tfrac{1}{3}(B_d - B_s)] \qquad & B_u &= \langle p|\bar{u}u|p\rangle \\ M_\Sigma - M_\Lambda &= (m_s - \hat{m})[\tfrac{2}{3}(B_d - B_s) - \tfrac{2}{3}(B_u - B_d)] & B_d &= \langle p|\bar{d}d|p\rangle \\ M_\Xi - M_\Sigma &= (m_s - \hat{m})(B_u - B_d) & B_s &= \langle p|\bar{s}s|p\rangle, \end{aligned} \tag{8}$$

with the current quark masses m_u, m_d, m_s and $\hat{m} = \frac{1}{2}(m_u + m_d)$. The corresponding quark contents in the proton are denoted by B_u, B_d, and B_s. By comparing the spectrum calculated from the standard Skyrme model (a) with the experimental spectrum in fig.1, it may be concluded that the $B_u - B_d$ content is too small in the Skyrme model.

The same conclusion is obtained from the $SU(2)$ mass splits of the octet baryons:

$$\begin{aligned} M_n - M_p &= (m_d - m_u)(B_u - B_d) \\ M_{\Sigma^-} - M_{\Sigma^+} &= (m_d - m_u)(B_u - B_s) \\ M_{\Xi^-} - M_{\Xi^0} &= (m_d - m_u)(B_d - B_s) \end{aligned} \tag{9}$$

Again, $B_u - B_d$ is too small in the standard Skyrme model (table 1). With $m_d - m_u \simeq 3.3$ MeV[8] it may be estimated $B_u - B_d \simeq 0.6-0.7$ from the np split. The $B_u - B_d$ content is also related to the Gottfried sum rule of deep inelastic μN scattering[9]. The EMC experiment determines $B_u - B_d = .76 \pm .11$ in accordance with the values deduced from the np split. This issue is discussed in a forthcoming letter[10]. These experimental numbers are in variance with $B_u - B_d = 0$ for the $SU(2)$ and $B_u - B_d = 0.26$ ($m_d - m_u \simeq 3.3$ MeV) for the $SU(3)$ standard Skyrme model. The standard symmetry breakers (6) are incapable to enhance the $SU(3)$ contribution substantially.

Table 1. Contributions to the $SU(2)$ mass splits for the octet baryons, (a) from the standard symmetry breakers and (b) from the additional symmetry breaker (10). The experimental values, corrected for electromagnetic contributions are taken from ref.8.

	(a)	(b)	sum	experiment e.m. corrected
$M_n - M_p$	.87	1.18	2.05	2.05±.30
$M_{\Sigma^-} - M_{\Sigma^+}$	6.61	.91	7.52	7.90±.31
$M_{\Xi^-} - M_{\Xi^0}$	6.13	−.32	5.81	~5.5

The next to leading order in a $1/N_C$ expansion adds contributions proportional to the angular velocities to the symmetry breakers (6)

$$\begin{aligned} L^{SB'} &= -\Delta_3 \sum_{a=1}^{3} D_{8a}\Omega_a^R - \frac{\sqrt{3}}{2}\Delta_2 \sum_{a=1}^{3} D_{3a}\Omega_a^R\,, & \Delta_3/\Delta_2 = 40 \\ H^{SB'} &= -\frac{\Delta_3}{\Theta_\pi} \sum_{a=1}^{3} D_{8a}R_a - \frac{\sqrt{3}}{2}\frac{\Delta_2}{\Theta_\pi} \sum_{a=1}^{3} D_{3a}R_a\,, & \frac{\Delta_3}{\Theta_\pi} = 70\text{MeV}\,. \end{aligned} \tag{10}$$

There are definitely also corresponding terms with the kaonic angular velocities. These are ignored here, because only the pionic contributions are closely related to the $B_u - B_d$ content which has problems in the standard Skyrme model.

Inclusion of the symmetry breaker (10) with the choice $\Delta_3/\Theta_\pi = 70$MeV results in a reasonable value $B_u - B_d \simeq .62$ ($m_d - m_u \simeq 3.3$ MeV). With that number and with a readjustment of the Skyrme parameter $e = 4.05$ to the $N\Delta$ split, the $\{8\}$ baryon spectrum agrees excellently with experiment (fig.1) and so do the $SU(2)$ mass splits (table 1). In all three cases the isospin symmetry breaking is improved by the additional symmetry breaker towards the experimental values.

Of course the symmetry breaker of the form (10) is not new and may be obtained in vector meson models[3,4] as well as in the NJL-model[11]. Before looking more closely into vector meson models (section 5) we will see whether the Gasser-Leutwyler lagrangian, obtained in chiral pertubation theory contains terms which go beyond the standard Skyrme model (1) and which are of importance in the $B = 1$ sector.

4. Gasser-Leutwyler lagrangian

In this section we will investigate the question, whether the Gasser-Leutwyler lagrangian can provide terms of a different structure, not contained in the standard Skyrme model. This lagrangian is obtained by a gradient expansion of the fields $U \in SU(3)$ and a simultaneous expansion in

$$\chi = \chi^\dagger = 2C\mathcal{M} = \begin{pmatrix} \mathring{m}_\pi^2 & & \\ & \mathring{m}_\pi^2 & \\ & & \mathring{m}_K^2 - \mathring{m}_\pi^2 \end{pmatrix}, \tag{11}$$

which is proportional to the quark mass matrix $\mathcal{M}$. The quantities $\mathring{m}_\pi$ and $\mathring{m}_K$ may deviate from the pion and kaon masses if corrections of order $\mathcal{M}^2$ are taken into account. Up to forth order the lagrangian is given by

$$\begin{aligned} L &= \frac{F_0^2}{4}\int d^3r\ [tr\ (\partial_\mu U \partial^\mu U^\dagger) + tr\ (U\chi^\dagger + \chi U^\dagger)] \\ &+ L_1 \int d^3r\ [tr\ (\partial_\mu U \partial^\mu U^\dagger)]^2 + L_2 \int d^3r\ [tr\ (\partial_\mu U \partial_\nu U^\dagger) tr\ (\partial^\mu U \partial^\nu U^\dagger)] \\ &+ L_3 \int d^3r\ [tr\ (\partial_\mu U \partial^\mu U^\dagger \partial_\nu U \partial^\nu U^\dagger)] + L_4 \int d^3r\ [tr\ (\partial_\mu U \partial^\mu U^\dagger) tr\ (U\chi^\dagger + \chi U^\dagger)] \\ &+ L_5 \int d^3r\ [tr\ (\partial_\mu U \partial^\mu U^\dagger (U\chi^\dagger + \chi U^\dagger))] + L_6 \int d^3r\ [tr\ (U\chi^\dagger + \chi U^\dagger)]^2 \\ &+ L_7 \int d^3r\ [tr\ (U\chi^\dagger - \chi U^\dagger)]^2 + L_8 \int d^3r\ [tr\ (\chi^\dagger U \chi^\dagger U + \chi U^\dagger \chi U^\dagger)]. \end{aligned} \tag{12}$$

Couplings to external fields which may become important if e.g. electromagnetic baryon properties are studied[12] are not considered here. The low-energy constants F_0 and L_ν , $\nu = 1 - 8$ depend on the chiral scale μ and are determined for $\mu = m_\eta$ by an analysis of experimental data in the mesonic sector. Because of this scale dependence and also because we want to use the effective lagrangian for low energies as well as in the soliton sector, we have to be careful about the numerical values of these constants.

On the other hand it should be noted, that if we just use the terms from the lagrangian (12) blindly (with $L_4 = L_6 = 0$ because they are suppressed by the Zweig rule) and include the singlet η-meson, then the experimental meson masses $m_\pi, m_K, m_\eta, m_{\eta'}$ and decay constants f_π, f_K fix the six parameters

$$\begin{aligned} F_0 = 91.2\text{MeV},\ \ \mathring{m}_\pi = 138\text{MeV},\ \ \mathring{m}_K = 498\text{MeV} \\ L_5 - 0.0022,\ \ L_7 = -0.0005,\ \ L_8 = 0.0011 \end{aligned} \tag{13}$$

exactly at the values given in ref.8. There remain the constants of the three different four-derivative terms, which may be debatable. The choice

$$L_1 = \frac{1}{2}L_2 = -\frac{1}{6}L_3 = \frac{1}{32e^2} \qquad e \simeq 6\,, \tag{14}$$

compatible with the Gasser-Leutwyler values, lets the Skyrme term survive alone, however with a strength much too weak to be used in the soliton sector (the same finding as from the local approximation on vector mesons, $e = 2g_\rho = 5.6$, discussed in the next section). For that reason, the usual 6th order term with mesonic parameters (next section) was added in the actual calculation to give the soliton a reasonable size. Deviations from (14) which were also considered, lead to $(\Omega_a^R)^4$ terms in the lagrangian. The corresponding pionic and kaonic right angular momenta R_a are then related to the angular velocities Ω_a^R by two coupled cubic equations, which posses a unique real solution and which may be solved numerically in order to perform the Legendre tranformation to the hamiltonian.

The result in the soliton sector does not depend sensitively on the numerical values employed for the constants L_ν and may be comprised in the following statement. The additional terms, which go beyond the standard Skyrme model alter in the first place the quantities $M, \Theta_\pi, \Theta_K, \Gamma_2$, and Γ_3 in the collective lagrangian (3). This does not introduce new aspects and may be compensated by altering the Skyrme parameter e, in particular the spacings of the $\{8\}$ and $\{10\}$ baryon spectra remain essentially unaffected if the $N\Delta$ split and the total splittings of the multiplets are kept fixed. In the second place new terms of a different operator structure are added to the collective lagrangian which are of the type

$$\begin{array}{cc} (\Omega_a^R)^4 & N_C^{-4} \\ D_{8a}d_{abc}\Omega_b^R\Omega_c^R & N_C^{-1}\chi \\ D_{88}(\Omega_a^R)^2 & N_C^{-1}\chi \\ D_{88}^2 & N_C\chi^2, \end{array} \tag{15}$$

and which all together turn out to be numerically small if they are used with reasonable strengths as e.g. with the Gasser-Leutwyler parameters. With these parameters the $\{8\}$ and $\{10\}$ baryons masses are changed by less than 10MeV only. It should be noted that the second and third operator in (15) appear already in the standard Skyrme model (1) due to the kinetic symmetry breaker, but usually are neglected.

The explanation why the terms (15) indeed should be small is as follows. All operators which contain the angular velocities are of order $1/N_C^2$ compared to the leading terms in (3). The terms proportional to χ^2 are negligible because the symmetry breaking in the $\{8\}$ and $\{10\}$ baryon wave functions is small. This latter statement is no longer true for the exotic states with their main components in the $\{\overline{10}\}$ and $\{27\}$ baryon multiplets. Their masses react more sensitively to symmetry breakers quadratic in the quark masses.

The conclusion of this section is, that a lagrangian of the Gasser-Leutwyler type in 4th order chiral perturbation theory does not introduce qualitatively new features to the baryon sector. In particular there is no term with the operator structure (10) which is only of order $1/N_C$ compared to the leading term. This of course was quite clear from the beginning: because the symmertry breaker (10) contains one time derivative only, it is due to an antisymmetric term in the lagrangian which goes along with the Levi-Cevita tensor. In order not to vanish identically, it must contain at least 6 derivatives. Being proportional to χ, such a term would therefore arise in chiral pertubation theory not until 8th order.*

This seems to be a very strong argument in favor of vector meson lagrangians, where such terms arise naturally through the anomaly. In the next section, a local approximation on the vector mesons is considered in order to isolate a term with the required properties.

5. Local approximation on vector mesons

As we have noticed in the last section the terms up to 4th order chiral perturbation theory are insufficient in two respects: the missing repulsion of a too weak Skyrme term with mesonic strength and the missing symmetry breaker (10), which is able to explain the spacing of the {8} baryon spectrum and the *np* split. Because in higher order chiral perturbation theory the number of terms increases rapidly, their handling, in particular the determination of their strength from meson phenomenology is considered as a serious obstacle in proceeding that way. As an alternative, we choose here a vector meson lagrangian of the type discussed by N. Kaiser[13] and H. Weigel[4] as starting point and investigate the pseudoscalar terms obtained by local approximation.

The equations of motion for the vector meson nonet $V_\mu \in U(3)$ which comprises ρ, K^*, ω and ϕ fields may be solved in a derivative expansion which is equivalent to a large mass expansion

$$\begin{aligned} V_\mu &= \frac{i}{2g_\rho}(\xi^\dagger\partial_\mu\xi - \partial_\mu\xi\xi^\dagger) + \frac{1}{m_\rho^2}\xi^\dagger(C_\mu - g_\omega N_C B_\mu)\xi + \ldots, \qquad \xi = \sqrt{U} \\ B_\mu &= \frac{1}{12\pi^2}\epsilon_{\mu\nu\rho\sigma}\alpha^\nu\alpha^\rho\alpha^\sigma \ , \qquad C_\mu = -\frac{i}{8g_\rho}(2\partial^\sigma[\alpha_\mu,\alpha_\sigma] - [\alpha^\sigma,[\alpha_\mu,\alpha_\sigma]]). \quad (16) \end{aligned}$$

The coupling constants $g_\rho = 2.8$ and $g_\omega = 3.3$ (linear combination of c_1 and c_3 of ref.14) are determined by experiment. In this way, the vector meson fields contained in the lagrangian may be replaced by pseudoscalar fields. This procedure is not without problems, but at least it was checked that the equations of motion for the pseudo-scalars are correct to 6th order in the derivative expansion. The result

$$L^0 = -\frac{f_\pi^2}{4}\int d^3r\, tr\alpha_\mu\alpha^\mu + \frac{1}{32(2g_\rho)^2}\int d^3r\, tr[\alpha_\mu,\alpha_\nu][\alpha^\mu,\alpha^\nu]$$

*Because of its pionic structure the term must survive in $SU(2)$ where antisymmetric four-derivative terms vansish $\epsilon^{\mu\nu\rho\sigma}\alpha_\mu\alpha_\nu\alpha_\rho\alpha_\sigma = 0$.

$$-\frac{1}{4m_\rho^2}\int d^3r\, tr(C_\mu - g_\omega N_C B_\mu)^2 \tag{17}$$

provides the Skyrme term with $e = 2g_\rho$ and a more complicated 6th order term. In $SU(2)$, where the static soliton is defined, the cross term vanishes, $trB_\mu C^\mu = 0$, and the familiar $B_\mu B^\mu$-term with the baryon current B_μ is obtained together with a new term $C_\mu C^\mu$. The new term is not positive definite, contains higher radial derivatives and therefore destroys the soliton. This is of course an artefact of the expansion (16) which is cut off; the vector meson model possesses stable solitons. For that reason the chiral angle $F(r)$ is calculated from (17) treating the new term $C_\mu C^\mu$ as a perturbation. This is justified from table 2, which contains the contributions of the individual terms to the soliton mass and the pionic moment of inertia: the contribution from the new term to the soliton mass is small. Alternatively, the exact chiral angle from the vector meson model would yield quite similar numbers.

Table 2. Contributions to the soliton mass and the pionic moment of inertia from the individual terms of the lagrangian (17) obtained in local approximation for $g_\rho = g_\omega = 2.925$ with a pion mass term included. The sum is compared to the exact results of the vector meson model.

	NLσ	Skyrme	$B_\mu B^\mu$	$C_\mu C^\mu$	mass term	sum	exact result
M (MeV)	920	409	283	-118	60	1554	1474
Θ_π (GeV^{-1})	2.8	2.1	2.0	-.20		4.9	4.2

The sum of the contributions up to the 6th order in the gradient expansion is already close to the exact vector meson model results. This finding is in contrast to the prevailing opinion that derivative expansions cannot work in the soliton sector because the gradients of the profiles are large.

In the vector meson model chiral symmetry breakers are usually introduced as

$$\begin{aligned} L^{SB} \quad = \quad & d_1 \int d^3r\, tr\chi(A_\mu^L U A^{R\mu} + A^{R\mu} U^\dagger A_\mu^L) \\ & + d_2 \int d^3r\, tr\chi(F_{\mu\nu}^L U F^{R\mu\nu} + F^{R\mu\nu} U^\dagger F_{\mu\nu}^L) \\ & + d_3 \int d^3r\, tr\chi(U + U^\dagger - 2) \\ & + d_4 \int d^3r\, tr\chi(\partial_\mu U \partial^\mu U^\dagger U + U^\dagger \partial_\mu U \partial^\mu U^\dagger) \end{aligned} \tag{18}$$

with $A_\mu^L = \xi(V_\mu + \frac{i}{g_\rho}\partial_\mu)\xi^\dagger$, $A_\mu^R = \xi^\dagger(V_\mu + \frac{i}{g_\rho}\partial_\mu)\xi$ and the corresponding field strength tensors $F_{\mu\nu}^L, F_{\mu\nu}^R$. The first two terms fix the vector meson masses and decay constants at their physical values and the next two terms do the same for the pseudoscalars. Accordingly the constants $d_\nu, \nu = 1-4$ are determined.

We do not want to discuss the whole variety of terms obtained from (18) by insertion of the local approximation (16). Instead we are interested in the missing symmetry breaker (10) in order to obtain an estimate for its strength Δ_3. Therefore we isolate all antisymmetric terms with 6 derivatives in lowest non-vanishing order which must contain B_μ and C_μ, namely

$$L^{\rho\omega} = (\frac{d_1}{m_\rho^2} + 2d_2)\frac{g_\omega N_C}{m_\rho^2} \int d^3r \, tr\{B_\mu, C^\mu\}(U\chi + \chi U^\dagger) \,. \tag{19}$$

As is noticed from (18), these terms are closely related to $\rho\omega$ mixing[15]. They are active in the soliton sector on the premises that the soliton contains non-vanishing ρ and ω-profiles. The ω-profile is due to the anomaly and is approximated here by B_μ, C_μ approximates the ρ-profiles in higher order. The operator structure of the expression (19) is precisely of the form (10), so it is a possible candidate for the explanation of the {8} baryon spacing and the np split. With d_1 expressed by the vector meson masses and $d_2 = 0$, the strength

$$\frac{\Delta_3}{\Theta_\pi} = -\frac{2g_\omega(m_{K^*}^2 - m_\rho^2)}{\sqrt{3}g_\rho m_\rho^4 \Theta_\pi} \int d^3r \; B_0(r) \left(\frac{2F' \sin F}{r} + \frac{\sin^2 F \cos F}{r^2}\right)$$

$$\frac{\Delta_3}{\Theta_\pi} = 70\text{MeV}, \qquad g_\rho = 2.8, \quad g_\omega = 3.3 \tag{20}$$

may be obtained as a functional of the chiral angle. The calculated strength (20) agrees with the phenomenological value (10) obtained by the fit to the {8} baryon spectrum. The $\rho\omega$ mixing mechanism explains nicely the np split. This does of course not exclude other explanations. The numerical coincidence with the fitted value may be partly accidential. There may be also contributions from the higher derivative terms. This statement is of course already true for the vector meson model itself, where only the lowest order symmetry breakers are included in (18).

An alternative explanation would require rotationally induced η-components in a $U(2)$ or $U(3)$ soliton, which may be driven by the anomalous terms in the vector meson lagrangian, or equivalently by the $B_\mu C^\mu$-term in (17) which does not vanish in $U(2)$ and $U(3)$ respectively. Regardless of some scepticism concerning the treatment of these induced components we may model this procedure within the local approximation. The result is again the same integral as in (20), but with a smaller factor depending on the $\eta\eta'$ mixing mechanism. This needs further clarification.

6. Conclusions

Starting from the standard Skyrme model, the goal was to construct a purely pseudoscalar soliton model with mesonic parameters, which is able to describe the low-energy properties in the $B = 1$ sector. The {8} baryon spectrum, the np split and related quantities require a symmetry breaker of a different structure not contained in the standard Skyrme model. The symmetry breaking terms of the Gasser-Leutwyler lagrangian in 4th order chiral perturbation theory, whose coefficients are determined by meson masses and decay constants do not introduce new aspects. This finding, and also the stabilisation mechanism require higher order terms in the soliton sector.

The desired terms may be obtained from a local approximation on the vector meson action, which is cut off at six derivatives. In addition to the familiar and frequently used 6th order $B_\mu B^\mu$ term there arises a new term with the unpleasant feature that it destroys the soliton (an artefact of the truncation of the local approximation). When treated perturbatively, this term turns out to be small, and from the various contributions to the soliton mass we may suspect that gradient expansions should in principle also work in the soliton sector. Because of the ill-behaved 6th order term, the construction of a purely pseudoscalar soliton model with mesonic parameters remains however incomplete.

It was demonstrated, that the missing symmetry breaker requires an antisymmetric term in the lagrangian which is proportional to the Levi-Cevita tensor and simultaneously proportional to the quark mass matrix. The vector meson anomaly in combination with the vector meson symmetry breakers supply the correct operator structure. The $\rho\omega$ mixing mechanism provides a reasonable strength for the explanation of the $\{8\}$ baryon spectrum and the np split. Other explanations are not excluded. This issue needs further clarification.

7. Acknowledgement

The author wishes to thank F. Meier for contributing to section 5.

8. References

1. B. Moussalam, talk at this workshop
2. J. Gasser, H. Leutwyler *Nuc. Phys.* **B250** (1985) 465
3. J. Schechter, talk at this workshop
4. H. Weigel, talk at this workshop
5. H. Yabu, K. Ando *Nuc. Phys.* **B301** (1988) 601
6. B. Schwesinger, H. Weigel *Phys. Lett.* **B267** (1991) 438
7. Particle Data Group *Phys. Lett.* **170B** (1986) 289, *Phys. Rev.* **D45** (1992) 1
8. J. Gasser, H. Leutwyler *Phys. Rep.* **87** (1982) 77
9. E. Reya *Phys. Rep.* **69** (1981)221, and private comunication
10. G. Holzwarth, H. Walliser, Gottfried sum rule in the Skyrme model, in preparation
11. M. Praszalowicz, talk at this workshop
12. F. Meier, talk at this workshop
13. N. Kaiser, talk at this workshop
14. P. Jain, R. Johnson, U.-G. Meissner, N.W. Park, J. Schechter *Phys. Rev.* **D37** (1988) 3252
15. L.N. Epele, H. Fanchiotti, C.A.G. Canal, R.M. Galain *Phys. Rev.* **D39** (1989) 1473

COLLECTIVE COORDINATE METHOD FOR SU(n) SOLITONS

H. WALLISER
Fachbereich 7, Universität Siegen
Adolf-Reichwein-Str., W-5900 Siegen, Germany

ABSTRACT

The collective coordinate method, utilized extensively in SU(3) soliton models is extended to general SU(n) flavor groups. The extension is nontrivial because for $n \geq 4$ qualitatively new constraints appear in addition to the familiar Wess-Zumino-Witten constraints. Baryon states, which obey all the constraints, are constructed in a systematical and definite procedure and may be used as a basis for the exact diagonalisation of the symmetry breaking. The standard symmetry breakers lead to the correct order of magnitude of the baryon multiplet splitting.

1. Introduction

The extension of the familiar SU(3) soliton models to higher flavor groups SU(n) is in the first place of fundamental interest. Especially the selection of the baryon multiplets by the Wess-Zumino-Witten constraints in combination with additional constraints appearing for SU(n), $n \geq 4$ is achieved group theoretically in an elegant way.

In principle SU(n) soliton models should also allow for a unified description of strange, charmed and bottom mesons and baryons. Because the chiral symmetry breaking in charmed and bottom direction is larger than the chiral scale, a chirally symmetric lagrangian as starting point with the chiral symmetry breakers added explicitly for these flavors, may be regarded with suspicion. However, as we have learned from Mannque Rho's talk this morning[1], there is a smooth connection from a chirally symmetric lagrangian to the heavy meson lagrangian of Isgur and Wise based on heavy quark spin symmetry[2] in the limit where the masses of the heavy pseudoscalar meson and its vector meson partner tend to infinity. Starting from a hidden gauge lagrangian[3,4], integrating out the ω and ρ fields[5], essentially the heavy meson lagrangian of Isgur and Wise with its couplings to the light mesons is recovered. Consequently there seems to be no serious argument against putting the heavy meson masses to their physical values by adding the appropriate symmetry breakers explicitly to a chirally invariant lagrangian.

Supposed, the lagrangian is well established, the further question which approach should preferably be chosen for the description of charmed and bottom baryons in the $B = 1$ sector to my opinion is decoupled from the discussion concerning the lagrangian. Different choices, as for example bound state approach[6] versus collective coordinate method[7], are discussed in the conclusions. In this talk, the extension of the collective coordinate method to higher flavor groups is presented, which is quite general and may be applied to a purly pseudoscalar as well as to e.g. a hidden gauge symmetric lagrangian which includes vector mesons.

2. Collective coordinate method

Let us specify on SU(4) for our discussion, the generalisation to an arbitrary SU(n) group is straight forward. Then, the 15 pseudoscalar fields are parametrised by the SU(4) matrix U , and analogously for the vector fields. The pseudoscalar meson {15} is illustrated in fig.1, where the Z-quantum number related to charm $Z = -C$ is plotted against hypercharge $Y = S - \frac{1}{3}C$ and the third component of isospin. In addition to the SU(3) {8}, this multiplet contains a SU(3) {3} and $\{\overline{3}\}$ with the $D(1867)$ and $D_s(1969)$ mesons and the heavy $\eta_c(2980)$ which is considered a $c\bar{c}$ state in the quark picture.

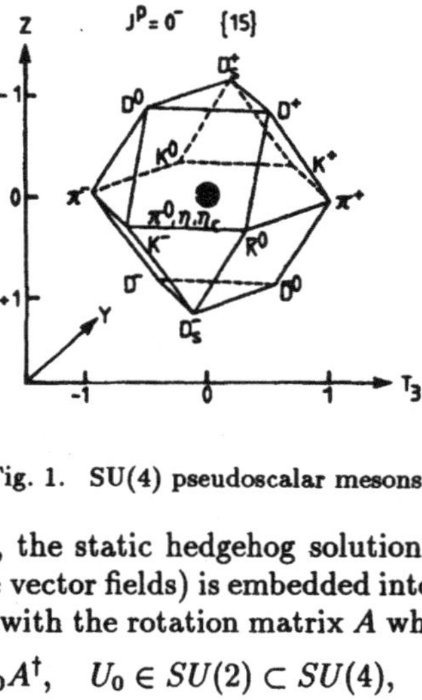

Fig. 1. SU(4) pseudoscalar mesons.

In the soliton sector, the static hedgehog solution U_0 for the pseudoscalar fields (and analogously for the vector fields) is embedded into the isospin subgroup of SU(4) and rotated collectively with the rotation matrix A which depends on 15 Euler angles

$$U = AU_0A^\dagger, \quad U_0 \in SU(2) \subset SU(4), \quad A \in SU(4)$$

$$D_{ab} = \frac{1}{2} tr \lambda_a A \lambda_b A^\dagger, \qquad A^\dagger \dot{A} = \frac{i}{2} \sum_{a=1}^{n} \lambda_a \Omega_a^R. \tag{1}$$

Consequently there are 15 right angular velocities Ω_a^R . Let us first consider the chirally symmetric part of the collective lagrangian, which is valid for every effective action with at most two time derivatives, in particular also for a hidden gauge symmetric action which includes vector mesons

$$L^0 = -M + \frac{1}{2}\Theta_\pi \sum_{a=1}^{3} (\Omega_a^R)^2 + \frac{1}{2}\Theta_K \left(\sum_{a=4}^{7} (\Omega_a^R)^2 + \sum_{a=9}^{12} (\Omega_a^R)^2 \right)$$

$$-\frac{N_C B_0}{2\sqrt{3}}\left(\Omega_8^R+\frac{1}{\sqrt{2}}\Omega_{15}^R\right). \tag{2}$$

It is interesting that all quantities which appear in that part of the lagrangian, the soliton mass M and the moments of inertia Θ_π and Θ_K are well known and the same as in SU(3). The contributions linear in the angular velocities are due to the Wess-Zumino term. As in SU(3), after defining right and left angular momenta

$$R_a=-\frac{\partial L^0}{\partial \Omega_a^R}, \qquad L_a=\sum_{b=1}^{15} D_{ab}R_b \tag{3}$$

the hamiltonian is obtained via a Legendre transformation under constraints and expressed by the SU(4) and SU(2) Casimir operators

$$H^0=M+\frac{1}{2\Theta_\pi}C(SU(2)_R)+\frac{1}{2\Theta_K}\left(C(SU(4)_R)-C(SU(2)_R)-\frac{9}{8}\right). \tag{4}$$

In addition to the well-known Wess-Zumino-Witten constraints stemming from the anomaly

$$R_8=\frac{N_C B_0}{2\sqrt{3}}=\sqrt{\frac{3}{4}}, \qquad R_{15}=\frac{N_C B_0}{2\sqrt{6}}=\sqrt{\frac{3}{8}} \tag{5}$$

$$R_{13}=0, \qquad R_{14}=0 \tag{6}$$

qualitatively new constraints appear, which are due to the angular velocities missing in the lagrangian, namely Ω_{13}^R and Ω_{14}^R in the SU(4) case. These additional constraints appear for SU(n) with $n\geq 4$ only and play an important role for the selection of the higher lying multiplets.

After quantisation, the right and left angular momenta become operators which obey the familiar commutation relations

$$[R_a,R_b]=-i\sum_c f_{abc}R_c, \qquad [L_a,L_b]=i\sum_c f_{abc}L_c. \tag{7}$$

From these equations it is noticed, that the constraints (5),(6) mutually commute and commute also with the hamiltonian (4), which is a necessary condition for the eigenstates of H^0 to obey the constraints simultaneously. Accordingly, such eigenstates may be labelled by the SU(4) irreducible representation $\{N\}$ and two sets of right and left quantum numbers $(\lambda\mu)_R J-J_3 Y_R Z_R$ and $(\lambda\mu)TT_3YZ$ with the SU(3) subgroup labels $(\lambda\mu)$, spin J–J_3 (isospin TT_3), hypercharge Y and Z- quantum number related to strangeness and charm $Y=S+B_0-\frac{1}{3}C$, $Z=\frac{3}{4}B_0-C$.

The Wess-Zumino-Witten constraints (5) fix the right hypercharge and Z-quantum number

$$Y_R=1, Z_R=\frac{3}{4} \qquad \text{or} \qquad S_R=C_R=0. \tag{8}$$

The additional constraints may be expressed by shift operators

$$W_{\pm} = R_{13} \pm iR_{14}, \tag{9}$$

which enhance respectively lower right strangeness and charm by one unit. These shift operators applied to a baryon state must yield zero according to (6).

The multiplets with the smallest Casimir energy are uniquely fixed by the Wess-Zumino-Witten constraints (8) to the $\{20_M\}$ with spin $J = \frac{1}{2}$ and the $\{20_S\}$ with spin $J = \frac{3}{2}$. These two lowest multiplets are illustrated in fig.2, again by plotting the Z-quantum number or charm against hypercharge and third component of isospin.

The additional constraints are fulfilled automatically for these two multiplets because $W_{\pm}$ applied to states with $S_R = C_R = 0$, the "nucleon" and "delta", yields zero. There are no states with $S_R = \pm 1$ and $C_R = \pm 1$ in this multiplet. The left quantum numbers may be equal to the 20 possibilities in these multiplets defining the 20 different baryons contained in each of them. The lowest baryon states may be represented by SU(4) D-functions which carry these quantum numbers.

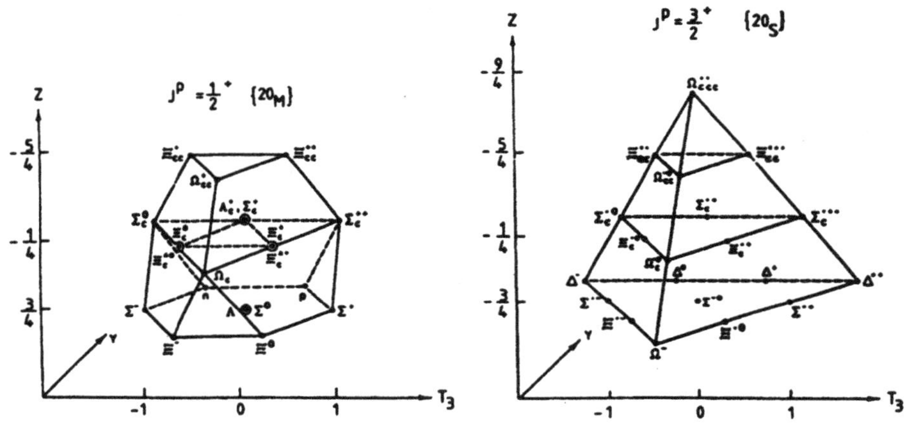

Fig. 2. The Wess-Zumino-Witten constraints select the lowest multiplets with the correct quantum numbers.

For the higher lying multiplets the situation complicates. Nevertheless all baryon states compatible with the constraints may be constructed in a systematical way. For this purpose it is convenient to introduce a SU(n) group representation with so-called Cartan-Weyl generators and Gel'fand-Zetlin patterns, which is described in detail in ref.8. The result is the following: the eigenstates of H^0 which obey all the constraints are in general represented by linear combinations of D-functions with the same left quantum numbers but different right quantum numbers i.e. different

SU(3)$_R$ subgroup labels. For example the lowest $J = \frac{1}{2}$ states are given by

$$\begin{aligned}
\langle\alpha|\{20_M\}\rangle &\sim D^{\{20_M\}}_{(\lambda\mu)TT_3YZ,(11)\frac{1}{2}-J_3 1\frac{3}{4}}(\alpha) \\
\langle\alpha|\{\ \overline{60}\ \}\rangle &\sim \sqrt{\frac{2}{3}} D^{\{\overline{60}\}}_{(\lambda\mu)TT_3YZ,(03)\frac{1}{2}-J_3 1\frac{3}{4}}(\alpha) - \sqrt{\frac{1}{3}} D^{\{\overline{60}\}}_{(\lambda\mu)TT_3YZ,(02)\frac{1}{2}-J_3 1\frac{3}{4}}(\alpha) \\
\langle\alpha|\{140\}\rangle &\sim \sqrt{\frac{3}{5}} D^{\{140\}}_{(\lambda\mu)TT_3YZ,(22)\frac{1}{2}-J_3 1\frac{3}{4}}(\alpha) - \sqrt{\frac{2}{5}} D^{\{140\}}_{(\lambda\mu)TT_3YZ,(11)\frac{1}{2}-J_3 1\frac{3}{4}}(\alpha).
\end{aligned} \tag{10}$$

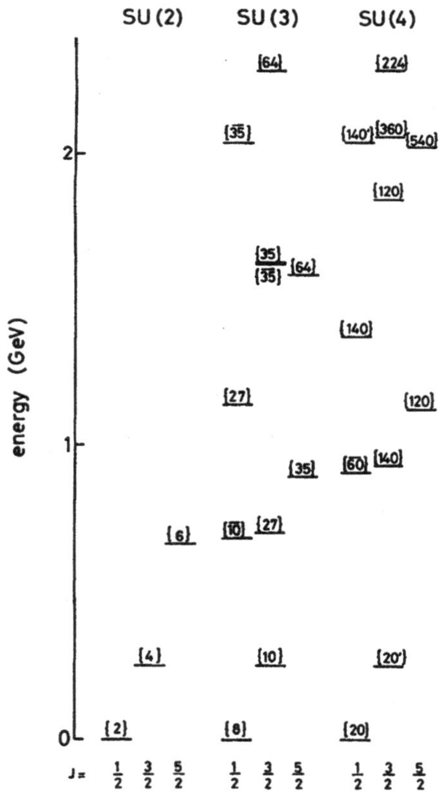

Fig. 3. Unperturbed SU(4) baryon spectrum for $J = \frac{1}{2}, \frac{3}{2}, \frac{5}{2}$ baryons compared to well-known SU(2) and SU(3) results. When the symmetry breakers are included the baryon masses split off in strange and charmed directions.

This possibility exists, because positions in the interior of higher multiplets are multiply occupied. The spectrum of the $J = \frac{1}{2}, \frac{3}{2}, \frac{5}{2}$ unperturbed baryon states is depicted in fig.3 for SU(2),SU(3) and SU(4). It is noticed that the split between the lowest $J = \frac{1}{2}$ and $J = \frac{3}{2}$ states is independent from the SU(n) group under consideration. In nature, of course, there is no good SU(n) symmetry for $n \geq 3$. The baryon masses attached to the same multiplet split off by about 150 MeV per unit of strangeness and even roughly by 1.5 GeV per unit of charm. The strong symmetry breaking makes its consideration indispensible for a reproduction of the experimental baryon spectrum.

3. Symmetry breaking

As already discussed, the heavy pseudoscalar and vector meson masses may be put to their experimental values by adding explicit symmetry breakers such that in the limit where these masses tend to infinity the essentials of the Isgur-Wise lagrangian are recovered. With the standard symmetry breakers for pseudoscalars and vectors, proportional to the quark mass matrix

$$H^{SB} = -L^{SB} = \frac{1}{2}\Gamma_3(1 - D_{88} - \frac{1}{\sqrt{3}}D_{815}) + \frac{1}{2}\Gamma_4(1 - \sqrt{3}D_{158} - D_{1515}), \qquad (11)$$

the strength of symmetry breaking is measured by one parameter Γ_ν for each flavor, which is a functional of the soliton profiles. The ratio $\Gamma_4/\Gamma_3 \simeq f_D^2 m_D^2 / f_\eta^2 m_\eta^2 \simeq 20$ provides the correct order of magnitude for the baryon mass splitting in charmed direction. However, for higher flavor groups SU(n), $n \geq 4$ the standard symmetry breakers (11) are considered to be insufficient because the Gell'man-Oakes-Renner relation does of course not hold for heavy mesons. Further symmetry breakers at least quadratic in the quark mass matrix are important to get a reasonable fit to the heavy meson masses. In the baryon sector these additional symmetry breakers lead to products of D-functions, which may be treated technically in the same way as the standard symmetry breaker (11).

For the calculation of the baryon spectrum and other baryon properties, the symmetry breakers have to be diagonalised in the basis of the eigenstates $|B\rangle$ of H^0

$$|\psi_B\rangle = \sum_B c_B|B\rangle, \qquad \sum_{B'}\langle B|H^0 + H^{SB}|B'\rangle c_{B'} = Ec_B. \qquad (12)$$

Consequently, the unperturbed baryon states of different multiplets mix and the energies of the individual baryons split off in strange, charmed and so on directions. In contrast to SU(3), where the wave functions are almost round, we expect extremely distorted wave functions for the higher flavor groups, which aquire components of quite high lying multiplets. For the calculation of the matrix elements in (12), integrals over D-functions have to be evaluated by Clebsch-Gordan coefficients, which despite of many promises are still not available for high lying irreducible representations of SU(n) with $n \geq 4$.

4. Conclusions

The collective coordinate method, which works remarkably well in SU(3) may be extended to general SU(n) groups, but the extension is nontrivial because for $n \geq 4$ in addition to the Wess-Zumino-Witten constraints, which supply the correct quantum numbers for the two lowest multiplets, qualitatively new constraints appear. The additional constraints fix the eigenstates of higher lying multiplets in a systematical and consistent way. These baryon states may serve as a suitable basis for the diagonalisation of the symmetry breaking piece of the lagrangian to finally obtain spectrum and wave functions of the physical baryons.

Provided that the lagrangian, e.g. a hidden gauge symmetric lagrangian, is compatible with the Isgur-Wise limit, when the heavy meson masses tend to infinity, the way how these masses tend to infinity has to be fixed by the symmetry breakers. Not only meson masses and decay constants should be reproduced, there is further information like for example the decays $\eta_c(2980) \longrightarrow \eta\pi\pi, K\bar{K}\pi$ which tests the symmetry breaking part of the lagrangian. This is an interesting task by its own.

Once the appropriate lagrangian is fixed, the best possible ansatz to solve the equations of motion in the $B = 1$ sector should be chosen. The following approaches will be briefly compared:

(a) SU(2) rigid rotator + bound state approach to strange, charmed and bottom mesons

(b) SU(3) rigid rotator + bound state approach to charmed and bottom mesons only

(c) SU(n) rigid rotator

(d) SU(n) slow rotator

The simple bound state approach (a) for strange, charmed and bottom mesons[9] bound to a rotating SU(2) soliton is already quite successful, but by construction it cannot account for the coupling of different flavors which may become important for baryons which carry for example strangeness and charm simultaneously. Furthermore, the strange quark is still considered as light in the sense that its mass is small compared to the chiral scale. The SU(3) symmetry breaking in strange direction is therefore weak, which is confirmed e.g. by the {8} baryon wave functions in the Skyrme model, which are almost spherical (less than 10 percent admixtures of higher representations). Thus, the approach (b) of binding the heavy mesons to a rotating SU(3) soliton would certainly be much more appropriate. This project, however, gets already quite involved because the equations of motion to be solved for the heavy mesons become Euler angle dependent through the SU(3) symmetry breakers. Also the charmed quark mass, although quite heavy, is still not too far from the chiral scale of roughly 1 GeV. The alternative (c), to start with a rotating SU(n) soliton as described in this contribution, has the appealing feature that the baryons are automatically attached to the right SU(n) multiplets from the very beginning. A serious disadvantage is the rigid rotator approach, which keeps the soliton profiles fixed such that the symmetry breakers just hinder the soliton to rotate say in charmed direction but may not influence the soliton itself. Consequently in the heavy meson limit, the bound state approach is not recovered. A way out would be the slow rotator (d) of Schwesinger and Weigel [10], which is supposed to comprise the approaches (b) and (c), and which may be treated with the techniques described in this contribution. In the end, there could be a unified description of strange, charmed and bottom mesons and baryons within one effective theory using a common approach in the $B = 1$ sector.

5. Acknowledgements

The author wishes to thank G. Holzwarth and B. Schwesinger for numerous discussions and helpful advices.

6. References

1. M. Rho, talk at this conference
2. N. Isgur and M. B. Wise *Phys. Rev. Lett.* **66** (1991) 1130
3. M. Bando, T. Kugo, K. Yamawaki *Phys. Rep.* **164** (1988) 317
4. T. Fujiwara, T. Kugo, H. Terao, S. Uehara, K. Yamawaki *Prog. Theor. Phys.* **73** (1985) 926
5. N. N. Scoccola, D.-P. Min, H. Nadeau, M. Rho *Nuc. Phys.* **A505** (1989) 497
6. C. G. Callan, I. Klebanov *Nuc. Phys.* **B262** (1985) 365
7. H. Yabu, K. Ando *Nuc. Phys.* **B301** (1988)601
8. H. Walliser, Siegen preprint 1992, to be published in *Nuc. Phys.*
9. M. Rho, D. O. Riska, N. N. Scoccola *Z. Phys.* **A341** (1992) 343
10. B. Schwesinger, H. Weigel *Phys. Lett.* **B267** (1991) 438, and talk at this conference

IV. Meson-Baryon System

CASIMIR ENERGY IN THE SKYRME MODEL

B. MOUSSALLAM
Division de physique théorique, Institut de physique nucléaire*
Université Paris-Sud, 91406 Orsay, France

ABSTRACT
The so called Casimir energy embodies the $O((N_c)^0)$ contribution to the skyrmion mass according to the semi-classical expansion theory of solitons. We claim that this contribution can be accurately estimated despite the fact that some of the counterterms, provided in principle by chiral perturbation theory, are not known in practice. Using ζ function techniques we show that $E_{cas} = E_{cas}(\mu) + E_{ct}(\mu)$ where $E_{cas}(\mu) >> E_{ct}(\mu)$ because it incorporates all the zero-mode contributions and it can be exactly calculated. Our results confirm that the fourth order Skyrme lagrangian does not seem to provide a correct description of the lightest baryons as solitons. We show that a simple extension to order six gives, on the contrary, good results without tuning the parameters of the chiral lagrangian.

1. Introduction

It is a common experience to most skyrmion practitioners that the mass tends to come out too high. Even if one uses rather sophisticated chiral lagrangians, incorporated vector and axial vector mesons (e.g. [1]) one finds the predicted nucleon mass to lie nearly 50% too high and the situation becomes much worse in SU(3) extensions. Most other observables, in contrast to that, seem to be predicted with a much better accuracy. This has led to the belief that the mass may have a somewhat special status.

In this talk, I would like to try to convince you that this is not the case and that the problem with the mass has a very natural explanation once one tries to perform the semi-classical (which is also the $1/N_c$) expansion in a systematic way. This, in fact, is not usually done in the context of the skyrmion. Following ref.[2], one identifies collective coordinates and quantization is effected only at the level of these coordinates. If one does this for the mass, one finds besides the classical $O(N_c)$ contribution a $O(1/N_c)$ one. This procedure gives no contribution of order N_c^0. This contribution is precisely what one calls the "Casimir energy" (CE) and its evaluation requires that one deals with the non-collective coordinates.

What one has to do in principle can be inferred from the semi-classical soliton theory which was developped a long time ago (see e.g.[3]). In practice, however, one seems to face the difficulty that in early work, solitons were taken to live in 1+1 dimensions with a renormalizable lagrangian, while the skyrmion lives in 3+1 dimensions and the lagrangian is not renormalizable. A safe approach to this problem is to use a lagrangian involving only chiral field degrees of freedom. This is after all what

*Unité de recherche des universités Paris 11 et Paris 6 associée au CNRS

Skyrme originally did and we nowadays have a theory, Chiral perturbation theory (ChPT)[4][5], which precisely gives us such a lagrangian and furthermore tells us how to make sense out of loops. Quite remarkably, recent determinations of the values of the parameters appearing in the lagrangian of ChPT turn out to be surprisingly close to those which Skyrme has guessed thirty years ago.

2. Basic ideas and formulae

Vibrational degrees of freedom for the skyrmion were first considered in ref.[6] who did not attempt, however, a complete evaluation of their role. One approximation scheme to actually calculate the CE was proposed by Schnitzer[7] who found a large positive value (of the order of 500 MeV). Later on, the presence of vector mesons in the lagrangian were found (in the same approximation) to considerably reduce this effect to a negligible 50 MeV[8]. Concurrently, an estimate was made in ref.[9] which makes use of the relationship between the CE and the effective action to one-loop. They recognized that an ultra-violet divergence was present and they used a derivative expansion approximation of the effective action. A somewhat better approximation was proposed later on [10] which seems nevertheless to yield a rather similar result of approximately −200 MeV.

How come that all these results are different from each other? The answer is that one must be very careful with the approximations that one makes. Most approximations will in fact kill the most important part of the effect. To show this, we will start from a formula which is exact and which relates the CE to the pion-skyrmion phase shifts. The correct size of the effect will be controlled by the fact that the phase-shifts are very large at the origin because there are six zero-modes (associated with three translations and three rotations). Any approximation of the Born type for the phase-shifts will be essentially incorrect. An approximation which respects the low-energy behaviour of the phases was proposed in ref.[11].

Let us start with a simple 1+1 dimensional situation as a warm up. Many features are exactly similar to the 3-dimensional case. Consider, for instance, the action:

$$S = \frac{1}{2}\int d^2x \left((\partial_\mu\phi)^2 + m^2\phi^2 - \frac{\lambda}{2}\phi^4\right) \tag{1}$$

It is convenient to rescale the coordiates and the fields:

$$\phi \to \frac{m}{\sqrt{\lambda}}\phi \qquad x \to \frac{1}{m}x$$

so the action now looks like:

$$S = \frac{m^3}{2\lambda}\int d^2x \left((\partial_\mu\phi)^2 + \phi^2 - \frac{1}{2}\phi^4\right) \tag{2}$$

We see that $1/\lambda$ appears in front of the action so the semi-classical expansion is identical to the weak coupling expansion. We have exactly the same situation in three dimensions with $N_c$replacing $1/\lambda$. The classical solution (the "kink") and the classical mass are easily found:

$$\phi_{class} = \tanh(\frac{x - x_0}{\sqrt{2}}), \qquad M_{class} = \frac{2\sqrt{2}m^3}{3\lambda} \tag{3}$$

Next, one has to consider fluctuations around the classical solution and it is not difficult to show that the leading correction to the classical mass (i.e. of order λ^0) has the following formal expression [3]:

$$M^{(0)} = \frac{1}{2}\left[\mathrm{Tr}\left(- \partial_x^2 + 3\tanh^2(\frac{x}{\sqrt{2}}) - 1\right)^{\frac{1}{2}} - \mathrm{Tr}\left(- \partial_x^2 + 2\right)^{\frac{1}{2}}\right] \equiv \frac{1}{2}(\mathrm{Tr}H^{\frac{1}{2}} - \mathrm{Tr}H_0^{\frac{1}{2}}) \tag{4}$$

Obviously if we rewrite the trace in terms of eigenvalues we obtain that

$$M^{(0)} = \sum \omega_n - \sum \omega_n^0 \tag{5}$$

which is why this contribution is called the Casimir energy, by analogy with the classic QED effect [12]. The operator involved in (4) is obtained by expanding the action to second order in powers of the fluctuation. The corresponding operator for the skyrmion is obtained by exactly the same procedure. A careful derivation can be found in ref.[13].

As was mentioned already, an extremely useful formula for practical purposes arises upon expressing (4) in terms of phase shifts[14]. A simple direct derivation, valid for any space dimension, is as follows. One first uses the identities:

$$\mathrm{Tr}H^{\frac{1}{2}} = 2\mathrm{Tr}\int_M^\infty dE\, E^2\delta(E^2 - H) = -\frac{1}{i\pi}\mathrm{Tr}\int_M^\infty dE\, E\left(\frac{E}{E^2 - H + i\epsilon} - \frac{E}{E^2 - H - i\epsilon}\right) \tag{6}$$

where M is the lower bound of the continuous spectrum (in the kink example $M = \sqrt{2}m$). Next one recognizes that:

$$\mathrm{Tr}\left(\frac{E}{E^2 - H + i\epsilon} - \frac{E}{E^2 - H_0 + i\epsilon}\right) = \frac{d}{dE}\ln\Delta^+, \tag{7}$$

where

$$\Delta^+ = \det\left(1 - \frac{1}{E^2 - H_0 + i\epsilon}(H - H_0)\right) \tag{8}$$

So we obtain a formula:

$$M^{(0)} = \frac{1}{4i\pi}\int_M^\infty dE\, E\frac{d}{dE}(\ln\Delta^+ - \ln\Delta^-) \tag{9}$$

(This takes care of the continuous part of the spectrum. Obviously, one must also add the discrete part in the trace if there is any). Now Δ^+ is nothing but the Fredholm determinant. In one space dimension it is a well know fact from scattering theory that its phase is the phase-shift [15]:

$$\Delta^{\pm} = |\Delta| \exp(\pm i\delta(E)) \tag{10}$$

so our expression (4) becomes:

$$M^{(0)} = \frac{1}{2\pi} \int_M^{\infty} dE E\, \delta'(E) \tag{11}$$

In order to generalize to higher dimensions we must return to (9). If $H - H_0$ is a radial potential for instance one can again express the phase of the Fredholm determinant in terms of the phase-shifts of the radial operators $\delta(E) \equiv \text{phase}(\Delta^+) = \sum(2j+1)\delta_j(E)$. The generalization to the skyrmion case is obvious: for every value of the grand-spin quantum number, J, we have three eigen-phase-shifts[16] so we must sum over these before summing over J.

At this point it would seem that we are running into trouble. A look at refs.[16] reveals that (except for J=0) the phase-shifts are linearly diverging functions of E and one can show that after performing the J sum things become worse: one ends up with a cubic divergence. Even in one dimension, in fact, the integral (11) is logarithmically divergent. In that case, however , it is enough to remember that at order one in $\hbar$ there is an extra contribution to the energy which comes from the one-loop counterterm in the lagrangian. Once this is taken into account the divergence disappears. We will see later how one can generalize this mechanism to the skyrmion. To begin with, we must discuss the lagrangian.

3. Chiral perturbation theory and the skyrmion:

In the limit where the masses of the $N_f$light quarks are set to zero the QCD lagrangian is invariant under the chiral SU(N_f)×SU(N_f) group. This invariance is spontaneously broken by the QCD vacuum down to SU(N_f). The spectrum thus consists of $N_f$zero-mass Goldstone bosons and there is a mass gap of around $\Lambda = 1$ GeV above which one finds meson resonances, baryons, etc.. . ChPT is a systematic framework for describing low energy phenomena (with typical energy E) as an expansion in powers of E/Λ. In fact, it is a more general expansion which involves quark masses, external fields etc... but let us assume that these are zero for the moment. It is convenient to encode the goldstone bosons in a unitary matrix on which the chiral group operates linearly: $U = \exp(i\vec{\tau}.\vec{\pi})$ then the most general dynamics can be expressed with the aid of a chiral lagrangian:

$$\mathcal{L}(U) = \mathcal{L}_2 + \mathcal{L}_4 + \mathcal{L}_6 + ... \tag{12}$$

where the subscripts denote the numbers of derivatives of the matrix field U. As shown by Weinberg[4] if one wants to expand an amplitude to order E^n with $n > 2$ one must compute loops. There are simple counting rules, for example that one loop made out of two $\mathcal{L}_2$ vertices contribute at order 4, one loop with one $\mathcal{L}_2$ vertex and one $\mathcal{L}_4$ contributes at order 6, a two-loop amplitude with $\mathcal{L}_2$ vertices is also of order 6, etc... The chiral lagrangian contains the counterterms which render the loops finite.

If we are able to construct the skyrmion out of the chiral lagrangian (12) then, at least in principle, one should be able to renormalize the Casimir energy, which is a one-loop effect. Note that it is not clear that the chiral expansion does apply for the soliton but, after all, one might expect the average energy of a pion "inside" a skyrmion to be of the order of 200 MeV since the skyrmion size is expected to be of the order of 1 fm. So why shouldn't we try?

The chiral lagrangian has so far been computed up to order four[5] and it reads (external sources being switched off):

$$\mathcal{L}_{ChPT} = \frac{F^2}{2}\vec{A}^{\mu}.\vec{A}_{\mu} + \frac{1}{96\pi^2}(\bar{l}_1 - 1 + \ln\frac{M^2}{\mu^2})\left(\vec{A}^{\mu}.\vec{A}_{\mu}\right)^2 + \frac{2}{96\pi^2}(\bar{l}_2 - 1 + \ln\frac{M^2}{\mu^2})\left(\vec{A}^{\mu}.\vec{A}_{\nu}\right)^2 \tag{13}$$

where

$$i\vec{\tau}.\vec{A}^{\mu} = U^{\dagger}\partial^{\mu}U, \qquad M \simeq m_{\pi}, \qquad F \simeq f_{\pi}.$$

We see that a scale μ appears in (13). This is because it incorporates counterterms. If one uses a regularization prescription like dimensional regularization, one gets rid of the 1/(n-4) pole and one is left with a scale dependence. A similar but more convenient scheme for our purposes is the zeta function one[17] as we will see later. In order to discuss solitons we must perform an $N_c$expansion so we might start by assuming:

$$\mathcal{L}^{(1)} = \mathcal{L}_{ChPT}(\mu = m_{\rho}) \tag{14}$$

where the superscript designates the $N_c$order. This is a meaningful assumption provided subleading terms in $N_c$are very small for this particular scale value $\mu = m_{\rho}$. This seems indeed to be the case as follows from the recent work of ref.[18] who have analysed Kl4 decays and showed that this provides a quantitatively neat test for this suppression. Note that the fact that the terms of order N_c^0 in the chiral lagrangian are found to be small by no means implies that the correction of the same order to the skyrmion mass should also be small. However, this will turn out to be important for the accuracy of the CE determination, as we will see.

Let us now rewrite the fourth order lagrangian using notations which are familiar in the skyrmion context:

$$\mathcal{L}_4^{(1)} = \frac{1}{4e^2}(\vec{A}^{\mu} \wedge \vec{A}^{\nu}).(\vec{A}_{\mu} \wedge \vec{A}_{\nu}) + \frac{\gamma}{2e^2}(\vec{A}^{\mu}.\vec{A}_{\mu})^2 \tag{15}$$

The first question that one might ask is whether the values of the parameters e (Skyrme parameter) and γ that are determined from ChPT are compatible with a stable soliton solution. Remember in this context that the γ term tends to destabilize the soliton. An upper bound for stability was found to be $\gamma < 0.12$[19]. If we look at the 1984 values of Gasser and Leutwyler we find:

$$e = 6.8 \pm 4.15 \qquad \gamma = -0.06 \pm 0.2$$

These numbers are not very conclusive because of the large error bars. Let us now consider the more recent determination of Riggenbach et al.[18]:

$$e = 7.1 \pm 2.30 \qquad \gamma = 0.03 \pm 0.03$$

The error bars are considerably smaller and we see that ChPT now seems perfectly compatible with a stable skyrmion. The results are in fact strikingly similar with the numbers proposed by Skyrme 30 years ago:

$$e = 6.28 \qquad \gamma = 0$$

Unfortunately, chiral order four is not enough for a consistent description of the skyrmion. Firstly it is not consistent with the chiral expansion because the virial theorem implies that the contribution of $\mathcal{L}_2$ is identical to that of $\mathcal{L}_4$ instead of being much larger. Secondly it it is not consistent with the large $N_c$expansion either, because, as we will soon discover, the $N_c$contribution turns out to be nearly identical to the N_c^0 one instead, again, of being significantly larger.

The situation is perhaps not desperate. In fact I claim that a mere extension to chiral order six is sufficient to cure, apparently, all the difficulties. We need however to have a model in order to make a guess for the coefficients of the terms appearing in $\mathcal{L}_6$ since they have not yet been worked out from ChPT. A reasonable starting point seems to be provided by the observation[20] that all of the 10 parameters which appear in the chiral order four lagrangian can be saturated to a very good approximation by the contribution of the low-lying vector, axial vector, scalar and pseudo-scalar resonances. In particular since the rho meson saturates the Skyrme term it is natural to consider the omega meson as well which contributes at sixth order:

$$\mathcal{L}_{6,\omega} = -\frac{1}{2}\frac{\beta^2}{M_\omega^2} B_\mu B^\mu, \qquad B_\mu = \frac{\epsilon_{\mu\nu\lambda\sigma}}{12\pi^2}(\vec{A}_\nu, \vec{A}_\lambda, \vec{A}_\sigma) \tag{16}$$

The value of the coupling parameter β can be estimated to be rather large $\beta \simeq 9.3$[21] so it is tempting to assume that $\mathcal{L}_{6,\omega}$ is the dominant sixth order term. This is supported by estimates by Walliser[22] who showed that the rho induced sixth-order term is much smaller than the omega one. One also notices that the rho and scalar contributions tend to cancel each other.

If one includes the contribution (16) in the chiral lagrangian and looks for a soliton solution, the result is found to be in much better agreement with the chiral expansion than before. Due to the repulsive effect of the omega induced sixth order term the profile size is much larger than before and, as a result, the contribution of $\mathcal{L}_2$ becomes five times larger than that of $\mathcal{L}_4$. We will see in the sequel that the $N_c$expansion also seems to become coherent, but before we can actually evaluate the CE in order to check that, we must discuss the ζ function method of regularization which is a very important technical ingredient in the calculation.

4. ζ function regularization method

The basic idea is to introduce instead of a sum like $\sum(\omega^2)^{\frac{1}{2}}$ the sum $\sum(\omega^2)^{\frac{1}{2}-s}$ depending on the complex parameter s. One can eventually show that the function is analytic in s and further define the limit of interest, $s = 0$, by analytic continuation. The non trivial part is to actually perform this analytic continuation in a situation like ours where the eigenvalues ω_n are known only numerically. As we will see, the phase-shift representation provides a simple solution to this problem. Before we turn to that, however, we must make sure that the regularization procedure that we use for the Casimir energy is the same as the one which is used to regularize Green's functions (leading to the counterterms as they appear in (13)). Now Green's functions to one loop are generated by an effective action which can be written as a trace log of a four dimensional operator. If we call O such an operator then the corresponding ζ function is defined as[23]:

$$\zeta_O(s) = \frac{1}{\Gamma(s)}\mathrm{Tr}\int_0^\infty d\tau\, \tau^{s-1}\exp\left(-\frac{\tau O}{\mu^2}\right) \tag{17}$$

the scale μ is introduced at this level in order to make τ dimensionless. The regularized form of the trace log is then given by:

$$\mathrm{Tr}\log(O) \equiv -\zeta'(s=0). \tag{18}$$

We can relate the Casimir energy to a four dimensional operator via the identity:

$$\mathrm{Tr}(H^{\frac{1}{2}}) = \lim_{T\to\infty}\frac{1}{T}\mathrm{Tr}\log\left(-\partial_t^2 + H\right) \tag{19}$$

which holds for a time independent operator H. If we let $O = (-\partial_t^2 + H)$ in formula (17) a simple calculation shows that the ζ function regularization of the Casimir energy which matches the one used for the effective action is:

$$\mathrm{Tr}(H^{\frac{1}{2}}) \equiv -\zeta'(0) \quad \text{with} \quad \zeta(s) = -\frac{\mu^{2s}\Gamma(s-\frac{1}{2})}{\Gamma(s)\Gamma(-\frac{1}{2})}\frac{1}{2}\mathrm{Tr}H^{\frac{1}{2}-s} \tag{20}$$

The same derivation as before yields the appropriate phase-shift formula:

$$\mathrm{Tr}H^{\frac{1}{2}-s} - \mathrm{Tr}H_0^{\frac{1}{2}-s} = \frac{1}{\pi}\int_0^\infty dp\,\delta'(p)(p^2+M^2)^{\frac{1}{2}-s} \tag{21}$$

where a finite pion mass M was introduced. It can be shown that the large momentum behavior of the phase function is of the form:

$$\delta(p) = \bar{a}_0 p^3 + \bar{a}_1 p + \frac{\bar{a}_2}{p} + \ldots \tag{22}$$

where the $\bar{a}_i$'s are numbers which are simply related to the heat kernel expansion of the operator H. One can subtract and add this leading asymptotic behavior and perform the analytic continuation in s by using the well-known integrals[24]

$$\int_0^\infty dp\,p^m(p^2+M^2)^{\frac{1}{2}-s} = \frac{1}{2}M^{m+2-2s}\frac{\Gamma(\frac{m+1}{2})\Gamma(s-1-\frac{m}{2})}{\Gamma(s-\frac{1}{2})} \tag{23}$$

Finally, one obtains the finite and closed form formula for the regularized CE:

$$\begin{aligned} E_{cas}(\mu) =& \frac{1}{2\pi}\Big\{ \int_0^\infty dp \Big[-\frac{p}{\sqrt{p^2+M^2}}(\delta(p) - \bar{a}_0 p^3 - \bar{a}_1 p) + \frac{\bar{a}_2}{\sqrt{p^2+\mu^2}}\Big] \\ & - \frac{3\bar{a}_0}{8} M^4 (\frac{3}{4} + \frac{1}{2}\ln\frac{\mu^2}{M^2}) + \frac{\bar{a}_1}{4} M^2 (1 + \ln\frac{\mu^2}{M^2}) - M\delta(0) \Big\} \end{aligned} \tag{24}$$

Note that it is well defined for zero as well as finite pion mass M. It is also clear that, by construction, only the low-energy behavior of the phase shift is important in the integration. Now, in a way similar to the 1+1 dimensional case we must add counterterms to (24), i.e. the $O((N_c)^0)$ part of the chiral lagrangian:

$$E_{ct}(\mu) = -(\mathcal{L}_2^{(1)} + ... + \mathcal{L}_n^{(1)} - \mathcal{L}_{ChPT}) \tag{25}$$

Actually, since $\mathcal{L}^{(1)}$ is being truncated at chiral order n (in practice $n = 4$ or 6) E_{ct} contains terms of order $O(N_c)$ and of chiral order $n+2, .., 2n$. Consistency requires that these should be of the same order in magnitude (or smaller) that the terms of lower chiral order but subleading in N_c. In this respect already, $n = 6$ is more satisfactory than $n = 4$. Next, when we add the two pieces:

$$M^{(0)} = E_{cas}(\mu) + E_{ct}(\mu) \tag{26}$$

the scale dependence should disappear (at least up to $O(1/N_c)$ terms). In practice, of course, it does not since we do not know enough terms in $\mathcal{L}_{ChPT}$. One would therefore like to argue that

$$E_{ct}(\mu) << E_{cas}(\mu) \tag{27}$$

This, of course, cannot hold for arbitrary values of μ. Now as we have seen, it follows from ref.[18] (and also from ref.[20]) that for the particular value $\mu = m_\rho$ terms of order N_c^0 are strongly suppressed in the chiral lagrangian (by a factor of 10 or so) so it seems a good idea for us to choose $\mu = m_\rho$. According to formula (25) this will suppress the contributions to E_{ct} up to chiral order n. Those of order $n+2, n+4, ...$ are of order $N_c$so one must assume that they are suppressed because of their high chiral order. The main reason why (27) should hold, though, is that $E_{cas}(\mu)$ is enhanced because it incorporates the zero-mode contributions. In our formalism, they show up via Levinson's theorem forcing the phase-shift at the origin to be fairly large (=6π). An approximate way to estimate the CE by singling out the zero-mode contributions was imagined recently by Holzwarth[25]. Let us now illustrate these points on some examples.

5. Some results and conclusions

Let us first consider the Skyrme lagrangian $\mathcal{L} = \mathcal{L}_2 + \mathcal{L}_{4,sk}$ (see (15)) with physical values of f_π and m_π and

a)e=5.5 (by analogy with ref.[2]). Following the method described above one finds:

$$M^{(0)} = -957 + (-72) = -1029 \text{ MeV}$$

where the number in parenthesis is the counterterm contribution (which is indeed small). This is to be compared with the leading $N_c$contribution:

$$M^{(1)} = 1263\,\mathrm{MeV}$$

Clearly, the $1/N_c$expansion seems to be in trouble and furthermore, the nucleon mass is found to be much too small. In fact, better phenomenological results are expected from the Skyrme lagrangian if one takes a smaller value for e. Let us consider then: b)e=4.0 (which gives for example a reasonable delta-nucleon splitting of 250 MeV). In this case, one finds:

$$M^{(0)} = -805 + (-452) = -1297\mathrm{MeV}$$

while the leading contribution is

$$M^{(1)} = 1761\mathrm{MeV}$$

This is slightly better than before and one could eventually arrive at a reasonable nucleon mass, but an unsatisfactory feature now is that the counterterm contribution is rather large. This is because of the large mismatch between the value of e and the one compatible with ChPT (see sec. (3)). In this situation one can no longer argue that the unknown contributions of E_{ct} are necessarily small.

The only way out of this dilemma seems to include higher order terms in the chiral lagrangian. Let us add an omega induced sixth order term then, and consider:$\mathcal{L}^{(1)} = \mathcal{L}_2 + \mathcal{L}_{4,sk} + \mathcal{L}_{6,\omega}$ with the parameters $e = 7.22$ (from the meson saturation fit of [20]) and $\beta = 9.3$ from [21]. The calculation gives, in that case:

$$M^{(0)} = -604 + (+153) = -451\mathrm{MeV}$$

while the classical value

$$M^{(1)} = 1553\mathrm{MeV}$$

The striking feature is the strong reduction of the Casimir contribution. Now the large $N_c$expansion looks much more reasonable than before (the correction is a factor of 3 smaller than the classical mass). We argued in sec. (3) that the chiral expansion was also more justified. The two things are in fact related: the reason why E_{cas} is smaller is that the phase-shift function $\delta(p)$ drops faster as a function of the momentum, and this is because the classical profile function has a larger extension in space. Note that the counterterm contribution is positive now (because the choice of e is slightly larger than that of ChPT) but the Casimir energy itself is always found to be negative. If we add $M^{(1)}$ and $M^{(0)}$ we find that the nucleon mass is correctly predicted to within 20% while we did not attempt to fit any parameter in the chiral lagrangian.

In conclusion, a reasonable picture seems to emerge provided one incorporates besides the fourth order rho induced term a sixth order omega meson one. This is not really a surprise. In fact, the most succesfull skyrmion phenomenology seems to require that one has these resonances (and a few others) explicitely in the lagrangian.

The rationale for including resonances is that this should extent the range of validity of the effective lagrangian from $E << \Lambda$ to $E \simeq \Lambda$. It is likely that one could extend the calculation of the CE to that case but the value of the "optimal scale" should perhaps be larger. An open question at the moment is how to estimate loop corrections for other nucleon observables. In particular, it is not clear (to me) whether the zero-mode enhancement which lead to a sizeable effect in the case of the mass will also operate for some other observables and how.

References

1. N. Kaiser, these proceedings.
2. G.S. Adkins, C.R. Nappi and E. Witten, *Nucl. Phys.* **B228** (1983) 552.
3. S. Coleman, *"Aspects of symmetry"* Cambridge University Press (1985) Chap. 6.
4. S. Weinberg, *Physica* **A96** (1979) 327.
5. J. Gasser and H. Leutwyler, *Ann. Phys.* **158** (1984) 142.
6. L.C. Biedenharn, Y. Dothan and M. Tarlini, *Phys. Rev.* **D31** (1985) 649.
7. H. Schnitzer, *Nucl. Phys.* **B261** (1985) 546.
8. M. Chemtob, *Nucl. Phys.* **A473** (1987) 613.
9. I. Zahed, A. Wirzba and U-G. Meissner,*Phys. Rev.* **D33** (1986) 830.
10. A. Dobado and J. Terron, *Phys. Lett.* **B247** (1990) 581.
11. B. Moussallam and D. Kalafatis, *Phys. Lett.* **B272** (1991) 196.
12. C. Itzykson and J-B Zuber, *"Quantum field theory"* Mc Graw-Hill (1980) p.138.
13. B. Moussallam, preprint IPNO/TH 92-75, submitted to *Annals of Physics.*
14. R.F. Dashen, B. Hasslacher and A. Neveu, *Phys. Rev.* **D10** (1974) 4114, 4130.
15. R.G. Newton, *"Scattering theory of waves and particles"*, Mc Graw-Hill (1982).
16. H. Walliser and G. Eckart, *Nucl. Phys.* **A429** (1984) 514.
    M.P. Mattis and M. Karliner, *Phys. Rev.* **D31** (1985) 2833.
17. D.G.C. McKeon and T.N. Sherry, *Phys. Rev.* **D35** (1987) 3854.
18. C. Riggenbach, J. Gasser, J.F. Donoghue and B.R. Holstein, *Phys. Rev.* **D43** (1991) 127.
19. T.N. Pham and T.N. Truong, *Phys. Rev.* **D31** (1985) 3027.
20. G. Ecker, J. Gasser, A. Pich, E. de Rafael, *Nucl. Phys.* **B321** (1989) 311.
21. M. Lacombe, B. Loiseau, R. Vinh Mau and W.N. Cottingham, *Phys. Rev.* **D38** (1988) 1491.
22. H. Walliser, these proceedings.
23. P. Ramond, *"Field theory: a modern primer"*, Addison-Wesley, New York (1981).
24. *"Handbook of mathematical functions"*, ed. M. Abramowitz and I.A. Stegun, Dover pub., p258.
25. G. Holzwarth, these proceedings.

LOW-ENERGY πN SCATTERING S AND P WAVES AND QUANTUM CORRECTIONS TO THE SOLITON MASS

G. HOLZWARTH
Fachbereich Physik, Universität-GH-Siegen
Adolf-Reichweinstr., W-5900 Siegen, Germany

ABSTRACT

Recent results within the soliton model for low-energy S- and P-wave πN elastic scattering are presented. The P33-resonance is analysed in terms of the dominating $\mathcal{O}(1)$ zero mode contribution and the linear time-derivative coupling. The isorotation and translation zero modes are used to obtain a lower limit for the self-energy corrections to the soliton mass. The resulting nucleon mass is near 1.1 GeV which still leaves room for other small quantum corrections. Additional $\mathcal{O}(1/N_c)$ mass shifts are estimated from the linear time-derivative interaction.

1. Meson-baryon scattering

A decisive advantage of the Skyrme-soliton approach to baryons is the ease with which it naturally incorporates meson-baryon scattering. Asymptotically ($r \to \infty$) time-dependent fluctuations $\eta(x,t)/f_\pi$ of the soliton represent free mesons and thus supply a natural basis for extracting the meson-baryon S-matrix. Serious complications arise from the fact that the soliton performs collective rotations in coordinate and flavor space with angular velocities Ω' (which supply spin, isospin, strangeness, etc. to the baryon). The corresponding moments of inertia are $\mathcal{O}(f_\pi^2)$, therefore the rotational velocities Ω' for the lowest multiplets are of the order $\mathcal{O}(1/f_\pi^2)$.

This allows an ordering of the soliton-meson coupling terms in powers of $1/f_\pi$, or, equivalently, in powers of $1/\sqrt{N_c}$ where N_c is the number of colors of the underlying fundamental field theory. Each power of the fluctuations η carries a factor $1/\sqrt{N_c}$, so the expansion of an effective lagrangian $\mathcal{L}$ (which contains time derivatives only up to second order) in terms of η has the general form

$$\begin{array}{rlll|ll} \mathcal{L}(\Omega',\eta,\dot\eta) = & - & M_s & & = & \mathcal{L}_{soliton}[\mathcal{O}(N_c)] \\ & + & \frac{1}{2f_\pi^2}(\int \dot\eta T \dot\eta - \int \eta V \eta) & & + & \mathcal{L}_{vib}[\mathcal{O}(1)] \\ & + & \frac{1}{2}\Theta\Omega'^2 & & + & \mathcal{L}_{rot}[\mathcal{O}(1/N_c)] \\ & + & \frac{1}{f_\pi}\Omega' \int W_1 \dot\eta & & + & \mathcal{L}^{(1)}_{coupl}[\mathcal{O}(1/\sqrt{N_c})] \\ & + & \frac{1}{f_\pi^2}\Omega' \int \eta W_2 \dot\eta & & + & \mathcal{L}^{(2)}_{coupl}[\mathcal{O}(1/N_c)] \\ & + & \cdots & & + & \cdots \end{array} \tag{1}$$

This is a very formal notation, in which M_s is the static soliton mass and η comprises the fluctuations of all fields in the theory $(\pi, \varrho, \omega, \ldots)$. The tensoroperators $T, V, W_1, W_2, \ldots$ are $\mathcal{O}(N_c)$ and fixed through the original $\mathcal{L}(U, \varrho, \omega, \ldots)$; their explicit form, however, depends sensitively on the choice of parametrizing the matrix U in terms of the pionic fluctuations.

Attempts to deal with all degrees of freedom in $\mathcal{L}(\Omega', \eta, \dot{\eta})$ in a complete coupled channels approach are confronted with a number of serious difficulties: The flavor rotations of the soliton and the fluctuations parametrize the original meson fields in an overcomplete way, which imposes constraints on the corresponding canonical variables. Quantization therefore (apart from the usual ambiguities in the ordering of operators) has to deal properly with these constraints and with the corresponding gauge conditions. This leads to extremely tedious procedures, while approximations may cause admixtures of spurious states which are difficult to identify. The Legendre transformation may lead to additional self-coupling of the fluctuations. All these problems have as yet prevented a fully consistent treatment of $\mathcal{L}(\Omega', \eta, \dot{\eta})$, even for πN scattering only.

The fact that $\mathcal{L}$ is formulated in terms of two types of dynamical degrees of freedom, rotations and fluctuations, creates two different types of baryon resonances: The $\mathcal{O}(1/N_c)$ rotational energy (i.e. the Casimir operators of the corresponding flavor group) separates different flavor multiplets. Together with the flavor symmetry breaking terms this accounts for the hyperon spectra and properties. On the other hand the $\mathcal{O}(1)$ fluctuational energy $\mathcal{L}_{vib}$ creates those baryon resonances which are based on a specific flavor multiplet member as a "ground state", and it provides the continuum in which the Δ and other hyperons are embedded as bound states.

The $\mathcal{O}(1/N_c)$ time-derivative (TD) coupling $\mathcal{L}^{(2)}_{coupl}$ is quadratic in the η's, so we consider it as a perturbation to the $\mathcal{O}(1)$ fluctuational part $\mathcal{L}_{vib}$. For this purpose we use for the η's in $\mathcal{L}^{(2)}_{coupl}$ the quantized distorted scattering waves of $\mathcal{L}_{vib}$. We shall see that this DWBA procedure gives excellent account of low-energy S-wave scattering and provides the right amount of isospin splitting for the P13 and P31 P-channels.

The $\mathcal{O}(1/\sqrt{N_c})$ linear (in η) time-derivative (TD) coupling $\mathcal{L}^{(1)}_{coupl}$ cannot be treated in the same way because it vanishes identically if we insert for η the distorted scattering solutions of $\mathcal{L}_{vib}$. This is due to the fact that $\mathcal{L}^{(1)}_{coupl}$ has the form of a scalar product between a scattering wave and the (rotational) zero mode, i.e. between two eigenmodes of $\mathcal{L}_{vib}$ with different energies. Still, $\mathcal{L}^{(1)}_{coupl}$ plays an important role for the coupled equations of motion or, in a Hamilton approach, for the definition of canonical momenta and their constraints. After all, we expect it to provide the width for Δ and hyperon states which causes these bound states to appear as resonances in meson-baryon scattering. Verschelde[1] has shown that in the "nonrigid gauge" $\mathcal{L}^{(1)}_{coupl}$ survives in the form of a linear time-derivative coupling. Here, we shall simply estimate its influence on πN scattering in second-order perturbation theory for incident and final plane-wave meson states.

Several years ago the investigation of the $\mathcal{O}(1)$ fluctuation term $\mathcal{L}_{vib}$ has led to a large variety of results for elastic and inelastic πN ($\bar{K}N$, KN, ϱN etc.) T-matrices covering energy ranges from threshold to more than 1 GeV, yielding at the same time important resonances and background scattering in all possible channels. These results have certainly contributed substantially to establish the soliton picture as a serious alternative description of low-energy hadron physics; they have been reviewed previously[2–6] and I will not repeat that here.

I will only remind you of one peculiar aspect of those $\mathcal{O}(1)$ results. Due to the hedgehog structure of the soliton the fluctuations in the intrinsic (i.e. rotating body-fixed) frame separate according to the quantum numbers I of the phonon- (or grand-) spin $\vec{I} = \vec{j} + \vec{t}$ (i.e. the sum of meson total spin and isospin). This means that certain groups of resonances in different JT-channels have a very similar internal structure, and as a consequence it leads to linear "magic relations" between S-matrix elements in different channels. For instance, for elastic πN scattering the scattering amplitudes $a_{L\,2T\,2J}$ in the JT-channel with pion angular momentum L must satisfy[7]

$$2(2L+1)\, a_{L\,3\,2L+1} = 3L\, a_{L\,1\,2L-1} + (L+2)\, a_{L\,1\,2L+1} \tag{2}$$
$$2(2L+1)\, a_{L\,3\,2L-1} = (L-1)\, a_{L\,1\,2L-1} + 3(L+1)\, a_{L\,1\,2L+1} \tag{3}$$

These relations are surprisingly accurate in many cases[8], particularly for higher partial waves. However, they seriously fail for S and P waves in elastic πN scattering, especially in the low-energy region. For S-waves they constrain the scattering amplitudes $a_{2T}(k)$ in the two isospin channels $T = 1/2,\ 3/2$ to satisfy

$$a_1(k) \equiv a_3(k) \tag{4}$$

(k: pion momentum). This relation excludes isospin splitting in the S-channels and, in fact, contradicts the Tomozawa-Weinberg current algebra result which states for the isoscalar and isovector amplitudes $b_0(k)$ and $b_1(k)$ in the the K-matrix for S-wave πN scattering[9]

$$K = \frac{\tan\delta}{k} = b_0(k) + b_1(k)\vec{\tau}\cdot\vec{t} \tag{5}$$

(with $\vec{\tau}/2$ and $\vec{t}$ for target and pion isospin) in the limit $k \to 0$

$$\begin{aligned} b_0 &= \tfrac{1}{3}(a_1 + 2a_3) &\to\quad & 0 + \mathcal{O}(m_\pi^2) &\quad (a)\\ b_1 &= \tfrac{1}{3}(a_3 - a_1) &\to\quad & -\tfrac{m_\pi}{8\pi f_\pi^2} + \mathcal{O}(m_\pi^2) &\quad (b) \end{aligned} \tag{6}$$

in very good agreement with the experimentally observed values $b_0 = -.010(3)m_\pi^{-1}$, $b_1 = -.091(2)m_\pi^{-1}$.

For P-waves Eqs. 2 and 3 require for the scattering amplitudes $a_{2T\,2J}(k)$ in the four different JT-channels

$$\begin{aligned} a_{31}(k) &\equiv a_{13}(k), &\quad (a)\\ a_{33}(k) &\equiv \tfrac{1}{2}\left(a_{11}(k) + a_{13}(k)\right). &\quad (b) \end{aligned} \tag{7}$$

Relation 7b evidently excludes a pronounced resonance in the $P33$-channel in an energy region where the T =1/2 amplitudes are nonresonant. These conspicuous deficiencies of the purely fluctuational $\mathcal{O}(1)$ part of $\mathcal{L}(\Omega', \eta, \dot{\eta})$ indicate the crucial role of $\mathcal{L}^{(1)}_{coupl}[\mathcal{O}(1/\sqrt{N_c})]$ and $\mathcal{L}^{(2)}_{coupl}[\mathcal{O}(1/N_c)]$ which couple the fluctuations (i.e. the mesons) to the collective (i.e. flavor) variables of the baryons.

2. Low-energy S-wave scattering

For S-waves the situation simplifies because the linear coupling $\mathcal{L}^{(1)}$ vanishes due to parity, and because S-waves are not affected by constraints. The static parts of the soliton characterized by the soliton profile functions $F(r), G(r)$, and $W(r)$ for the π, ρ, ω-meson fields serve as background for the scattering of the time-dependent fluctuations $\vec{\eta}, \vec{u}_\nu$ and w_μ:

$$
\begin{aligned}
U &= e^{i\vec{\tau}\cdot(\hat{r}F(r) + \vec{\eta}/f_\pi)} \\
\omega_\mu &= (f_\pi W(r) + w_0, \vec{w}) \\
\rho^a_\mu &= (u^a_0, \frac{G(r)}{2gr}\varepsilon_{ika}\hat{r}_k + u^a_i) \,.
\end{aligned}
\tag{8}
$$

The constants $f_\pi = \mathcal{O}(\sqrt{N_c})$ and $g = \mathcal{O}(1/\sqrt{N_c})$ are carried along explicitly in order to have the fluctuation fields of $\mathcal{O}(1)$ and to keep trace of the various orders of $1/N_c$. In the decomposition of the fluctuations into vector and tensor spherical harmonics the S-waves are of pure electric dipole (E1) character and, with the notation for the radial functions defined in ref.5, are written in terms of spectral amplitudes $\vec{q}_k$

$$
\begin{aligned}
\vec{\eta} &= \int dk_B \, \{p_r(k,r)\hat{r}(\hat{r}\vec{q}_k) + p_\nabla(k,r)[\vec{q}_k - \hat{r}(\hat{r}\vec{q}_k)]\} \\
\vec{w} &= \int dk_B \left\{\omega_r(k,r)\hat{r}(\hat{r}\dot{\vec{q}}_k) + \omega_\nabla(k,r)\{\dot{\vec{q}}_k - \hat{r}(\hat{r}\dot{\vec{q}}_k)\}\right\} \\
\vec{u}_0 &= \int dk_B \;\; v_\Lambda(k,r)\hat{r}\times\dot{\vec{q}}_k \\
u^a_i &= \int dk_B \Big\{t_4(r)\hat{r}_i(\hat{r}\times\vec{q}_k)_a + (t_5(k,r) - t_7(k,r) - t_8(k,r))\,\hat{r}_a(\hat{r}\times\vec{q}_k)_i \\
&\qquad + (t_5(k,r) + t_7(k,r) - t_8(k,r))\,\varepsilon_{aij}\hat{r}_j(\hat{r}\vec{q}_k)\Big\}
\end{aligned}
\tag{9}
$$

where the abbreviation $\int dk_B = \int d^3k/((2\pi)^3 2\omega_k)$ with $\omega_k = \sqrt{m_\pi^2 + k^2}$ is used. The lagrangian is expanded up to quadratic order in these fluctuations and diagonalization then leads to a set of coupled differential equations for all radial functions. From the boundary condition for the S-wave pionic fluctuations (all other fluctuations decrease exponentially below the ρ- and ω-thresholds)

$$
\lim_{r\to\infty} p_r \sim \lim_{r\to\infty} p_\nabla \sim \frac{\sin(kr+\delta)}{kr}
\tag{10}
$$

the $\mathcal{O}(1)$ background phase shifts are extracted.

In order to determine the DWBA scattering amplitude to the order $\mathcal{O}(1/N_c)$ the E1 modes (9) are inserted into the lagrangian

$$\begin{aligned} L &= -M_s + \frac{1}{2}\int dk_B \left[\dot{\vec{q}}_k^{\,2} - \omega_k^2 \vec{q}_k^{\,2}\right] \\ &+ \frac{1}{2}\Theta\vec{\Omega}'^2 - \vec{\Omega}' \int dk_B \int dk'_B \sqrt{2\omega_k}\sqrt{2\omega_{k'}} B_{kk'} (\vec{q}_k \times \dot{\vec{q}}_k) + \ldots \end{aligned} \tag{11}$$

The symmetrical matrix $B_{kk'}$ is a very lengthy spatial integral which involves the static soliton profiles and the radial functions for all fluctuations. Via the conjugate momenta

$$\begin{aligned} \vec{S} = \frac{\partial L}{\partial \vec{\Omega}'} &= \Theta\vec{\Omega}' - \int dk_B \int dk'_B \sqrt{2\omega_k}\sqrt{2\omega_{k'}} B_{kk'} (\vec{q}_k \times \dot{\vec{q}}_{k'}) \\ \vec{p}_k = \frac{\delta L}{\delta \dot{\vec{q}}_k} &= \dot{\vec{q}}_k - \vec{\Omega}' \times \int dk'_B \sqrt{2\omega_k}\sqrt{2\omega_{k'}} B_{kk'} \vec{q}_{k'} \end{aligned} \tag{12}$$

the hamiltonian then is obtained in the form[10]

$$\begin{aligned} H &= M_s + H^0 + H^1 + \ldots \\ H^0 &= \frac{1}{2}\int dk_B \left[\vec{p}_k^{\,2} + \omega_k^2 \vec{q}_k^{\,2}\right] = \int dk_B\, \omega_k \vec{a}_k^{\,+} \vec{a}_k \\ H^1 &= \frac{1}{2\Theta}\left[\vec{J} - \int dk_B \int dk'_B \sqrt{2\omega_k}\sqrt{2\omega_{k'}} A_{kk'} \vec{q}_k \times \vec{p}_{k'}\right]^2 + \mathcal{O}(q^4/f_\pi^2) \\ H^2 &= \mathcal{O}(\frac{S^2 q^2}{\Theta^2}, \ldots) \end{aligned} \tag{13}$$

In H^1 the total angular momentum $\vec{J}$

$$\vec{J} = \vec{S} + \vec{I} \tag{14}$$

with phonon spin

$$\vec{I} = \int dk_B \vec{q}_k \times \vec{p}_k \tag{15}$$

has been introduced which causes the replacement of $B_{kk'}$ by

$$A_{kk'} = \frac{2\pi^2}{k^2}\delta(k - k') - B_{kk'}. \tag{16}$$

(A drastic simplification may be obtained by approximating the vector meson fields in $A_{kk'}$ by their local pionic counterparts

$$V_\mu \approx -\frac{gN_c}{m_\omega^2} B_\mu + \frac{i}{2g}(l_\mu - r_\mu) \tag{17}$$

After this replacement, the expressions for $A_{kk'}$ and for Θ depend on the pionic profile F and the pionic fluctuations p_r and p_∇ only, and as expected, become identical to the corresponding expressions for a Skyrme lagrangian with 4th- and 6th-order terms.)

Following Walliser[11] we consider in H^1 all terms which have their origin in the collective rotation and neglect other $\mathcal{O}(1/f_\pi^2)$ terms which are due to anharmonicities in the vibrations. The relevant perturbation then is

$$\begin{aligned} H^1 = & - \frac{1}{\Theta}\int dk_B \int dk'_B \sqrt{2\omega_k}\sqrt{2\omega_{k'}} A_{kk'} \vec{J}\cdot(\vec{q}_k \times \vec{p}_{k'}) + \\ & + \frac{1}{2\Theta}\left[\int dk_B \int dk'_B \sqrt{2\omega_k}\sqrt{2\omega_{k'}} A_{kk'}(\vec{q}_k \times \vec{p}_{k'})\right]^2 \end{aligned} \tag{18}$$

proportional to $1/\Theta$. For one-pion on-shell matrix elements of H^1 (momentum k, isospin components a, b) only one-boson intermediate states contribute in the second term because of the antisymmetric product of q_k and p_k, and the exact result is

$$\langle k, b|H^1|k, a\rangle = -\frac{2\omega_k}{\Theta}A_{kk}\cdot i\epsilon_{bca}J_c + \frac{2\omega_k}{\Theta}\int dk'_B 2\omega_{k'} A_{kk'}A_{k'k}\,\delta_{ab}. \tag{19}$$

The K-matrix in DWBA may directly be read off from this expression

$$K^1 = \frac{\omega_k}{4\pi\Theta}A_{kk}\cdot 2\vec{J}\cdot\vec{I} - \frac{2\omega_k}{4\pi\Theta}\int dk'_B 2\omega_{k'}A_{kk'}A_{k'k}, \tag{20}$$

where the representation of the phonon spin $\langle b|I_c|a\rangle = i\epsilon_{bca}$ for $I = 1$ was used. Because of $\vec{J}=\vec{S}+\vec{I}$, $\vec{S}^2=\vec{T}^2$ and $\vec{T}=\vec{t}+(\vec{\tau}/2)$ we have

$$2\vec{J}\cdot\vec{I} = -\vec{\tau}\cdot\vec{t} \quad \text{for S-waves.} \tag{21}$$

So, evidently, it is the first term in (20) which creates the isospin splitting for the S-waves. But, most interestingly, the second term in (20) is an isospin independent rotationally induced $\mathcal{O}(1/N_c)$ contribution to the background scattering. It is repulsive, and eventually will compensate the attractive background phase-shift.

By comparing the K-matrix (20) with (5) we find the coefficients

$$b_0(k) = b_0^{BG}(k) - \frac{2\omega_k}{4\pi\Theta}\int dk'_B 2\omega_{k'}A_{kk'}A_{k'k} \tag{22}$$

and

$$b_1(k) = -\frac{\omega_k}{4\pi\Theta}A_{kk} \tag{23}$$

where $b_0^{BG}(k)$ is the $\mathcal{O}(1)$ isoscalar amplitude due to the meson scattering off the static soliton background.

In this $1/N_c$ order (23) now can be shown to satisfy the Low Energy Theorem (6b): In the chiral limit the Distorted Waves for $k \to 0$ approach a superposition of (translational and chiral) zero modes leading to

$$\lim_{k\to 0} A_{kk} = \frac{\Theta}{2f_\pi^2} \tag{24}$$

so (6b) is established. For (6a) we note that the translational zero mode is bound and the scattering phase shift of the chiral zero mode is $\delta = 0$ therefore $b_0^{BG}(k)$ vanishes in

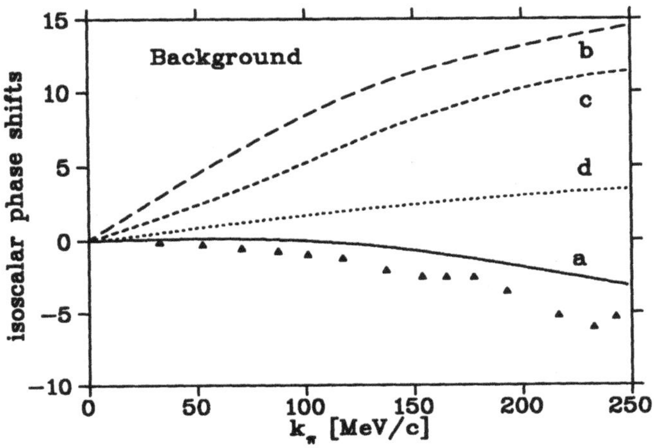

Fig. 1. The $\mathcal{O}(1)$ isoscalar background phase shifts $\delta_0^{BG}(k)$ calculated for models with different choices for the anomalous part of the effective action as described in ref.10. Triangles denote the Karlsruhe[12] data.

this limit. Although the LET thus is formally observed, agreement with experiment depends on the actual numerical magnitude of the $\mathcal{O}(m_\pi^2)$ terms.

Unfortunately, satisfactory results so far had been prevented by the fact that $b_0^{BG}(k \to 0)$ turned out to be large, positive and extremely sensitive to the choice of the effective lagrangian, while the experimental number is very small and negative. This is shown in fig.1 where the isoscalar background S-phase shifts $\delta_0^{BG}(k) = \arctan k b_0^{BG}(k)$ are plotted for some different choices of the anomalous part of the effective action, together with Karlsruhe[12] data. Evidently, the resulting background scattering generally is much too attractive. However, this attraction is virtually compensated by the rotationally induced $1/N_c$ contribution in (22). This remarkable fact (first noticed by Walliser[11]) is demonstrated in fig.2 which also shows that for the Bardeen-subtracted form[10] of the anomalous action (dashed lines in figs.1 and 2) which severely violates chiral symmetry, this compensation does not occur.

Altogether, fig.3 contains the results for both S-wave channels and shows agreement of comparable quality with the Karlsruhe data for different choices of effective lagrangians (the Bardeen-subtracted form is omitted from fig.3). Although the details of these calculations are very tedious they clearly demonstrate that the soliton model does indeed provide an accurate account of low-energy πN S-wave scattering.

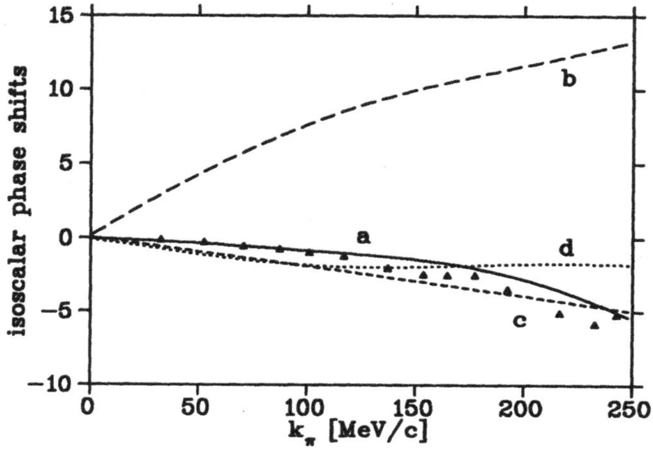

Fig. 2. The isoscalar phase shifts δ_0 including the $\mathcal{O}(1/N_c)$ effects for the same models as in fig.1

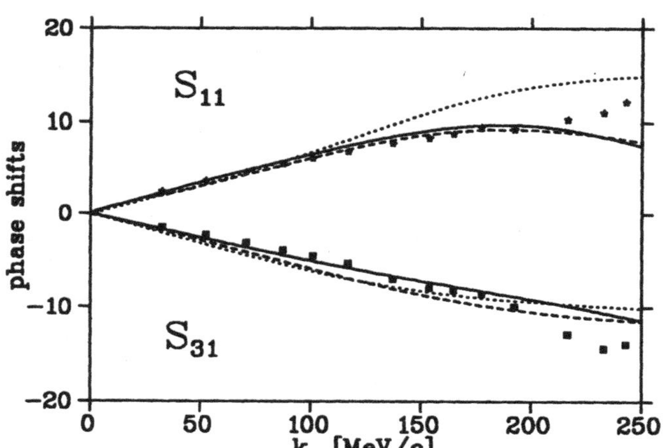

Fig. 3. The phase shifts δ_{S11} and δ_{S31} for the models labelled a, c and d as in fig.1, together with the Karlsruhe data.

3. *P*-waves in elastic πN-scattering

For *P*-waves the fluctuation-rotation coupling adds to the $\mathcal{O}(1)$ background scattering additional contributions from the linear and quadratic (time-derivative) coupling $\mathcal{L}^{(1)}[\mathcal{O}(1/\sqrt{N_c})]$ and $\mathcal{L}^{(2)}[\mathcal{O}(1/N_c)]$. The *S*-matrix $S^{2T\,2J}_{\ell s;\,\ell' s'}$ contains the πN *P*-waves in the channels with isospin-spin labels $2T$=1, 3; $2J$=1, 3; they are coupled to other channels with pion orbital angular momentum $\ell = 1,3$ and target spin-isospin $s = \frac{1}{2}, \frac{3}{2}, \frac{5}{2}$. To order $\mathcal{O}(1)$ it is given by[7]

$$S^{2T\,2J}_{\ell s;\,\ell' s'} = \sum_I C_{TJ\ell sI} S^I_{\ell\ell'} C_{TJ\ell' s' I} \tag{25}$$

with recoupling coefficients

$$C_{TJ\ell sI} = (-)^{\ell+s+J}((2I+1)(2s+1))^{1/2} \begin{Bmatrix} J & \ell & s \\ 1 & T & I \end{Bmatrix} \tag{26}$$

and intrinsic *S*-matrix elements $S^I_{\ell\ell'}$ calculated for the non-rotating soliton. The complexity of the full problem becomes apparent, if we note that the thresholds for final channels with $s' > \frac{1}{2}$ and the energies of intermediate states with $s > \frac{1}{2}$ (coupled to initial and final channels through $\mathcal{L}^{(1)}$) are also of order $\mathcal{O}(1/N_c)$. Although at very low energies we may hope to safely truncate inelastic closed channels, any such approximation scheme affects the position of zero modes which are crucial for *P*-waves and requires intricate projection techniques to eliminate spurious contributions[3,6]. Therefore the full problem has not been satisfactorily treated as of now; however, the essential physics can be piecewise extracted and discussed.

Different choices for the parametrization of the time dependence of U in terms of pionic fluctuations lead to different explicit forms of $\mathcal{L}_{vib}$. Specifically, in the kinetic part $\int \dot{\eta} T \dot{\eta}$ the longitudinal and transverse components of η (with respect to the static soliton field)

$$\vec{\eta} = \eta_L \hat{x} + \vec{\eta}_T, \qquad \eta_L = (\hat{x} \cdot \vec{\eta}), \quad \vec{\eta}_T \cdot \hat{x} = 0 \tag{27}$$

appear with metric functions $M_L(r)$ and $M_T(r)$

$$T = \frac{1}{2f_\pi^2} \left(\int \dot{\eta}_L M_L(r) \dot{\eta}_L + \dot{\vec{\eta}}_T M_T(r) \dot{\vec{\eta}}_T \right) \tag{28}$$

which are quite different for different parametrizations. The standard form of the Schroedinger equation for the scattering waves is obtained by absorbing $\sqrt{M_L(r)}$ and $\sqrt{M_T(r)}$ into the definition of the longitudinal and transverse components which at the same time makes the scattering potentials independent of the parametrization.

At low energies the I-sum in (25) is dominated by the I=1 (Magnetic Dipole) contribution. The radial dependence of the M1 partial wave is determined by one single differential equation which in terms of $k^2 = \omega_k^2 - m_\pi^2$ has the form of a nonrelativistic Schrödinger equation for *P*-wave potential scattering[3]

$$\left(-\partial_r^2 - \frac{2}{r}\partial_r + \frac{2}{r^2} + V_1(r) - k_n^2 \right) \Phi_n(r) = 0. \tag{29}$$

(The potential will in general contain further gradient terms). Appropriate boundary conditions determine the phase shifts $\delta_1(k)$ and the partial scattering amplitude

$$f_1(k^2) = \frac{1}{k}\sin\delta_1 e^{i\delta_1}. \tag{30}$$

The general expression for this partial amplitude is[13]

$$f_1(k^2) = -\int (j_1(kr))^2 V_1(r) r^2 dr - \sum_n \frac{(\int j_1(kr)V_1(r)\Phi_n(r)r^2dr)^2}{k^2 - k_n^2 + i\varepsilon}, \tag{31}$$

where the n-sum picks up all normalized bound states of (29) and implies an integral over the scattering states. At low energies this n-sum is dominated by the contribution from the M1 zero modes

$$\boldsymbol{\Phi}_i^{(0)} = \frac{f_\pi}{\sqrt{\Theta}} i(\hat{\boldsymbol{a}}_i \times \hat{\boldsymbol{x}}) \sin F \sqrt{\mathcal{M}(r)}, \tag{32}$$

which describe global rotations in isospace of the Skyrme $SU(2)$ hedgehog

$$U_0 = \exp(i\boldsymbol{\tau}\cdot\hat{\boldsymbol{x}}F(r)) \tag{33}$$

around three orthogonal axes $\hat{\boldsymbol{a}}_i$. The metric $\mathcal{M}(r)$ arises from the kinetic part in $\mathcal{L}_{vib}[\mathcal{O}(1)]$. It is

$$\mathcal{M}(r) = 1 + \frac{1}{e^2 f_\pi^2}\left(F'^2 + \frac{\sin^2 F}{r^2}\right) \tag{34}$$

for the standard Skyrme model and it has been absorbed into the definition of $\boldsymbol{\Phi}_i^{(0)}$ in order to have the orthogonality

$$\int \boldsymbol{\Phi}_i^{(0)*}\cdot\boldsymbol{\Phi}_j^{(0)} d^3x = \delta_{ij} \tag{35}$$

with the moment of inertia Θ as normalization constant. In this form the radial part of the zero modes (32) satisfies Eq. 29 at $k_n^2 = -m_\pi^2$, therefore their contribution to the scattering amplitude (31) is

$$f_1^{(ZM)}(k^2) = -\frac{8\pi f_\pi^2}{3}\frac{\left((k^2+m_\pi^2)\int j_1(kr)\sin F\sqrt{\mathcal{M}(r)}r^2dr\right)^2}{(k^2+m_\pi^2)\Theta} = -\frac{3}{2\pi\Theta}\frac{k^2}{m_\pi^2}\frac{f_{\pi NN}^2(k^2)}{\omega_k^2} \tag{36}$$

with the πNN form factor defined as

$$f_{\pi NN}(k^2) = \frac{m_\pi}{k} f_\pi \frac{4}{3}\pi\omega_k^2 \int \sin F(r)\sqrt{\mathcal{M}(r)}j_1(kr)r^2dr\,. \tag{37}$$

Combined with the appropriate recoupling coefficients the M1 zero mode contribution to the $\mathcal{O}(1)$ background (BG) scattering T-matrix is (for plane wave states $|\boldsymbol{k}, a>$ for the incoming pion (momentum $\boldsymbol{k}$, isospin orientation $\hat{\mathbf{e}}_a$) and $< \boldsymbol{k}', b|$ for the outgoing ($k^2 = k'^2$) pion)

$$T^{M1}_{(BG)\,ab}(\boldsymbol{k},\boldsymbol{k}') = \frac{3}{\omega_k^2 \Theta m_\pi^2} f^2_{\pi NN}(k^2)$$
$$\times < jb|4P_{11} + P_{13} + P_{31} + \frac{5}{2}P_{33}|ia > k_i k'_j \qquad (38)$$

with $< jb|P_{2T\,2J}|ia >$ matrixelements for the projectors on the TJ isospin-spin channels.

Through $\mathcal{M}(r)$ the form factor (37) depends explicitly on the form of the lagrangian. For $\mathcal{M} = 1$ it coincides with the πNN form factor derived from the static pion cloud as suggested by Cohen[14] and further discussed in ref.15. Its value at the pole $k^2 = -m_\pi^2$, the πNN coupling constant $f = f_{\pi NN}(-m_\pi^2)$ (with experimental value of $f^2/4\pi = 0.088$), picks up only the asymptotic amplitude of $F(r)$, and the deviation of $\mathcal{M}(r)$ from 1 is irrelevant. However, as can be seen from fig.4 the actual form of $f_{\pi NN}(k^2)$ for the same profile function $F(r)$ differs appreciably for $\mathcal{M}(r)$ given

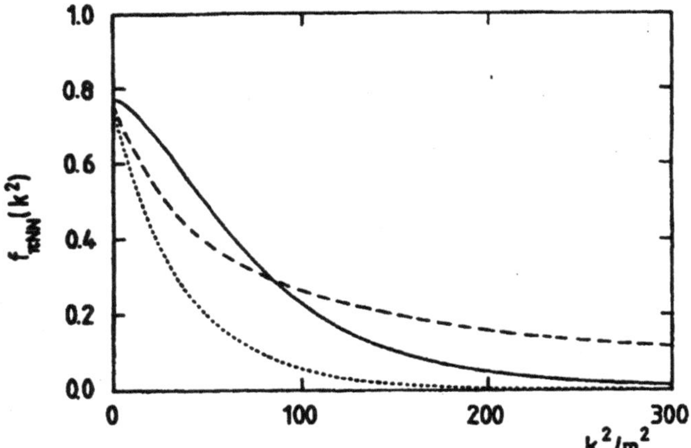

Fig. 4. The pion-nucleon form factor $f_{\pi NN}(k^2)$ as defined in eq.(37) for the Skyrme model with e=4 and f_π=93MeV (full line). The dotted line shows the same form factor with $\mathcal{M} = 1$. The dashed line is the standard monopole form factor with cutoff Λ=1GeV. The abscissa is the square of the pion three-momentum in units of m_π^2.

either by (34) or for $\mathcal{M}(r) \equiv 1$. From the origin of $\mathcal{M}$ in (28) it is evident that purely static considerations as in refs.14,15 cannot account for this specific feature of the dynamical form factor (37).

The same metrical factor appears again in the linear (time-derivative) coupling $\mathcal{L}^{(1)}[\mathcal{O}(1/\sqrt{N_c})]$ because this also derives from the kinetic part of the original lagrangian. In the Schnitzer representation[16]

$$U = SU_0(\alpha)S, \quad S = \exp(i\boldsymbol{\tau}\cdot\boldsymbol{\xi}/(2f_\pi)) \;, \qquad (39)$$

$$U_0(\alpha) = \exp(i\boldsymbol{\tau} \cdot \hat{\boldsymbol{\Phi}}(\alpha)F(r)) \,, \tag{40}$$

$\mathcal{L}^{(1)}$ couples the fluctuations $\boldsymbol{\xi}$ to the baryonic axial charge density $\boldsymbol{A}_0$ of the hedgehog rotating with angular velocitiy $\boldsymbol{\Omega}'$

$$\mathcal{L}^{(1)} = f_\pi \int \sin F \cos F \mathcal{M}(r)(\dot{\hat{\boldsymbol{\Phi}}} \cdot \dot{\boldsymbol{\xi}}) d^3x \tag{41}$$

$$\dot{\hat{\boldsymbol{\Phi}}} = (\boldsymbol{\Omega}' \times \hat{\boldsymbol{\Phi}}) \,. \tag{42}$$

The transverse components of $\dot{\boldsymbol{\xi}}$ contained in (41) appear in the kinetic energy for the Schnitzer fluctuations with a metrical factor $\cos^2 F\ \mathcal{M}(r)$. For a plane-wave Born approximation we therefore substitute plane waves for the transverse normalized modes $\cos F \sqrt{\mathcal{M}(r)}\, \boldsymbol{\xi}_T$.

We use the naive quantization rule

$$\boldsymbol{\Omega}' = \boldsymbol{T}/(2\Theta) \tag{43}$$

(where T_i (i=1,2,3) are the generators of isospin $SU(2)$) and symmetrize (42) with respect to the noncommuting $\boldsymbol{\Omega}'$ and $D^\dagger(\alpha)$ to bring $\boldsymbol{A}_0$ into hermitean form. Then with $\boldsymbol{T} = \boldsymbol{\tau}$ for nucleon states and

$$< N'|D^\dagger_{aj}|N >= -\frac{1}{3} < N'|\tau_a\sigma_j|N >, \tag{44}$$

we have

$$\epsilon_{abc}(\Omega'_b\tau_c + \tau_c\Omega'_b) = 0, \tag{45}$$

so $\mathcal{L}^{(1)}$ creates no πNN vertex. However, with

$$< \Delta|D^\dagger_{aj}|N >= -\frac{1}{\sqrt{2}} < \Delta|T^{(+)}_a S^{(+)}_j|N > \tag{46}$$

(where matrix elements of $T^{(+)}$ and $S^{(+)}$ are equal to the corresponding $< \Delta|\pi N >$ Clebsch-Gordan coefficients) and

$$\epsilon_{abc}\frac{1}{2}(\Omega'_b T^{(+)}_c + T^{(+)}_c\Omega'_b) = i\frac{3}{2\Theta}T^{(+)}_a \tag{47}$$

the interaction (41) connects baryon states with spin(=isospin) s, s' differing by one unit. The vertex for emission or absorption of a pion with momentum $\boldsymbol{k}$ and isospin orientation $\hat{\boldsymbol{e}}$ then is given by

$$< s'|\mathcal{L}^{(1)}|\boldsymbol{k}, \hat{\boldsymbol{e}}\,; s >= -i\frac{3}{2\Theta\omega_k} f_{\pi s s'}(k^2) < s'|(\hat{\boldsymbol{e}}\cdot\boldsymbol{T}^{(\pm)})(\boldsymbol{k}\cdot\boldsymbol{S}^{(\pm)})|s > \,, \quad s' = s\pm 1 \,. \tag{48}$$

It contributes to the elastic πN-scattering K-matrix through an intermediate Δ state as

$$K^{(1)}(k^2) = -\frac{1}{12\pi}\left(\frac{3}{2\Theta}\frac{k}{\omega_k m_\pi} f_{\pi N\Delta}(k^2)\right)^2 \left(\frac{P_{33}}{\omega_k - \Delta M} - \frac{P_{33} + 4(P_{13} + P_{31}) + 16P_{11}}{9(\omega_k + \Delta M)}\right). \tag{49}$$

The form factor $f_{\pi N\Delta}$ has the same k-dependence as $f_{\pi NN}$ and differs from it only by the geometrical factor which arises from the different baryonic matrix elements

$$f_{\pi N\Delta}(k^2) = 3\sqrt{\frac{1}{2}} f_{\pi NN}(k^2). \tag{50}$$

The nucleon-delta mass difference is given by $\Delta M = 3/(2\Theta)$.

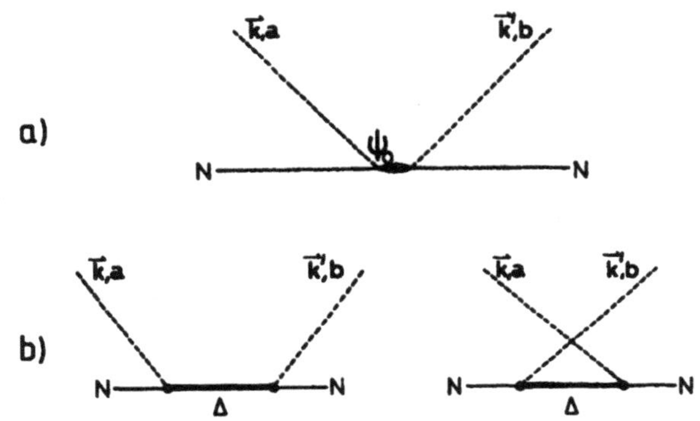

Fig. 5. a) The zero-mode contribution (38) to $\mathcal{O}(1)$ background scattering; b) the direct and crossed diagram for the $\mathcal{O}(1/N_c)$ contribution of the linear time-derivative interaction (48) to πN scattering.

The sum of the (M1) zero mode background term (38) (fig.5a) and the linear time-derivative term (49) (fig.5b) yields for the K-matrix elements $K_{2J\,2T}$ in the individual JT channels

$$\begin{aligned} K_{11} &= C(k^2)\left(-\frac{12}{\omega_k^2\Theta} - (\frac{3}{2\omega_k\Theta})^2 \cdot \frac{9}{2} \cdot (-\frac{16}{9})\frac{1}{\omega_k + \Delta M}\right) \\ &= -8C(k^2)\left(\frac{\Delta M}{\omega_k(\omega_k + \Delta M)}\right) \end{aligned} \tag{51}$$

$$K_{13} = K_{31} = K_{11}/4 \tag{52}$$

$$\begin{aligned} K_{33} &= C(k^2)\left(-\frac{3}{\omega_k^2\Theta} \cdot \frac{5}{2} - (\frac{3}{2\omega_k\Theta})^2 \cdot \frac{9}{2} \cdot \left(\frac{1}{\omega_k - \Delta M} - \frac{1/9}{\omega_k + \Delta M}\right)\right) \\ &= C(k^2)\frac{\Delta M(4\Delta M + 5\omega_k)}{\omega_k(\Delta M^2 - \omega_k^2)} \end{aligned} \tag{53}$$

with

$$C(k^2) = \frac{1}{12\pi}\frac{k^2}{m_\pi^2} f_{\pi NN}^2(k^2). \tag{54}$$

Verschelde[17] noticed that this result coincides with the well-known and successful static Δ-isobar model[5] where the K-matrix is obtained as the sum of (direct and crossed) nucleon and Δ-exchange contributions if we use relation (50) for the πNN and $\pi N\Delta$ pseudovector coupling constants:

$$K_{11} = C(k^2)\left(-\frac{8}{\omega_k}+\frac{9}{2}\cdot\frac{16/9}{\omega_k+\Delta M}\right) \quad (55)$$

$$K_{31} = K_{13} = K_{11}/4 \quad (56)$$

$$K_{33} = C(k^2)\left(\frac{4}{\omega_k}+\frac{9}{2}\cdot\left(\frac{-1}{\omega_k-\Delta M}+\frac{1/9}{\Delta M+\omega_k}\right)\right) . \quad (57)$$

It should be noted that this equivalence holds only for the sum of all terms; it is not possible to identify individual nucleon or delta direct or crossed terms for the usual linear pseudovector coupling with any of the diagrams of fig.5. In this sense the transparency of the familiar fermion-boson picture is lost in the soliton description. However, the latter has the advantage of supplying the formfactor from the theory.

For low k^2 other (E0,M1,E2) contributions to the I-sum in (25) and to (31) do not seriously disturb this result[18], although the E0 part is too large due to the Roper (breathing mode) resonance which appears in the Skyrme model too low. This causes too much attraction in the P11 channel.

Similarly as for S-waves the quadratic (time derivative) coupling $\mathcal{L}^{(2)}$ provides also for P-waves the $\boldsymbol{t}\cdot\boldsymbol{\tau}$ interaction between pion and target isospins and leads to the following contribution to the elastic πN T-matrix

$$T^{(2)} = -\frac{\omega_k}{3\Theta}\int j_1(kr)^2 \sin^2 F \mathcal{M}(r) r^2 dr \left(-2(P_{11}+P_{13})+(P_{31}+P_{33})\right) . \quad (58)$$

It turns out be small for low energies and provides the correct splitting between the P_{13} and P_{31} phase shifts[18].

4. The P_{33} resonance, the πNN-formfactor, and g_A

The shape of the static πNN-formfactor $G_{\pi NN}(q^2)$ defined through

$$i\, q_i\, \chi^\dagger \sigma_i \tau_a \chi \frac{G_{\pi NN}(q^2)}{2M_N} = \int d^3x e^{i\boldsymbol{q}\cdot\boldsymbol{x}} < \boldsymbol{p}'|(\Delta - m_\pi^2)\pi_a(\boldsymbol{x})|\boldsymbol{p} > \quad (59)$$

(with spinors χ, nucleon mass M_N, nucleon momenta $\boldsymbol{p},\boldsymbol{p}'$) depends of course on the definition of the interpolating pion field $\pi_a(\boldsymbol{x})$. A natural definition is suggested by PCAC

$$\partial_\mu \boldsymbol{A}^\mu = -f_\pi m_\pi^2 \boldsymbol{\pi} . \quad (60)$$

The so-defined pion field then is related to the axial form factor $G_A(q^2)$ at $q^2 = 0$, the axial coupling constant $g_A = G_A(q^2 = 0)$, by

$$g_A\, q_i\, \chi^\dagger \sigma_i \frac{\tau_a}{2}\chi = \lim_{q^2\to 0}\int d^3x e^{i\boldsymbol{q}\cdot\boldsymbol{x}} < \boldsymbol{p}'|\, m_\pi^2 f_\pi \pi_a|\boldsymbol{p} > . \quad (61)$$

In the limit $q^2 \to 0$ the πNN formfactor defined through (59) therefore satisfies

$$\frac{G_{\pi NN}(q^2=0)}{M_N} = \frac{g_A}{f_\pi}\,. \tag{62}$$

This identity is a consequence of the above definitions and thus it holds for all values of m_π. In the chiral limit it coincides with the Goldberger Treiman (GT) relation. For $m_\pi \neq 0$ the πNN-coupling constant $g_{\pi NN} = G_{\pi NN}(q^2 = -m_\pi^2)$ differs only slightly from $G_{\pi NN}(q^2 = 0)$, so Eq. 62 implies a small violation of the GT relation. As we mentioned already, at $q^2 = -m_\pi^2$ the dynamical formfactor $f_{\pi NN}(q^2)$ which appears in Eq. 36 coincides with the static $G_{\pi NN}(q^2)$ defined in Eq. 59 (up to a kinematical factor)

$$G_{\pi NN}(q^2 = -m_\pi^2) = \frac{m_\pi}{2M_N} f_{\pi NN}(q^2 = -m_\pi^2) \tag{63}$$

because for $q^2 \to -m_\pi^2$ both pick up only the asymptotic tail of $\boldsymbol{\pi}$, i.e. the metric $\mathcal{M} \to 1$.

As a consequence of the equations of motion for U the pionfield $\boldsymbol{\pi}$ in (60) is determined by the chiral symmetry breaking mass term $\mathcal{L}_{mass}$ alone

$$-m_\pi^2 f_\pi \boldsymbol{\pi} = tr\left(\frac{\delta\mathcal{L}_{mass}}{\delta U}\{i\frac{\boldsymbol{\tau}}{2}, U\} + h.c.\right)\,. \tag{64}$$

With the standard choice of $\mathcal{L}_{mass}$

$$\mathcal{L}_{mass} = m_\pi^2 \frac{f_\pi^2}{4} tr(U + U^\dagger - 2) \tag{65}$$

and $U = \exp iA^\dagger \boldsymbol{\tau}\cdot\boldsymbol{\Phi} A$ this leads to

$$\pi_a = \frac{f_\pi}{2} \sin\Phi\; tr(\tau_a A^\dagger \boldsymbol{\tau}\cdot\hat{\boldsymbol{\Phi}} A)\,. \tag{66}$$

To lowest order in a $1/N_c$ expansion we replace matrixelements in one-soliton states by the classical soliton field[19]

$$\int d^3x e^{i\boldsymbol{q}\cdot\boldsymbol{x}} < \boldsymbol{p}'|U(\boldsymbol{x})|\boldsymbol{p} >= \delta(\boldsymbol{q} - \boldsymbol{p}' + \boldsymbol{p}) \int U_{soliton}(\boldsymbol{x}) e^{i\boldsymbol{q}\cdot\boldsymbol{x}} d^3x\,. \tag{67}$$

Therefore to order N_c g_A is obtained from (61) as

$$g_A = \frac{8\pi}{9} m_\pi^2 f_\pi^2 \int \sin F r^3 dr\,. \tag{68}$$

In the standard Skyrme model ($f_\pi = 93$ MeV, $m_\pi = 138$ MeV) the Skyrme parameter e=3.5 leads to g_A=1.28, which due to (62) guarantees also good values for $f_{\pi NN}$. In fact, one obtains for e=3.5

$$f = f_{\pi NN}(q^2 \to -m_\pi^2) = 1\,. \tag{69}$$

Like (68) this is also a $\mathcal{O}(N_c)$ result. On the other hand, the $\mathcal{O}(1/N_c)$ nucleon-delta mass difference $\Delta M = 3/(2\Theta)$ is much too small ($\sim$ 175 MeV) for e=3.5. We have obtained $f_{\pi NN}(q^2)$ from analysing the zero mode graph in fig. 5a. For $\mathcal{O}(1)$ corrections to $f_{\pi NN}$ (and correspondingly to g_A) contributions of the graph in fig. 9a should be evaluated.

To demonstrate the influence of the different form factors we show in fig. 6 a comparison with experimental elastic πN P33 phase shifts. Such a comparison, however, is only feasible if we replace (57) by the relativistically improved expression[9]

$$K_{33} = \frac{1}{12\pi}\frac{q^2}{m_\pi^2}\frac{M_N}{\sqrt{s}} f_{\pi NN}^2(q^2)\left(\frac{8M_N}{M_N^2-\bar{u}} + M_\Delta\left(\frac{9}{M_\Delta^2-s}+\frac{1}{M_\Delta^2-\bar{u}}\right)\right) \tag{70}$$

(with $s=(E_q+\omega)^2$, $\bar{u} = M_N^2+m_\pi^2-2\omega E_q$, $E_q^2 = M_N^2+q^2$) which takes due account of recoil effects. For this demonstrative purpose we choose e=4 (which leads to f=

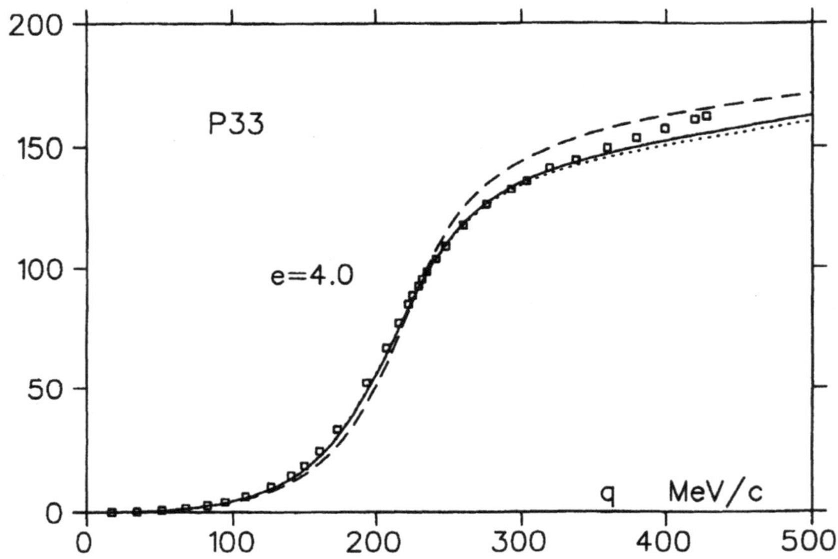

Fig. 6. The phaseshifts (in degrees) for the P33 resonance as obtained from (70) with three different formfactors: full line: with $f_{\pi NN}$ given by (37) and the metric (34); dashed line: the same with the metric $\mathcal{M}$=1; dotted line: with constant $f_{\pi NN} = f$. All formfactors are normalized to $f = f_{\pi NN}(-m_\pi^2) = 1$. The squares are the data points of ref. 20.

0.78), rescale the formfactors with 1/0.78 and put M_N and M_Δ to their experimental values 938 and 1232 MeV. It can be seen that the influence of the difference in the form factors is noticeable already in the resonance region: the rapid decrease with

q^2 of the static form (59) (which implies $\mathcal{M} \equiv 1$ in Eq. 37) (cf.fig. 4) makes the resonance too narrow and leaves no space for the additional attractive background effects which begin to contribute above resonance energy (especially the influence of the Roper resonance). The dynamical form factor (37) stays closer to the coupling constant f for small q^2 and the resulting phaseshifts nearly coincide with the dotted line in fig. 6 which is calculated with $f \equiv 1$ for all q^2.

5. Quantum corrections to the soliton mass

The fact that the $\mathcal{O}(1)$ contribution of the zero mode alone plus the $\mathcal{O}(1/N_c)$ diagrams of $\mathcal{L}^{(1)}_{coupl}$ are sufficient for a quite accurate description of the low-energy scattering T-matrix suggests that also $\mathcal{O}(1)$ and $\mathcal{O}(1/N_c)$ corrections to nucleon and delta mass could be reliably obtained in the same approximation.

The expression quoted by Cahill, Comtet, and Glauber[21] for the renormalized mass shift of a soliton

$$\delta M_s = -\frac{1}{4}\mathrm{tr}[(h_s - h_0)^2/h_0] \tag{71}$$

where h_s or h_0 is the meson hamiltonian with or without the soliton, can be trivially rewritten in the form

$$\delta M_s = -\frac{1}{2}\mathrm{tr}\frac{(h_s^2 - h_0^2)^2}{2h_0(h_s + h_0)^2}. \tag{72}$$

This form suggests that the $\mathcal{O}(1)$ corrections to nucleon and delta mass could be obtained by formally closing the loop in the diagram fig.5a, which describes the zero mode contribution to the background scattering amplitude (incoming and outgoing pion with momenta $\boldsymbol{k}, \boldsymbol{k}'$, isospin orientation $\hat{e}^{(a)}, \hat{e}^{(b)}$, respectively)

$$t^{(ZM)}_{ab}(\boldsymbol{k}, \boldsymbol{k}') = \frac{<\boldsymbol{k}', b|V|\psi_n><\psi_n|V|\boldsymbol{k}, a>}{(k^2 + m_\pi^2) - (k_n^2 + m_\pi^2)}\,|_{\psi_n = \psi_0}\,. \tag{73}$$

Here V is the potential exerted through the classical soliton field on the mesonic waves; the zero modes ($k_0^2 = -m_\pi^2$) satisfy

$$<\psi_0|V|\boldsymbol{k}> = -<\psi_0|\boldsymbol{k}>\omega_k^2, \quad \omega_k^2 = k^2 + m_\pi^2\,. \tag{74}$$

Closing the loop implies to replace the energy denominator in (73) by $-(\sqrt{k^2 + m_\pi^2} + \sqrt{k_n^2 + m_\pi^2}\,)^2$, summation and integration, and an additional factor of 1/2. This is represented by fig.7a.

The resulting mass correction then is

$$\delta M^{(ZM)} = -\frac{1}{2}\int \frac{d^3k}{(2\pi)^3\, 2\omega_k}\sum_a t^{(ZM)}_{aa}(\boldsymbol{k}, \boldsymbol{k}). \tag{75}$$

With the relation between the T-matrix (73) and the P-wave scattering amplitude (36)

$$\sum_a t^{(ZM)}_{aa}(\boldsymbol{k}, \boldsymbol{k}) = -12\pi f_1^{(ZM)}(k^2) \tag{76}$$

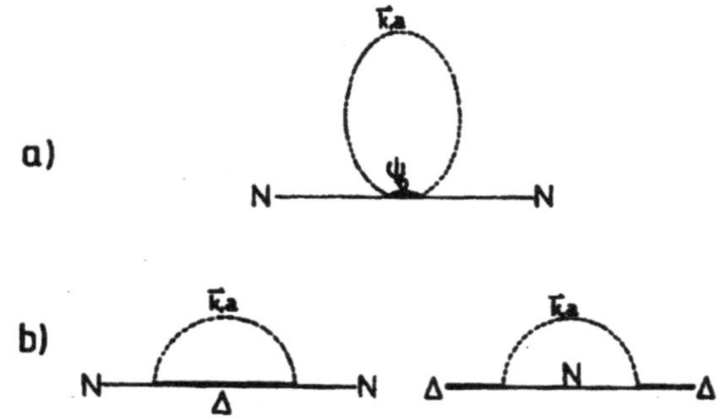

Fig. 7. a) The zero-mode contribution (75) to the $\mathcal{O}(1)$ background mass correction; b) the self-energy diagrams for the $\mathcal{O}(1/N_c)$ contributions of the linear time-derivative interaction (48) to nucleon and Δ mass.

we obtain for the mass shift

$$\begin{aligned}\delta M^{(ZM)} &= \frac{3}{2\pi}\int k f_1^{(ZM)}(k^2)\, d\omega_k \\ &= -4\frac{f_\pi^2}{\Theta}\int k^2 dk\, \omega_k f^2(k^2)\end{aligned} \tag{77}$$

$$\text{with} \quad f(k^2) = \int \sin F\sqrt{\mathcal{M}} j_1(kr) r^2 dr\,. \tag{78}$$

This mass shift is of $\mathcal{O}(1)$ and it is the same for nucleon and delta states.

One might expect another sizable contribution to the background loop graph in fig.7a from the translational (Electric dipole E1) zero mode, which enters into the low-energy S-wave phase shifts δ_{10M}

$$\tan\delta_{10M}(k^2) = -k b_0(k^2) \tag{79}$$

with a corresponding mass shift

$$\delta M^{(trans)} = \frac{3}{2\pi}\int k b_0^{(trans)}(k^2)\, d\omega_k. \tag{80}$$

The contribution from the translational zero mode ψ_0^{trans} to $b_0(k^2)$ is[22]

$$b_0^{trans}(k^2) = -\omega_k^2 \,|<\boldsymbol{k}\,|\,\psi_0^{trans}>|^2 \tag{81}$$

$$= -\frac{\omega_k^2}{4\pi}\frac{f_\pi^2}{\mathcal{M}_S}\left(\frac{4\pi}{3}\int j_0(kr)\left(\sqrt{\mathcal{M}_L}F' + 2\sqrt{\mathcal{M}}\frac{\sin F}{r}\right) r^2 dr\right)^2. \tag{82}$$

Here $\mathcal{M}$ again is the transverse metric (34) while $\mathcal{M}_L$ is the longitudinal metric

$$\mathcal{M}_L(r) = 1 + \frac{2}{e^2 f_\pi^2}\left(\frac{\sin^2 F}{r^2}\right) \tag{83}$$

and $\mathcal{M}_S$ is the translational (soliton) mass

$$\mathcal{M}_S = \frac{4\pi}{3} f_\pi^2 \int \left(\mathcal{M}_L F'^2 + 2\mathcal{M}\frac{\sin^2 F}{r^2} \right) r^2 dr. \tag{84}$$

The resulting functions $kf_1^{(ZM)}$ and $kb_0^{(trans)}$ are plotted in fig.8. The contributions

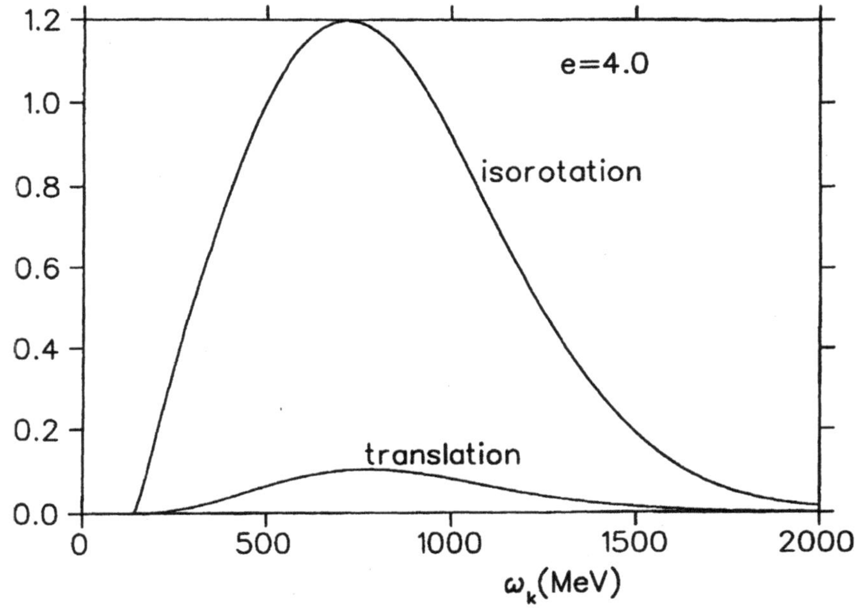

Fig. 8. The zero-mode contribution $kf_1^{(ZM)}(k^2)$ of the isorotation (36) and $kb_0^{(trans)}(k^2)$ of the translation (82) to background scattering for the Skyrme model with f_π=93 MeV and e=4.0 as functions of the pion energy ω_k.

$\delta M^{(ZM)}$ and $\delta M^{(trans)}$ to the masses are listed below. The latter is less than 10% of the rotational contribution, it is also of order $\mathcal{O}(1)$ and it is the same for nucleon and delta states.

Let us now consider the $\mathcal{O}(1/N_c)$ corrections to nucleon and delta masses which are due to the linear time-derivative interaction (48). Closing the loop in fig.5b leads to the two diagrams of fig.7b. The corresponding mass shift for the delta is

$$\delta\mathcal{M}_\Delta^{(1)} = \int \frac{d^3k}{(2\pi)^3 2\omega_k} \sum_{i,N} \frac{|<\Delta|\mathcal{L}^{(1)}|\boldsymbol{k}, \hat{\boldsymbol{e}}^{(i)}; N>|^2}{M_\Delta - (\omega_k + M_N)} \tag{85}$$

Strictly, the sum over the intermediate states in (85) should also include the coupling to collective $\frac{5}{2}\frac{5}{2}$ states. However, in contrast to the N-Δ split the position of the 55-state is subject to a considerable increase when going from SU(2) to SU(3). Within

SU(2) we do not expect a reliable estimate for the influence of this hypothetical 55-resonance on $\delta\mathcal{M}_\Delta^{(1)}$ and prefer to omit it altogether. The result then is

$$\delta\mathcal{M}_\Delta^{(1)} = \frac{3}{2\Theta^2} f_\pi^2 \int k^2 dk \frac{\omega_k}{-\omega_k + M_\Delta - M_N} f^2(k^2) \, . \tag{86}$$

For the mass shift of the nucleon the rôle of N and Δ in (85) is interchanged. Therefore the CG - coefficients in (46) have to be rewritten as

$$< \Delta | T^{(+)} | N > = (-)^a \sqrt{2} (\frac{3}{2} m_\Delta 1 - a | \frac{1}{2} m_N) \tag{87}$$

before the sum over the intermediate $\pi\Delta$ states is performed.

This leads to an overall factor of 4 as compared to (86):

$$\delta\mathcal{M}_N^{(1)} = \frac{6}{\Theta^2} f_\pi^2 \int k^2 dk \frac{\omega_k}{-\omega_k - M_\Delta + M_N} f^2(k^2) \, . \tag{88}$$

The imaginary part of the integral (86) is not affected by the coupling to the 55-state and it provides the width Γ of the Δ-resonance

$$\Gamma(\omega_k) = 3\pi k \left(\frac{f_\pi \, \omega_k}{\Theta} f(k^2) \right)^2 \, . \tag{89}$$

The following numerical results are calculated for a Skyrme lagrangian with f_π=93 MeV and for two values of the Skyrme parameter: e=4.0 and (in brackets: e=4.25). For these cases the soliton mass $\mathcal{M}_S$ (Eq. 84) is 1756 (1647) MeV. The N-Δ mass split $\Delta^{(0)}=3/(2\Theta)$ is 249 (293) MeV. The "background" mass correction from the isorotational zero mode (Eqns. 77, 78) is $\delta M^{(ZM)}$ = -501 (-522) MeV; the corresponding translational contribution is $\delta M^{(trans)}$ = -37 (-39) MeV. The $\mathcal{O}(1/N_c)$ correction (88) is

$$\delta\mathcal{M}_N^{(1)} = -130 \, (-147) \, \text{MeV} \tag{90}$$

leading to the nucleon mass

$$M_N = \mathcal{M}_S + \delta M^{(ZM)} + \delta M^{(trans)} + \delta\mathcal{M}_N^{(1)} + \Delta^{(0)}/4 = 1150 \; (1012)\text{MeV}.$$

Keeping in mind that the principal value of the integral (86) can only give a rough estimate for the $\mathcal{O}(1/N_c)$ correction to the Δ mass

$$\delta\mathcal{M}_\Delta^{(1)} = -89 \, (-103) \, \text{MeV}$$

we obtain the total N-Δ mass split

$$\Delta^{(0)} + \delta\mathcal{M}_\Delta^{(1)} - \delta\mathcal{M}_N^{(1)} = 290 \; (337) \, \text{MeV}. \tag{91}$$

The width (89) of the Δ-resonance at the resonance energy (91) is $\Gamma(290)$ = 91 MeV ($\Gamma(337)$ = 128 MeV).

Although there certainly will be other small quantum contributions to the masses the very simple formulae presented here at least provide a lower limit for the effects to be expected from the quantization of the soliton. Due to the formfactors (78) and (82) which are naturally supplied by the theory all the occurring k-integrals are finite. A very natural extension and check would be to include in the trace in (72) not only the zero modes but the complete set of distorted waves. In the 1+1 dimensional sine-Gordon or Φ^4-kink models this is known to increase the mass corrections by about 10% (see Appendix).

6. Conclusion

In this lecture we have pointed out that some of the most persistent difficulties of Skyrme type soliton models can be resolved by including terms beyond the leading order in a $1/N_c$ expansion of the lagrangian in terms of its dynamical coordinates or fluctuations. That $1/N_c$ terms restore the Tomazawa-Weinberg LET for the isovector part of S-wave scattering has been suggested very early by Uehara and Kondo[25]. That they also correct the isoscalar part was first observed by Walliser[11]. For P-waves Verschelde[17] noticed that $1/N_c$ corrections lead to expressions which are identical with the static isobar model, and we have seen here that the resulting dynamical πNN formfactor allows for an excellent reproduction of the P33 phase shifts. Finally, we have presented here a lower limit for the $\mathcal{O}(1)$ quantum corrections which have to be subtracted from the $\mathcal{O}(N_c)$ soliton mass. It turns out that the resulting nucleon mass is still slightly above its experimental value, leaving room for further small corrections. We have shown that the important zero mode contributions to these $\mathcal{O}(1)$ loop graphs are finite and very easy to calculate. This strongly suggests to reconsider the analyses of the πNN coupling constant f, the closely related normalization of the axial current g_A, and the nucleon magnetic moments. All of these have persistently resisted improvement by more sophisticated extensions of the Skyrme model, and we may expect sizeable contributions from the graphs in fig. 9 to these quantities. If

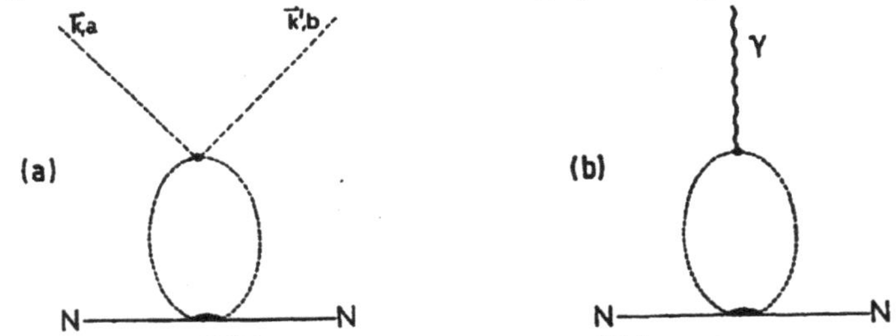

Fig. 9. Graphs for the $\mathcal{O}(1)$ contributions to (a) the πNN coupling constant f and (b) to the electromagnetic properties of the nucleons.

not, then there still remains the possibility to choose the Skyrme parameter e=3.5, which leads to satisfactory $\mathcal{O}(N_c)$ results f=1.0, g_A=1.28 and[26] μ_p=2.82 and (not satisfactory) μ_n=-2.52. The reason for the deficiency in this last number is discussed in ref. 26 and it shows that in any case we should expect important $\mathcal{O}(1)$ contributions from the graph in fig. 9b. With e=3.5, however, we need large $\mathcal{O}(1/N_c)$ corrections to the $N\Delta$-split, of the order of 100 MeV. Presently, these questions have not yet been investigated, but the answers may well provide new evidence for our experience that the dynamical content of Skyrme-type soliton models is much better than the approximation methods which so far have been used for extracting it.

7. Acknowledgements

The material collected in this lecture has grown out of numerous discussions with B. Schwesinger, G. Pari, B. K. Jennings, M. Robilotta, and especially with H. Walliser. Ref.21 was pointed out to me by B. Moussalam.

8. References

1. H. Verschelde, *Phys. Lett.* **B215** (1988) 444.
2. M. P. Mattis and M. Karliner, *Phys. Rev.* **D31** (1985) 2833.
3. G. Eckart, A. Hayashi and G. Holzwarth, *Nucl. Phys.* **A448** (1986) 732.
4. M. Karliner and M. P. Mattis, *Phys. Rev.* **D34** (1986) 1991.
5. B. Schwesinger, H. Weigel, G. Holzwarth and A. Hayashi, *Phys. Rep.* **173** (1989) 173.
6. B. Schwesinger, *Nucl. Phys.* **A537** (1992) 253.
7. A. Hayashi, G. Eckart, G. Holzwarth and H. Walliser, *Phys. Lett.* **B147** (1984) 5.
8. M. P. Mattis and M. E. Peskin, *Phys. Rev.* **D32** (1985) 58.
9. T. E. O. Ericson and W. Weise, *Pions and Nuclei* (Clarendon, Oxford, 1988).
10. D. Masak, H. Walliser and G. Holzwarth, *Nucl. Phys.* **A536** (1992) 583.
11. H. Walliser, *Nucl. Phys.* **A524** (1991) 706.
12. G. Hoehler, F. Kaiser, R. Koch and E. Pietarinen, *Physics Data*, Universität Karlsruhe (1979),
    R. Koch and E. Pietarinen, *Nucl. Phys.* **A336** (1980) 331.
13. A. Messiah, *Quantum Mechanics*, (Wiley, New York, 1960) ch. XIX. p. 830.
14. T. D. Cohen, *Phys. Rev.* **D34** (1986) 2187.
15. N. Kaiser, U. G. Meissner and W. Weise, *Phys. Lett.* **B198** (1987) 319,
    N. Kaiser, U. Vogel, W. Weise and U. G. Meissner, *Nucl. Phys.* **A484** (1988) 593.
16. H. J. Schnitzer, *Phys. Lett.* **B139** (1984) 217.
17. H. Verschelde, *Phys. Lett.* **B232** (1989) 15.
18. G. Holzwarth, G. Pari and B. K. Jennings, *Nucl. Phys.* **A515** (1990) 665.
19. J. Goldstone and R. Jackiw, *Phys. Rev.* **D11** (1975) 1486.
20. R. Koch and M. Sararu, Universität Karlsruhe, *TKP 84-6* (1984).
21. K. Cahill, A. Comtet and R. J. Glauber, *Phys. Lett.* **B64** (1976) 283.
22. G. Holzwarth, H. Walliser and G. Pari, *Z. Phys.* **A339** (1991) 195.
23. R. F. Dashen, B. Hasslacher and A. Neveu, *Phys. Rev.* **D10** (1974) 4130.

24. R. F. Dashen, B. Hasslacher and A. Neveu, *Phys. Rev.* **D11** (1975) 3424.
25. M. Uehara and H. Kondo, *Prog. Theor. Phys.* **75** (1986) 981.
26. F. Meier, Contribution to these Proceedings.

Appendix

Evaluating the loop graph of fig.7a for the normalized translational zero mode

$$\psi_0(x) = \sqrt{\frac{m}{2}}\frac{1}{\cosh(mx)} \tag{92}$$

in the 1+1 dimensional sine-Gordon model leads to (with $\omega_k^2 = k^2 + m^2$)

$$\begin{aligned}\delta M &= \frac{1}{2}\int \frac{dk}{(2\pi)\,2\omega_k}\frac{|\omega_k^2 \int e^{ikx}\psi_0(x)dx|^2}{-\omega_k^2} \\ &= -\frac{m}{4\pi}\int_{-\infty}^{+\infty} dq\sqrt{q^2+(\pi/2)^2}\frac{1}{\cosh^2 q} = -0.89\frac{m}{\pi}\end{aligned} \tag{93}$$

to be compared with the value $\delta M_C = -m/\pi$ quoted in ref.23 for the renormalized Casimir energy.

The corresponding consideration for the Φ^4-kink is slightly complicated by the existence of two bound states, the zero mode with $\omega_0^2=0$,

$$\psi_0(x) = c_0 \cosh^{-2}\left(\frac{mx}{\sqrt{2}}\right), \quad c_0^2 = \frac{3m}{4\sqrt{2}} \tag{94}$$

and a bound state at $\omega_1^2 = 3m^2/2$

$$\psi_1(x) = c_1 \sinh\left(\frac{mx}{\sqrt{2}}\right)\cosh^{-2}\left(\frac{mx}{\sqrt{2}}\right), \quad c_1^2 = \frac{3m}{2\sqrt{2}}. \tag{95}$$

Including both bound states in the loop graph (fig.7a) leads to (with $\omega_k^2 = k^2 + 2m^2$)

$$\begin{aligned}\delta M &= \frac{1}{2}\int \frac{dk}{(2\pi)\,2\omega_k}\frac{|\omega_k^2 \int e^{ikx}\psi_0(x)dx|^2}{-\omega_k^2} \\ &\quad + \frac{1}{2}\int \frac{dk}{(2\pi)\,2\omega_k}\frac{|(\omega_k^2-\omega_1^2)\int e^{ikx}\psi_1(x)dx|^2}{-(\omega_k+\omega_1)^2} \\ &= -m\frac{3}{2\sqrt{2}\pi^3}(I_0 + \frac{1}{4}I_1).\end{aligned} \tag{96}$$

The integrals I_0 and I_1 are

$$\begin{aligned}I_0 &= \int_{-\infty}^{+\infty} dq\sqrt{q^2+\pi^2}\frac{q^2}{\sinh^2 q} = 11.2 \\ I_1 &= \int_{-\infty}^{+\infty} dq\frac{(\sqrt{q^2+4\pi^2}-\sqrt{3\pi^2})^2}{\sqrt{q^2+4\pi^2}}\frac{q^2}{(1+\cosh q)} = 3.3\end{aligned} \tag{97}$$

so we obtain δM=−0.41m to be compared with the renormalized Casimir energy quoted in ref.24

$$\delta M_C = \frac{m}{\sqrt{2}}\left(\frac{1}{2\sqrt{3}} - \frac{3}{\pi}\right) = -0.47m\,.$$

MESON-BARYON COUPLING IN A SOLITON PICTURE

A. HAYASHI *
Department of Applied Physics, Fukui University
910 Fukui, Japan

ABSTRACT

We quantize the rotational and translational motions of the skyrmion and the fluctuation fields around the skyrmion. For the motions of the skyrmion, the collective coordinates are used. We carefully investigate the consequences of the redundancy in the decomposition of the original fields into the skyrmion and fluctuation fields. Taking into account all the relevant terms at the tree level in the Hamiltonian, we show that pion-nucleon scattering amplitudes exhibit the Born amplitudes with the correct recoil energy, which are attributed to the Yukawa coupling of order $\sqrt{N_c}$. Furthermore, we show that the derivation of the Born amplitudes gets much simpler if one takes the original fields, which are the sum of the skyrmion and fluctuation fields, as the interpolating fields for the scattering matrices.

1. Introduction

It was almost ten years ago when the Skyrme model was revived [1] and applied to the static and dynamical properties of the baryons [2]. In this model, the baryons are identified with the solitonic solution of a nonlinear meson theory. This picture is very different from other conventional ones, like quark models for example. We now must ask what are nice features in regarding "baryons as solitons" and what are not.

Let us concentrate on the dynamical properties of the baryons, namely the problems on the pion-nucleon scattering and baryon resonances. In the Skyrme model, the pion fields are expressed as the fluctuations around the moving skyrmion. In the large N_c limit, the collective mass parameters of the skyrmion, the translational mass and the moment of inertia, are very large and consequently the motions of the skyrmion are supposed to be slow. In this case, the fluctuation fields are only distorted by the presence of the background field, the skyrmion field. This is the lowest-order adiabatic picture of pion-nucleon scattering in the Skyrme model. The lowest order adiabatic description of the pion-nucleon scattering has the following nice features [3,4,5]:

- The formalism gives not only resonance positions and widths but also the S matrices themselves as functions of energy, namely including the background.
- Linear relations between the S matrices in different channels can be derived and found to be well satisfied in the high-angular-momentum partial waves.
- Including vector mesons and fixing the parameters in the meson sector, one obtains a nice quantitative agreement between theoretical prediction and experimental data in the high partial waves.

In the lower partial waves (S-, P- and D-waves), however, the linear relations are badly violated experimentally. Especially, the delta resonance in the P_{33} channel is

*This talk is based on work in collaboration with S. Saito (Nagoya) and M. Uehara (Saga).

missing in the calculation. The reason is that our lagrangian lacks the conventional Yukawa coupling term, which is linear in the pion fields and therfore creates and annihilates the pion by the presence of the nucleon [6]. By definition, the static skyrmion is the configuration with the lowest energy. So only when the skyrmion is moving, we obtain linear terms in the pion fields in the lagrangian . Consequently, in order to improve resultes in the lower partial waves, we need to take into account the non-adiabatic effects, namely higher order terms in $1/N_c$ expansion.

There is yet another delicate but important issue concerning the redundancy in the decomposition of the original fields into the moving skyrmion and the fluctuation fields around it. The redundancy comes from the obvious fact that a little change in the position or the orientation of the skyrmion is also expressed by the appropriate small fluctuation of the skyrmion. As we will see in the following sections, the redundancy also produces the higher order effects in the $1/N_c$ expansion. This is also a problem in regarding "baryons as solitons".

Thus we conclude that the soliton model has the following unsatisfactory feature in the pion-nucleon scattering problem.

- The Yukawa couplings, or the Born amplitudes arising from it in the conventional picture, are hidden in the higher order effects in the $1/N_c$ expansion.

So far, many attempts to solve the problem have been reported in the literatures[7, 8,9,10,11]. In the following §2 and §3, we will report our results [12,13,14] which we believe are most straightforward and general. Taking into all the relevant terms at the tree level, we will show that pion-nucleon scattering amplitudes exhibit the P-wave Born amplitudes attributed to the Yukawa coupling of order $\sqrt{N_c}$. The derivation of the desired Born terms presented in §2 and §3 is straightforward but not simple. In §4, we will show that without decomposing the original fields one gets a much simpler derivation as far as the Born amplitudes are concerned. The concluding remarks are given in §5.

2. Quantization of Fluctuating Fields

In this section, we quantize the moving skyrmion and the fluctuating fields around it. This is an extension of the method developed in the $1+1$ dimensional soliton model [12] to the $3+1$ dimensional model, the Skyrme model. The important point is that there appear two kinds of zero modes, the rotational and translational zero modes. In a previous paper [13], we have considered only the rotational zero modes. Here, we properly treat both of the zero modes, and then derive the desirable $\pi - \mathrm{N}$ Born term with the recoil effect in the scattering amplitude.

Denoting the original iso-triplet fields by $\Psi^a (a = 1,2,3)$, we write the Skyrme Lagrangian:

$$L = \int d^3x \, \left(\frac{1}{2}\partial\Psi^a\partial\Psi^a - \frac{1}{2}m^2\Psi^a\Psi^a - \mathcal{V}(\Psi, \partial_i\Psi)\right), \tag{1}$$

where m is the pion mass and $\mathcal{V}$ is the interaction potential which is a complicated function of Ψ and its space derivative $\partial_i\Psi$ $(i = 1,2,3)$. Although the details of the potential part $\mathcal{V}$ are irrelevant to the following argument, we assume the metric tensor for the time derivatives is unity for the sake of simplicity. In the following, we use the letters $a, b, \ldots$ for isospin indices and $i, j \ldots$ for space indices.

The skyrmion, the static soliton solution, is given by the hedgehog form:

$$\phi^i(\mathbf{x}) = f_\pi \hat{\mathbf{x}}^i \sin F(r), \tag{2}$$

where $F(r)$ is the chiral angle and f_π is the pion decay constant. The motions of the skyrmion are represented by introducing two sets of collective coordinates, an orthogonal 3 by 3 matrix D_{ai} for the rotation and $\mathbf{R}$ for the position of the soliton. The skyrmion in the laboratory frame is then written as

$$\Phi^a(\mathbf{x}) = D_{ai}\phi^i(\mathbf{x} - \mathbf{R}). \tag{3}$$

In terms of six generators l_α ($\alpha = i$ or a) defined as

$$l_i \equiv \frac{d}{dx^i}, \quad , \quad l_a \equiv iL_a, \tag{4}$$

with $(L_a)_{bc} = -i\epsilon_{abc}$, the six zero mode solutions in the laboratory frame are conveinently given by $l_\alpha \Phi(\mathbf{x})$. The collective mass parameters are given by the integral

$$\int d^3x \; l_\alpha \Phi \cdot l_\beta \Phi = \delta_{\alpha\beta} M_\alpha, \tag{5}$$

where $M_a = \Lambda$ is the momoent of inertia and $M_i = M$ is the mass of the skyrmion.

We now decompose the total field Ψ into the moving and rotating skyrmion Φ and the fluctuation fields χ around it as follows;

$$\Psi^a(\mathbf{x}, t) = \Phi^a(\mathbf{x}, t) + \chi^a(\mathbf{x}, t). \tag{6}$$

Since the decomposition is redundant, the quantization in this frame work requires gauge fixing conditions in the Dirac canonical formalism. We first observe that the canonical momenta conjugate to the collective coordinates and the fluctuating fields are constrained by the relations

$$\phi_\alpha \equiv K_\alpha + \int d^3x \; l_\alpha \Phi \cdot \pi' = 0, \tag{7}$$

where the canonical momenta conjugate to the fluctuating fields and the collective coordinates are difined by

$$\pi'^a(\mathbf{x}) = \delta L / \delta \dot{\chi}^a(\mathbf{x}), \;\; K_i = \partial L / \partial \dot{R}_i, \;\; K_a = \partial L / \partial \omega_a,$$

with the angular velocity ω_a defined by

$$(\dot{D}D^{-1})_{ab} = -i(\omega \cdot \mathbf{L})_{ab}.$$

By calculating Poisson brackets, those constraints are readily shown to be of the first class, namely

$$\begin{aligned} \{\phi_a, \phi_b\}_P &= \epsilon_{abc}\phi_c, \quad \text{(otherwise zero)}, \\ \{H, \phi_\alpha\}_P &= 0, \end{aligned} \tag{8}$$

with $H = \int d^3x \ (1/2(\pi' \cdot \pi' + \nabla\Psi \cdot \nabla\Psi + m^2\Psi \cdot \Psi) + \mathcal{V}(\Psi, \partial_i\Psi))$. We take the following natural gauge fixing conditions;

$$\int d^3x \ l_\alpha \Phi \cdot \chi = 0, \tag{9}$$

which impose the fluctuations be orthogonal to all the zero mode solutions. Using the normalized zero modes solution $\Phi_\alpha = l_\alpha\Phi/\sqrt{M_\alpha}$, we decompose the momentum π' in the form

$$\pi'^a = \pi^a - \frac{K_\beta}{\sqrt{M_\beta}}\Phi^a_\beta, \tag{10}$$

where the first term π^a is transverse to and the second is londitudinal to the collective motions. With the transverse momenta π, the constraint Eq. 7 can be rewritten as

$$\phi_\alpha = \int d^3x \ l_\alpha \Phi \cdot \pi = 0. \tag{11}$$

Furthermore we define new collective momenta $\mathcal{P}_\alpha$ as

$$\mathcal{P}_\alpha = (1 + \Xi)_{\alpha\beta}\frac{K_\beta}{\sqrt{M_\beta}}, \tag{12}$$

where the symmetric 6 by 6 matrix Ξ is given by

$$\Xi_{\alpha\beta} = -\int d^3x \ \chi \cdot \Phi_{\alpha\beta}, \tag{13}$$

with

$$\Phi_{\alpha\beta}(\mathbf{x}) = \frac{1}{\sqrt{M_\alpha M_\beta}} l_\alpha l_\beta \Phi(\mathbf{x}).$$

In the above we have absorbed a factor of $1/\sqrt{M_\alpha}$ into the definition of $\mathcal{P}_\alpha$, namely

$$\mathcal{P}_a = \frac{\tilde{I}_a}{\sqrt{\Lambda}}, \quad \mathcal{P}_i = \frac{\tilde{P}_i}{\sqrt{M}}, \tag{14}$$

where as will be seen in the following $\tilde{I}_a$ is asymptotically identified with the isospin and $\tilde{P}_i$ the linear momentum operator of the skyrmion in the laboratory frame.

With the new momenta π^a and $\mathcal{P}_\alpha$, the Hamiltonian is now written as

$$\begin{aligned} H = &\int d^3x \ \left(\frac{1}{2}(\pi \cdot \pi + \nabla\Psi \cdot \nabla\Psi + m^2\Psi \cdot \Psi) + \mathcal{V}\right) \\ &+ \frac{1}{8}\left\{(1+\Xi)^{-1}_{\beta\alpha}, \mathcal{P}_\alpha\right\}\left\{(1+\Xi)^{-1}_{\beta\gamma}, \mathcal{P}_\gamma\right\}, \end{aligned} \tag{15}$$

where we adopted the Tomboulis symmetrization procedure[15]. Since we will calculate everything in the tree approximation ,we expand the Hamiltonian in power series of the fluctuation χ. Noting that Ξ is linear in the fluctuation, we obtain

$$H = M + H_{\rm meson} + H_{\rm coll} + H_1 + H_2, \tag{16}$$

where

$$H_{\rm meson} = \frac{1}{2}\int d^3x \left(\pi\cdot\pi + \chi_a(-\nabla^2 + m^2 + \mathcal{V}^{(2)}(\Phi))_{ab}\chi_b\right), \tag{17}$$

$$H_{\rm coll} = \frac{1}{2}\mathcal{P}_\alpha\mathcal{P}_\alpha = \frac{\tilde{\mathbf{P}}^2}{2M} + \frac{\tilde{\mathbf{I}}^2}{2\Lambda}, \tag{18}$$

$$H_1 = -\frac{1}{4}\left\{\mathcal{P}_\alpha, \{\mathcal{P}_\beta, \Xi_{\alpha\beta}\}\right\}, \tag{19}$$

$$H_2 = \frac{1}{8}\{\mathcal{P}_\alpha, \Xi_{\alpha\gamma}\}\{\mathcal{P}_\beta, \Xi_{\beta\gamma}\} + \frac{1}{4}\left\{\mathcal{P}_\alpha, \{\mathcal{P}_\beta, {\Xi^2}_{\alpha\beta}\}\right\}. \tag{20}$$

The quantum corection terms of order $\hbar^2$ and the terms of higher powers of χ are neglected as before. Note that $H_{\rm meson}$ is order N_c^0 ,$H_{\rm coll}$ order N_c^{-1}, H_1 order $N_c^{-3/2}$, and H_2 order N_c^{-2} , because the collective momentum $\mathcal{P}_\alpha$ is of order $N_c^{-1/2}$ and Ξ order $N_c^{-1/2}$.

It is tedious but straightforward to calculate the Dirac brackets between various quantities defined above. Multiplying the Dirac brackets by the imaginary unit i, we obtain the commutators of the corresponding quantized operators. The fluctuations χ and the transverse canonical momenta π defined in Eq. 10 obey the following equal-time commutation relations

$$[\chi^a(\mathbf{x}), \pi^b(\mathbf{y})] = i(\delta^{ab}\delta(\mathbf{x}-\mathbf{y}) - \Phi^a_\alpha(\mathbf{x})\Phi^b_\alpha(\mathbf{y})), \tag{21}$$

where the second term in the right hand side comes from the fact that χ's are constrained by Eq. 7 and π's by Eq. 11. The collective momentum $\mathcal{P}_\alpha$ acts on the skyrmion as follows:

$$[\mathcal{P}_\alpha, \Phi^a(\mathbf{x})] = i\Phi^a_\alpha(\mathbf{x}), \tag{22}$$

by which we may identify $\mathcal{P}_\alpha$ with the collective momentum of the skyrmion alone. However, $\mathcal{P}_\alpha$ is not quite independent of the fluctuations around the skyrmion. This can be seen from the following commutation relations

$$[\mathcal{P}_\alpha, \chi^a(\mathbf{x})] = i\Phi^a_\beta(\mathbf{x})\Xi_{\alpha\beta}, \quad [\mathcal{P}_\alpha, \pi^a(\mathbf{x})] = i\Phi^a_\beta(\mathbf{x})\Theta_{\alpha\beta}, \tag{23}$$

with

$$\Theta_{\alpha\beta} = -\int d^3x\ \pi\cdot\Phi_{\alpha\beta}.$$

Since the normalized zero mode solution $\Phi^a_\alpha(\mathbf{x})$ is non-vanishing only near by the skyrmion, the above identification for $\mathcal{P}_\alpha$ is valid when the fluctuations are far away from the skyrmion. By the similar reason we take the fluctuation fields χ^a for the interporation fields when we evaluate the S matrix in the following.

3. Born Amplitudes

In this section we calculate the meson-baryon scattering amplitudes. We have seen that the fluctuations χ^a and the skyrmion collective momenta $\mathcal{P}^a$ are asymptotically independent. Therfore we take χ^a for the interporation fields and write the reduction formula for the S matrix in a symmetrized form, by which the following calculation becomes simpler than that by the non-symmetrized one.

$$(S-1)_{B'\mathbf{p}'\mathbf{k}'a',B\mathbf{p}\mathbf{k}a} = i^2 \int d^3x'dt'\ d^3xdt\ f^*_{k'}(\mathbf{x}',t')f_k(\mathbf{x},t)$$
$$\langle\, B'\mathbf{p}' \mid T\left(J^{a'}(\mathbf{x}',t')J^a(\mathbf{x},t)\right)$$
$$+\frac{1}{2}\delta(t'-t)\left([\dot{\chi}^a(\mathbf{x},t), J^{a'}(\mathbf{x}',t)] + [\dot{\chi}^{a'}(\mathbf{x}',t), J^a(\mathbf{x},t)]\right)$$
$$+\frac{1}{2}\delta(t'-t)\left(-i\omega_{k'}[\chi^{a'}(\mathbf{x}',t), J^a(\mathbf{x},t)] + i\omega_k[\chi^a(\mathbf{x},t), J^{a'}(\mathbf{x}',t)]\right)\ \mid B\mathbf{p}\,\rangle, \quad (24)$$

where the in- and out-states are specified by the isospin a and the momentum $\mathbf{k}$ of the pion, and for the baryon by the spin-isospin quantum number B and the momentum $\mathbf{p}$. The source functions $J^a(\mathbf{x},t)$ in the reduction formula are defined by

$$J^a(\mathbf{x},t) = (\partial^2 + m^2)\chi^a(\mathbf{x},t). \quad (25)$$

The scattering amplitude T_{fi} is now defined as

$$(S-1)_{fi} = \frac{i(2\pi)^4\delta(E_{B',\mathbf{p}'} + \omega_{k'} - E_{B,\mathbf{p}} - \omega_k)\delta(\mathbf{p}'+\mathbf{k}'-\mathbf{p}-\mathbf{k})}{((2\pi)^3 2\omega_k(2\pi)^3 2\omega_{k'}(2\pi)^3(2\pi)^3)^{1/2}}\ T_{fi}, \quad (26)$$

where

$$E_{B,\mathbf{p}} = \frac{I_B(I_B+1)}{2\Lambda} + \frac{\mathbf{p}^2}{2M}, \quad \omega_k = \sqrt{\mathbf{k}^2+m^2},$$

with I_B the magnitude of baryon's isospin.

The time derivatives of the fields χ^a in the reduction formula Eq. 24 and the source functions Eq. 25 should be calculated by using the commutators with the Hamiltonian Eq. 15. Noting that the Hamiltonian is expanded in powers of $N_c^{-1/2}$, we can also expand the source functions in the similar manner:

$$J_a(\mathbf{x},t) = J_a^{(0)}(\mathbf{x},t) + J_a^{(1)}(\mathbf{x},t) + J_a^{(3/2)}(\mathbf{x},t) + J_a^{(2)}(\mathbf{x},t), \quad (27)$$

where $J_a^{(n)}$'s mean that they are of order N_c^{-n}. We will ignore the terms higher than order N_c^{-2} in the following, because they do not contribute to the scattering amplitude in the tree approximation. In the same way, the scattering amplitude is also expanded according to the order of $1/N_c$:

$$T = B^{(0)} + B^{(1)} + B^{(2)} + B^{(3)}. \quad (28)$$

Let us first calculate the source function $J_a^{(3/2)}$, which comes from the second time derivative of the fluctuation fields χ^a. Noting that $H_{\rm meson}$ is of order N_c^0 and H_1 of order $N_c^{-3/2}$, we obtain

$$J_a^{(3/2)}(\mathbf{x}) = -[H_{\rm meson}, [H_1, \chi^a(\mathbf{x})]] - [H_1, [H_{\rm meson}, \chi^a(\mathbf{x})]]. \quad (29)$$

In the tree approximation, we can neglect the terms which involve more than second power of the fluctuation fields. In the approximation, $J_a^{(3/2)}$ can be calculated as

$$\begin{aligned} J_a^{(3/2)}(\mathbf{x}) &= -\frac{1}{4}\{\mathcal{P}_\alpha, \{\mathcal{P}_\beta, \Phi^a_{\alpha\beta}(\mathbf{x})\}\} \\ &= \left[H_{\rm coll}, [H_{\rm coll}, \Phi^a(\mathbf{x})]\right]. \end{aligned} \tag{30}$$

In the above and in the following, we rewrite the anticommutators involving the momenta $\mathcal{P}$ and the soliton fields in terms of the commutators. The necessary identities for those quantities are given in the Appendix of Ref. [14]. In Eq. 30, the source function $J_a^{(3/2)}$ is expressed simply as the double commutator between the skyrmion field and the total collective Hamiltonian $H_{\rm coll}$, which involves both the translational and rotational energy. As we will see, this is the reason why we obtain the desired Born term, which includes the full recoil effect and the rotational excitation of the baryon. Substituting Eq. 30 for $J_a^{(3/2)}$ into the time-ordering part in the reduction formula Eq. 24, we obtain

$$\begin{aligned} B^{(3)} = \sum_{B''} & \left(\frac{(E_{B',\mathbf{p}'} - E_{B'',\mathbf{p}+\mathbf{k}})^2 (E_{B'',\mathbf{p}+\mathbf{k}} - E_{B,\mathbf{p}})^2}{E_{B'',\mathbf{p}+\mathbf{k}} - E_{B,\mathbf{p}} - \omega_k} \langle B' \mid \Phi^{*a'}(\mathbf{k}') \mid B''\rangle\langle B'' \mid \Phi^a(\mathbf{k}) \mid B\rangle \right. \\ & \left. + \frac{(E_{B',\mathbf{p}'} - E_{B'',\mathbf{p}-\mathbf{k}'})^2 (E_{B'',\mathbf{p}-\mathbf{k}'} - E_{B,\mathbf{p}})^2}{E_{B'',\mathbf{p}-\mathbf{k}'} - E_{B',\mathbf{p}'} + \omega_k} \langle B' \mid \Phi^a(\mathbf{k}) \mid B''\rangle\langle B'' \mid \Phi^{*a'}(\mathbf{k}') \mid B\rangle \right), \end{aligned} \tag{31}$$

where $\Phi^a(\mathbf{k})$ is the Fourier transform of the skyrmion field:

$$\Phi^a(\mathbf{k}) \equiv D_{ai} \int d^3x \exp(i\mathbf{k}\mathbf{x})\phi^i(\mathbf{x}). \tag{32}$$

The time deriatives of the fluctuating fields are given by

$$\begin{aligned} \dot\chi^a(\mathbf{x}) &= \dot\chi^{a(0)}(\mathbf{x}) + \dot\chi^{a(1)}(\mathbf{x}) + \cdots \\ &= \pi^a(\mathbf{x}) - \frac{1}{2}\{\mathcal{P}_\alpha, \Phi^a_\beta(\mathbf{x})\Xi_{\alpha\beta}\} + \cdots. \end{aligned} \tag{33}$$

Here the first term $\dot\chi^{a(0)} = \pi^a$ is of order N_c^0 and the second term $\dot\chi^{a(1)}$ is of order N_c^{-1}. The zeroth order of the source function $J_a^{(0)}$ is written as

$$J_a^{(0)}(\mathbf{x}) = -\left[H_{\rm meson}, [H_{\rm meson}, \chi^a(\mathbf{x})]\right] + (-\nabla^2 + m^2)\chi^a(\mathbf{x}) = -\mathcal{V}^{(2)}_{ab}(\Phi)\chi^b(\mathbf{x}). \tag{34}$$

The equal-time commutator of $\dot\chi^{a(0)}$ with $J_a^{(0)}$ in the reduction formula gives the zeroth order amplitudes $B^{(0)}$. This can be again expressed in terms of the total collective Hamiltonian $H_{\rm coll}$, for example:

$$[\dot\chi^{a(0)}(\mathbf{x},t), J_{a'}^{(0)}(\mathbf{x}',t)] = i\mathcal{V}^{(2)}_{a'a}(\mathbf{x})\delta(\mathbf{x}'-\mathbf{x}) - i(\nabla'^2 - m^2)\left[\Phi^{a'}(\mathbf{x}'), [H_{\rm coll}, \Phi^a(\mathbf{x})]\right]. \tag{35}$$

Noting the symmetric form of the reduction formula, we find the zeroth order scattering amplitude to be

$$\begin{aligned} B^{(0)} = &-\langle B' \mid \mathcal{V}^{(2)}_{\mathbf{k}'a',\mathbf{k}a} \mid B\rangle \\ &-\sum_{B''}\frac{\omega_k^2+\omega_{k'}^2}{2}(E_{B'',\mathbf{p}+\mathbf{k}}-E_{B,\mathbf{p}})\langle B' \mid \Phi^{*a'}(\mathbf{k}') \mid B''\rangle\langle B'' \mid \Phi^{a}(\mathbf{k}) \mid B\rangle \\ &-\sum_{B''}\frac{\omega_k^2+\omega_{k'}^2}{2}(E_{B'',\mathbf{p}-\mathbf{k}'}-E_{B',\mathbf{p}'})\langle B' \mid \Phi^{a}(\mathbf{k}) \mid B''\rangle\langle B'' \mid \Phi^{*a'}(\mathbf{k}') \mid B\rangle. \end{aligned} \tag{36}$$

The source function $J^{(1)}$ can be obtained by using the double commutator of the fluctuations χ^a with H_{meson} and H_{coll} as follows:

$$\begin{aligned} J_a^{(1)}(\mathbf{x}) &= -\Big[H_{\text{meson}},[H_{\text{coll}},\chi^a(\mathbf{x})]\Big]-\Big[H_{\text{coll}},[H_{\text{meson}},\chi^a(\mathbf{x})]\Big] \\ &= -\{\mathcal{P}_\alpha,\Phi^a_\beta(\mathbf{x})\Theta_{\alpha\beta}\}, \end{aligned} \tag{37}$$

where we retained only the linear term in the fluctuations. The scattering amplitude of order N_c^{-1} comes from the commutator between χ and the $J^{(1)}$ in the reduction formula. We find the commutator is calculated as follows:

$$\begin{aligned} [\chi^{a'}(\mathbf{x}',t),J_a^{(1)}(\mathbf{x},t)] &= i\{\mathcal{P}_\alpha,\Phi^a_\beta(\mathbf{x})\Phi^{a'}_{\{\alpha\beta\}}(\mathbf{x}')\} \\ &= \Big[\Phi^a(\mathbf{x}),\Big[H_{\text{coll}},[H_{\text{coll}},\Phi^{a'}(\mathbf{x}')]\Big]\Big], \end{aligned} \tag{38}$$

with $\Phi^a_{\{\alpha\beta\}} = 1/2(\Phi^a_{\alpha\beta}+\Phi^a_{\beta\alpha})$. Thus we obtain

$$\begin{aligned} B^{(1)} = &-\frac{1}{2}\sum_{B''}\Big(\omega_{k'}(E_{B',\mathbf{p}'}-E_{B'',\mathbf{p}+\mathbf{k}})^2+\omega_k(E_{B'',\mathbf{p}+k}-E_{B,\mathbf{p}})^2\Big) \\ &\quad\langle B' \mid \Phi^{*a'}(\mathbf{k}') \mid B''\rangle\langle B'' \mid \Phi^{a}(\mathbf{k}) \mid B\rangle \\ &+\frac{1}{2}\sum_{B''}\Big(\omega_{k'}(E_{B'',\mathbf{p}-\mathbf{k}'}-E_{B,\mathbf{p}})^2+\omega_k(E_{B',\mathbf{p}'}-E_{B'',\mathbf{p}-\mathbf{k}'})^2\Big) \\ &\quad\langle B' \mid \Phi^{a'}(\mathbf{k}) \mid B''\rangle\langle B'' \mid \Phi^{*a'}(\mathbf{k}') \mid B\rangle, \end{aligned} \tag{39}$$

The second order source function $J^{(2)}$ consists of two pieces, which we denote by $K^{(2)}$ and $K'^{(2)}$, and given by

$$\begin{aligned} K^{(2)}(\mathbf{x}) &= -\Big[H_{\text{meson}},[H_2,\chi^a(\mathbf{x})]\Big]-\Big[H_2,[H_{\text{meson}},\chi^a(\mathbf{x})]\Big] \\ K'^{(2)}(\mathbf{x}) &= -\Big[H_{\text{coll}},[H_{\text{coll}},\chi^a(\mathbf{x})]\Big]. \end{aligned} \tag{40}$$

The second order scattering amplitudes come from the commutator terms of $\dot{\chi}^a = \pi^a$ with $J^{(2)}$. Although the calculation is now more involved, we find that the commutators between these second order source functions and the π field are expressed in terms of $H_{\rm coll}$ as in the previous cases. For $K^{(2)}$, the commutator is calculated as follows:

$$
\begin{aligned}
& [\pi_a(\mathbf{x},t), K_{a'}^{(2)}(\mathbf{x}',t)] + [\pi_{a'}(\mathbf{x}',t), K_a^{(2)}(\mathbf{x},t)] \\
= & +\frac{i}{4}\{\{\mathcal{P}_\alpha, \Phi^{a'}_{\{\alpha\gamma\}}(\mathbf{x}')\}\{\mathcal{P}_\beta, \Phi^{a}_{\{\beta\gamma\}}(\mathbf{x})\}\} \\
& +\frac{i}{2}\{\mathcal{P}_\alpha, \{\mathcal{P}_\beta, \Phi^{a'}_{\{\alpha\gamma\}}(\mathbf{x}')\Phi^{a}_{\{\gamma\beta\}}(\mathbf{x}) + \Phi^{a}_{\{\alpha\gamma\}}(\mathbf{x})\Phi^{a'}_{\{\gamma\beta\}}(\mathbf{x}')\}\} \\
= & -i[\Phi^{a'}(\mathbf{x}'), [H_{\rm coll}, [H_{\rm coll}, [H_{\rm coll}, \Phi^a(\mathbf{x})]]]] \\
& -3i[[H_{\rm coll}, \Phi^{a'}(\mathbf{x}')], [H_{\rm coll}, [H_{\rm coll}, \Phi^a(\mathbf{x})]]]. \qquad (41)
\end{aligned}
$$

On the other hand, the commutator between π and $K'^{(2)}$ takes the following form:

$$
\begin{aligned}
& [\pi_{a'}(\mathbf{x}',t), K_a'^{(2)}(\mathbf{x},t)] + [\pi_a(\mathbf{x},t), K_{a'}'^{(2)}(\mathbf{x}',t)] \\
= & -i[H_{\rm coll}, [H_{\rm coll}, [\Phi^{a'}(\mathbf{x}'), [H_{\rm coll}, \Phi^a(\mathbf{x})]]]]. \qquad (42)
\end{aligned}
$$

In the previous paper [13], this term did not contribute to the scattering amplitudes, because we assumed that the final baryonic energy is the same as the initial one. Substituting those commutators into the reduction formula, we finally obtain the second order scattering amplitude $B^{(2)}$ as

$$
\begin{aligned}
B^{(2)} = & \sum_{B''}\Big(\frac{1}{2}(E_{B'',\mathbf{p+k}} - E_{B,\mathbf{p}})^3 + \frac{2}{3}(E_{B',\mathbf{p}'} - E_{B'',\mathbf{p+k}})(E_{B'',\mathbf{p+k}} - E_{B,\mathbf{p}})^2 \\
& +\frac{1}{2}(E_{B',\mathbf{p}'} - E_{B,\mathbf{p}})^2(E_{B'',\mathbf{p+k}} - E_{B,\mathbf{p}})\Big)\langle B' \mid \Phi^{*a'}(\mathbf{k}') \mid B''\rangle\langle B'' \mid \Phi^a(\mathbf{k}) \mid B\rangle \\
& +\sum_{B''}\Big(\frac{1}{2}(E_{B'',\mathbf{p-k}'} - E_{B',\mathbf{p}'})^3 + \frac{2}{3}(E_{B,\mathbf{p}} - E_{B'',\mathbf{p-k}'})(E_{B'',\mathbf{p-k}'} - E_{B',\mathbf{p}'})^2 \\
& +\frac{1}{2}(E_{B,\mathbf{p}} - E_{B',\mathbf{p}'})^2(E_{B'',\mathbf{p-k}'} - E_{B',\mathbf{p}'})\Big)\langle B' \mid \Phi^a(\mathbf{k}) \mid B''\rangle\langle B'' \mid \Phi^{*a'}(\mathbf{k}') \mid B\rangle \qquad (43)
\end{aligned}
$$

We still have to evaluate the commutator $[\dot{\chi}^{a(1)}(\mathbf{x}), J_{a'}^{(1)}(\mathbf{x}')]$ which is order of N_c^{-2}. However, by the same argument as in the 1+1 dimensional case, the contribution from this term is found to be of higher order by $\hbar^2$ than the amplitudes $B^{(2)}$ given above. Thus we can safely discard this term.

Let us now sum up all the amplitudes calculated so far. The calculation can be done almost in the same way as in the 1+1 dimensional case. We, however, find that the sum does not immediately give the expected Born term, because of the following unequality:

$$
\langle B' \mid \Phi^{*a'}(\mathbf{k}') \mid B''\rangle\langle B'' \mid \Phi^a(\mathbf{k}) \mid B\rangle \neq \langle B' \mid \Phi^{a'}(\mathbf{k}) \mid B''\rangle\langle B'' \mid \Phi^{*a'}(\mathbf{k}') \mid B\rangle. \qquad (44)
$$

Here we should notice that the Fourie transforms of the skyrmion fields $\Phi^a(\mathbf{k})$ defined in Eq. 32 carry the spin-isospin degrees of freedom only through the matrix D_{ai}, therefore they commute. This point was emphasized by Gervais and Sakita [16] in connection with the N_c power counting in the storong-coupling theory. Thus, summing over the intermediate spin-isosin state B'', we observe

$$\sum_{B''}\langle B' \mid \Phi^{*a'}(\mathbf{k}') \mid B''\rangle\langle B'' \mid \Phi^{a}(\mathbf{k}) \mid B\rangle = \langle B' \mid \Phi^{*a'}(\mathbf{k}')\Phi^{a}(\mathbf{k}) \mid B\rangle$$
$$= \sum_{B''}\langle B' \mid \Phi^{a}(\mathbf{k}) \mid B''\rangle\langle B'' \mid \Phi^{*a'}(\mathbf{k}') \mid B\rangle. \tag{45}$$

Using the above relation Eq. 45, we find

$$\begin{aligned} T &= B^{(0)} + B^{(1)} + B^{(2)} + B^{(3)} \\ &= -\langle B' \mid \mathcal{V}^{(2)}_{\mathbf{k}'a',\mathbf{k}a} \mid B\rangle \\ &- \sum_{B''}\left(\frac{\langle B' \mid j^{a'}(\mathbf{k}') \mid B''\rangle\langle B'' \mid j^{a}(\mathbf{k}) \mid B\rangle}{E_{B'',\mathbf{p}+\mathbf{k}} - E_{B,\mathbf{p}} - \omega_k} + \frac{\langle B' \mid j^{a}(\mathbf{k}) \mid B''\rangle\langle B'' \mid j^{a'}(\mathbf{k}') \mid B\rangle}{E_{B'',\mathbf{p}-\mathbf{k}'} - E_{B',\mathbf{p}'} + \omega_k}\right), \end{aligned} \tag{46}$$

where $j^a(\mathbf{k})$ is the Fourier transform of the classical source function $j^a(\mathbf{x})$ defined by

$$(-\nabla^2 + m^2)D_{ai}\phi^i(\mathbf{x}) = j^a(\mathbf{x}), \tag{47}$$

therefore we have the relation $j^a(\mathbf{k}) = \omega_k^2\Phi^a(\mathbf{k})$. Note that the first term in Eq. 46 represents the back ground scattering of order N_c^0 and the remaining amplitudes are just the disired Born amplitudes of order $N_c^{1/2}$, where the recoil and the excitation of the baryon are correctly included. Furthermore, noting that the matrix element $\langle B' \mid D_{ai} \mid B\rangle$ is given by $\Lambda_{B'B}(T_a)_{B'B}(S_i)_{B'B}$ with $\Lambda_{NN} = -1/3$, $\Lambda_{N\Delta} = 1/\sqrt{2}$, and $\Lambda_{\Delta\Delta} = -1/15$, we can rewrite $j^a(\mathbf{k})$ as

$$j^a(\mathbf{k}) = i\frac{G_{B'B\pi}(\mathbf{k}^2)}{2M_N}\mathbf{k}\cdot\mathbf{S}_{B'B}(T_a)_{B'B}, \tag{48}$$

where M_N is the nucleon mass and the $B'B\pi$ coupling form factor is given by

$$G_{B'B\pi}(\mathbf{k}^2) = -8\pi M_N f_\pi \Lambda_{B'B}\frac{\mathbf{k}^2 + m^2}{\mid \mathbf{k} \mid}\int_0^\infty r^2 dr\, j_1(kr)\sin F(r). \tag{49}$$

The integral in the avobe has the pion pole, owing to the pion tail of the chiral angle $F(r)$. The pseudoscalar coupling constant is then given by $G_{B'B\pi}(\mathbf{k}^2 = -m^2)$, which is finite and agrees with the coupling constant [1] determined by the pion tail of skyrmion. In the case of massless pions, it satisfies the Goldberger-Treiman relation.

4. Simpler Derivation

In the previous sections, we decomposed the original total fields into the soliton and fluctuating fields. And the fluctuation fields are used as the interpolating fields in the reduction formula of the S matrix, since it is clear that the solitonic parts have no contribution in the scattering. As we have seen, the calculation is rather involved owing to the redundancy in the docomposition. In this section, without decomposing the origianl fields, we will show the correct Born amplitudes can be obtained in a much simpler way.

Since we do not use the decompositon, the canonical quantization is straightforward. According to the Lagrangian Eq. 1, the conjugate fields $\Pi^a(\mathbf{x},t)$ to the original fields $\Psi^a(\mathbf{x},t)$ are just their time derivatives and the usual canonical commutators can be assumed between them;

$$\begin{aligned} \Pi^a(\mathbf{x},t) &= \dot{\Psi}^a(\mathbf{x},t), \\ \left[\Psi^a(\mathbf{x},t), \Pi^{a'}(\mathbf{x}',t)\right] &= i\delta_{aa'}\delta(\mathbf{x}-\mathbf{x}'). \end{aligned} \tag{50}$$

In terms of Ψ and Π, the Hamiltonian is simply given by

$$H = \int d^3x \, \left(\frac{1}{2}(\Pi^a\Pi^a + \nabla\Psi^a\nabla\Psi^a + m^2\Psi^a\Psi^a) + \mathcal{V}(\Psi, \partial_i\Psi)\right). \tag{51}$$

Now using the original total fields Ψ^a, we write the reduction formula,

$$\begin{aligned} (S-1)_{B'\mathbf{p}'\mathbf{k}'a', B\mathbf{p}\mathbf{k}a} &= i^2 \int d^3x' dt' \, d^3x dt \, f^*_{k'}(\mathbf{x}',t') f_k(\mathbf{x},t) \\ &\langle B'\mathbf{p}' \mid T\left(J^{a'}_{\mathrm{tot}}(\mathbf{x}',t') J^a_{\mathrm{tot}}(\mathbf{x},t)\right) + \delta(t'-t)[\dot{\Psi}^{a'}(\mathbf{x}',t), J^a_{\mathrm{tot}}(\mathbf{x},t)] \mid B\mathbf{p}\rangle, \end{aligned} \tag{52}$$

where J^a_{tot}'s are the source functions of the original total fields Ψ^a and defined by

$$J^a_{\mathrm{tot}}(\mathbf{x},t) = (\partial^2 + m^2)\Psi^a(\mathbf{x},t). \tag{53}$$

It should noted here that the commutator terms between the fields and the source functions vanish in this representation. Using the Hamiltonian Eq. 51 and the canonical commutators Eq. 50, we find the source functions J_{tot} are given by,

$$J^a_{\mathrm{tot}} = -[H\,[\,H, \Psi^a]] + (-\nabla^2 + m^2)\Psi^a = -\mathcal{V}^{(1)}_a(\Psi), \tag{54}$$

where $\mathcal{V}^{(1)}_a$ is the functional derivative of $\mathcal{V}$ with respect to Ψ^a.

Let us first evaluate the time ordering term in the reduction formuala Eq. 52. In the tree approximation, intermediate states are assumed to be single baryonic states without pions. Therfore relevant matirx element is of the form,

$$\langle B\mathbf{p} \mid \mathcal{V}^{(1)}_a(\Psi) \mid B'\mathbf{p}'\rangle \tag{55}$$

As noted sometime ago[17,18], we can safely replace $\mathcal{V}^{(1)}_a(\Psi)$ in the above by its classical counter part $\mathcal{V}^{(1)}_a(\Phi)$, where Φ is the soliton field. The $\mathcal{V}^{(1)}_a(\Phi)$ is, with an additional negative sign, just the classical source function j^a defined in Eq. 47. Thus after performing the integrations over space and time, we immediately obtain the same Born amplitudes as the one given in the previous section. As for the equal-time commutator term, we find that the commutator is given by the second functinal derivative of $\mathcal{V}$ with respect to Ψ^a and Ψ^b;

$$\left[\dot{\Psi}^{a'}(\mathbf{x}',t)\ ,\ J^a_{\mathrm{tot}}(\mathbf{x},t)\right] = i\delta(\mathbf{x}-\mathbf{x}')\mathcal{V}^{(2)}_{aa'}(\Psi), \tag{56}$$

where the original field Ψ can also be replaced by the soliton field Φ, if the expression is inserted in the reduction formula. Therefore the contribution from the commutator term is the back ground scattering amplitude. Together with the Born amplitudes arising from the time ordering term, we have reproduced the whole amplitudes given Eq. 46 in a much simpler way.

5. Concluding Remarks

When we regard " baryons as solitons " as in the Skyrme model, we find that the meson-baryon scattering is conveinently described in the adiabatic approximation, where N_c is assumed to be large and the skyrmion is supposed to move slowly compared with fluctuations. However, in order to have the conventional picture of the meson-baryon scattering, which is attributed to the Yukawa coupling, we realize that we should go beyond the adiabatic approximation.

It is natural and reasonable attempt to decompose the original fields into the soliton part and fluctuations, since the soliton fields have no propagating components therefore no contribution in the meson-baryon scattering. As we have seen in §2, however, going beyond the adiabatic approximation with the decomposition is not at all trivial. The trouble arises from the redundancy in the decomposition. We must carefully study the consequences of the redundancy and pick up all the relevant terms in the $1/N_c$ expansion. By doing so, we have shown in §3 that pion-nucleon scattering amplitudes exhibit the Born amplitudes with the correct recoil energy, which are attributed to the Yukawa coupling of order $\sqrt{N_c}$. Thus we successfully solved one the the unsatisfactory features in the Skyrme model.

Furthermore, we have shown in §4 that the decomposition is not necesarry to get the correct Born terms and the derivation of the Born amplitudes is even much simpler without the decomposition. This is a good news, since it may be possible to study non-adiabatic (higher order in $1/N_c$) effects in the soliton model in a more tranparent way[19].

Acknowledgements

The author would like to thank Dr. S. Saito and Dr. M. Uehara for the collaboration and usefull discussions.

References

1. G. S. Adkins, C. R. Nappi and E. Witten, *Nucl. Phys.* **B228**, 552 (1983).
2. For a general review, see I. Zahed and G. E. Brown, *Phys. Rep.* **142**, 1 (1986); G. Holzwarth and B. Schwesinger, *Rep. Prog. Phys.* **49**, 825 (1986).
3. A. Hayashi, G. Eckart, G. Holzwarth and H. Walliser, *Phys. Lett.* **147B**, 5 (1984); H. Walliser and G. Eckart, *Nucl. Phys.* **A429**, 541 (1984); G. Eckart, A. Hayashi and G. Holzwarth, *ibid.* **A448**, 732 (1986).
4. M. P. Mattis and H. Karliner, *Phys. Rev.* **D31**, 2833 (1985); M. P. Mattis and M. E. Peskin, *idid.* **D32**, 58 (1985).
5. B. Schwesinger, H. Weigel, G. Holzwarth and A. Hayashi, *Phys. Rep.* **173**,

173 (1989).
6. T. Otofuji, S. Saito and M. Yasuno, *Prog. Theor. Phys.* **73**, 520 (1985); C. Hajduk and B. Schwesinger, *Nucl. Phys.* **A453**, 620 (1986); M. Uehara, *Prog. Theor. Phys.* **75**, 212 (1986).
7. S. Saito, *Nucl. Phys.* **A463**, 169 (1986); G. Holzwarth, A. Hayashi and B. Schwesinger, *Phys. Lett.* **B191**, 27 (1987); M. Uehara, *Prog. Theor. Phys.* **78**, 984 (1997); C. Adami and I. Zahed, *Phys. Lett.* **213**, 373 (1988); H. Liu and M. Oka, *Phys. Rev.* **D40**, 883 (1989).
8. H. Verschelde, *Phys. Lett.* **B215**, 444 (1988); **232**, 15 (1989); H. Verschelde and H. Verbeke, *Nucl. Phys.* **A495**, 523 (1989).
9. K. Ohta, *Phys. Lett.* **B234**, 229 (1990); *Nucl. Phys.* **A511**, 621 (1990).
10. G. Holzwarth, *Phys. Lett.* **B241**,165 (1990).
11. M. Uehara, *Prog. Theor. Phys.* **80**, 768 (1988); **83**, 790 (1990).
12. A. Hayashi, S. Saito and M. Uehara, *Phys. Lett.* **B246**, 15 (1990).
13. A. Hayashi and S. Saito and M. Uehara, *Phys. Rev.* **D43**, 1520 (1991).
14. A. Hayashi S. Saito and M. Uehara, *Prog. Theor. Phys. Suppl.* **109**, 45 (1992).
15. E. Tomboulis, *Phys. Rev.* **D12**, 1678 (1975).
16. J.-L. Gervais and B. Sakita, *Phys. Rev.* **D30**, 1795 (1984).
17. J. Goldstone and R. Jackiw, *Phys. Rev.* **D11**, 2943 (1975).
18. J.-L. Gervais, A. Jevicki and B. Sakita, *Phys. Rev.* **D12**, 1038 (1975).
19. A. Hayashi, S. Saito and M. Uehara, *Phys. Rev.* **D46**, 12 (1992).

MESON-BARYON SCATTERING FROM SKYRMIONS: A CONSTRUCTION MADE OF LARGE N_C STONES

BERND SCHWESINGER
FB Physik, University Siegen, Postfach 101240
D-5900 Siegen, Germany

ABSTRACT

Baryon resonances and meson-baryon scattering are discussed in terms of small amplitude fluctuations of Skyrme's soliton in SU(2) and SU(3). Special attention has been given to the role of rotational degrees of freedom up to first order in the inverse number of colors which are of crucial importance for the whole program.

1. Introduction

It was Skyrme's conjecture that baryons may be described as meson clouds, differing from actual meson fields only through the topological structure of the configuration. Such a picture leads to a particulary simple approach to meson-baryon scattering[1-4,10-14] where scattered bosons are small amplitude fluctuations around the static minimum in the topological baryon number one sector. Quite in contrast to this principally simple access formidable efforts are required to make the picture work quantitatively. The tedious progress from the first attempts dating back eight years to now came from the insight into the importance of the rotational vibrational couplings of order $\mathcal{O}(N_C^{-1})$ and the way they had to be incorporated. The figures of the section 3. will just display this difference between the early calculations of the πN scattering amplitudes to the present state of art, which not only includes features of the rotating hedgehog but also a necessary extension to SU(3) flavor degrees of freedom.

In the first part of this contribution I will concentrate on a careful presentation of how to systematically include all $\mathcal{O}(N_C^{-1})$ contributions into such scattering calculations because they have been proven to be the cornerstone to the low-energy theorems of pion nucleon physics, fig.1. The next part then reviews the effects of the $\mathcal{O}(N_C^{-1})$ terms in an SU(3) calculation for πN and $\bar{K}$N scattering amplitudes omitting the case of P-wave scattering which is special in the sense that one has to account for a linear term in the fluctuations responsible for a coupling of the P33 channel of πN scattering to the rotional $J = I = \frac{3}{2}^+$ excitation of the soliton. The presentation of P-wave scattering up to $\mathcal{O}(N_C^{-1})$ will conclude this contribution.

Figure 1: Graphical representation of the relative importance of various building blocks when the construction of a bridge to QCD is attempted.

2. A construction made of large N_C stones

2.1. *Effective action*

The lowest energy configuration of the chirally invariant lagrangian

$$
\begin{aligned}
L_0 &= L^{(2)} + L^{(4A)} + L^{(6)} + L^{(WZ)} \quad , \\
L^{(2)} &= -\frac{f_\pi^2}{4}\int d^3r \quad tr\ell_\mu \ell^\mu \quad , \qquad \ell_\mu = U^\dagger \partial_\mu U \quad , \\
L^{(4A)} &= \frac{1}{128 g^2}\int d^3r \quad tr[\ell_\mu, \ell_\nu][\ell^\mu, \ell^\nu] \cdot \ , \qquad U \in SU(3) \quad , \\
L^{(6)} &= -\frac{\epsilon_6^2}{2}\int d^3r \, B_\mu(U) B^\mu(U), \quad B^\mu(U) = \frac{\epsilon^{\mu\nu\rho\sigma}}{24\pi^2} tr l_\nu l_\rho l_\sigma \\
\int dt L^{(WZ)} &= \frac{iN_c}{240\pi^2}\int_Q d\Sigma^{\mu\nu\lambda\rho\sigma} tr\ell_\mu \ell_\nu \ell_\lambda \ell_\rho \ell_\sigma
\end{aligned}
\tag{1}
$$

in the topological winding number one sector is the hedgehog solution

$$
\begin{aligned}
U_H &= \exp\left(i\chi(r)\sum_{i=1}^{3}\lambda_i \hat{r}_i\right) \in SU(2) \subset SU(3), \\
& \chi(0) - \chi(\infty) = \pi .
\end{aligned}
\tag{2}
$$

It represents a baryon in its ground state because the Wess-Zumino term $L^{(WZ)}$ in (1) enforces an identification of winding number with baryon number under the assumption that the lagrangian above effectively describes the colorless, bosonized degrees of freedom of QCD[5]. The Wess-Zumino term thus provides the link between meson fields and baryon structure.

Meson masses may readily be introduced[6–9] into the effective action by addition of

$$
\begin{aligned}
L^{(FSB)} &= -\frac{f_K^2 - f_\pi^2}{12} \int d^3r \quad tr(1-\sqrt{3}\lambda_8)(U\ell_\mu\ell^\mu + \ell_\mu\ell^\mu U^+) \\
&\quad + \frac{f_K^2 m_K^2 - f_\pi^2 m_\pi^2}{6} \int d^3r \quad tr(1-\sqrt{3}\lambda_8)(U + U^+ - 2) \qquad (3) \\
L^{(CSB)} &= \frac{f_\pi^2 m_\pi^2}{4} \int d^3r \quad tr(U + U^+ - 2)
\end{aligned}
$$

where now SU(3) flavor symmetry and chiral symmetry is broken. In the mesonic sector the lagrangian

$$L = L_0 + L^{(CSB)} + L^{(FSB)} \tag{4}$$

fixes the η mass by a modified Gell-Mann - Okubo relation

$$m_\eta^2 = \frac{4f_K^2 m_K^2 - f_\pi^2 m_\pi^2}{4f_K^2 - f_\pi^2} \tag{5}$$

since the SU(3) symmetry breaking also affects the weak decay constants.

2.2. *The coupling of rotations and fluctuations*

Any continuous deformation of the hedgehog conserves the winding number present in the fields and thus baryon number. In case such deformations obey the equations of motion governed by L they must change in time and have to be interpreted as excitations of the baryon. In the Skyrme model such excitations extend out infinitly far away from the soliton center and are viewed as mesons scattering off the soliton since they are unbound then. An algebrically convenient but numerically very inconvenient parametrization of these deformations puts the static soliton to the e.g. left of the matrix U_f containing the fluctuations

$$
\begin{aligned}
U &= A(\alpha_i(t))\, U_H(\mathbf{r})\, U_f(\mathbf{r},t)\, A^\dagger(\alpha_i(t)), \\
U_f &= e^{i\lambda\cdot\psi} \qquad (6)
\end{aligned}
$$

and both between time dependent global flavor rotations

$$A^\dagger \dot{A} = \frac{i}{2}\lambda \cdot \mathrm{C}(\alpha_i) \cdot \dot{\alpha} = -\frac{i}{2}\lambda \cdot \Omega. \tag{7}$$

We use a dot product to indicate a contraction over flavor indices and the symbol λ for the set of generators of the flavor group considered. Most parts of the derivation to be presented are independent of the specific number of flavors and we will explicitly point out those steps where differences do occur.

In left parametrization the flavor rotational zero modes acquire a very simple form:

$$U_f(\mathbf{r},t) = U_H^\dagger[\lambda_a, U_H]. \tag{8}$$

When all rotational velocities Ω are assumed to be zero the lagrangian only comprises very few terms up to quadratic order in the fluctuations ψ :

$$\begin{aligned} \mathcal{O}(\Omega^0): \quad \mathcal{L}^{(2)}(U^\dagger\partial_\mu U) \quad &\rightarrow \quad \mathcal{L}^{(2)}(U_H^\dagger\partial_\mu U_H - U_f\partial_\mu U_f^\dagger) \\ &= \quad -\frac{f_\pi^2}{4}\mathrm{tr}(U_H^\dagger\partial_\mu U_H + i\lambda\cdot\partial_\mu\psi - \frac{1}{2}[\lambda\cdot\psi, \partial_\mu\lambda\cdot\psi])^2. \end{aligned} \tag{9}$$

Since we wish to focus on the construction of the $\mathcal{O}(N_C^{-1})$ terms and since the higher order stabilizers don't add any new features we concentrate on the terms from the non-linear σ-model $\mathcal{L}^{(2)}$ alone. Up to first order in the rotational velocities we find:

$$\mathcal{O}(\Omega^1): \quad \mathcal{L}_{rot}^{(2)} = -\frac{f_\pi^2}{4}\mathrm{tr}(U_H^\dagger[\lambda\cdot\Omega, U_H]\lambda\cdot\dot{\psi} + U_H^\dagger\{\lambda\cdot\Omega, U_H\}\frac{i}{2}[\lambda\cdot\psi, \lambda\cdot\dot{\psi}]). \tag{10}$$

The introduction of fluctuations and global rotations leads to redundant coordinates and, in a hamiltonian description of the dynamics, generally necessitates the Dirac formalism for constrained systems . We avoid such complications by using the well known fact that the Euler Lagrange equations

$$\begin{aligned} \frac{\delta\int L dt}{\delta\lambda\cdot\psi} &= 0 \\ (\partial_t - \Omega\times)\frac{\delta\int L dt}{\delta\Omega} &= \mathrm{C}^{-1}\cdot\frac{\delta\int L dt}{\delta\alpha} \overset{\mathrm{SU(2)}}{=} 0 \end{aligned} \tag{11}$$

immediately lead to the correct equations of motion, even with redundant coordinates.

In the three flavor case the action generally depends on the orientation A of the soliton in flavor space due to the symmetry breaking terms $L^{(FSB)}$ causing drastic complications unless their contribution is ignored, as done here. The result of the the variation is

$$\begin{aligned} \text{(i)} \qquad \frac{\delta\int L^{(2)}dt}{\delta\lambda\cdot\psi} = -\frac{f_\pi^2}{2}(\,&\partial_\mu\partial^\mu\lambda\cdot\psi - [\partial_\mu\lambda\cdot\psi,\, U_H^\dagger\partial^\mu U_H] \\ &-\frac{1}{2}U_H^\dagger[\lambda\cdot\dot{\Omega},\, U_H] \\ &+\frac{i}{4}[\lambda\cdot\psi,\, U_H^\dagger\{\lambda\cdot\dot{\Omega}, U_H\}] + \frac{i}{2}[\lambda\cdot\dot{\psi},\, U_H^\dagger\{\lambda\cdot\Omega, U_H\}]) \end{aligned} \tag{12}$$

$$\begin{aligned} \text{(ii)} \qquad (\partial_t - \Omega\times)(\Theta\cdot\Omega - \frac{f_\pi^2}{4}\int &\mathrm{tr}(U_H^\dagger[\lambda, U_H]\lambda\cdot\dot{\psi} \\ &+ U_H^\dagger\{\lambda, U_H\}\frac{i}{2}[\lambda\cdot\psi, \lambda\cdot\dot{\psi}])) = 0. \end{aligned} \tag{13}$$

With the N_C counting rules

$$\begin{aligned} \psi &= \psi_{adiab} + \psi_{non-ad} = \mathcal{O}(N_C^{-\frac{1}{2}}) + \mathcal{O}(N_C^{-\frac{3}{2}}) \\ \Omega &= \mathcal{O}(N_C^{-1}) \end{aligned} \tag{14}$$

and the right angular momenta

$$\begin{aligned} R &= \frac{\delta L}{\delta \Omega} \\ &= (\boldsymbol{\Theta} \cdot \Omega - \frac{f_\pi^2}{4} \int \mathrm{tr}(U_H^\dagger[\lambda, U_H]\lambda \cdot \dot{\psi} + U_H^\dagger\{\lambda, U_H\}\frac{i}{2}[\lambda \cdot \psi, \lambda \cdot \dot{\psi}])) \\ &\rightarrow -\frac{1}{i}\mathbf{C}^{-1} \cdot \frac{\partial}{\partial \alpha} \end{aligned} \tag{15}$$

which for the baryon states considered are $\mathcal{O}(N_C^0)$ we deduce

$$\Omega = \boldsymbol{\Theta}^{-1} \cdot (R + \frac{f_\pi^2}{4} \int \mathrm{tr}(U_H^\dagger\{\lambda, U_H\}\frac{i}{2}[\lambda \cdot \psi, \lambda \cdot \dot{\psi}])) + \mathcal{O}(N_C^{-\frac{3}{2}}) \tag{16}$$

and

$$\dot{\Omega} = \boldsymbol{\Theta}^{-1} \cdot \frac{f_\pi^2}{4} \int \mathrm{tr}(\, U_H^\dagger[\lambda, U_H]\, \lambda \cdot \ddot{\psi}) + \mathcal{O}(N_C^{-2}) = \mathcal{O}(N_C^{-\frac{3}{2}}). \tag{17}$$

In the derivation of the N_C behaviour above I have used that the adiabatic fluctuations are orthogonal to the rotational zero mode

$$\int \mathrm{tr}(U_H^\dagger[\lambda, U_H]\lambda \cdot \dot{\psi}_{adiab} = 0 \tag{18}$$

since both are eigensolutions to the same hermitian differential operator, the zero mode having $\omega = 0$. This means that the vertex linear in the fluctuations coming from the lagrangian $\mathcal{L}_{rot}$ is $\mathcal{O}(N_C^{-\frac{3}{2}})$.

We can now gather all terms of $\mathcal{O}(N_C^{-1})$ in $\frac{\delta \int L^{(2)} dt}{\delta \lambda \cdot \psi}$:

$$\begin{aligned} &\frac{f_\pi^4}{16} U_H^\dagger[\lambda, U_H] \cdot \boldsymbol{\Theta}^{-1} \cdot \int \mathrm{tr}(U_H^\dagger[\lambda, U_H]\, \lambda \cdot \ddot{\psi}) \\ - &\frac{if_\pi^2}{4}[\lambda \cdot \dot{\psi},\, U_H^\dagger\{\lambda, U_H\}] \cdot \boldsymbol{\Theta}^{-1} \cdot (\frac{if_\pi^2}{8} \int \mathrm{tr}(\, U_H^\dagger\{\lambda, U_H\}\,[\lambda \cdot \psi, \lambda \cdot \dot{\psi}] + R) \end{aligned} \tag{19}$$

and find that they lead to a non-linear differential equation for ψ due to the corrections to the right angular momenta coming from the fluctuations. However, up to order $\mathcal{O}(N_C^{-1})$ it is sufficient to keep the contributions from ψ_{adiab} alone in these corrections, so we can use their completeness relation

$$\psi_a(\mathbf{x}, t)\dot{\psi}_b(\mathbf{x}', t) \rightarrow \frac{i}{2f_\pi^2}\delta(\mathbf{x} - \mathbf{x}')\mathbf{N}^{-1}(\mathbf{x})_{ab} \tag{20}$$

contracting the term cubic in the fluctuations to

$$-\frac{if_\pi^2}{32\Theta}\mathbf{N}_{ab}^{-1}\,[\lambda_a,\,2[\lambda_b,\lambda\cdot\dot{\psi}]+U_H^\dagger[\lambda_b,\lambda\cdot\dot{\psi}]U_H+U_H[\lambda_b,\lambda\cdot\dot{\psi}]U_H^\dagger\,]. \tag{21}$$

N is the metric of the adiabatic fluctuations which actually is just a unit matrix for the non-linear σ-model with left parametrization and also with the stabilizing terms, asymptotically.

In the course of this derivation we have tacitly assumed that the tensor for the moments of inertia can be inverted and later also that it is proportional to the unit tensor. Both assumptions only hold for SU(2) and in case of three flavors we thus have introduced a second uncontrolled approximation by the use of an average moment of inertia.

For SU(2) the sum of all terms up to $\mathcal{O}(N_C^{-1})$ in the equations of motion contains no Euler angle dependence and the right angular momenta $R=-J_A$ are constants of motion. J_A is the spin carried by the Euler angles. Thus, in SU(2) $\lambda\cdot\psi\sim e^{-i\omega t}$ solves for the time dependence of the fluctuation. For SU(3) this ansatz unfortunately is unjustified and introduces the next uncertainty.

2.3. *The equations of motion in the laboratory system*

Although we now have reached our goal: the equations of motion for the fluctuations complete up to $\mathcal{O}(N_C^{-1})$ it will be useful to proceed by rotating them from the soliton fixed flavor axes to physical ones. The latter define the physical channels and thus the asymptotic boundary conditions for the solutions. This may be done by multiplication of the soliton fixed e.o.m. from the left by A and from the right by $A^\dagger$ using

$$A\lambda\cdot\psi A^\dagger=\lambda\cdot\varphi. \tag{22}$$

In the laboratory system we couple the isospin and hypercharge of the generators λ carrying the mesonic charges with the isospin τ and hypercharge y of the target baryon carried by the Euler angle dependence of the laboratory fluctuation $\varphi(\mathbf{r},\alpha)$ to conserved total isospin T and total hypercharge Y:

$$\lambda\cdot\varphi=\sum_{\sigma\tau yl}(\lambda\circ\varphi^{l\sigma\tau y}(\mathbf{r},\alpha))_{JTY}e^{-i\omega t}. \tag{23}$$

A further coupling of the mesonic angular momentum l carried by the angular dependence on $\mathbf{r}$ and the spin σ of the target baryon carried by the Euler angle dependence, again leads to an equally conserved total reaction spin J. Time derivatives of the intrinsic fluctuations translate to

$$A\lambda\cdot\dot{\psi}A^\dagger=-i\omega\lambda\cdot\varphi \tag{24}$$

and further caution must be taken to correctly rotate the operator for the right angular momenta. Using $RA=-\frac{1}{2}A\lambda$ this rotation gives rise to additional spin contributions as may be seen e.g. from the following transformation of total spin J to the intrinsic system:

$$J\lambda\cdot\varphi = (l+J_A)\lambda\cdot\varphi = (l-R)A\lambda\cdot\psi A^\dagger = A((l+\frac{1}{2}[\lambda,\cdot]-R)\lambda\cdot\psi)A^\dagger \tag{25}$$

leading to a body-fixed total spin

$$J^{b.f.} = l+\frac{1}{2}[\lambda,\cdot]-R = K-R. \tag{26}$$

Here we have applied the usual definition of K-spin as a sum of intrinsic isospin and orbital angular momentum.

With the introduction of the left operators $L_a = D_{ab}(A)R_b$ which comprise isospin and hypercharge carried by the Euler angles and with the abbreviation $U_S = AU_HA^\dagger$ we finally find the e.o.m. in the laboratory system :

$$\begin{aligned}
&A\frac{\delta\int L^{(2)}dt}{\delta\lambda\cdot\psi}A^\dagger = \\
&-\frac{f_\pi^2}{2}\Big(-(\omega^2+\Delta)\lambda\cdot\varphi+[\nabla\lambda\cdot\varphi,U_S^\dagger\nabla U_S] \\
&+\omega^2\frac{f_\pi}{\sqrt{4\Theta}}U_S^\dagger[\lambda,U_S]\cdot\int\frac{1}{2}\mathrm{tr}(\frac{f_\pi}{\sqrt{4\Theta}}U_S^\dagger[\lambda,U_S]\,\lambda\cdot\varphi) \\
&-\frac{\omega}{2\Theta}\Big([U_S^\dagger\{\lambda_a,U_S\},L_a\lambda\cdot\varphi+\frac{1}{2}[\lambda_a,\lambda\cdot\varphi]] \\
&-\frac{1}{8}D_{ab}\mathrm{N}_{ac}^{-1}D_{cd}[\lambda_b,2[\lambda_d,\lambda\cdot\varphi]+U_S^\dagger[\lambda_d,\lambda\cdot\varphi]U_S+U_S[\lambda_d,\lambda\cdot\varphi]U_S^\dagger]\Big)\Big)
\end{aligned} \tag{27}$$

$$\begin{aligned}
&\overset{r\to\infty}{\to} -\frac{f_\pi^2}{2}\Big(-(\omega^2+\Delta)\lambda\cdot\varphi-\frac{\omega}{\Theta}[\lambda\cdot L,\lambda\cdot\varphi]-\frac{\omega}{4\Theta}[\lambda_a,[\lambda_a,\lambda\cdot\varphi]]\Big) \\
&= -\frac{f_\pi^2}{2}\Big(-(\omega^2+\Delta)\lambda\cdot\varphi-\frac{\omega}{\Theta}\Big((L+\frac{1}{2}[\lambda,\cdot])^2-L^2\Big)\lambda\cdot\varphi\Big).
\end{aligned}$$

In addition to the adiabatic terms in the second line of (27) we encounter in the third line a projection term of $\mathcal{O}(N_C^{-1})$ which is proportional to the overlap of the fluctuation with the rotational zero mode and thus only effective for mesonic p-waves. The fourth and fifth line provide the vibrational-rotational coupling in $\mathcal{O}(N_C^{-1})$. It will affect mesonic waves of any angular momentum and evidently introduces thresholds into different channels. As may be read off from the last line these thresholds are proportional to the differences of the Casimir energies of the combined meson-target system and of the target alone. In SU(2), together with the chiral symmetry breaking mass term this leads to an asymptotic dispersion relation

$$-\omega^2+k^2+m_\pi^2-\frac{\omega}{\Theta}(T(T+1)-\tau(\tau+1)) = 0 \tag{28}$$

for a reaction of total isospin T and a specific reaction channel of target spin τ. The threshold in channel τ obtained from

$$\omega^2 + \frac{\omega}{\Theta}T(T+1) - \frac{\omega}{\Theta}\tau(\tau+1) = m_\pi^2 \tag{29}$$

is the expected result up to $\mathcal{O}(N_C^{-1})$: the channel frequency is shifted by the differences of the rotational target energies $\frac{\tau(\tau+1)}{2\Theta}$.

In SU(3) the neglect of the symmetry breaking and the use of a common moment of inertia has lead to target energies which are degenerate within a fixed multiplet.

2.4. *The Tomozawa - Weinberg relation*

Finally, we would like to show that for πN S-wave channels the rotational term in the equation of motion (27) leads to the low energy theorem of Tomozawa and Weinberg. It follows under the very general assumptions that the lagrangian leading to the hedgehog solution is atmost quadratic in the time derivatives and that the sum of all terms containing two time derivatives

$$L_{tt}[U] = \frac{1}{2}\int d^3r \sum_{A,B} \mathrm{tr}\Lambda_A(U)\dot{U}\Lambda_B(U)\dot{U} \tag{30}$$

is Lorentz invariant, invariant under parity transformations $PU(\mathbf{r},t)P = U^\dagger(-\mathbf{r},t)$ flavor and chirally symmetric, i.e. the $\Lambda_A(U)$ transform like $U^\dagger$ under global chiral and flavor rotations.

Let us insert a hedgehog configuration that has been shifted in space by an infinitesimal distance ϵ, rotated in flavor space by the matrix A and has undergone a global chiral rotation by an infinitesimal angle γ:

$$U(\mathbf{r},t) = e^{i\gamma\cdot\lambda/2}AU_H(\mathbf{r}+\epsilon)A^\dagger e^{i\gamma\cdot\lambda/2}. \tag{31}$$

A, ϵ, γ may depend on time but not on $\mathbf{r}$. The two time derivatives in $\mathcal{L}_{tt}$ differentiate these parameters and give rise to the following contributions:
(i) contributions with both time derivatives acting on the flavor rotations leading to the rotational energy $\frac{1}{2}\theta_{ab}\Omega_a\Omega_b$ of the hedgehog.
(ii) contributions where no time derivative acts on the flavor rotation leading to a quadratic form in the velocities of the infinitesimal parameters. They will be connected by the adiabatic norm matrix of the soliton fluctuations.
(iii) contributions linear in the rotational velocities

$$\dot{U} = e^{i\gamma\cdot\lambda/2}A[-\frac{i}{2}\Omega\cdot\lambda, U_H(\mathbf{r}+\epsilon)]A^\dagger e^{i\gamma\cdot\lambda/2} \tag{32}$$

together with either the translational velocity (case a)

$$\dot{U} = e^{i\gamma\cdot\lambda/2}A\dot{\epsilon}\cdot\nabla U_H(\mathbf{r}+\epsilon)A^\dagger e^{i\gamma\cdot\lambda/2} \tag{33}$$

or the chiral rotational velocity (case b)

$$\begin{aligned}\dot{U} &= e^{i\gamma\cdot\lambda/2}\{(e^{-i\gamma\cdot\lambda/2}\partial_t e^{i\gamma\cdot\lambda/2})AU_H(\mathbf{r}+\epsilon)A^\dagger \\ &\quad -AU_H(\mathbf{r}+\epsilon)A^\dagger(e^{i\gamma\cdot\lambda/2}\partial_t e^{-i\gamma\cdot\lambda/2})\}e^{i\gamma\cdot\lambda/2}.\end{aligned} \tag{34}$$

By assumption Λ_A can only involve spatial derivatives. Consequently

$$\Lambda_A(U) = e^{-i\gamma\cdot\lambda/2}A\,\Lambda_A\,(U_H(\mathbf{r}+\epsilon))\;A^\dagger e^{-i\gamma\cdot\lambda/2}. \tag{35}$$

Therefore translation and chiral invariance reduce the contribution of case (a) to

$$\int d^3r \sum_{A,B} \mathrm{tr}\Lambda_A(U_H(\mathbf{r}))\dot{\epsilon}\cdot\nabla U_H(\mathbf{r})\Lambda_B(U_H(\mathbf{r}))[-\frac{i}{2}\Omega\cdot\lambda, U_H(\mathbf{r})] \tag{36}$$

which vanishes by parity of the integrand since $PU_H(\mathbf{r})P = U_H(\mathbf{r})$.

In case (b) the shift parameter ϵ drops out of the whole expression because of the translational invariance of L_{tt} leaving a linear and a quadratic contribution in γ up to second order. Again, the linear term in γ may be shown to vanish by parity. Thus all terms from the time derivatives are given by

$$\begin{aligned}&L_{tt}[U] = \frac{1}{2}\int d^3r \sum_{A,B} \\ &\Big\{\mathrm{tr}\Lambda_A(U_H(\mathbf{r}))\,[-\frac{i}{2}\Omega\cdot\lambda, U_H(\mathbf{r})]\,\Lambda_B(U_H(\mathbf{r}))\,[-\frac{i}{2}\Omega\cdot\lambda, U_H(\mathbf{r})]+ \\ &\mathrm{tr}\Lambda_A(U_H(\mathbf{r}))\,\Big[\frac{1}{8}A^\dagger[\gamma\cdot\lambda,\dot{\gamma}\cdot\lambda]A,\,U_H(\mathbf{r})\Big]\,\Lambda_B(U_H(\mathbf{r}))\,[-\frac{i}{2}\Omega\cdot\lambda, U_H(\mathbf{r})]\Big\} \\ &+\mathcal{O}(\dot{\epsilon}^2,\dot{\gamma}^2,\dot{\epsilon}\dot{\gamma}).\end{aligned} \tag{37}$$

Since γ was spatially constant the two integrals both involve the tensor for the moments of inertia giving

$$L_{tt}[U] = \frac{1}{2}\theta_{ab}\Omega_a\Omega_b + \frac{1}{2}\theta_{ab}\Omega_b\frac{i}{4}\mathrm{tr}\lambda_a[A^\dagger\gamma\cdot\lambda A, A^\dagger\dot{\gamma}\cdot\lambda A] + \mathcal{O}(\dot{\epsilon}^2,\dot{\gamma}^2,\dot{\epsilon}\dot{\gamma}), \tag{38}$$

where we find the sum of rotational energy and the rotation-vibration coupling quadratic in the chiral rotation. Up to $\mathcal{O}(N_C^0)$) we have $R_a = \Theta_{ab}\Omega_b$ (15) and the rotation-vibration coupling amounts to

$$\frac{i}{8}\mathrm{tr}\lambda\cdot R[A^\dagger\gamma\cdot\lambda A, A^\dagger\dot{\gamma}\cdot\lambda A] = -\frac{i}{8}\mathrm{tr}\gamma\cdot\lambda[\lambda\cdot L,\dot{\gamma}\cdot\lambda] \tag{39}$$

If the total lagrangian is chirally symmetric - i.e. not only the parts involving derivatives - then fluctuations of the type considered here are the S-wave zero modes of the adiabatic equations of motion:

(i) the bound infinitesimal translation: $\epsilon = \frac{1}{\sqrt{M}}\mathbf{e}_a$, $\gamma = 0$,

(ii) the unbound chiral rotation $A^\dagger\lambda\cdot\gamma A = \frac{1}{f_\pi}\lambda_a$, $\epsilon = \frac{3g_A}{2f_\pi M}e_a$ which is orthogonal on the bound solution (i) and has a vanishing phase shift.

Thus the rotational-vibrational term is just the DWBA expression $\frac{2\pi}{k}\sin\Delta\delta = \Delta L$ for the change of the S-wave phase shifts due to its own presence in the equations of motion. Insertion of the properly normalized chiral rotation (ii) together with the little trick replacing Euler angle matrix elements of commutators of $\lambda\cdot L$ by the difference of the Casimir operator of the combined meson-baryon system to the ones of the individual systems, see eq(27) and following remarks, finally leads to

$$\frac{2\pi}{k}\sin\delta^S = -\frac{\omega}{4f_\pi^2}\langle(L_a+[\frac{1}{2}\lambda_a,\cdot])^2 - L_a^2 - [\frac{1}{2}\lambda_a,[\frac{1}{2}\lambda_a,\cdot]]\rangle. \tag{40}$$

For the scattering lengths $a = \frac{1}{k}\tan\delta$ we recover in SU(2) the Tomozawa-Weinberg relation from this expression: total isospin is $T^2 = (L+[\frac{1}{2}\lambda,\cdot])^2$ and S-waves have $L^2 = R^2 = J^2$ such that at πN threshold for the two different isospin channels

$$\frac{1}{3}(a_{31}-a_{11}) = -\frac{m_\pi}{8\pi f_\pi^2} + \mathcal{O}(m_\pi^2). \tag{41}$$

The threshold $\omega = m_\pi$ is introduced by the chiral symmetry breaking term $\sim m_\pi^2$ which also gives rise to additional DWBA contributions to the background scattering lengths but not to the difference above[19].

3. The status of πN and $\bar{K}$N scattering calculations

In contrast to the SU(2)-case there exist only few attempts to derive the meson-baryon resonances in the SU(3) Skyrme model[10–14]. The approaches differ by the degree to which SU(3) symmetry breaking is taken into account.

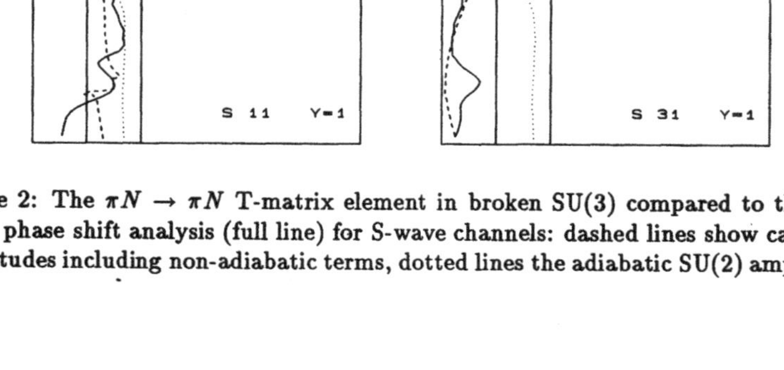

Figure 2: The $\pi N \to \pi N$ T-matrix element in broken SU(3) compared to the Karlsruhe phase shift analysis (full line) for S-wave channels: dashed lines show calculated amplitudes including non-adiabatic terms, dotted lines the adiabatic SU(2) amplitudes.

Karliner and Mattis[10] consider the case of exact SU(3) symmetry where their lagrangian (1) without the symmetry breaking terms only describes massless pseudoscalar mesons.

In the Callan and Klebanov approach[15] collective rotations A of the soliton are constrained to the SU(2) subgroup and strange baryons are formed by bound kaonic fluctuations. Scoccola[11] tied his investigations of $\bar{K}$N-scattering to the "bound state approach" exploring the continuum solutions of these kaonic fluctuations.

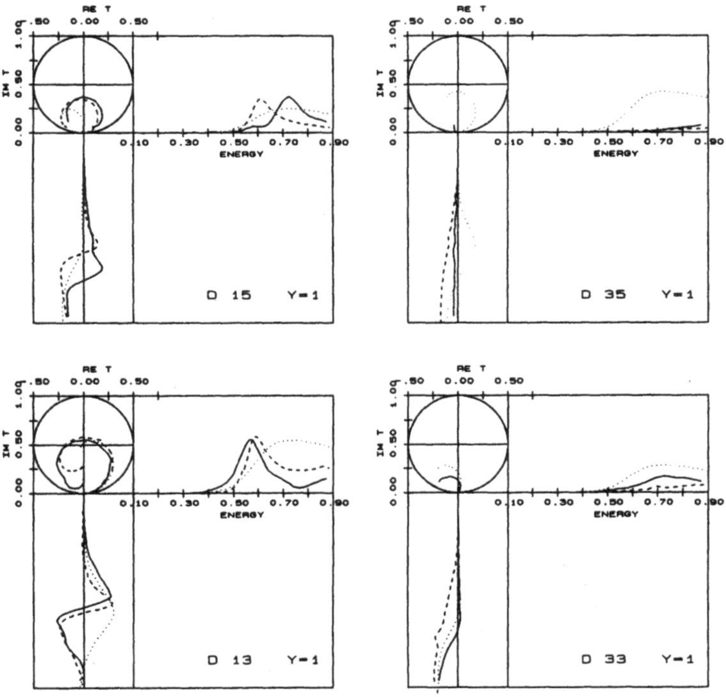

Figure 3: The $\pi N \to \pi N$ T-matrix element in broken SU(3) compared to the Karlsruhe phase shift analysis (full line) for D-wave channels: dashed lines show calculated amplitudes including non-adiabatic terms, dotted lines the adiabatic SU(2) amplitudes.

Both attempts mentioned oversimplify the problem for different reasons. Since the SU(3) symmetry breaking terms of $L^{(FSB)}$ (3) can bind kaons to the soliton[15] the additional potential from $L^{(FSB)}$ must strongly affect the hyperon resonances. Thus the SU(3) symmetric calculation of Karliner and Mattis is missing an important term. The Callan-Klebanov description of hyperon resonances[11] improves on this point, however, at the high price of having to neglect all scattering processes where strangeness is transferred from the meson to the baryon since the fluctuation is always a kaonic one. Thus important excitation and decay channels are missing.

In order to find out what the SU(3) soliton model has to say about the meson-baryon system one has to take account of the full complexity of the problem and let the equations of motion decide on the size of SU(3)-breaking effects in scattering[12–14].

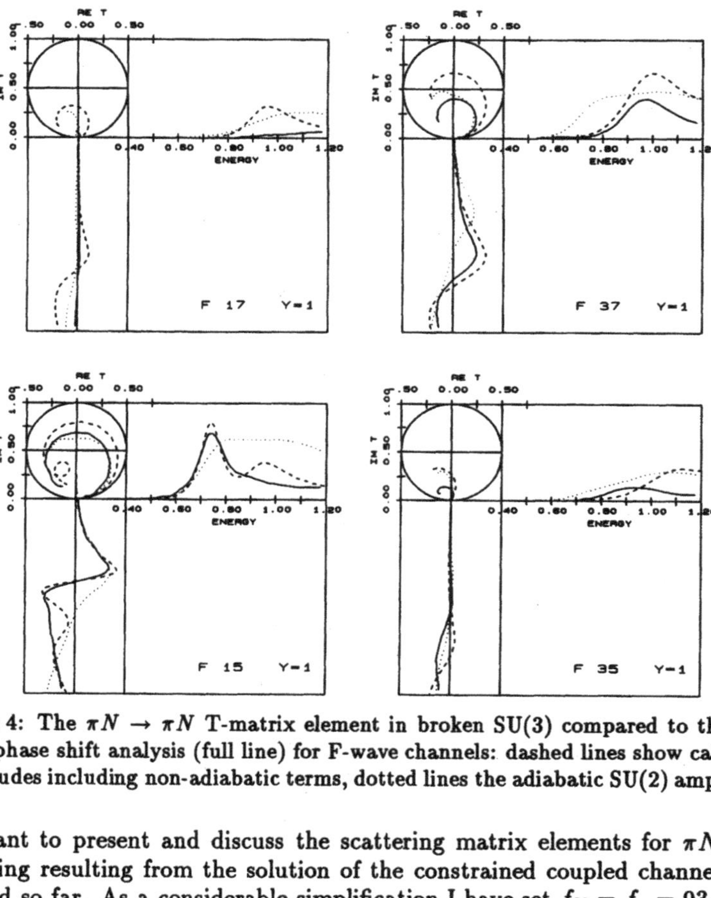

Figure 4: The $\pi N \to \pi N$ T-matrix element in broken SU(3) compared to the Karlsruhe phase shift analysis (full line) for F-wave channels: dashed lines show calculated amplitudes including non-adiabatic terms, dotted lines the adiabatic SU(2) amplitudes.

I want to present and discuss the scattering matrix elements for πN and $\bar{K}N$ scattering resulting from the solution of the constrained coupled channels problem outlined so far. As a considerable simplification I have set $f_K = f_\pi = 93MeV$. The meson masses are fixed to their physical values by $m_\pi = 138MeV$ and $m_\eta = 549MeV$ giving $m_K = 480MeV$. As a further simplification I have assumed SU(3) symmetric target wave functions for the moment. Data will be presented using an effective action the parameters of which correspond to the local approximation of the vector mesons in the action used in ref.4: $g = 2.8, \epsilon_6 = .0075MeV^{-1}$. As explained in the preceeding section various approximations, specific to the SU(3), case lead to degenerate target energies within the baryon octet and decuplet, a situation which turns out to be very unsatisfactory. We have chosen an ad hoc remedy which might account for some of the unavoidable approximations made in SU(3) by fixing the thresholds to their empirical positions.

All results will be compared to the Karlsruhe analysis[17] of πN scattering for total hypercharge $Y = 1$ and to the $\bar{K}N$ phase shift analysis by the Rutherford Laboratory-Imperial College Collaboration[18] for $Y = 0$. For further comparison I also include the results for adiabatic SU(2) and the strong symmetry breaking case, however, only with the adiabatic terms of (27) contributing to the equations of motion. Thus for πN scattering I can demonstrate the progress relative to the old SU(2) Skyrme model results and for $\bar{K}N$ I can compare to results of the Callan-Klebanov approach.

Figures 2 - 4 show the T-matrix elements for $Y = 1$ for increasing pionic angular momentum up to the F-waves.

The lowest partial S-wave has been a problem in chiral soliton models for two reasons:
(i) the low-energy scattering lengths satisfied an adiabatic geometrical relation $a_1 = a_3$ instead of the the Weinberg-Tomozawa relation as dictated by chiral symmetry and current algebra.
(ii) it was impossible to obtain two resonances one of them decaying strongly into the ηN channel.
Walliser[19] was the first to notice that within SU(2) the current algebra result follows exactly by taking all $1/N_C$ contributions from the Skyrme model lagrangian into account. Within SU(3) analogous terms are present and one may deduce from the figure that the current algebra result is also closely followed in the broken SU(3) approach here despite of some uncontrolled approximations involved.

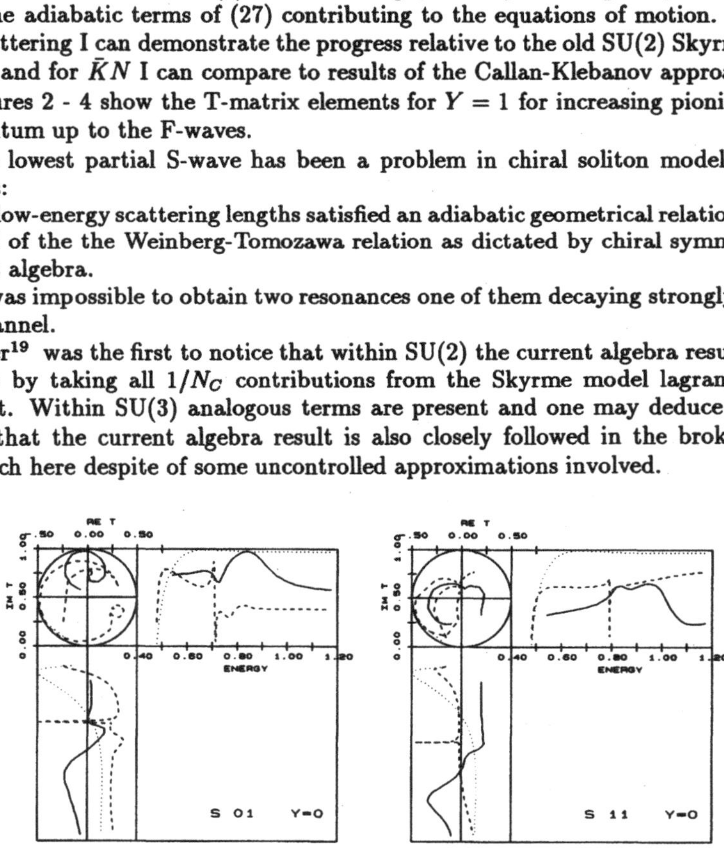

Figure 5: The $\bar{K}N \to \bar{K}N$ T-matrix element in broken SU(3) compared to the Rutherford Laboratory- Imperial College Collaboration phase shift analysis (full line) for S-wave channels: dashed lines for realistic symmetry breaking, dotted lined for the strong symmetry breaking limit, i. e. Callan Klebanov approach.

As for the resonances the drastic impact of the SU(3) symmetry breaking mass term manifests itself in the appearence of several very sharp S-wave resonances related to quasibound $K\Lambda$ and $K\Sigma$ S-states which couple weakly to the πN channel. In the adiabatic calculation the hyperons are degenerate with the nucleon, thus the states appear at energies slightly below the kaon mass of $480MeV$. The implementation of physical channel thresholds shifts these spikes by the corresponding hyperon-nucleon mass difference; in the S11 channel this mass difference moves the quasibound $K\Lambda$ S-state above ηN threshold, a channel into which it decays preferentially. Thus the resonant S11 structures from the non-adiabatic calulation in fig. 2 have a large flux into the ηN channel (60%). We will see later, that bound or resonant S-wave states of heavy mesons with the octet baryons are more a rule than an exception in the soliton approach and that they offer an understanding of why some of the nucleon or hyperon S-wave resonances have strong branching ratios into channels with the η meson present. The flaw in the results displayed in fig. 2 comes from the missing broad second S11 and first S31 resonance, which I think will appear once scalar mesonic degrees of freedom are taken into account, too[4,20] .

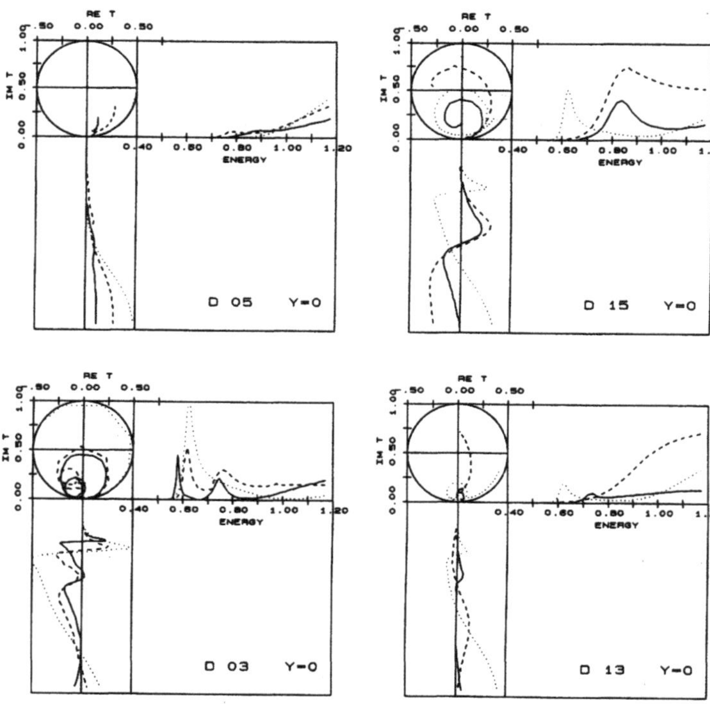

Figure 6: The $\bar{K}N \rightarrow \bar{K}N$ T-matrix element in broken SU(3) compared to the Rutherford Laboratory- Imperial College Collaboration phase shift analysis (full line) for D-wave channels: dashed lines for realistic symmetry breaking, dotted lined for the strong symmetry breaking limit, i. e. Callan Klebanov approach.

SU(3) extension and rotational effects combine to improve the overall agreement of the higher angular momentum partial waves which traditionally already have been satisfactory. Generally, existing resonances become more pronounced as may be seen from figures 3,4 in case of the partial waves D13, D15 and F15. Also, quite pleasingly, a D35 resonance which has disturbed previous Skyrme model calculations[3] has disappeared due to the extension to SU(3) now, fig. 3.

Extension to SU(3) and the presence of a sixth order term in the effective action invariably indicate the existence of a second F15 resonance. In context with the Skyrme model including vector mesons, ref.21 , we have argued that the decay channels of the ρ meson will damp this second resonance, but possibly it deserves a more serious consideration.

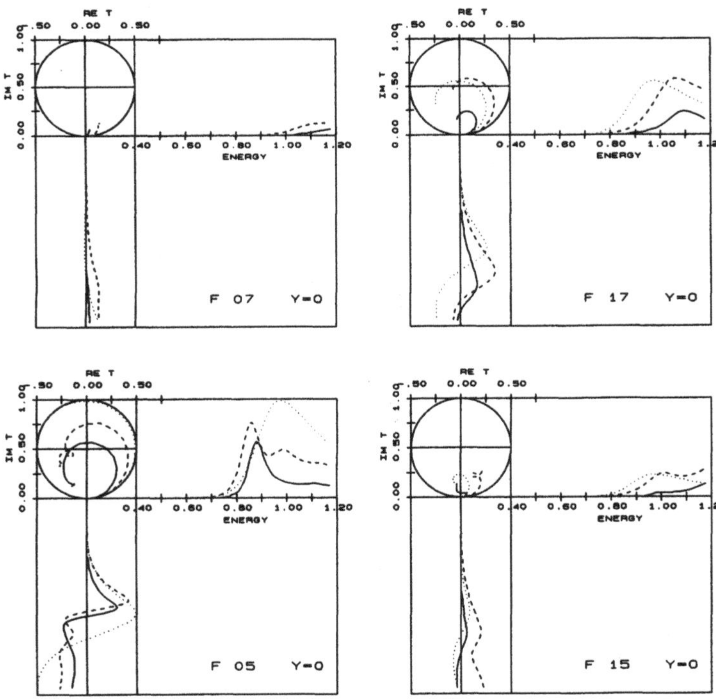

Figure 7: The $\bar{K}N \rightarrow \bar{K}N$ T-matrix element in broken SU(3) compared to the Rutherford Laboratory- Imperial College Collaboration phase shift analysis (full line) for F-wave channels: dashed lines for realistic symmetry breaking, dotted lined for the strong symmetry breaking limit, i. e. Callan Klebanov approach.

Upon inspection one can immediately see from the figures 5-7 that the general outcome for the calculated $\bar{K}N$ scattering amplitudes is very similar to the results just discussed for the $Y = 1$ sector. Firstly, the structure of the S-wave amplitudes (fig. 5) is dominated by the appearence of very sharp resonances. In the isospin zero channel there is an additional resonance just below $\bar{K}N$ threshold visible here only by the sharp clockwise motion of the $\bar{K}N$ T-matrix element in the complex plane. This resonance is due to a quasibound $\bar{K}N$ state coupled to the open $\pi\Sigma$ channel. This resonance corresponds, of course, to the experimentally observed $\Lambda(1405)$. The visible spikes in the S01 channel and the S11 channel arise through the coupling of the $\bar{K}N$ channel to a quasibound $K\Xi$ state. It therefore appears that the broken SU(3) soliton model likes to bind kaons to octet baryons B of appropriate hypercharge. In all cases calculated these bound or resonant states had a strong coupling to the channel containing the η meson with its corresponding octet baryon B'. In the adiabatic approximation the $\eta B'$ channel has a higher threshold energy than the KB channel because there all octet baryons are degenerate. So through the inclusion of the target energies in the coupled channels equations the bound KB states decay strongly into the now open $\eta B'$ channel.

Existing resonance data[22] on N^*, Λ^* and Σ^* do not contradict such a picture. This means that in all cases there is a resonance 50 - 100 MeV below the threshold energy of KB (or $\bar{K}B$ if applicable) which has the strong decay to the open $\eta B'$ channel.

Table 1 lists all known S-wave resonances with more than two stars from the tables of the Particle Data Group together with their decay probability into the corresponding $\eta B'$ channel, an upper bound on the energy of a bound KB state of appropriate hypercharge and a threshold energy for a decay to $\eta B'$. In case of the N^* resonances there exists the further possibility of a bound $K\Sigma$ system: for isospin $\frac{1}{2}$ this state will lie above $K\Lambda$ threshold and for isospin $\frac{3}{2}$ it can only have η decays to $\eta\Delta \to \eta\pi N$ leading to a three-body decay. These reasons might explain why the effects of a bound $K\Sigma$ system are not seen in the data.

Table 1. Kaon-Baryon bound states and their η decay.

	N^*		$m_K + M_\Lambda$		$m_\eta + M_N$
****	S11(1535)	$\to$ $\eta N(45-55\%)$	1606	$\to$	1550
****	S11(1650)	$\not\to$ ηN			
	Λ^*		$m_K + M_\Xi$		$m_\eta + M_\Lambda$
****	S01(1405)	below $\bar{K}N$ threshold			
****	S01(1670)	$\to$ $\eta\Lambda(15-35\%)$	1808	$\to$	1665
***	S01(1800)	$\not\to$ $\eta\Lambda$			
	Σ^*		$m_K + M_\Xi$		$m_\eta + M_\Sigma$
***	S11(1750)	$\to$ $\eta\Sigma(15-55\%)$	1808	$\to$	1742

Therefore, it appears that the broken SU(3) soliton model offers a new interpretation of some S-wave resonances. There remains, however, the problem of their narrow width in the calculation.

The D- and F-wave $\bar{K}N$ amplitudes, finally, show the same qualitative agreement as do the πN amplitudes. The most striking effect comes from the non-adiabatic terms which with their rotational target energies split the D03 resonance into two pieces (fig. 6) in nice agreement with the data. As in the $Y = 1$ sector there is a weak second F-wave resonance for $J = \frac{5}{2}$ (fig. 7). Apart from this additional structure the calculation again displays a rather impressive similarity to the phaseshift analyses, and maybe it should be stressed here, despite of the few really free parameters which enter the calculation - their actual number may be debatable, but it is somewhere in the region around two or three -. From the figures 6,7 one can also notice how grossly the strong symmetry breaking approach underestimates the inelasticities at least for $T = 0$ in lack of important physical reaction channels.

4. The P-wave channels up to $\mathcal{O}(N_C^{-1})$

We now address the special case of πN P-wave amplitudes. The structure of the equations of motion for the laboratory fluctuations (27) is given in a simplified notation

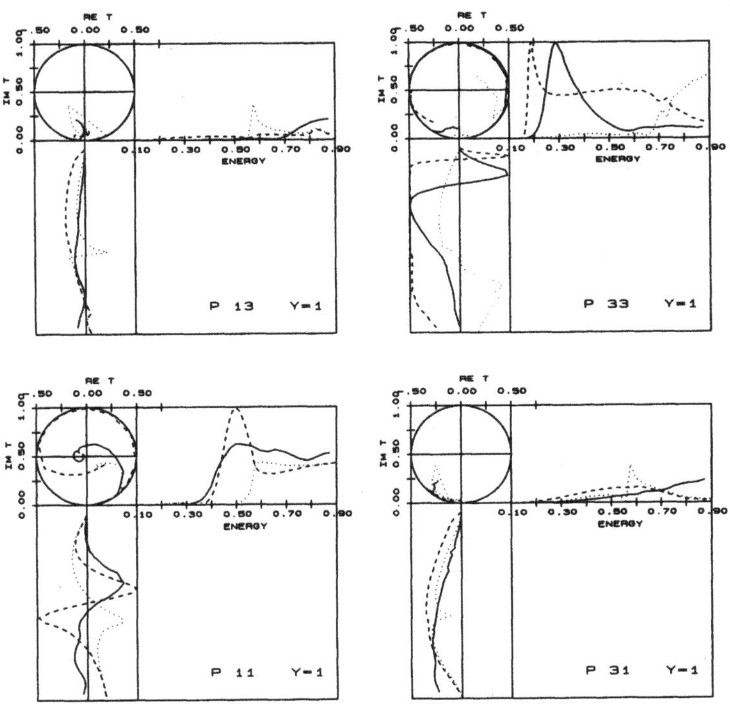

Figure 8: The $\pi N \to \pi N$ T-matrix element in SU(2) compared to the Karlsruhe phase shift analysis (full line) for P-wave channels: dotted lines show calculated amplitudes including all non-adiabatic terms, dashed lines with the dummy parameter $\epsilon = -1$.

by

$$(\mathbf{D} - \omega^2\mathbf{N} - \omega\mathbf{R}) \cdot \varphi = -\epsilon\omega^2\mathbf{N} \cdot \xi \int \xi \cdot \mathbf{N} \cdot \varphi \qquad \epsilon = +1. \tag{42}$$

Here, $\mathbf{D}$ is the adiabatic differential operator, $\mathbf{N}$ the corresponding norm matrix, $\mathbf{R}$ the rotational-vibrational coupling active for all partial waves. On the r.h.s. we find the projection term which is special for p-waves and which measures the overlap of the laboratory fluctuation φ with the rotational zero mode ξ. It is multiplied by a dummy parameter ϵ which up to order $\mathcal{O}(N_C^{-1})$ came out to be one. In order to be free of the uncertainties inherent in the SU(3) case we restrict ourselves to SU(2). There, eq. (27) gives all contributions to the e.o.m. up to $\mathcal{O}(N_C^{-1})$.

We first concentrate on the effects of the projection term and ignore the rotational-vibrational coupling for a moment. Without this coupling the scattering amplitudes of the coupled channels problem are identical to the adiabatic case:

- if φ is not the rotational zero mode, the l.h.s. of the equation (42) vanishes but so does the r.h.s. because the projection integral is zero.
- if φ is the rotational zero mode, the l.h.s. of the equation amounts to $-\omega^2\mathbf{N}\cdot\xi$, but so does the r.h.s. because the projection integral is one. Thus the projection term allows the rotational zero mode to appear anywhere in the spectrum. However, because the rotational zero mode is localized at the soliton's center it does not contribute to the scattering amplitudes in any way. The linear coupling term - which has lead to the

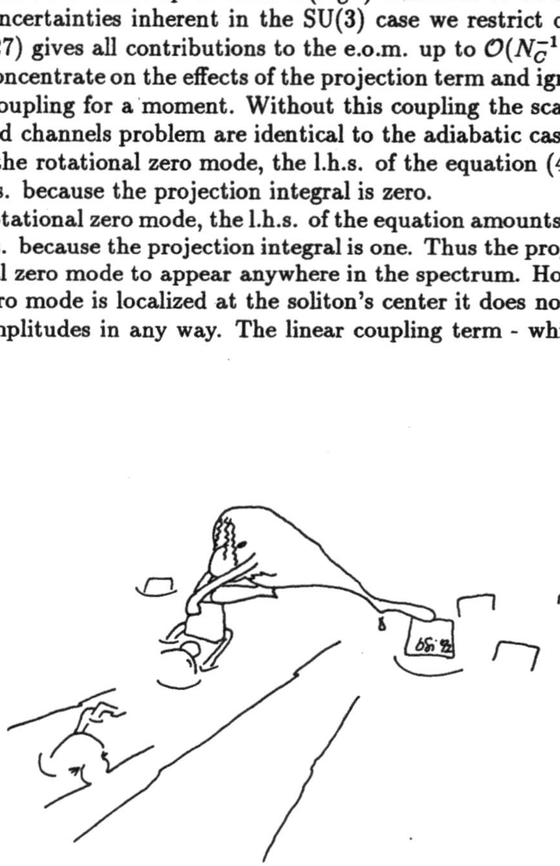

Figure 9: The art of building a bridge to QCD requires more than stones just down to $\frac{1}{N_C}$-size.

projection term in the e.o.m. - cannot affect the scattering amplitudes alone. It is left to the rotational-vibrational terms now to place the zero mode at a definite position in the spectrum. There, this bound state in the continuum will couple to the πN channels. We may estimate its position by insertion of $\varphi = \xi$ into the differential equation and integrate after multiplication with ξ again. We find:

$$(1-\epsilon)\omega^2 = -\omega \int \xi \cdot \mathbf{R} \cdot \xi. \tag{43}$$

Up to order $\mathcal{O}(N_C^{-1})$ the zero mode which should be responsible for the appearence of the P33 resonance resides at energy zero and does not show up as a resonance in the continuum. This feature is also borne out by the calculation (dotted lines). In case we set our dummy parameter to a value different from one (dashed lines, $\epsilon = -1$) we see that the resonance in the P33 channel indeed emerges. The other channels then also turn out to be qualitatively correct since no resonances occur in the P13 and P31 channels leaving an enhanced Roper resonance in the P11 channel.

Possible deviations of ϵ from one are at least $\mathcal{O}(N_C^{-\frac{3}{2}})$ and this fact leads to the conclusion of this contribution summarized in figure 9.

5. Acknowledgement

Since it is not perfectly visible I would like to reveal that the other person partially submerged in figure 9. is Hans Walliser, with whom large parts of this contribution have been worked out.

6. References

1) H. Walliser and G. Eckart, *Nucl. Phys.* **A429** (1984) 514
2) A. Hayashi, G. Eckart, G. Holzwarth and H. Walliser, *Phys. Lett.* **B147** (1984) 5
3) G. Eckart, A. Hayashi and G. Holzwarth, *Nucl. Phys.* **A448** (1986) 732
4) B. Schwesinger, H. Weigel, G. Holzwarth and A. Hayashi, *Phys. Rep.* **173(4)** (1989) 173
5) E. Witten, *Nucl. Phys.* **B223** (1983) 422,433
6) J. Gasser and H. Leutwyler, *Nucl. Phys.* **B250** (1985) 465
7) N. W. Park, J. Schechter and H. Weigel, *Phys. Rev.* **D41** (1990) 2836
8) G. Pari, B. Schwesinger and H. Walliser, *Phys. Lett.* **B255** (1991) 1
9) B. Schwesinger and H. Weigel, *Phys. Lett.* **B67** (1991) 438, B. Schwesinger and H. Weigel, *Nucl. Phys.* **A540** (1992) 461
10) M. Karliner and M. P. Mattis, *Phys. Rev.* **D34** (1986) 1991
11) N. N. Scoccola, *Phys. Lett.* **B236** (1990) 245
12) B. Schwesinger and H. Walliser, *Phys. Lett.* **B231** (1989) 11, *Phys. Lett.* **B234** (1990) 543
13) B. Schwesinger and H. Walliser, *Phys. Lett.* **B242** (1990) 139
14) B. Schwesinger, *Nucl. Phys.* **A537** (1992) 253
15) C.G. Callan and I. Klebanov, *Nucl. Phys.* **B262** (1985) 365 C.G. Callan, K. Hornbostel and I. Klebanov, *Phys. Lett.* **B202** (1988) 269

17) G. Höhler, F. Kaiser, R. Koch and E. Pietarinen, *Physics Data*, Universität Karlsruhe (1979), Nr 12-1
18) G.P. Gopal et al., *Nucl. Phys.* **B119** (1977) 362
19) H. Walliser, *Nucl. Phys.* **A524** (1991) 706
20) B. Schwesinger and H. Weigel, *Nucl. Phys.* **A465** (1987) 733
21) H. B. Geyer and B. Schwesinger, *Phys. Lett.* **B252** (1990) 319
22) Particle Data Group, *Phys. Lett.* **B239** (1990) 1

V. Systems with $B \neq 1$

THE NUCLEON-NUCLEON INTERACTION IN THE SKYRME MODEL

TIMOTHY S. WALHOUT
Physics Department, The Ohio State University
174 West 18th Avenue, Columbus, Ohio 43210, USA

ABSTRACT
The nucleon-nucleon potential is derived from the Skyrme model in a semiclassical framework via exact numerical calculations with lagrange constraints. All the usual terms — including spin-orbit and momentum-dependent terms — arise naturally in a Born-Oppenheimer approach. The results show that the product ansatz is a poor approximation at short and intermediate ranges, and that the Skyrme model predicts a potential which compares much better with realistic potentials than had been previously thought.

1. Introduction

Despite that quantum field theories have an infinite number of degrees of freedom, some phenomena in certain such theories may be accurately described by a finite set of dynamical variables. If one thinks of a path integral description, this would occur if the major contribution to the action comes from fields which lie along the bottom of a steep valley in configuration space. For a theory which contains solitons, such a truncation of the dynamics will be justified when there is a small parameter which may be scaled out of the lagrangian through a rescaling of the fields; for then this parameter will multiply $\hbar$ in the path integral, and a semiclassical expansion ensues.[1] Thus to lowest order, the important configurations will be the classical solitons, and the important degrees of freedom will be the collective coordinates which describe the zero modes of these solitons. The other degrees of freedom — the fluctuations of the fundamental field about the soliton — produce effects which are suppressed by a power of the small parameter.

One such theory where this approach is thought to be valid is the Skyrme model: the fundamental field is the pion, the expansion parameter is $1/N_c$, and upon quantization of their collective coordinates the solitons of the theory — the *skyrmions* — become generalized nucleons.[a] If we look at phenomena which involve only skyrmions, then the $1/N_c$ expansion tells us that we may effectively truncate the degrees of freedom to those collective coordinates which describe adiabatic motion of these solitons. The semiclassical approach consists in first approximation of quantizing just these collective degrees of freedom. Expressed in a different way, the collective coordinates of the skyrmions are "slow" variables which describe large amplitude motion, whereas

[a] A *generalized nucleon* has spin and isospin $j=t=\frac{1}{2},\frac{3}{2},\ldots$; so that its ground state is the nucleon, its first isorotational excitation is the delta, and so on.

the fluctuations of the intrinsic pion field about the solitons are "fast," small amplitude degrees degrees of freedom which adjust quickly to the adiabatic motion of the skyrmions.

Skyrmions, then, are characterized by inertial masses, rotational and isorotational moments of inertia, and interactions which have $\mathcal{O}(N_c)$ contributions from the configurations which are the classical solutions to the equations of motion. If $1/N_c$ is a good expansion parameter, then one can expect that the inclusion of the meson degrees of freedom will necessitate a renormalization of the coupling constants and mass of the fundamental field, but will produce higher order corrections to the skyrmion masses, moments of inertia, and interactions which will not alter qualitatively the features of these quantities which are found at lowest order. Since the lowest order calculation of nucleon and delta properties in the one-baryon sector of the Skyrme model is qualitatively successful, then, one can expect to find a similiar success when looking at B=2 properties in the same approximation. In particular, if the $\mathcal{O}(N_c)$ features of the nucleon-nucleon (NN) interaction as found from the Skyrme model are not in qualitative agreement with realistic potentials, the effectiveness of a solitonic picture of baryons must be questioned.

Unfortunately, the first calculations of the NN interaction in the Skyrme model have shown a qualitative discrepancy with realistic potentials: use of the popular *product ansatz* (defined below) results in a potential with no intermediate range attraction in the central channel.[2] This is a serious shortcoming because in realistic models this attraction is responsible for the binding of nuclei; and if nuclei in the Skyrme model are not bound at leading-order, then the $1/N_c$ expansion will not be very useful. Many proposed resolutions to the problem proceed by extending the original Skyrme lagrangian by introducing terms with more derivatives of the pion field, heavier mesons, or scalar fields into the model. The assumption here is that the Skyrme model cannot describe the B=2 sector with the same accuracy as the B=1 sector. There is little reason, however, to see why this would be so; and one should hope that such a complication would not be necessary. A second proposed resolution to the problem of the missing attraction is that it is the result of a higher order effect in $1/N_c$; namely, one cannot see the attraction unless one includes pion fluctuations. Although this is certainly possible, and in the B=1 sector $\mathcal{O}(1)$ effects appear to be significant (see the contributions from Holzwarth and Moussallam in this volume), it would nevertheless be particularly damaging to the utility of the $1/N_c$ expansion if $\mathcal{O}(1)$ contributions were capable of qualitatively altering $\mathcal{O}(N_c)$ results. Of course, one needs to perform much more extensive calculations to include such effects, and I shall have little to say about $\mathcal{O}(1)$ corrections in the following.

Another possible resolution stems from the observation that when one starts from an ansatz instead of the true classical solution to the equations of motion, one runs the risk of messing up $1/N_c$ counting. If for some range of the collective coordinates the ansatz is much different from the true solution, then pion fluctuations in this range will be large amplitude and no longer suppressed by $1/N_c$: they will give $\mathcal{O}(N_c)$ corrections to compensate for the ineffectiveness of the ansatz in approximating the true solution.[3] In the following, I shall describe a proper semiclassical approach to deriving the NN potential in the Skyrme model: that is, one based on numerical solutions to the equations of motion. The results show that the product ansatz is in fact a bad approximation to the true solution at small and intermediate separations between the skyrmions, and so it provides a poor starting point for understanding the NN potential in these ranges.

Now there is of course one final possibility: namely, that the idea that baryons are solitons is simply unrealistic. There is a lot of evidence against such an assertion, and I hope to show in the following that a successful semiclassical description of the NN potential is a new item which may be added to the now large list of evidence which supports the utility of the picture of baryons as solitons.

2. Collective Coordinates and Truncated Dynamics

The Skyrme model can be understood as a low-energy, large-N_c approximation to two-flavor QCD: namely, as an effective meson lagrangian where the degrees of freedom corresponding to mesons more massive than the pion have been integrated out and terms with more than four derivatives of the pion field which result from this process have been dropped. Using the familiar $SU(2)$ notation

$$U(\mathbf{x},t) = \frac{1}{f_\pi}\Big(\sigma(\mathbf{x},t) + i\boldsymbol{\tau}\cdot\boldsymbol{\pi}(\mathbf{x},t)\Big); \quad \sigma^2 + \boldsymbol{\pi}^2 = f_\pi^2, \tag{1}$$

and defining the left and right currents as

$$\mathcal{L}^a_\mu(U) = -\frac{i}{2}\mathrm{Tr}[\tau_a(\partial_\mu \mathrm{U})\mathrm{U}^\dagger], \quad \mathcal{R}^a_\mu(\mathrm{U}) = -\frac{i}{2}\mathrm{Tr}[\tau_a \mathrm{U}^\dagger(\partial_\mu \mathrm{U})], \tag{2}$$

the Skyrme lagrangian can be written as a sum of "kinetic" and "potential" parts:

$$\mathcal{L} = \mathcal{T} - \mathcal{V}; \tag{3}$$

where

$$\begin{aligned} \mathcal{T} &= \frac{1}{2}\mathcal{R}^a_i \mathcal{M}_{ab} \mathcal{R}^b_i, \\ \mathcal{M}_{ab} &= [f_\pi^2 + \frac{1}{e^2}(\mathcal{R}^c_i)^2]\delta_{ab} - \frac{1}{e^2}\mathcal{R}^a_i\mathcal{R}^b_i, \\ \mathcal{V} &= \frac{1}{4}\mathcal{R}^a_i[f_\pi^2\delta_{ab} + \mathcal{M}_{ab}]\mathcal{R}^b_i + \frac{f_\pi^2 m_\pi^2}{2}\mathrm{Tr}(1-U). \end{aligned} \tag{4}$$

The constants f_π and $1/e$ scale as $\sqrt{N_c}$ and the pion mass as 1, so we can scale N_c out of the lagrangian $\mathcal{L} = N_c\tilde{\mathcal{L}}$, with $\tilde{\mathcal{L}}$ of $\mathcal{O}(1)$.

2.1. *B=1 collective coordinates*

In the B=1 sector there is a class of hedgehog solutions to the equations of motion which have the form

$$U_0(\mathbf{x}) = AU_H(\mathbf{x} - \mathbf{R})A^\dagger \tag{5}$$

with

$$U_H(\mathbf{x}) = \cos F(r) + i\boldsymbol{\tau}\cdot\hat{\mathbf{x}}\sin F(r), \tag{6}$$

where $\mathbf{R}$ is the center of the hedgehog — that is, the point where $U = -1$, often refered to as the hedgehog "singularity" — and A is its isospatial orientation. From Eq. (1) one sees that $\boldsymbol{\pi}_H \propto f_\pi$, so the classical hedgehog solution is of order $\sqrt{N_c}$. In contrast, fluctuations $\boldsymbol{\eta}$ about the skyrmion, $\boldsymbol{\pi} = \boldsymbol{\pi}_H + \boldsymbol{\eta}$, are $\mathcal{O}(1)$. The truncated dynamics consists of promoting the collective coordinates $\mathbf{R}$ and A to time-dependent variables and ignoring the fluctuations $\boldsymbol{\eta}$.

Let us define the rotation matrix

$$D_{ab}(A) = \frac{1}{2}\mathrm{Tr}[\tau_a A \tau_b A^\dagger] \;=\; D^{-1}_{ab}(A^\dagger). \tag{7}$$

Note the identity

$$\mathcal{L}^a_\mu(A) = D_{ab}(A)\mathcal{R}^a_\mu(A). \tag{8}$$

Since the $SU(2)$ matrix A can also be parameterized by three Euler angles as

$$A = \exp(-\frac{i}{2}\tau_3\alpha)\exp(-\frac{i}{2}\tau_2\beta)\exp(-\frac{i}{2}\tau_3\gamma), \tag{9}$$

one can see that $\mathbf{D}(A)\cdot\mathbf{x}$ just describes a general rotation of the vector $\mathbf{x}$. Now the hedgehog has the symmetry

$$AU_H(\mathbf{x})A^\dagger = \cos F(r) + i\boldsymbol{\tau}\cdot\mathbf{D}(A)\cdot\hat{\mathbf{x}}\sin F(r) = U_H(\mathbf{D}(A)\cdot\mathbf{x}); \tag{10}$$

and therefore rotating a hedgehog in space is equivalent to rotating it in isospace, which means a skyrmion has just six zero modes. Substituting Eq. (5) into Eqs. (2) and (3) gives

$$\mathcal{R}^a_i(U_0) = -D_{ab}(A)\Big\{[\delta_{bc} - D_{bc}(U_H)]\mathcal{L}^c_i(A) + \mathcal{R}^b_i(U_H)\dot{R}_i\Big\} \tag{11}$$

and

$$L = \int d^3x\mathcal{L} = \frac{1}{2}M_H\dot{R}_i^2 + \frac{1}{2}\lambda_H[\mathcal{R}^a_i(A)]^2 - M_H, \tag{12}$$

with M_H the hedgehog's mass and λ_H its moment of inertia. The canonical momenta are the isospin $T_a = -2\lambda_H\mathcal{L}^a_i(A)$, the spin $J_a = 2\lambda_H\mathcal{R}^a_i(A)$, and the linear momentum $P_i = M_H\dot{R}_i$. Eq. (8) ensures that $\mathbf{J}^2 = \mathbf{T}^2$ for the quantized B=1 skyrmion, the generalized nucleon. The angular momentum operator $\mathbf{J}$ generates a right shift of A — or, equivalently, a spatial rotation of the hedgehog; and the isospin operator $\mathbf{T}$ generates a left shift of A — or, equivalently, an isospatial rotation of the hedgehog. The spin and isospin eigenstates of the system are the generalized nucleon states

$$\psi_{\mathcal{N}}(A) = \langle A|\mathcal{N}\rangle = \langle A|t,i,s\rangle = \sqrt{\frac{2t+1}{8\pi^2}}(-1)^{t+i}\mathcal{D}^{(j=t)}_{-i,s}(A), \tag{13}$$

where $\mathcal{D}$ is a Wigner rotation matrix, i is the third component of isospin and s is the third component of spin. I shall use $\mathcal{N}$ to denote a generalized nucleon state and N to denote the usual nucleon in what follows.

Let us now move to the B=2 sector of the Skyrme model. Ultimately, we will want to extract a nucleon-nucleon potential using methods similar to those introduced here. This will require an understanding of the dependence of the system on the relative position and orientation of the two skyrmions. These are dynamical collective variables corresponding to nonzero modes, and even on the level of naive semiclassical quantization we must go beyond the simple static calculation which determined the minimum-energy solution in the B=1 sector. As an introduction to the B=2 sector, however, I will first describe the static solutions.

2.2. *Onions, donuts and hedgehogs*

There are three known unconstrained, time-independent solutions to the equations of motion. First, there is a B=2 hedgehog solution

$$U_{H2}(\mathbf{x}) = \cos F_2(r) + i\boldsymbol{\tau}\cdot\hat{\mathbf{x}}\sin F_2(r), \tag{14}$$

where the label 2 distinguishes this from the B=1 solution. The boundary conditions $F_2(0) = 2\pi$ and $F_2(\infty) = 0$ guarantee the baryon number is two. The energy of this solution is $M_{H2} \sim 3M_H$, so this hedgehog is unstable. It has a "onion" structure, consisting of one spherical B=1 object wrapped by another. One can see, in fact, that on the sphere where $F_2 = \pi$, the baryon number is zero, but that there is total baryon number one both inside $(2\pi \geq F_2 \geq \pi)$ and outside $(\pi \geq F_2 \geq 0)$ this sphere. The B=2 hedgehog represents an extreme in the two-skyrmion interaction, and should be of little consequence for low energies. It is hard to imagine a gradual evolution from two well-separated skyrmions to such a configuration, and as a matter of fact nothing of the sort has been seen in numerical two-skyrmion scattering simulations.

Second, there is the minimum-energy energy solution for B=2, which may be obtained from numerical relaxation of the full static equations of motion on a discrete spatial lattice. This stable solution, whose energy is 1660 MeV (for parameters $f_\pi = 108MeV$, $e = 4.84$, and $m_\pi = 138MeV$), was discovered independently by Verbaarschot, et al.[4] and by Kopeliovich and Shtern.[5] As conjectured by Manton,[6] who had studied a two-monopole system, the minimum-energy B=2 solution has the shape of a donut.[7] More precisely, the solution has the following axial symmetry

$$U(\mathbf{x}) = e^{i\tau_3\alpha/2}U(\mathbf{D}(e^{-i\tau_3\alpha/2})\cdot\mathbf{x})e^{-i\tau_3\alpha/2}, \tag{15}$$

where $\mathbf{D}(e^{i\tau_3\alpha/2})$ sends the cylindrical coordinate ϕ to $\phi+\alpha$. Since this is the classical minimum-energy B=2 configuration, it will surely be important for any semiclassical study of the NN interaction based upon the Skyrme model. We shall see this solution again later.

The third static B=2 solution is the trival solution

$$U_\infty(\mathbf{x}) = A_1U_H(\mathbf{x}-\mathbf{R}_1)A_1^\dagger A_2U_H(\mathbf{x}-\mathbf{R}_2)A_2^\dagger, \tag{16}$$

where $U_H(\mathbf{x})$ is the B=1 hedgehog solution, the A_n are $SU(2)$ matrices giving the orientation of the nth hedgehog in isospace, and the relative separation $|\mathbf{R}_1 - \mathbf{R}_2|$ is infinite. At infinite separation the fields $U_n = A_n U_H(\mathbf{x}-\mathbf{R}_n)A_n^\dagger$ commute since $U_2 = 1$ in the neighborhood of U_1, and vice versa. This is an asymptotically stable solution. Its energy is that of two hedgehogs, $2M_H = 1730MeV$ for the above parameters. Clearly, one may quantize the asymptotic solution (16) as in the previous section — that is, by promoting the collective degrees of freedom $\{A_n, \mathbf{R}_n\}$ to dynamical variables — for these twelve collective coordinates here describe zero modes of non-interacting hedgehogs. In the limit of infinite separation, then, the quantized two-skyrmion system consists of independent generalized nucleon-nucleon states

$$\psi_{\mathcal{N}_1\mathcal{N}_2}(A_1, A_2) = \langle A_1 A_2 | \mathcal{N}_1 \mathcal{N}_2 \rangle = \langle A_1 | t_1 i_1 s_1 \rangle \langle A_2 | t_2 i_2 s_2 \rangle, \tag{17}$$

with the generalized nucleon states $|t_n, i_n, s_n\rangle$ given in Eq. (13).

2.3. *B=2 collective coordinates*

It is natural to describe the two-skyrmion system in terms of the twelve collective variables $A_1, A_2, \mathbf{R}_1$ and $\mathbf{R}_2$ so that the skyrmions may be quantized as individual nucleons.[8–10] Since the difference in energy between the toroidal solution (15) and the asymptotic solution (16) is only four percent, these collective variables describe nearly-zero mode oscillations about the classical solution. If the semiclassical expansion is valid, the degrees of freedom may be truncated to only these twelve. The truncated dynamics is more complicated than for the B=1 sector, however, because it cannot be derived from simple static solutions to the equations of motion. At finite separation, $R = |\mathbf{R}_1 - \mathbf{R}_2|$ and $C = A_1^\dagger A_2$ describe non-zero modes. There have been several different approaches to this problem. One method uses an ansatz for the field U which contains an explicit dependence on the collective variables. A second method numerically solves the static equations of motion, modified so that the collective variables may be constrained to given values. Finally, there have been some numerical simulations of time-dependent skyrmion-skyrmion scattering.

First, let me here discuss how to identify the collective coordinates for a general B=2 configuration. As with the hedgehog above, for all separations the system is invariant with respect to spatial translations $\mathbf{R}_T$, spatial rotations B, and isospatial rotations A of the field U:

$$U(\mathbf{x}, t) \rightarrow AU(\mathbf{D}(B)\cdot(\mathbf{x} - \mathbf{R}_T), t)A^\dagger. \tag{18}$$

These nine global collective variables, which describe zero modes, are then easy to identify. They move the system from the body-fixed to the lab frame. When promoted to dynamical variables, the canonical momenta corresponding to these global variables will be conserved. These are the total momentum $\mathbf{P}$, the total angular momentum $\mathbf{J}$, and the total isospin $\mathbf{T}$. In general, spatial and isospatial rotations of the field describe independent zero modes, and there will be no relation between $\mathbf{J}$ and $\mathbf{T}$ as for the hedgehog.

Now let's choose the body-fixed frame so that the two skyrmions are separated along the z-axis and so that the skyrmion in the $z > 0$ half of space (skyrmion 1) is unrotated. Then

$$\begin{aligned} A_1^{bf} &= 1, \\ A_2^{bf} &= C, \\ \mathbf{R}_1^{bf} &= \frac{1}{2}R\hat{\mathbf{e}}_3, \\ \mathbf{R}_2^{bf} &= -\frac{1}{2}R\hat{\mathbf{e}}_3. \end{aligned} \tag{19}$$

Moving to the lab frame via global translations, rotations, and isorotations, gives

$$\begin{aligned} A_1 &= AB, \\ A_2 &= ACB, \\ \mathbf{R}_1 &= \mathbf{R}_T + \mathbf{D}^{-1}(B)\cdot\frac{1}{2}R\hat{\mathbf{e}}_3, \\ \mathbf{R}_2 &= \mathbf{R}_T - \mathbf{D}^{-1}(B)\cdot\frac{1}{2}R\hat{\mathbf{e}}_3. \end{aligned} \tag{20}$$

In this formulation of the collective coordinates the reference to individual skyrmions is clearly more apparent. However, the momenta conjugate to these variables, the isospins and linear momenta of each skyrmion, are not good eigenoperators of the full quantized system.

In the body-fixed frame, then, the pion field depends on only the relative coordinates as $U_{2B}^{bf}(\mathbf{x}; R, C)$. R is just the separation between the skyrmion centers — that is, the distance between the points where the pion field $U = -1$. The relative isospatial orientation C may be identified in this frame as follows. For two skyrmions separated at a given time along the z axis, the fields in the $z > 0$ and $z < 0$ halves of space will be related by the rotation defined by C. For each C one may define a vector $\hat{\mathbf{n}}$ perpendicular to $\hat{\mathbf{e}}_3$ such that $C(\boldsymbol{\tau}\cdot\hat{\mathbf{n}}) = (\boldsymbol{\tau}\cdot\hat{\mathbf{n}})C^\dagger$;[b] and the field will have the symmetry[10,11]

$$U_{2B}^{bf}(\mathbf{x}, C) = C(\boldsymbol{\tau}\cdot\hat{\mathbf{n}})U_{2B}^{bf}(\mathbf{D}(i\boldsymbol{\tau}\cdot\hat{\mathbf{n}})\cdot\mathbf{x}, C)(\boldsymbol{\tau}\cdot\hat{\mathbf{n}})C^\dagger, \tag{21}$$

where $\mathbf{D}(i\boldsymbol{\tau}\cdot\hat{\mathbf{n}})$ is a rotation of π about the $\hat{n}$-axis. Under this symmetry transformation, the two skyrmions – that is, their positions and orientations — are simply interchanged. Eq. (21) is respected by two infinitely separated hedgehogs, Eq. (16), and by the Skyrme lagrangian, and so will also be respected by the field configuration resulting via the equations of motion from the time-evolution of two well-separated hedgehogs.

Configurations with $\hat{\mathbf{n}}$'s related by a rotation about the z-axis are simply related by simultaneous global rotations in space and isospace about the 3-axes:

$$U_{2B}^{bf}(\mathbf{x}, C) = e^{-i\tau_3\theta/2}U_{2B}^{bf}(\mathbf{D}(e^{i\tau_3\theta/2})\cdot\mathbf{x}, e^{i\tau_3\theta/2}Ce^{-i\tau_3\theta/2})e^{i\tau_3\theta/2}. \tag{22}$$

[b]If we write $C = \exp(i\boldsymbol{\tau}\cdot\boldsymbol{\gamma})$, then $\hat{\mathbf{n}} = \hat{\gamma}\times\hat{\mathbf{e}}_3$.

This just says that we may get from C to $C' = e^{i\tau_3\theta/2}Ce^{-i\tau_3\theta/2}$ simply by simultaneously re-defining the 1- and 2-axes (via rotations about the 3-axis) in space and isospace, which is a solitonic realization of the usual cylindrical symmetry of a two-particle system. Let us now anticipate quantization of the collective coordinates, for which purpose we can express Eq. (22) in the lab frame as

$$\begin{aligned} U_{2B}(\mathbf{x},t) &= AU_{2B}^{bf}(\mathbf{D}(B)\cdot(\mathbf{x}-\mathbf{R}_T);C)A^\dagger \\ &= (Ae^{-i\tau_3\theta/2})U_{2B}^{bf}(\mathbf{D}(e^{i\tau_3\theta/2}B)\cdot(\mathbf{x}-\mathbf{R}_T);C')(Ae^{-i\tau_3\theta/2})^\dagger. \end{aligned} \tag{23}$$

Writing the generator of left shifts of the $SU(2)$ matrix A as $\hat{\mathcal{L}}_k(A)$ and the generator of right shifts as $\hat{\mathcal{R}}_k(A)$,[c] from Eq. (23) we conclude that

$$\hat{\mathcal{L}}_3(B) - \hat{\mathcal{R}}_3(A) = \hat{\mathcal{R}}_3(C) - \hat{\mathcal{L}}_3(C). \tag{24}$$

The eigenstates of the system must satisfy this relation, and C describes only two independent degrees of freedom.

Assume now we know the (adiabatically varying) field in the body-fixed frame $U_{2B}^{bf}(\mathbf{x};R(t),\boldsymbol{\gamma}(t))$, where $C=\exp(i\boldsymbol{\tau}\cdot\boldsymbol{\gamma})$. Promoting also the global coordinates to dynamical variables, we find using Eq. (18) that

$$\begin{aligned} \mathcal{L}_t^a\big(U_{2B}(\mathbf{x},t)\big) &= D_{ab}(A)\big\{\mathcal{R}_t^b(A) - D_{bc}(U_{2B}^{bf})\mathcal{R}_t^c(A) \\ &\quad +[2\epsilon_{ikc}D_{kl}(B)x_l\mathcal{L}_t^c(B) + \dot{R}_{Ti}]D_{ij}(B)\mathcal{L}_j^b(U_{2B}^{bf}) \\ &\quad +\dot{R}\mathcal{L}_R^b(U_{2B}^{bf}) + \dot{\gamma}_i\mathcal{L}_{\gamma_i}^b(U_{2B}^{bf})\big\}, \end{aligned} \tag{25}$$

where

$$\begin{aligned} \mathcal{L}_R^a(U) &= -\frac{i}{2}\mathrm{Tr}[\tau_a\frac{\partial U}{\partial R}U^\dagger] \\ \mathcal{L}_{\gamma_i}^a(U) &= -\frac{i}{2}\mathrm{Tr}[\tau_a\frac{\partial U}{\partial \gamma_i}U^\dagger]. \end{aligned} \tag{26}$$

For notational simplicity let's combine the collective coordinates into a single vector $\dot{\mathbf{Q}} = (2\mathcal{R}_t(A), 2\mathcal{L}_t(B), \dot{\mathbf{R}}_T, \dot{\gamma}, \dot{R})$, so that Eq. (25) becomes $\mathcal{L}_t^a(U_{2B}) = S_{ak}(U_{2B})\dot{Q}_k$. Substituting this into the lagrangian (3) and integrating over space, we arrive at a total lagrangian in terms of just the collective variables

$$L = \frac{1}{2}\dot{Q}_k M_{kl}(R,C)\dot{Q}_l - \mathcal{U}(R,C). \tag{27}$$

Here the mass matrix is

$$M_{kl}(R,C) = \int d^3x S_{ak}(U_{2B}^{bf})\mathcal{M}_{ab}(U_{2B}^{bf})S_{bl}(U_{2B}^{bf}), \tag{28}$$

[c] $\hat{\mathcal{R}}_k(A)$ is the body-fixed total isospin operator, and $-\hat{\mathcal{L}}_k(B)$ is the body-fixed total angular momentum operator.

where $\mathcal{M}_{ab}$ is given in Eq. (4). The two-skyrmion potential is simply

$$\mathcal{U}(R,C) = \int d^3x \mathcal{V}(U_{2B}^{bf}). \tag{29}$$

Now we may convert to hamiltonian form via the canonical momenta $P_i = M_{ij}\dot{Q}_j$, finding then

$$H = \frac{1}{2\sqrt{m}} P_i \Big(\sqrt{m}\, M_{ij}^{-1}(R,C)\Big) P_j + \mathcal{U}(R,C), \tag{30}$$

where $m = \det(\mathbf{M})$. This is the truncated hamiltonian for two interacting skyrmions. The above operator ordering of the kinetic term assures that the Schrödinger equation involves the covariant laplacian on the manifold which defines the truncated dynamics.

According to the $1/N_C$ expansion, the mass $\mathbf{M}(R,C)$ and the potential $\mathcal{U}(R,C)$ are of order N_C, while the momenta $\mathbf{P}$ are of order 1. Thus the NN potential, which is determined by diagonalizing the hamiltonian H in Eq. (30), will have $\mathcal{O}(N_C)$ "static" contributions from $\mathcal{U}$ and $\mathcal{O}(1/N_C)$ "kinetic" contributions from $\mathcal{T}$. Then the system may be quantized under the adiabatic assumption, namely, the collective coordinates are considered to be "slow" variables. Moreover, the rotational kinetic energy is characterized by $1/\lambda_H$, and the translational kinetic energy is characterized by the smaller quantity $1/M_H$; so C may be considered a "fast" variable with respect to R.[9,10] This amounts to calculating a generalized $\mathcal{NN}$ potential for R under the Born-Oppenheimer approximation. The natural set of basis functions to use are combinations of the states (17) with good total spin and isospin

$$|(t_1,t_2)TM_TSM_S\rangle = \sum_{\{i_n,s_n\}} \langle t_1 i_1 t_2 i_2 | TM_T\rangle \langle t_1 s_1 t_2 s_2 | SM_S\rangle |t_1 i_1 s_1\rangle |t_2 i_2 s_2\rangle. \tag{31}$$

Because the hamiltonian depends upon the relative orientation $C = A_1^\dagger A_2$, however, the state $|(1/2,1/2)TM_TSM_S\rangle$ with the lowest asymptotic energy will not be an exact eigenstate of the system; there will be an admixture of $|(1/2,3/2)TM_TSM_S\rangle$ and $|(3/2,3/2)TM_TSM_S\rangle$ states — that is, intermediate deltas.

3. Use of Ansätze

Because of the nonlinearity of the Skyrme model, the equations of motion are difficult to solve. One has to rely on numerical techniques even to find the hedgehog angle F. As we go to the B=2 sector, the numerical calculations become even more tedious. It would thus be a great advantage if we could somehow guess an accurate analytic form for the field U_{2B}. In particular, if we knew how the field depended on the collective coordinates — and therefore the derivatives of U_{2B} in Eq. (25) with respect to R and C — the evaluation of Eq. (27) would be greatly facilitated. This has led to the use of ansätze for determining the NN potential.

3.1. *The product ansatz*

Skyrme[12] first suggested employing the product

$$U_P(\mathbf{x}) = A_1 U_H(\mathbf{x}-\mathbf{R}_1) A_1^\dagger A_2 U_H(\mathbf{x}-\mathbf{R}_2) A_2^\dagger \tag{32}$$

for all separations $R = |\mathbf{R}_1 - \mathbf{R}_2|$ as an ansatz for the B=2 sector. Considering Eq. (16), this is clearly a good approximation at large R, for then one soliton does not feel the presence of the other. Moreover, the baryon number is additive $B = B_1 + B_2$, and the product ansatz (32) has baryon number equal to two for all separations and orientations. For finite R, the matrices $U_n(\mathbf{x}) = A_n U_H(\mathbf{x} - \mathbf{R}_n) A_n^\dagger$ in Eq. (32) no longer commute, and the symmetry (21) is not satisfied. At smaller separations, then, we can expect that this configuration will be much different from the exact solution. For example, it cannot reproduce the minimum-energy toroidal solution. On the other hand, the product ansatz is intuitively appealing since the skyrmions retain their individual properties. The main advantage is its compact form, which reduces the numerical difficulty in extracting a NN potential from skyrmions.

Let's write the collective coordinates as $\dot{\mathbf{Q}}_P = \left(2\mathcal{R}_t(A_1), 2\mathcal{R}_t(A_2), \dot{\mathbf{R}}_1, \dot{\mathbf{R}}_2\right)$, which is slightly different than in the previous section. The canonical momenta are then $\mathbf{P}_P = \mathbf{M} \cdot \dot{\mathbf{Q}}_P = (\mathbf{T}_1, \mathbf{T}_2, \mathbf{P}_1, \mathbf{P}_2)$; and the hamiltonian is

$$H = \frac{1}{2\sqrt{m}} \mathbf{P}_P \cdot \left(\sqrt{m}\, \mathbf{M}_P^{-1}(R, C)\right) \cdot \mathbf{P}_P + \mathcal{U}(R, C). \tag{33}$$

I will concentrate first on the two-skyrmion potential $\mathcal{U}(R, C)$, which generates the leading-order contribution to the NN potential. As a first approximation, this potential is

$$\mathcal{U}_{NN}(\mathbf{R}) = \langle N_1' N_2' | \mathcal{U}(\mathbf{R}, A_1^\dagger A_2) | N_1 N_2 \rangle - 2M_H, \tag{34}$$

where from Eq. (17)

$$|N_1 N_2\rangle = |1/2, i_1, s_1\rangle |1/2, i_2, s_2\rangle \tag{35}$$

I will consider the inclusion of deltas presently.

Using Eq. (32), one finds a potential of the form

$$\mathcal{U} = 2M_H + \mathcal{U}^{(0)}(R) + D_{ab}(C)\mathcal{U}_{ab}^{(1)}(\mathbf{R}) + \ldots, \tag{36}$$

where the dots stand for terms proportional to $\mathcal{D}^{(2)}_{mm'}(C)$ which correspond to isospin-2 exchange between the skyrmions and so do not contribute to Eq. (34). The potential $\mathcal{U}_{ab}^{(1)}$ has the form

$$\mathcal{U}_{ab}^{(1)}(\mathbf{R}) = \mathcal{U}_{\tau\sigma}(R)\delta_{ab} + \mathcal{U}_{\tau T}(R)(3\hat{R}_a\hat{R}_b - \delta_{ab}). \tag{37}$$

Actually, explicit calculation[2] has shown that the terms proportional to $\mathcal{D}^{(2)}(C)$ in Eq. (36) are negligible. Using now the identity[13]

$$\langle N_1' N_2' | D_{ab}(C) | N_1 N_2 \rangle = \frac{1}{9}(\boldsymbol{\tau}_1 \cdot \boldsymbol{\tau}_2)\sigma_{1a}\sigma_{2b}, \tag{38}$$

the contribution to the NN potential given by Eq. (34) can be written in terms of three spin-isospin structures

$$\mathcal{U}_{NN}(R) = \mathcal{U}^{(0)}(R) + \frac{1}{9}(\boldsymbol{\tau}_1 \cdot \boldsymbol{\tau}_2)[(\boldsymbol{\sigma}_1 \cdot \boldsymbol{\sigma}_2)\mathcal{U}_{\tau\sigma}(R) + S_{12}\mathcal{U}_{\tau T}(R)], \tag{39}$$

where $S_{12} = 3(\boldsymbol{\sigma}_1 \cdot \hat{\mathbf{R}})(\boldsymbol{\sigma}_2 \cdot \hat{\mathbf{R}}) - \boldsymbol{\sigma}_1 \cdot \boldsymbol{\sigma}_2$ is the usual tensor structure. Interestingly enough, these three terms are known from phenomenological potentials to give the strongest contributions to the NN interaction.

At large separation R the potentials in Eq. (39) can be calculated by expanding in $F(|\mathbf{x} - \mathbf{R}_1|)$ — which has a Yukawa form asymptotically with strength[2] related to the coupling $g_{\pi NN}$ — in the neighborhood of $\mathbf{R}_2$, and *vice versa*. Then one finds

$$\mathcal{U}_{NN}(R) \sim \frac{1}{4\pi}\left(\frac{g_{\pi NN}}{2M_N}\right)^2 (\boldsymbol{\tau}_1 \cdot \boldsymbol{\tau}_2)\sigma_{1a}\sigma_{2a}\frac{\partial}{\partial R_a}\frac{\partial}{\partial R_b}\left(\frac{e^{-m_\pi R}}{R}\right), \tag{40}$$

which is the familiar one-pion-exchange potential. Since the product ansatz is accurate in this limit, the long-range part of the NN potential as derived from the Skyrme model is correct. Further calculation[13] shows that the central term is of order $e^{-2m_\pi R}$ at large separation and so is due to two-pion-exchange. With relatively small numerical effort, the product ansatz can also be used for smaller separation. In Refs. [2] and [14], a general agreement was found with conventional phenomological potentials such as the Paris. A striking difference is that no intermediate range attraction was found in the central channel. Since in conventional potentials it is this attraction which is responsible for the binding of nuclei, this could be a major failure of the Skyrme model. In fact, much work has gone into correcting for this supposed shortcoming.

One improvement upon the above analysis is to consider also the kinetic term in Eq. (33). At infinite separation the system consists of two independent skyrmions. We may write the mass matrix in Eq. (33) as $\mathbf{M}_P = \mathbf{M}_0 + \boldsymbol{\mu}(\mathbf{R}, A_1^\dagger A_2)$, where

$$\mathbf{M}_0 = \mathrm{diag}(\lambda_\mathrm{H}, \lambda_\mathrm{H}, \mathrm{M_H}, \mathrm{M_H}) \tag{41}$$

and $\boldsymbol{\mu} \to 0$ at large separation. In general, then, when we evaluate $\langle N_1' N_2' | H | N_1 N_2 \rangle$, there will be additional contributions to $\mathcal{U}_{NN}$ in Eq. (39) coming from the kinetic terms. Writing

$$T_0 = \frac{1}{2}\mathbf{P}_P \cdot \mathbf{M}_0^{-1} \cdot \mathbf{P}_P = \frac{1}{2\lambda_H}(\mathbf{J}_1^2 + \mathbf{J}_2^2) + \frac{1}{2M_H}(\mathbf{P}_1^2 + \mathbf{P}_2^2), \tag{42}$$

we have

$$\begin{aligned} \langle N_1' N_2' | H | N_1 N_2 \rangle &= \frac{\mathbf{P}_1^2}{2M_H} + \frac{\mathbf{P}_2^2}{2M_H} + 2M_N + \mathcal{U}_{NN}(\mathbf{R}) \\ &\quad + \langle N_1' N_2' | \frac{1}{2\sqrt{m}}\mathbf{P}_P \cdot \sqrt{m}\,\mathbf{M}_P^{-1} \cdot \mathbf{P}_P - T_0 | N_1 N_2 \rangle, \end{aligned} \tag{43}$$

where $M_N = M_H + 3/8\lambda_H$. The NN potential can then be identified in this approximation as $\mathcal{U}_{NN} + \langle N_1' N_2' | T - T_0 | N_1 N_2 \rangle$. All possible terms in the NN potential, including momentum-dependent and spin-orbit terms, are generated form this. As a matter of fact, there are also parity violating terms such as $\mathbf{P} \cdot \mathbf{S}$ which arise from the product ansatz. This is a direct consequence of the fact that the product ansatz does not obey the symmetry (21), and such terms will not be present when the exact solution is used. The contribution from the kinetic terms to the central channel is attractive, as noted originally in Ref.[15]. There the calculation was performed for closer separations, but this term was not found to provide a strong enough source of attraction to overcome the repulsion found from the two-skyrmion potential. This is what we expect from N_c counting: all these kinetic terms are much smaller in magnitude.

Another obvious improvement over the above analysis — which also provides a second possible source of intermediate range attraction in the central channel — is to expand the basis used for the isorotational variables. As mentioned above, the NN states (35) are not exact eigenstates of the hamiltonian for finite separation because $\mathcal{U}(\mathbf{R}, C)$ mixes generalized nucleon states with different total spin. By restricting the basis to only nucleon states, we are ignoring $\mathcal{O}(N_c)$ contributions to the NN interaction. If we think of a perturbative expansion in $H' = H - H_0$, with $H_0 = T_0 + 2M_H$ and H given by Eq. (33), this amounts to the inclusion of intermediate isorotational excitations of the generalized nucleon. Then the NN potential becomes

$$V_{NN} = \langle N_1' N_2' | H' | N_1 N_2 \rangle - \sum_{\mathcal{N}_1 \mathcal{N}_2 \neq N_1 N_2} \frac{\langle N_1' N_2' | H' | \mathcal{N}_1 \mathcal{N}_2 \rangle \langle \mathcal{N}_1 \mathcal{N}_2 | H' | N_1 N_2 \rangle}{E_{\mathcal{N}_1 \mathcal{N}_2} - 2M_N} + \ldots, \quad (44)$$

where $\mathcal{N}$ is a generalized nucleon state as given by Eq. (13). This just describes the inclusion of intermediate states. The energy differences $E_{\mathcal{N}\mathcal{N}'} - 2M_N \propto 1/\lambda_H$ are of the order $1/N_c$, however, so we cannot expect the above perturbative expansion to work very well when the mixing becomes appreciable.

If we only consider nucleons and deltas,[d] though, it does not require a great deal of numerical effort to simply diagonalize the matrix $\langle \mathcal{N}_1 \mathcal{N}_2 | H' | \mathcal{N}_3 \mathcal{N}_4 \rangle$. Such a calculation has been performed using the product ansatz in Ref. [17]. One important result of this inclusion of intermediate states is that all six allowed local terms in the NN interaction are generated now at $\mathcal{O}(N_c)$, so that the potential becomes

$$\begin{aligned} V_{NN} &= V_c(R) + V_\sigma(R)\boldsymbol{\sigma}_1 \cdot \boldsymbol{\sigma}_2 + V_T(R) S_{12} \\ &\quad + \boldsymbol{\tau}_1 \cdot \boldsymbol{\tau}_2 \Big(V_\tau(R) + V_{\tau\sigma}(R)\boldsymbol{\sigma}_1 \cdot \boldsymbol{\sigma}_2 + V_{\tau T}(R) S_{12} \Big). \end{aligned} \quad (45)$$

Iteration of H' through state mixing thus produces three new terms and corrections to the three terms which already exist from first approximation. Considering Eq. (44), one can expect that the inclusion of intermediate states will produce an attractive contribution to the central channel; and in fact a noticable attractive contribution was found. For the product ansatz, however, this attraction is not enough to overcome the repulsion found from the term $\langle N_1' N_2' | \mathcal{U} | N_1 N_2 \rangle$.

There have also been many other suggested improvements upon these product ansatz calculations. Kälbermann and Eisenberg[18] considered coupling to the Roper

[d]Witten[16] has argued that the only allowed generalized nucleon states are those with isospin $t \leq N_c/2$, which for N_c=3 means only the nucleon and the delta.

resonance — the breathing mode of the hedgehog — as a possible source of attraction. Jackson and Jackson[19] have considered one-loop corrections from pion fluctuations about the product ansatz. Others have argued that one should consider extensions of the Skyrme model, either by including new higher derivative terms in the lagrangian,[20] or by explicitly adding mesons[21] or a scalar "glueball" field.[22] Of course, all these preclude the possibility — which indeed was mentioned in the first paper by Jackson, *et al.*[2] — that the product ansatz simply gives inaccurate results at short and intermediate ranges and that the NN potential obtained from exact solutions in the basic Skyrme model might, in fact, be much different.

3.2. *Other ansätze*

Clearly the use of an ansatz with explicit dependence on collective variables facilitates the calculations. Numerical calculations, however, show that the product ansatz is a poor approximation to the true solution at small and intermediate separations. There have therefore been several attempts to use more general ansätze. The simplest extension is the variational product ansatz.[23] This generalizes the chiral angle $F(|\mathbf{x}-\mathbf{R}_n|)$ in Eq. (32) to $F(\mathbf{x},\mathbf{R}_n)$ — that is, it allows for distortions of the chiral angles F of the individual skyrmions from the hedgehog solutions. This is clearly an improvement over the usual product ansatz, and some differences have been seen. However, these differences are not enough to affect the conclusions one reaches from the product ansatz: there is still no attraction in the central channel. The symmetrized product ansatz[24], another generalization of the product ansatz, is

$$U_{SP}(\mathbf{x}) = \frac{1}{\chi}\Big[U_1(\mathbf{x})U_2(\mathbf{x}) + U_2(\mathbf{x})U_1(\mathbf{x})\Big], \tag{46}$$

where the factor $\chi^2 = (U_1U_2 + U_2U_1)(U_1U_2 + U_2U_1)^\dagger$ ensures that U_{SP} is an $SU(2)$ matrix. This ansatz is designed to respect the symmetry (21) of the exact solution (which the product ansatz respects only in the asymptotic limit), and indeed a small attraction was found with this ansatz at large range in the central central. As shown in Ref. [25], however, the symmetrized product ansatz is inadequate at short range since for sufficiently small R the factor χ equals zero at some values of $\mathbf{x}$ and $A_1^\dagger A_2$. So neither of these extended product ansätze give the skyrmions enough freedom to distort from their asymptotic hedgehog form.

There have also been some more complicated ansätze proposed.[26,27] The latter, the Atiyah-Manton ansatz, which expresses the field U as a time-ordered integral over an instanton solution of the $SU(2)$ Yang-Mills theory, is particularly interesting. Hosaka, *et al.*[28] have described a choice of the instanton parameters suitable for modeling the B=2 sector of the Skyrme model. In particular, this ansatz reproduces the energies of the known static solutions to within a few percent. (Quantities such as the moment of inertia, however, are less satisfactorily reproduced.) The Atiyah-Manton ansatz is thus much more powerful than the product ansatz. This ansatz was used in Ref. [28] to study the NN potential numerically; significantly less repulsion at intermediate range in the central channel was found, but there was still no attraction. Unfortunately, a different definition of the separation was chosen, which makes direct comparison with existing numerical calculations difficult. Moreover, since the magnitude of the central attraction is only a few percent of $2M_N$, it could easily be missed by even a relatively accurate ansatz.

3.3. *Including pion fluctuations*

A full quantum theoretical treatment of the Skyrme model requires a consideration of pion fluctuations about the solitons. As mentioned above, such fluctuations will be higher order in $1/N_C$, and their contribution to the soliton-soliton interaction should in general be small. The most important fluctuations will be those corresponding to zero and nearly-zero modes of the solitons. These are isolated by the identification of collective coordinates, and other modes can then be ignored in first approximation. When this is done through the use of an ansatz, however, these statements must be altered.[3] A problem arises with the usual N_C counting when one uses a solitonic ansatz instead of the true solution to the equations of motion. This is because terms which would vanish if one expanded about the true solution to the classical equations of motion will not vanish when one expands about an ansatz. The result is that pion fluctuations about the ansatz will produce contributions of order N_C which are proportional to the deviation of the ansatz from the true solution.

Thorsson and Zahed[29] have presented a first calculation of the effects of pion fluctuations about the product ansatz in the ω-stabilized Skyrme model. Also, in the context of the Skyrme model, Otofuji, *et al.*[30] have done a calculation for the central channel of the NN interaction. Both these calculations involve a double expansion in $1/N_C$ and $e^{-m_\pi R}$, and so are valid for long range. The results are by no means conclusive but do give strong indication that the addition of pion fluctuations about the product ansatz will produce the missing attraction in the central channel. This procedure, however, becomes more and more difficult computationally as higher orders in $e^{-m_\pi R}$ are included. So the initial gain in computational ease from the simplicity of the product ansatz is lost because one can no longer ignore higher order pion fluctuations. Perhaps the insight to be gained from this work is that one would be better off solving the equations of motion and expanding about the constrained or full time-dependent classical solution so that the usual $1/N_C$ counting will apply. In particular, a truncation of the degrees of freedom so as to exclude mesonic fluctuations can only then be justified.

4. Exact Numerical Calculations

From the results discussed in the previous section, it seems clear that one must analyze the Skyrme equations of motion numerically. At the very least, this is necessary to determine the reliability of the various ansätze. As a matter of fact, it would seem that in practice if one wants to obtain reliable results the path of least effort is to start from the classical solutions to the equations of motion. There have been two sets of approaches to the problem. One method is to solve the full time-dependent equations of motion. This is quite a difficult task if one wants to include all the important degrees of freedom. Nevertheless, some progress has been made in this direction. A second approach is to introduce constraints so as to fix the collective coordinates, and then to solve the resulting static equations of motion. The numerical effort involved in this second approach is considerably less. Its validity relies on the assumption of adiabaticity: that the time-dependent motion can be parameterized in terms of a few collective variables, and the the paths of the time-dependent solutions lie on (or at least close to) the manifold of constrained static solutions.

4.1. *Time-dependent numerical simulations*

Time-dependent solutions of the classical field equations are useful in visualizing the dynamics as well as guiding suitable approximations to represent it. In addition, such solutions are essential in any semiclassical approach trying to incorporate leading-order quantum effects. In a nonlinear field theory, however, it is quite difficult to obtain time-dependent solutions, for one has to rely on costly numerical methods. Some simulations of the scattering of two skyrmions have been performed,[31,32] however, and a bound state has also been found.[33] I cannot comment on the details of these studies here, but will simply refer to Ref. [13] and the contribution to this volume from Snyderman. These simulations yield a picture consisting of particle-like scattering of the soliton centers with the dynamics of the scattering depending on the relative orientations of the pion fields which "dress" these centers. They confirm — and have at the same time helped to guide — the constrained calculations which I shall describe next.

4.2. *Constrained static solutions*

Static calculations are numerically more stable and less costly than time-dependent simulations. Several relaxation algorithms have been used for static B=2 skyrmion calculations.[4,10,34,35] To employ techniques such as these which find minimum-energy static configurations, one must introduce constraints that fix the collective coordinates of the two skyrmions. In the body-fixed frame, the relative orientation can be fixed via boundary conditions at the z=0 plane; and the separation can be fixed via Lagrange multipliers[4] or by explicitly pinning down the points where $U_{2B} = -1$, the topological centers.[34] In this way an explicit dependence of the two-skyrmion system upon the collective coordinates may be obtained.

Now the truncated hamiltonian (30) for the interacting $B = 2$ system contains a potential $\mathcal{U}(R, C)$ which depends on only the separation R, taken here to be along the z-axis, and the relative isospatial orientation C. The latter can be expressed as Euler angles via the $SU(2)$ matrix $C = e^{-i\tau_3\alpha/2}e^{-i\tau_2\beta/2}e^{-i\tau_3\gamma/2}$. Since the $\mathcal{D}$-functions form a complete set over $SU(2)$ for half-integer j, a general form for the two-skyrmion potential is then given by

$$\mathcal{U}(R,C) = \sum_{j=0}^{\infty} \sum_{m=-j}^{j} \sum_{n=-j}^{j} V_{jmn}(R)\mathcal{D}^{(j)}_{m,n}(C). \tag{47}$$

However, as discussed above, configurations with orientations C differing only by a rotation about the 3-axis are degenerate in energy: they are simply related by a redefinition of the 1- and 2-axes in space and isospace, Eq. (22). Thus the potential is independent of $\alpha - \gamma$. Furthermore, one expects $\mathcal{U}(R, C^\dagger) = \mathcal{U}(R, C)$; that is, that the classical potential is symmetric under skyrmion interchange $A_1 \leftrightarrow A_2$. Indeed, this follows from the discrete symmetries of the exact solution

$$\begin{aligned} U_{2B}^{bf}(x,y,z;-C) &= U_{2B}^{bf}(x,y,z;C), \\ \tau_1 U_{2B}^{bf}(-x,y,z;C)^\dagger \tau_1 &= U_{2B}^{bf}(x,y,z;C^\dagger), \\ \tau_2 U_{2B}^{bf}(x,-y,z;C)^\dagger \tau_2 &= U_{2B}^{bf}(x,y,z;C^\dagger), \\ \tau_3 U_{2B}^{bf}(x,y,-z;C)^\dagger \tau_3 &= C U_{2B}^{bf}(x,y,z;C^\dagger) C^\dagger, \end{aligned} \tag{48}$$

which are symmetries of the product solution at $R \to \infty$ and of the Skyrme lagrangian and so remain symmetries of the field which results from the time-evolution of two initially well-separated skyrmions. Employing these symmetries, then, gives

$$\mathcal{U}(R,C) = \sum_{j=0}^{\infty} \sum_{m=0}^{j} V_{jm}(r) \frac{1}{2} [\mathcal{D}^{(j)}_{m,m}(C) + \mathcal{D}^{(j)}_{-m,-m}(C)], \tag{49}$$

where j is now restricted to integer values.

To first approximation the NN potential is obtained by sandwiching Eq. (49) between the asymptotic two-nucleon states as given by Eq. (35). The terms with $j \geq 2$ in Eq. (49) can contribute only through intermediate states, and so will not contribute in this first approximation. The product ansatz generates a potential $\mathcal{U}(R, C)$ which when expressed as in Eq. (49) truncates at $j = 2$. The $j = 2$ terms, however, are negligible;[2] and the potential is essentially determined by three functions of R — one function corresponding to an isospin-exchange of zero and two to an isopin-exchange of one between the skyrmions. It seems reasonable[8] — and is certainly numerically expedient — to also assume for the exact calculations that the potential (49) contains only terms with $j \leq 1$. At the very least this assumption will allow us to determine the reliability of the product ansatz in approximating the two-skyrmion potential, but it is nevertheless likely that $j \geq 2$ terms will be negligible except at very short range. We have then

$$\begin{aligned} \mathcal{U}(R,C) &= V_{00}(R)\mathcal{D}^{(0)}_{0,0}(C) + V_{10}(R)\mathcal{D}^{(1)}_{0,0}(C) + \frac{1}{2}V_{11}(R)[\mathcal{D}^{(1)}_{1,1}(C) + \mathcal{D}^{(1)}_{-1,-1}(C)] \\ &= V_{00}(R) + V_{10}(R)\cos\beta + V_{11}(R)\cos^2\frac{\beta}{2}\cos(\alpha+\gamma). \end{aligned} \tag{50}$$

The radial potentials V_{00}, V_{10}, V_{11} can be straightforwardly deduced by calculating $\mathcal{U}$ from Eqs. (29) and (3) for only three relative isospatial orientations, for which a convenient choice is $C = 1$, $i\tau_3$ and $i\tau_2$. (From the product ansatz results, we know that asymptotically these yield zero, the most repulsive, and the most attractive potentials, respectively.) We have

$$\begin{aligned} \mathcal{U}(R,1) &= V_{00}+V_{10}+V_{11}, \\ \mathcal{U}(R,i\tau_3) &= V_{00}+V_{10}-V_{11}, \\ \mathcal{U}(R,i\tau_2) &= V_{00}-V_{10}. \end{aligned} \tag{51}$$

When sandwiched between NN states, Eq. (50) leads to the well-known result

$$\mathcal{U}_{NN} = V_c(R) + V_{\tau\sigma}(R)(\vec{\tau}_1\cdot\vec{\tau}_2)(\vec{\sigma}_1\cdot\vec{\sigma}_2) + V_{\tau T}(R)(\vec{\tau}_1\cdot\vec{\tau}_2)(\sigma_{1i}\sigma_{2j})(3\delta_{i3}\delta_{3j}-\delta_{ij}), \tag{52}$$

which is of the same form as Eq. (39), except that it is expressed in the body-fixed frame ($\hat{\mathbf{R}}$ along the z-axis).

The potential $\mathcal{U}(R,C)$ may be found from the exact field solution obtained through constrained minimization of the static lagrangian which holds the distance R fixed. Such a calculation has been performed[36] by discretizing the equations of motion on a spatial lattice and relaxing from an initial B=2 configuration . The calculations were done with the Skyrme lagrangian on a $20\times20\times40$ grid — the discrete symmetries (48) reduce the calculation to one-eighth of space — with the constants f_π and e taken to be $93MeV$ and 4.76. This parameter set guarantees that the long-range part of the potential — which automatically in the Skyrme model is due to one pion exchange — will agree with phenomenological results.[e] Fig. 1 displays the evolution of the baryon number density as the two solitons approach in a given isospatial orientation,[13,36] which is obtained by varying the separation R (actually, it is more convenient to fix the rms-radius r_0 with lagrange multipliers, so that R is then read off from the solutions). The cases $C=1$ and $C=i\tau_2$ are shown; the evolution for $C=i\tau_3$ looks much like that for $C=1$. When the solitons come close together significant distortions in shape are observed. As with the time-dependent simulations discussed above, the solitons shrink as they approach each other for $C=1$. As expected, the donut, the lowest-energy $B=2$ configuration, forms at R=0 for $C=i\tau_2$. This has also been observed in time-dependent simulations[33] where the skyrmions approach, form the donut and then separate at right angles to the initial axis of separation. Fig. 2 presents the interaction energies corresponding to the three orientations $C=1, i\tau_2$, and $i\tau_3$. (The coarseness of the lattice prevents the skyrmions from approaching closer than $r_0\sim .4fm$, for then numerical errors accumulate and baryon number "slips away.") While the product ansatz reproduces the defensive ($C=1$) and repulsive ($C=i\tau_3$) channels fairly well, it fails to reproduce the strong attraction for $C=i\tau_2$, which leads to a medium-range central attraction in the NN potential already in first approximation [36]. Indeed, the central, tensor, and spin-spin potentials obtained via Eq. (52) from the exact calculations[36] compare much better with the Paris,[37] for example, than do the corresponding potentials calculated using the product ansatz. This is shown for the central part $V_c(R)$ in Fig. 3.

The above calculation is easily improved by allowing for mixing through intermediate states,[13,38] which corrects the potential (52) still to leading order in $1/N_c$. This entails the diagonalization of the hamiltonian $H'=T_0+\mathcal{U}$ in the space of asymptotic

[e]To compare with Ref. [2] I have included the finite-N_c corrections used there, which are determined by a comparison with quark hedgehog models.

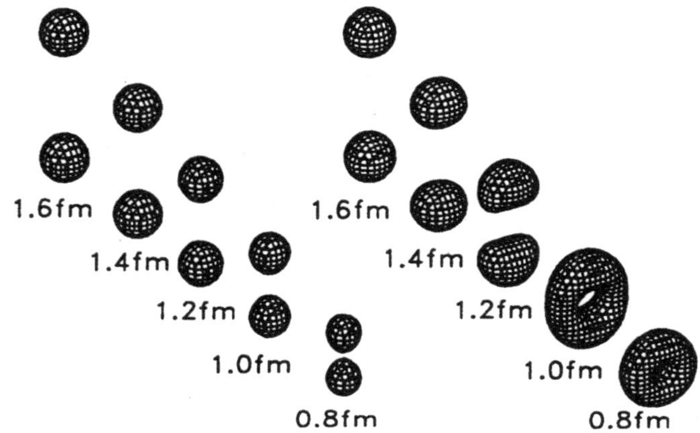

Figure 1: The baryon density for different rms radii r_0 of two interacting skyrmions. The defensive (left) and attractive (right) orientations are shown.

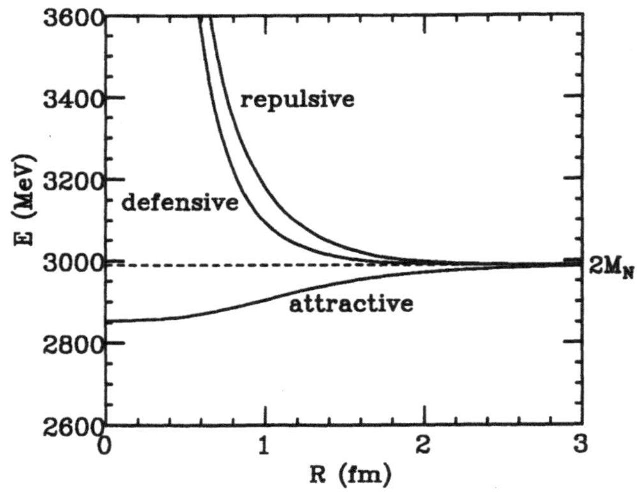

Figure 2: The exact two-soliton interaction energies for the three isospatial orientations, from top to bottom, $C = i\tau_3, 1, i\tau_2$ which determine the classical two-skyrmion potential.

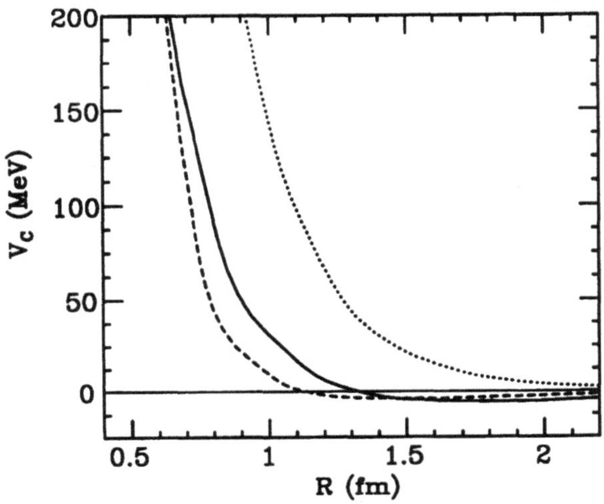

Figure 3: The central part of the NN potential in the product ansatz (dotted line), the exact calculation (full line) and the Paris potential (dashed line).

NN, $N\Delta$ and $\Delta\Delta$ states. These are conveniently written in an $LSJT$-representation labeled by the total isospin T, the total angular momentum J, the orbital angular momentum L and the spin S, with the quantization axis taken to be along $\hat{\mathbf{R}}$. The states split into subspaces that are either symmetric or antisymmetric under the exchange of identical particles. In accordance with the Pauli principle we must select the antisymmetric ones, which satisfy the standard selection rule $L+S+T=$ odd. Indeed, insistence that each skyrmion is quantized as a fermion[4] leads to the so-called Finkelstein-Rubinstein constraints,[39] which yield the above selection rule. The states $|(t_1t_2)LSJT\rangle$ are obtained by coupling the states (31) to orbital angular momentum states $|LM\rangle$; and evaluation of the interaction matrix elements proceeds by using the $\mathcal{D}$-function expansion (49) of $\mathcal{U}(R,C)$. One may then follow a state at large R, where it is made up purely of nucleons, to small R, where there is mixing. The energy eigenvalue of this state as a function of R then defines the adiabatic NN potential. At $R \sim 1fm$, the mixing amplitudes are typically 10%, which is similar to results from conventional meson-exchange potentials with intermediate isobars. As with the product ansatz calculation, this inclusion of intermediate states generates the three remaining allowed local terms in the NN potential — namely, $V_\tau(R)\boldsymbol{\tau}_1\cdot\boldsymbol{\tau}_2$, $V_\sigma(R)\boldsymbol{\sigma}_2\cdot\boldsymbol{\sigma}_2$, and $V_T(R)S_{12}$ — as well as corrections to the three other local terms in Eq. (52). The results[13] are shown [f] in Fig. 4 for the $^1S_0$ channel (upper part) and the

[f]The potentials from the Skyrme model are plotted as functions of $2r_0$ instead of R here; this causes

$^3S_1$ channel (lower part). They indicate that mixing leads to a considerable increase in the attraction, particularly for $^1S_0$, and hence an improvement when compared with phenomenological interactions (here the Reid soft core (RSC)[40] potential). Similar

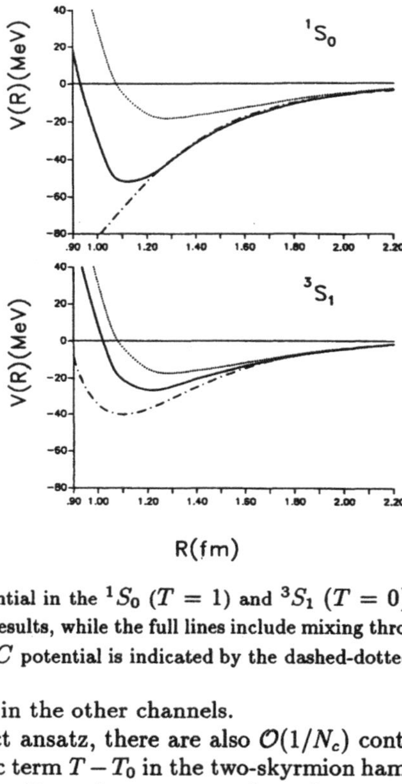

Figure 4: The NN potential in the $^1S_0$ ($T = 1$) and $^3S_1$ ($T = 0$) channels. The dotted lines indicate the lowest-order results, while the full lines include mixing through intermediate $NN, N\Delta$ and $\Delta\Delta$ states. The RSC potential is indicated by the dashed-dotted lines.

agreement is achieved in the other channels.

As with the product ansatz, there are also $\mathcal{O}(1/N_c)$ contributions to the NN potential from the kinetic term $T - T_0$ in the two-skyrmion hamiltonian (30). This gives rise to non-local spin-orbit and momentum-dependent terms, as well as to higher order corrections to the local terms obtained from the diagonalization of H'. Evaluation of the mass matrix $\mathbf{M}$, Eq. (28), however, is not a simple task since it involves derivatives with respect to the relative coordinates R and C. This means one must find field solutions for many values of these coordinates, which increases the computational effort needed. For R, this increase is not too great; and it has been found that the inertial mass of an interacting skyrmion does not differ from its asymptotic value M_H until it nears $R \sim 0.5fm$.[41] This results in a short-range momentum-dependent contribution to the NN potential. Determination of the C-dependence of $\mathbf{M}$ — and therefore the spin-orbit interaction — from exact numerical calculations, however, remains a problem for future study.

no change in the qualitative shape of the curves.

5. Discussion

To summarize, I have attempted to show how a realization of baryons as topological solitons of an effective meson theory leads to a nucleon-nucleon interaction which compares well with realistic potentials. Such effective meson theories are modeled on *QCD* and incorporate $1/N_c$ as the natural expansion parameter which characterizes the semiclassical quantization of their solitons. Calculations have been presented for the Skyrme model, the simplest such theory, under the assumption that only twelve zero and nearly-zero collective modes of the two-skyrmion system need be considered for quantization. From consideration of the product ansatz, we know that the theory produces the correct long-range OPEP form for the NN interaction. At closer range one must resort to numerical calculations. Then an intermediate range attraction in the central channel and a short range repulsion is found. The core is of similar range to that of the Paris potential. When intermediate states are included, all six allowed local terms in the NN interaction are present, with relative strengths which agree with phenomenological potentials. This is all at order N_c. When the order-$1/N_c$ kinetic terms which contain the inertial masses and moments of inertia of the two interacting skyrmions are considered, spin-orbit and momentum-dependent terms are also generated. The picture of nucleons as solitons thus yields a natural derivation of a qualititatively accurate NN potential, and the use of extended Skyrme models and the inclusion of $\mathcal{O}(1)$ effects from pion fluctuations seems likely to produce quantitative agreement with phenomenology.

So it seems that one can do a pretty good job of describing nucleons and their interactions without ever having to mention or use quarks. Although *QCD* is the underlying theory and quarks and gluons are the fundamental quanta of the strong interaction, at low energies the important degrees of freedom are the mesons — in particular, the pion. The Skyrme model is an elegant realization of this idea. In particular, I find the description of the NN potential in terms of interacting skyrmions at once simple and elegant and compelling. I want to emphasize, however, that this mesonic picture of the strong interaction is complementary, rather than contradictory, to the quark picture. Indeed, the predictions of nuclear properties coming from Skyrme models and from bag models are quite similiar. This observation is the basis of the Chesire cat philosophy, which models the nucleon as a quark bag surrounded by a field of mesons. Nuclear properties such as baryon number are then shared by the inner quark bag and the external solitonic field. According to the Chesire cat principle, physical quantities are insensitive to changes in the size of the bag. The skyrmion corresponds then to the limit where the bag shrinks to zero size, leaving behind just the topological singularity at the center of the hedgehog, $U_H(0) = -1$. One deficiency of meson exchange potentials with point nucleons has always been the description at short range. For the Bonn potential,[42] for example, *ad hoc* form factors must be introduced to fix the short range interaction. For the skyrmion, however, the point singularity is dressed with a cloud of mesons from the start; so nucleons are extended objects, and no such problem occurs. This is the power of effective meson theories: while quark models emphasize nuclear structure and meson exchange models emphasize nuclear interactions, the Skyrme model places structure and interactions on the same footing.

It would be interesting to push the connection between the Skyrme model and the meson exchange picture further. Establishing a precise correspondence, however, is a much more subtle and difficult problem than one might at first expect. Soliton methods are inherently nonlinear, with only physical pions resulting from a linearization about the solitonic background. But because nuclear structure and interactions are both mediated by mesons in the Skyrme model, two nuclei interact by distorting; and it is therefore difficult to extract from our calculations a description in terms of one-pion exchange, two-pion exchange, and so on. Clearly, the two skyrmions interact through multi-pion exchange, for pions are the only things available to be exchanged in the Skyrme model. So surely the intermediate range central attraction found in the Skyrme model is due primarily to two-pion exchange, but I do not know how to isolate this effect. Moreover, the usual soft pion theorems are not transparently reproduced by the semiclassical $1/N_c$ expansion.[3] This is because soft pions correspond to nearly-zero mode fluctuations, and an adiabatic quantization of the soliton is not good for this regime (see Ref. [43] and the contribution from Hayashi). Nevertheless, as we have seen here, the adiabatic approximation and the $1/N_c$ expansion seem to provide an appropriate scheme for describing the NN interaction.

The calculations described here, although quite encouraging, are still rather preliminary. From a technical standpoint, it will be necessary — and certainly numerically feasible — to perform the calculations on lattices of differing size so as to extrapolate to the continuum limit and to study shorter range behavior in the repulsive channels. A numerical technique such as that used by Shtern [34], where the skyrmion centers are fixed by hand and a variable grid — more dense near these centers — is employed, seems necessary for small separations. The calculations should also be performed for other relative isospatial orientations so as to allow for the determination of terms in the classical two-skyrmion potential corresponding to exchange of isospin two or even greater. This will also allow a determination of the relative moment of inertia and thus the spin-orbit interaction. There are many other ways by which the calculations presented here can be improved, such as the use of more realistic, extended effective meson lagrangians and the inclusion of higher order mesonic fluctuations. Another interesting, and ultimately necessary, extension of the Skyrme lagrangian is to account for the other flavors — in particular, strangeness (see the contribution from Kopeliovich). We can expect, however, that improvements such as these will just bring relatively small quantitative corrections to the qualitatively accurate results of the simple Skyrme model presented here.

Acknowledgements

This work was supported by NSF grants PHY-88-58250 and PHY-91-02922. I have benefited from discussions with Neal Snyderman and Jochen Wambach, and I thank Thomas Waindzoch for help in some of the numerical work.

References

1. R. Rajaraman, *Solitons and Instantons*, (North-Holland, Amsterdam, 1982).
2. A. Jackson, A.D. Jackson and V. Pasquier, Nucl. Phys. **A432**, 567 (1985).
3. H. Yamagishi and I. Zahed, Phys. Rev. **D43**, 891 (1991).
4. J.J.M. Verbaarschot, T.S. Walhout, J. Wambach and H.W. Wyld, Nucl. Phys. **A468**, 520 (1987).
5. V.B. Kopeliovich and B.E. Shtern, JETP Lett. **45**, 203 (1987).
6. N.S. Manton, Phys. Lett. **B192**, 177 (1987).
7. J.J.M. Verbaarschot, Phys. Lett. **195B**, 450 (1987).
8. D.I. D'yakonov and A.D. Mirlin, Sov. J. Nucl. Phys. **47**, 421 (1988).
9. N.S. Manton, Phys. Rev. Lett. **60**, 1916 (1988).
10. T.S. Walhout, Nucl. Phys. **A531**, 596 (1991).
11. J.J.M. Verbaarschot, *Workshop on Skyrmions and Anomalies* (World Scientific, Krakow, 1987).
12. T.H.R. Skyrme, Nucl. Phys. **31**, 556 (1962).
13. T.S. Walhout and J. Wambach, "From Skyrmions to the Nucleon-Nucleon Potential," to appear in Int. J. Mod. Phys. E.
14. R. Vinh Mau, M. Lacombe, B. Loiseau, W.N. Cottingham, and P. Lisboa, Phys. Lett. **150B**, 259 (1985).
15. E.M. Nyman and D. O. Riska, Phys. Lett. **175B**, 392 (1986).
16. E. Witten, Nucl. Phys. **B223**, 422,433 (1983).
17. H. DePace, H. Müther and A. Faessler, Nucl. Phys. **A457**, 541 (1986).
18. G. Kälbermann and J.M. Eisenberg, Nucl. Phys. **A500**, 589 (1989).
19. A. Jackson and A.D. Jackson, Nucl. Phys. **A457**, 687 (1986).
20. A. Jackson, A.D. Jackson, A.S. Goldhaber, G.E. Brown and L.C. Castillejo, Phys. Lett. **154B**, 101 (1985).
21. M. Lacombe, B. Loiseau, R. Vinh Mau and W.N. Cottingham, Phys. Lett. **161B**, 31 (1985); **169B** 121 (1986); Phys. Rev. Lett. **57**, 170 (1985).
22. H. Yabu, B. Schwesinger and G. Holzwarth, Phys. Lett. **224B** (1989) 25.
23. U.B. Kaulfuss and U.G. Meissner, Phys. Rev. **D31**, 3024 (1985).
24. E.M. Nyman and D.O. Riska, Phys. Lett. **203B**, 13 (1988).
25. M. Sternheim and G. Kälbermann, Phys. Lett. **225B**, 27 (1989).
26. E. Sorace and M. Tarlini, Phys. Rev. **D33**, 253 (1986).
27. M.F. Atiyah and N.S. Manton, Phys. Lett. **222B**, 438 (1989).
28. A. Hosaka, M. Oka, and R.D. Amado, Nucl. Phys. **A530**, 507 (1991).
29. V. Thorsson and I. Zahed, Phys. Rev. **D45**, 965 (1992).
30. T. Otofuji, Y. Kondo, S. Saito and R. Seki, Phys. Rev. D45, 2538 (1992).
31. J.J.M. Verbaarschot, T.S. Walhout, J. Wambach and H.W. Wyld, Nucl. Phys. **A461**, 603 (1986).
32. A.E. Allder, S.E. Koonin, R. Seki, H.M. Sommermann, Phys. Rev. Lett. **59**, 2836 (1987).
33. W.Y. Crutchfield, N. Snyderman and V.R. Brown, Phys. Rev. Lett. **68**, 1660 (1992).

34. B.E. Shtern, Sov. J. Nucl. Phys. **49**, 332 (1989).
35. E. Braaten, S. Townsend and L. Carson, Phys. Lett. **235B**, 147 (1990).
36. T.S. Walhout and J. Wambach, Phys. Rev. Lett. **67**, 314 (1991).
37. M. Lacombe, B. Loiseau, J. M. Richard, R. Vinh Mau, J. Cote, D. Pires and R. DeToureil, Phys. Rev. **C21** 861 (1980).
38. N.R. Walet, R.D. Amado and A. Hosaka, Phys. Rev. Lett. **68**, 3849 (1992).
39. D. Finkelstein and J. Rubinstein, J. Math. Phys. **9**, 1762 (1968).
40. R.V. Reid, Ann. Phys. **50**, 411 (1968).
41. T.S. Walhout, in preparation.
42. R. Machleidt, K. Holinde and C. Elster, Phys. Rep. **149**, 1 (1987).
43. H. Verschelde and H. Verbeke, Nucl. Phys. **A495** (1989) 523.

BOUND SKYRMIONS: THE PROPERTIES AND PHYSICS CONSEQUENCES OF THEIR EXISTENCE

VLADIMIR B. KOPELIOVICH
Institute for Nuclear Research of Academy of Sciences, Moscow 117312, Russia.

ABSTRACT
The properties of bound states of skyrmions with baryon number $B \geq 2$ are discussed. These bound states possess definite topology, specific for every baryon number. Torus-like configurations are considered in more details. The increase of dimensions of solitons and their inertia with increasing baryon number leads to the decrease of zero modes quantum corrections to the energy and, as a result, to the additional stabilization of quantized baryonic systems. The whole soliton approach becomes more selfconsistent with increasing B. $SO(3)$-solitons with $B = 2$ and 4 are briefly discussed.

1. Introduction

The Skyrme model of baryons as solitons of chiral fields allows to describe mesons (pions) and baryons (nucleons) starting from one and the same Lagrangian. Essential advantage of the model and its extensions is that systems with different values of baryon number B can be considered on equal footing: in many cases of interest B is defined by the boundary condition on the profile function F. The first example of bound skyrmion is the $SO(3)$-soliton with $B = 2$ and strangeness $S = -2$, discovered in the Skyrme soliton model by A.P.Balachandran et al [1],[2]. In $SU(2)$ soliton models which are suitable for studies of nucleons and nuclei all advantages of this approach had not been realized until 1987.

The so called hedgehog configurations considered previously possess unpleasant property: their energy grows rapidly with increasing B, like $B(B+0.9)$ [3]. It is clear that the role of such configurations in the case of nucleon-nucleon interactions or in ordinary nuclei is negligible by this reason. On the other side, when the nucleon-nucleon interaction was considered in the model of two rigid $B = 1$ nondeformable hedgehogs it was too much repulsion at small distances [4],[5].

Further investigations have shown that, first, the hedgehog form of the $B = 1$ skyrmions is changed as a result of interaction between them [6],[7]. Second, it was shown that the change of the Ansatz allows to lower considerably the energy of the $B = 2$ system: it equals to only 2.14 of that of $B = 1$ skyrmion for the Ansatz with winding number $n = 2(B = n)$, i.e. for the Ansatz with chiral field rotating 2 times faster than radius-vector R [8], but described by one function of one variable R.

More general axially symmetrical Ansatz was investigated in [9] for $B = 2-5$ and independently for $B = 2$ in [10]- [14]. It was found that states with $B = 2-4$ are bound relative to the decay into B unit skyrmions [9]. The calculations performed in [10] were made in general 3-dimensional form, and it was not clear that the mass and baryon number distributions of the state revealed here have pronounced torus-like shape. This fact was established later in [11], [12]. More complicated Ansatz for $B = 4$ configuration was proposed in [15] which allowed to get the energy considerably lower than that of $B = 4$ torus.

A very important step was made in paper [16] where direct minimization of the energy functional for $B \leq 6$ was made in 3 dimensions in most general form. It was found that for every baryon number the configuration of minimal energy possesses specific topological properties: it has tetrahedral form for $B = 3$, cubic form for $B = 4$, etc. Each of these configurations can be obtained by means of fusion of several basic deformed torus-like distributions.

The discovery of bound skyrmions has shown that chiral fields can be packed in such a way that the mass of the state grows more slowly than the baryon number. It was clear that this opens new prospects for the application of soliton models in nuclear physics.

Further efforts have been made in different directions.

In a number of papers the consequences of the existence of bound states have been investigated in $SU(2)$-model for $B = 2$ and 3 configurations [14],[17],[18]. Nucleon-nucleon interactions also were studied [19]-[21], as well as nuclear matter (see also the talk by A.Wirzba at the present workshop).

The other direction was the study of the consequences of the $SU(3)$-extension of the model [22]-[26], [41]. The quantization of zero modes of $SU(2)$-solitons in $SU(3)$-collective coordinates space leads to the spectrum of dibaryons, tribaryons etc. with large values of strangeness, down to $S/B = -3$. The additional binding of quantized states comes from the quantum corrections to their energy. The fact that quantum corrections have the property to stabilize systems is well known, its nature is the suppression of quantum fluctuations of $B = 1$ skyrmions when they are united into bigger soliton. The stabilization due to quantum correction associated with isorotations was discussed in [27]. The effect due to rotation into "strange" direction proportional to $1/\Theta_s$, Θ_s being strange, or kaonic inertia is numerically greater. There is a remarkable property of this correction: it stabilizes the system the stronger the greater its baryon number is [26].

After discussion of the results obtained in above directions some comments will be made also on $SO(3)$-solitons, dihyperon and tetrahyperon, which are probably the lowest in energy states with $S/B = -1$.

2. Lagrangian and Ansatz

Here I draw the Lagrangian of the model in the modified form which allows to describe the mass splittings of the octet and decuplet of baryons and gives the value of the axial constant g_A close to the experimental value [28]-[31]. The characteristic feature of this variant of the model is the presence of considerable $SU(3)$- and chiral symmetry breaking terms and a relatively small value of the constant $e = 3.46$ (in comparison with the $SU(2)$-value $e = 4.84$ [32] or $e = 2\pi$ which follows from QCD [33],[34].

$$
\begin{aligned}
L &= L^{(2)} + L^{(4A)} + L^{(WZ)} + L^{(SB)} \quad , \\
L^{(2)} &= -\frac{F_\pi^2}{16} \int d^3r \quad tr\ell_\mu \ell^\mu \quad , \qquad \ell_\mu = U^\dagger \partial_\mu U \quad , \\
L^{(4A)} &= \frac{1}{32e^2} \int d^3r \quad tr[\ell_\mu, \ell_\nu][\ell^\mu, \ell^\nu] \quad , \qquad U \in SU(3) \quad , \\
L^{(SB)} &= \frac{F_\pi^2 m_\pi^2}{16} \int d^3r \quad tr(U + U^\dagger - 2) \\
&- \frac{F_K^2 - F_\pi^2}{48} \int d^3r \quad tr(1 - \sqrt{3}\lambda_8)(U\ell_\mu \ell^\mu + \ell_\mu \ell^\mu U^\dagger) \\
&+ \frac{F_K^2 m_K^2 - F_\pi^2 m_\pi^2}{24} \int d^3r \quad tr(1 - \sqrt{3}\lambda_8)(U + U^\dagger - 2)
\end{aligned}
\tag{1}
$$

The Wess-Zumino-Witten action

$$
\int L^{(WZW)} dt = \frac{iN_c}{240\pi^2} \int_Q d\Sigma^{\mu\nu\lambda\rho\sigma} tr\ell_\mu \ell_\nu \ell_\lambda \ell_\rho \ell_\sigma \tag{2}
$$

defines the topological properties of the model and is important for the quantization of the soliton. It makes also some contribution to the observables, e.g. to the axial constant g_A [29].

$L^{(SB)}$ is responsible for all chiral and flavor symmetry breaking present in the mesonic sector and, for higher winding numbers, also in the baryonic sector. This form of the symmetry breaking terms is dictated by requirement that they transform as quark mass terms present in the underlying QCD-lagrangian. For $F_K = F_\pi$ we get from (1) the lagrangian of original Skyrme model with the mass term [32]. The generalized axially symmetrical Ansatz in $SU(2)$ is defined by

$$
U_T(\vec{r}) = c_F + is_F \vec{n}\vec{\tau} \tag{3}
$$

with Pauli matrices $\tau_i, i = 1, 2, 3$ and a unit vector $\vec{n}$,

$$
\begin{aligned}
&n_z = c_\alpha, \quad n_x = s_\alpha c_{n\varphi}, \quad n_y = s_\alpha s_{n\varphi}, \\
&s_\chi = \sin\chi, \quad c_\chi = \cos\chi, \quad s_\alpha = \sin\alpha.
\end{aligned}
\tag{4}
$$

$F(r,z), \alpha(r,z)$ are the profile and angular functions defining the field configuration in the soliton, they depend on 2 variables, r and z. ϕ is the azimuthal angle (n is integer). The total baryon (winding) number of the system

$$
B = n\frac{F(0,0) - F(\infty)}{\pi} \tag{5}
$$

The torus-like configuration of minimal energy is obtained at $F(0,0)-F(\infty)=\pi$. One takes usually for simplicity $F(\infty)=0$. Note, that for $n=1$ and $\alpha=\arctan(r/z)$ the so called hedgehog configuration is obtained which is believed to possess minimal energy in the $B=1$ sector. For $n=2$ the axially symmetrical configuration gives the minimal known till now energy of the system. As it was shown in [35] it is at least the local minimum. But there is no known proof that the the axially symmetrical configuration for $B=2$ gives the absolute minimum of the energy.

It should be noted that most general Ansatz without any symmetry constraints is described by 3 functions of 3 variables: $F(x,y,z), n_x, n_y, n_z(x,y,z); n_x^2+n_y^2+n_z^2=1$. This Ansatz was used in [19].

The extension to $SU(3)$ which I also shall discuss here is made by embedding of $SU(2)$ Ansatz (3) into $SU(3)$-collective coordinate space:

$$U(\mathbf{r},\mathbf{t}) = A(t)\begin{pmatrix} U_T(R(t)\cdot\mathbf{r}) & 0 \\ 0 & 1 \end{pmatrix} A^\dagger(t). \tag{6}$$

The $SU(3)$ rotation matrix depends generally on 8 rotation angles and may be chosen in the form

$$A = I(\alpha,\beta,\gamma)e^{-i\nu\lambda_4}I(\alpha',\beta',\gamma')e^{-i(\rho/\sqrt{3})\lambda_8} \tag{7}$$

The matrices $I(\alpha,\beta,\gamma)$ here denote rotations in the $SU(2)$ (isospin) subgroup of $SU(3)$.

The Lagrangian (1) can be rewritten in terms of the static characteristics of the system, its mass M and inertia relative to isotopic rotations (Θ_N,Θ_3), spatial rotations (Θ_J and also Θ_3), inertia relative to the rotation into "strange" direction (strange inertia) Θ_S. In the most general case also the interference inertia Θ_i enters (see below). So, in the most general case the configuration is characterized by 5 independent moments of inertia. In terms of Θ-s and angular velocities of rotation defined in the following way:

$$\begin{aligned} A^\dagger\dot{A} &= \frac{i}{2}\sum_{a=1}^{8}\lambda_a\Omega_a \\ (\dot{R}R^\dagger)_{ik} &= \sum_{l=1}^{3}\epsilon_{ikl}\omega_l \end{aligned} \tag{8}$$

the Lagrangian takes the form:

$$\begin{aligned} L &= -M(\nu)+\frac{1}{2}\Theta_N(\nu)\sum_{p=1}^{2}\Omega_p^2+\frac{1}{2}\Theta_J(\nu)\sum_{p=1}^{2}\omega_p^2 \\ &\quad +\frac{1}{2}\Theta_3(\nu)(\Omega_3+n\omega_3)^2+\Theta_i(\nu)\sum_{p=1}^{2}\Omega_p\omega_p+\frac{1}{2}\Theta_S(\nu)\sum_{k=4}^{7}\Omega_k^2 \\ &\quad -\frac{N_c}{2\sqrt{3}}\Omega_8. \end{aligned} \tag{9}$$

3. Static characteristics of bound states

The static mass of the solitonic configurations described by Ansatz (3), (4) for the fixed angle ν of rotation into "strange" direction equals to:

$$\begin{aligned} M(\nu) &= \int d^3r \left[\frac{F_\pi^2}{8}((F,F) + s_F^2((\alpha,\alpha) + \frac{n^2}{r^2}s_\alpha^2)) \right. \\ &+ \frac{s_F^2}{2e^2}([F,\alpha]^2 + \frac{n^2 s_\alpha^2}{r^2}((F,F) + s_F^2(\alpha,\alpha))) + m_\pi^2 F_\pi^2(1-c_F)/4 \\ &+ \sin^2\nu \left(\frac{1}{8}(F_K^2 - F_\pi^2)((F,F) + s_F^2(\alpha,\alpha) + \frac{n^2}{r^2}s_F^2 s_\alpha^2)c_F \right. \\ &\left.\left. + (m_K^2 F_K^2 - m_\pi^2 F_\pi^2)(1-c_F)/4\right)\right], \qquad (10) \\ (f,f) &= (\frac{\partial f}{\partial r})^2 + (\frac{\partial f}{\partial z})^2, \quad [f,g] = \frac{\partial f}{\partial r}\frac{\partial g}{\partial z} - \frac{\partial g}{\partial r}\frac{\partial f}{\partial z} \end{aligned}$$

and $s_F = \sin F, c_F = \cos F$.

This mass does not depend on other rotation angles in (6) for the $SU(2)$ solitons embedded into $SU(3)$. The terms in (10) proportional to n^2 , are damped by readjusting of profile F and α when n is not large. But they dominate at large n and make the state unbound relative to the decay into n unit skyrmions already at $n = 5$ [9].

One should use the most general Ansatz to find the configuration with minimal energy . For this Ansatz M takes the form ($SU(2)$-case):

$$M = \int d^3r[\frac{F_\pi^2}{8}((d_iF)^2 + s_F^2((d_i\alpha)^2 + s_\alpha^2(d_i\beta)^2))$$

$$\frac{s_F^2}{2e^2}([\vec{d}F \times \vec{d}\alpha]^2 + s_\alpha^2[\vec{d}F \times \vec{d}\beta]^2 + s_F^2 s_\alpha^2[\vec{d}\alpha \times \vec{d}\beta]^2) + m_\pi^2 F_\pi^2(1-c_F)/4] \qquad (11)$$

where $n_x = s_\alpha c_\beta, n_y = s_\alpha s_\beta$.

The calculations made in [14] have shown that really configurations found by means of direct minimization are considerably lower in energy than torus-like configurations, for $B = 3 - 6$ [14], especially for $B = 4$ (see Table 1). The inertia also depend on the angle ν only:

$$\begin{aligned} \Theta_N &= \frac{1}{2}\int d^3r \left[\frac{1}{4}F_\pi^2(\nu)(1+c_\alpha^2) + \right. \\ &\left. + \frac{1}{e^2}((F,F)(1+c_\alpha^2) + s_F^2 c_F^2(\alpha,\alpha) + \frac{n^2}{r^2}s_F^2 s_\alpha^2)\right], \\ \Theta_S &= \frac{1}{8}\int d^3r(1-c_F)\left[F_\pi^2\frac{2-c_F}{2}\sin^2\nu + F_K^2(1 - \frac{2-c_F}{2}\sin^2\nu) + \right. \\ &\left. + \frac{1}{e^2}((F,F) + s_F^2(\alpha,\alpha) + \frac{n^2}{r^2}s_F^2 s_\alpha^2)\right], \end{aligned}$$

$$
\begin{aligned}
\Theta_3 &= \int d^3r \left[\frac{1}{4}F_\pi^2(\nu) + \frac{1}{e^2}((F,F) + s_F^2(\alpha,\alpha))\right] s_F^2 s_\alpha^2 \\
\Theta_J &= \frac{1}{2}\int d^3r \left[\frac{1}{4}F_\pi^2(\nu)([\vec{r},F]^2 + s_F^2[\vec{r},\alpha]^2 + n^2\frac{z^2}{r^2}s_F^2 s_\alpha^2) + \right. \\
&\left. + \frac{s_F^2}{e^2}((z^2+r^2)\,[F,\alpha]^2 + \frac{n^2}{r^2}s_\alpha^2([\vec{r},F]^2 + s_F^2[\vec{r},\alpha]^2 + z^2((F,F) + s_F^2(\alpha,\alpha))))\right] \\
2\Theta_i &= \delta_{n1}\int d^3r s_F^2 \left[\frac{F_\pi^2(\nu)}{4}([\vec{r},\alpha] + \frac{z}{r}c_\alpha s_\alpha) + \frac{1}{e^2}(\frac{s_F^2 s_\alpha^2}{r^2}[\vec{r},\alpha] - \right. \\
&\left. -[F,\alpha](zF' + r\dot{F}) - \frac{z}{r}c_\alpha s_\alpha((F,F) + s_F^2(\alpha,\alpha)))\right]; \qquad (12)
\end{aligned}
$$

$$
F_\pi^2(\nu) = F_\pi^2 + c_F\sin^2(\nu)(F_K^2 - F_\pi^2), \quad [\vec{r},F] = z\frac{\partial F}{\partial r} - r\frac{\partial F}{\partial z}.
$$

It can be easily checked that for the case of hedgehog the inertia $\Theta_N, \Theta_3, \Theta_J$ and Θ_i are equal to each other and to the only inertia Θ which characterizes the $SU(2)$-hedgehog. The rotation energy takes the simple form

$$
E_{rot} = \frac{\Theta}{2}(\vec{\omega} + \vec{\Omega})^2 + \frac{\Theta_S}{2}\sum_{\alpha=4}^{7}\Omega_\alpha^2 \qquad (13)
$$

what is expected since isotopical and spatial rotations are equivalent for hedgehog.

The expressions (12) are valid for any configuration which possesses the axial symmetry, in particular for the case of two (or more) equally oriented interacting skyrmions [7]. In this case $n = 1$ and profile $F = \pi$ in two (or more) points on the z-axis. As it was shown in [36] for two skyrmions on the large distance between their centers $2R$ the orbital inertia Θ_J goes over into $2M_1R^2$, where M_1 is the mass of unit skyrmion. This rather natural value can be obtained after applying Derrick theorem. The interference moment $\Theta_i = 0$ when $R \to \infty$. For the case $n \geq 2$ the interference inertia $\Theta_i = 0$ after integration over azimuthal angle ϕ i.e. rotations in iso- and usual space (around x, y -axes) decouple and rotation energy depends on 4 inertia only. Some static characteristics of bound skyrmions with $B = 2, 3, 4$ are shown in Table 1. For $B = 3$ and 4 solitons with topology of tetrahedron and cube only few quantities are drawn calculated in [14],[18]. This computation was made on finite 3-dimensional lattice, as a result, baryon number is smaller by 3.5% than the true value, and other quantities are also smaller by several percents. A nice feature of $B = 3$ state is that its tensors of inertia a proportional to the unit matrix. Probably, the $B = 4$ configuration which has the form of a cube possesses the same property. The important feature is monotonic increase with B of all quantities shown in Table 1. R_B^2, Θ_S increase almost proportional to B, $\Theta_J \sim B^2$. The Σ-term also increases almost linearly with B for torus-like configurations.

The numbers in Table 1 are given in natural units, however, there is no scaling because of the presence of the mass term in the Lagrangian. Recent calculations have shown that in the chiral limit, $m_\pi = 0$,$M_2 = (69.9 \pm 0.1)F_\pi/e$.

Some configurations described by relatively simple Ansaetze have been considered in [15],[37] which are intermediate between torus-like and true one for $B \geq 3$. E.g. in [15] double-torus configuration for $B = 4$ was proposed. For this configuration polar angle α also rotates twice faster then radius-vector. These configurations can be used for approximate description of spectra of few-baryon systems.

Table 1. The masses of configurations with $B = 2 - 4$ —in units F_π/e, R_B^2 — in $1/(F_\pi e)^2$; inertia and Σ-term — in $1/(F_\pi e^3)$. For $B = 3, 4$ the upper line contains numbers for torus-like configurations, the lower line - for solitons of more complicated topology [14]. All numbers are obtained at $F_\pi = 108$ MeV, $e = 4.84$.

B	M	R_B^2	Q_B/R_B^2	Θ_N	Θ_J	Θ_3	Θ_S	Σ
2	74.2	6.20	0.095	129	201	88	46.4	191
	73.4	5.2						
3	113.1	9.52	0.137	206	464	111	76	315
	107.4	6.9		136	435			
4	154.5	13.5	0.160	290	888	131	108	460
	140.5	8.2						

The fact of existence of torus-like bound skyrmion in the $B = 2$ sector has been checked in several different variants of the model:
a) in the model with the Skyrme term replaced by baryon number density squared [9].
b) in the model with an explicit scalar meson modifying one of the gluon degrees of freedom [26].
c) in the model with vector meson (ω-meson) stabilized soliton [20].

In view of this the existence of bound skyrmions seems to be rather general property of chiral soliton model which does not depend on its particular variant.

4. Quantization of bound skyrmions ($SU(2)$- and $SU(3)$- cases)

The physical results can be obtain after ascribing to baryonic systems definite quantum numbers. This can be made by means of quantization procedure in the collective coordinates space, according to procedure performed first in [39] and for $B = 2$ systems in [41],[9],[14]. In comparison with hedgehog case, for few-baryon systems one should use 2 or more sets of collective coordinates.

The imposing of canonical commutation relations between collective coordinates a_i in $SU(2)$-collective coordinates space and conjugate momenta $\pi_i = \partial L/\partial \dot{a}_i$ is equivalent to definition that corresponding (body-fixed) momenta are equal to $J_i = \partial E_{rot}/\partial \omega_i$ and $N_i = \partial E_{rot}/\partial \Omega_i$. After quantization of J_i and N_i the Hamiltonian of the system is obtained as a function of J_i and N_i .

In $SU(3)$ -space the differential operators R_α are defined by relation $AR_\alpha A^\dagger = A\gamma_\alpha A^\dagger$. It follows then that $[R_\alpha, R_\beta] = -if_{\alpha\beta\gamma}R_\gamma$ The explicit form of operators $R_\alpha, (\alpha = 1, ...8)$ was given in [29].

The Hamiltonian of the system equals to

$$H = M(\nu) + \frac{1}{2}\{\frac{1}{2\Theta_S(\nu)}, C_2\,[SU(3)]\,\} + \frac{1}{2\Theta_J(\nu)}J(J+1)$$
$$+ (\frac{1}{2\Theta_N(\nu)} - \frac{1}{2\Theta_S(\nu)})N(N+1)$$
$$+ (\frac{1}{2\Theta_3(\nu)} - \frac{1}{2\Theta_N(\nu)} - \frac{n^2}{2\Theta_J(\nu)})L^2 - \frac{3}{8\Theta_S(\nu)}B^2 \tag{14}$$

Here $C_2(SU(3)) = \sum_{i=1}^{8} R_\alpha^2$ is the Casimir operator of $SU(3)$. For definite $SU(3)$-representation (p,q) $C_2(SU(3)) = \frac{1}{3}(p^2+q^2+pq)+p+q$. $N = p_0/2$ is the right isospin, where (p_0, q_0) is the representation of minimal triality present in the mixture. nL is 3-d projection of body fixed angular momentum J, L defines also the parity of the state, $P = (-1)^L$. The relation $J_3 = nL$ follows from generalized axial symmetry of configuration.

For $n \geq 2$ $J_{x,y}$ are defined by spatial rotations only, for $n = 1$ there is contribution to $J_{x,y}$ due to isorotations, and, vice versa, spatial rotations contribute to $N_{x,y}$ (or body-fixed isospin in $SU(2)$-case):

$$J_{x,y} = \Theta_J \omega_{x,y} + \Theta_i \Omega_{x,y}$$
$$N_{x,y} = \Theta_N \Omega_{x,y} + \Theta_i \omega_{x,y} \tag{15}$$

The case $n = 1$ was investigated in [36], but some questions remain to be cleared up.

For ν-dependent inertia Θ_s and $C_2(SU(3))$ do not commute and there is an ordering ambiguity in the second term of (14). In [41] we have checked two simplest variants for this term,

$$\frac{1}{2}\{\frac{1}{\Theta_S}, C_2(SU(3))\} \tag{16}$$

and

$$\frac{1}{\sqrt{\Theta_S}}C_2(SU(3))\frac{1}{\sqrt{\Theta_S}} \tag{17}$$

and have found that the results do not differ considerably.

5. $SU(2)$-quantized baryonic systems

Let us consider first the consequences of the existence of bound skyrmions in $SU(2)$-variant of the model.

The Hamiltonian of quantized system can be obtained from (13) setting $\Theta_S \to \infty$ and $N = T$-isospin operator of the system [36], [14], [9]

$$H = M_{cl} + \frac{T(T+1)}{2\Theta_T} + \frac{J(J+1)}{2\Theta_J} + \frac{J_3^2}{2}(\frac{1}{n^2\Theta_3} - \frac{1}{n^2\Theta_T} - \frac{1}{\Theta_J}) \tag{18}$$

Because of constraint on body-fixed components $J_3 = nT_3$ the last term in (18) does not contribute in both cases of main interest: $T = 0, J = 1$ (deuteron-like system) and $T = 1, J = 0$ (singlet scattering state). It follows immediately from the mass formula (18) that the energy of deuteron-like system is smaller than that of S_0-state by amount $\frac{3}{8}(\frac{1}{\Theta_T} - \frac{1}{\Theta_J})$ since $\Theta_T < \Theta_J$. It is in remarkable qualitative agreement with the fact that there exists deuteron in the nature but not dineutron. This difference, $\cong 30$ MeV, is too large in comparison with experimental value of deuteron binding energy, and this is another argument against straightforward identification of bound skyrmions with nuclei.

In Table 2 some characteristics of the nucleon and lightest nuclei (^{2}D, ^{3}He, ^{3}H) are shown in comparison with theoretical results obtained in papers [14],[18], [23]. These results should not be taken too seriously but can be useful for orientation.

Table 2. Mass M, (in MeV), magnetic moment μ (in nuclear magnetons), electric radius r_e and magnetic radius r_m (in fm) for nucleon and lightest nuclei. Data (e) in comparison with theoretical results (t) taken from [14],[18], [32]. F_π = 108 MeV, e = 4.84. For the nucleon N magnetic moments are given for the proton and neutron separately, r_e and r_m are isoscalar radii.

	$M(t)$	$M(e)$	$\mu(t)$	$\mu(e)$	$r_e(t)$	$r_e(e)$	$r_m(t)$	$r_m(e)$
N	939	939	1.97	2.79	0.68	0.72	0.95	0.81
			-1.24	-1.91				
^2D	1717	1876	0.74	0.857	0.94	2.095		
^3He	2463	2809	-1.92	-2.13	1.02	1.97	0.88	1.99
^3H	2463	2809	2.72	2.98	0.87	1.63	0.89	1.72

The axial constant for ^{3}He - ^{3}H transition obtained in [18], $g_A = -0.58$, is about twice smaller than experimental value, $g_A = -1.21$, quite similar to the nucleon case. However, one can obtain greater value of g_A by changing the parameter e. The price to be paid is greater masses of bound skyrmions.

The spectrum of dibaryons can be easily obtained from the mass formula (18). The most interesting is dibaryon with isospin $T = 2$ because it cannot decay strongly into two nucleons. Its energy, 2.01 Gev, differs from the energy of lowest states greater than by one pion mass [14] and it can decay into pion and two nucleons.

The important problem to be solved in any model is, of course, the description of nucleon-nucleon interaction (NN-potential). Outside product approximation first attempts were made for equally oriented skyrmions in [6],[7]. In [19] it was shown that torus-like $B = 2$ bound skyrmion can be separated into two separate unit skyrmions in most profitable orientation. This adiabatic process can be described only by the most general Ansatz (see expression (11)). The scattering plane coincides with the plane of the torus formed in intermediate state.

Recently the attempts to get NN-potential in $SU(2)$-Skyrme model have been made in papers [20],[21]. It was shown in [20] that central, tensor and spin-spin NN-potential can be reproduced in qualitative agreement with phenomenological Paris potential. A delicate question is the definition of distance between two interacting skyrmions. In [20] this distance was defined as a mean square radius of baryon number distribution.

The direct calculation of NN-scattering phase should give results which do not depend on the definition of the distance between interacting and deforming skyrmions. This is 4-dimensional problem, and it is a challenging task to solve it.

The two-nucleon dynamics in the Skyrme model is considered also by N.Snyderman and T.Walhout at the present workshop.

6. Baryonic systems with unusual properties

The natural thing is to quantize the $SU(2)$-bound skyrmions in $SU(3)$-collective coordinates space and to obtain the spectrum of dibaryons, tribaryons etc. with large values of strangeness. As it was shown in [24],[22], this procedure allows to obtain states with the ratio S/B down to -3. All such states may be interpreted in the framework of quark models as consisting of minimal amount of quarks, $3B$.

The important peculiarity of $SU(3)$-quantization procedure is that it leads to additional binding of all states relative to $SU(2)$-case.

The coefficient before $1/(2\Theta_S)$ in formula (14) equals for representation (p,q), ($N = p/2$, $Y' = B = (p+2q)/3$:

$$C_2(SU(3)) - N(N+1) - 3Y'^2/4 = (p+2q)/2 = 3B/2 \tag{19}$$

It does not depend on particular value of p and is the same for the whole family of "minimal" $SU(3)$-multiplets which possess the only property: the right hypercharge Y' equal to $N_cB/3$ due to quantization condition [40] coincides with the maximal hypercharge of the multiplet (triality) $Y_{max} = (p+2q)/3$. For $B=1$ this family consists of the octet $(1,1)$ and decuplet $(3,0)$; for $B=2$ - of $\bar{10}(0,3), 27(2,2), 35(4,1), 28(6,0)$ - multiplets. The total number of such representations equals to $[3B/2]+1$. This nice property of cancellation between Casimir operators of $SU(3)$ - and $SU(2)$-groups was known previously for $B=1$ [2], since coefficients before $1/\Theta_S$ coincide for the nucleon and isobar $\Delta(1232)$.

In flavor symmetric case ($\sin\nu = 0$) the contribution to the binding energy per unit baryon number connected with strange inertia equals to:

$$\varepsilon = \frac{3}{4}\left(\frac{1}{\Theta_S(B=1)} - \frac{1}{\Theta_S(B)}\right) \tag{20}$$

and increases with increasing baryon number [26].

In the case $\sin\nu = 0$ the eigenfunctions of rotational Hamiltonian coincide with the product of $SU(2)$- and $SU(3)$-rotational functions:

$$\begin{aligned}\Psi^{NL}_{YTT_3,JJ_3}(A,R) &= \sum_{M_L M_R} D^{(T)^*}_{T_3 M_L}(\alpha,\beta,\gamma) d^{(p,q)}_{TY\,M_L\,N\,2M_R}(\nu) \\ &\quad D^{(N)^*}_{M_R L}(\alpha',\beta',\gamma') e^{iY_R\rho} D^{J*}_{T_3\,-nL}(R)\end{aligned} \tag{21}$$

The right lower indices in two last D-functions differ by the factor $-n$ because of axial symmetry of basic soliton configuration.

In the case of symmetry breaking ($\sin\nu \neq 0$) the sum over different $SU(3)$-representations (p,q) enters into (21) instead of single $SU(3)$-rotation function $d^{(p,q)}$ with fixed p and q. All possible multiplets enter in the mixture which contain isomultiplets (I,Y) and (N,Y_R), $N = p_0/2$, p_0, q_0 is the representation of minimal triality present in the mixture.

The eigenfunctions of Hamiltonian (14) have been found, similar to the approach of Yabu and Ando [42], in the slow rotation approximation proposed in [31]. In this approximation the ν- dependent inertia in (14) are represented as polynomials of $\sin^2\nu$ or $D^{(1,1)}_{0,0}$. It corresponds to the picture that for every value of ν there is time enough for mesonic excitations inside the soliton to change its form according to the forces induced by the symmetry breaking terms in the Lagrangian. Simple estimates show that this assumption is probably better than rigid rotator approximation when it is supposed that soliton has the same shape and dimensions at all values of Euler angles in (21) [39].

Calculations performed in the $B = 1$ sector [31] have shown however that results obtained in both approximations - rigid rotator and slow rotator - do not differ very much. Slow rotator approach gave the possibility to describe mass splittings of the octet and decuplet of baryons with good accuracy with the value of parameters of the model F_π = 186 Mev, F_K = 236 Mev (a bit higher than empirical value 226 Mev), $e = 3.46$, which is fixed by $\Delta - N$ -splitting. Since this splitting is roughly proportional to $F_\pi e^3$, larger value of F_π demands smaller value of e in comparison with $SU(2)$-values $F_\pi = 108$ Mev, $e = 4.84$.

Some results for the most bound states with different values of strangeness S are shown below.

Table 3. The mass M (in GeV), binding energy ϵ, strangeness content $\langle\sin^2\nu\rangle/2$ and probability w of the basic representation (p,q) in the mixture for dibaryon states with maximal binding for a given strangeness S.

S,T (p,q)	0,1 (2,2)	-1,$\frac{1}{2}$ (0,3)	-2,1 (0,3)	-3,$\frac{1}{2}$ (2,2)	-4,1 (2,2)	-5,$\frac{1}{2}$ (4,1)	-6,0 (6,0)
M	4.72	4.92	5.2	5.41	5.63	5.87	6.21
ϵ	0.82	0.79	0.72	0.68	0.67	0.78	0.79
$\frac{1}{2}\langle\sin^2\nu\rangle$	0.085	0.150	0.240	0.295	0.390	0.415	0.425
w	0.70	0.75	0.79	0.88	0.94	0.97	0.98

The characteristic features of the quantities presented in Table 3 are:
the large values of binding energy calculated in the model, $0.7 - 0.8$ Gev which do not variate very much for different S;
monotonic decrease of admixture probability $1 - w$ with increasing $-S$. States with $S = -4, -5, -6$ are almost pure;
monotonic increase of strangeness content (s.c.) $\langle\sin^2\nu\rangle/2$ which is very natural (note that s.c. cannot exceed 0.5). S.c. of the mixture is always lower than s.c. of pure state

defined by some combination of $SU(3)$ Klebsh-Gordon coefficients. This difference is especially large (a factor ~ 2) for nonstrange components of multiplets.

According to the concept developed recently the soliton model can define mass splittings inside the multiplets of baryonic systems, not the absolute values of masses [29], [30], [31].

Some mechanism - the Casimir energy estimated in [43] is probably this mechanism - must shift the energy of the $T = 1, J = 0$ NN-scattering state to a position at the nucleon-nucleon threshold. If so, we should subtract the value of binding energy of the singlet state, 820 Mev, from the value of ϵ in Table 3. The immediate result of this procedure is that all strange dibaryon states become unbound. The accuracy of calculation is not better than 30 Mev, since the binding energy of the deuteron equals to this value instead of the observed 2.23 Mev.

Qualitatively the same result can be obtained also in the original version of the Skyrme model -in rigid rotator approximation and without mixing, for dibaryons [25] and tribaryons [23].

The spectrum of states with additional quark-antiquark pairs (nonminimal states according to our terminology) can be easily obtained from (14) [41]. In this case the right isospin $N = (p + m)/2$ and the contribution to the energy of the state is proportional to m which can be interpreted as a number of additional quark-antiquark pairs:
$\delta E = m((3B/2 + 1 + m - N)/(2\Theta_S) + (2N + 1 - m)/(2\Theta_N))$. Simple estimate allows to conclude that all such states have energy enough for the decay into m kaons and corresponding baryonic system in the final state.

7. $SO(3)$-solitons

As it was mentioned already, the soliton with lowest possible baryon number, $B = 2$, on $SO(3)$-subgroup of $SU(3)$ is chronologically the first bound skyrmion discovered theoretically [1]. It was rather straightforward and simple idea to try to generalize the $SO(3)$-Ansatz [1], similar to $SU(2)$-case. The unit vector $\vec{n}$ defining the direction of the field in $SO(3)$-space was written as in (4). The energy of the static configuration is then the function of two profiles, F and G, and angular function α, all are function of cylindrical coordinates (r, z); and also winding number n. Baryon number for configurations of minimal energy $B = 2n$. The functional for the mass has the form [44]:

$$\begin{aligned}
M &= \frac{1}{8}\int d^3\vec{r}\Big[F_\pi^2[(F,F) + \frac{1}{3}(G,G) + (1 - c_F c_G)(2(\alpha,\alpha) + \frac{n^2}{r^2}(1 - c_{2\alpha}))] \\
&+\frac{1}{e^2}\big[2(1 - c_F c_G)([F,\alpha]^2 + [G,\alpha]^2)4 s_F s_G[F,\alpha][G,\alpha] \\
&+(n^2/r^2)(1 - c_{2\alpha})\,((1 - c_F c_G)((F,F) + (G,G))2 s_F s_G(F,G) \\
&+2(\alpha,\alpha)((1 - c_F c_G)^2 + 3 s_F^2 s_G^2))\big] + F_\pi^2\mu^2(3 - c_{2G/3} - 2 c_F c_{G/3})\Big], \\
\mu^2 &= (m_\pi^2 + m_\eta^2)/2 \qquad (22)
\end{aligned}$$

For $n = 1, \alpha = \arctan(r/z)$ the $SO(3)$-hedgehog is obtained.

It was found that in the chiral limit the state with $n = 2$, $B = 4$, tetrahyperon, is bound relative to the decay into two $B = 2$ skyrmions, or H-particles, the binding energy $\epsilon = 2.5F_\pi/e$, [44]. The introduction of mass terms makes the binding energy smaller. For $F_\pi = 108MeV, e = 4.84$ it was obtained $\epsilon \simeq 0$.

In view of the fact that masses of torus with $B = 2$ and $SO(3)$-soliton with $B = 2$ are very close to each other in chiral limit the question has been raised recently by B.Schwesinger that probably these states are degenerate and can be transformed one to another by some transformation. To check this, calculations were performed for both states in the vicinity of chiral limit with accuracy exceeding that of previous calculations. The results for $\mu^2 = 0.5m_\pi^2$ and greater are shown in Table 4.

Table 4. Masses and radii of baryon number distribution for the H-particle ($SO(3)$-dibaryon) and torus-like biskyrmion near chiral limit. M_H and r_H^2 are calculated by B.Schwesinger. $F_\pi = 186MeV, e = 3.46$; M_T, M_H - in MeV, $8.44 = (m_\pi^2 + m_\eta^2)/(2m_\pi^2)$. r^2 - in fm^2.

μ^2/m_π^2	0.5	1.0	1.5	3.0	8.44
M_H	3868.2	3933.3	3989.4	4112.2	4454.2
M_T	3849.3	3925.3	3987.9	4140	4519
r_H^2	0.70	0.54	0.40	0.33	0.19
r_T^2	0.676	0.613	0.575	0.493	0.493

The result obtained in chiral limit is $M_H = 70.35F_\pi/e, M_T = (69.9 \pm 0.1)F_\pi/e$. So, the H-particle and torus are different objects. Especially big difference is in behaviour of radii. This is connected probably with the fact that for $SO(3)$-solitons two profile functions, F and G, contribute into the mass term.

8. Summary

Several lessons can be drawn from the results obtained till now in framework of chiral soliton models.

First, the $SU(2)$-solitons with $B \geq 2$ should not be interpreted directly as corresponding nuclei (deuteron, ^3He, ^3H). As a rule, skyrmions have considerably smaller dimensions than real nuclei, even when F_π and e fit the masses of nucleon and Δ-isobar.

On the other hand, we may wonder that many results obtained in the simplest variant of the model are in qualitative agreement with data, especially for magnetic moments. For $B = 2, 3$ they agree with data better than for the nucleons. The general impression is that soliton approach works better with increasing baryon number. The theoretical argument is that quantum corrections to the energy (per unit baryon number) of the state decrease with increasing B.

Bound skyrmions should be considered as some "sceleton" configurations of fields. Quantum corrections to physical observables due to nonzero modes should be taken into account. Probably, this will produce results in better agreement with nature.

It is clear that fine tuning of the model will be necessary to describe data with good accuracy, e.g. the binding energy of the deuteron.

As for exotic states with high value of strangeness they are predicted to be bound in all known variants of the model. The large value of binding energy (relative binding $\epsilon/\Sigma M \simeq 0.2$) is connected with rather natural property of solitons: the increase with B of their dimensions and inertia.

However, after the procedure of subtraction (renormalization) which is necessary to produce the energy of NN-scattering state on the right place all strange dibaryon states become unbound. It will be difficult to observe them if they are far above threshold. In view of uncertainties of the model there is a chance that some of states are bound, probably, states with $S = -5, -6$. In any case, the flavor dependence, or independence of such contributions as Casimir energy should be investigated carefully.

Further studies of baryonic systems (nuclei) in the framework of chiral soliton models seem to be very promising.

9. Acknowledgement

I am indebted for useful discussions to G.Holzwarth, B.Schwesinger, H.Walliser and A.Wirzba. The program for numerical calculations performed in several papers which the present talk is based upon was written by B.E.Stern to whom I am thankful especially.

10. References

1. A. P. Balachandran, A. Barducci, F. Lizzi, V. G. J. Rodgers and A. Stern, *Phys. Rev. Lett.* **52** (1984) 887.
2. A. P. Balachandran, et al, *Nucl. Phys.* **B256** (1985) 525.
3. E. B. Bogomolny and V. A. Fateev, *Yad. Fiz.* **37** (1983) 228.
4. A. Jackson, A. D. Jackson and V. Pasquier, *Nucl. Phys.* **A432** (1985) 367.
5. R. Vinh Mau, M. Lacombe, B. Loiseau, W. N. Cottingham and P. Lisboa, *Phys. Lett.* **150B** (1985) 259.
6. H. H. Sommermann, H. W. Wyld and C. J. Pethick, *Phys. Rev. Lett.* **55** (1985) 476.
7. S. V. Zenkin, V. B. Kopeliovich and B. E. Stern *Yad. Fiz.* **45** (1987) 165.
8. H. Weigel, B. Schwesinger and G. Holzwarth, *Phys. Lett.* **168B** (1986) 321.
9. V. B. Kopeliovich and B. E. Stern, *Pis'ma v ZhETF* **45** (1987) 165.
10. J. J. M. Verbaarschot, T. S. Walhout, J. Wambach and H. W. Wyld, *Nucl. Phys.* **A468** (1987) 520.
11. N. S. Manton, *Phys. Lett.* **192B** (1987) 177.
12. J. J. M. Verbaarschot, *Phys. Lett.* **195B** (1987) 235.
13. A. J. Schramm, Y. Dothan and L. C. Biedenharn, *Phys. Lett.* **205B** (1988) 151.
14. E. Braaten and L. Carson, *Phys. Rev.* **D38** (1988) 3525.
15. E. Sorace and M. Tarlini, *Phys. Lett.* **232B** (1989) 154.

16. E. Braaten, L. Carson and S. Townsend, *Phys. Lett.* **235B** (1990) 147.
17. L. Carson *Phys. Rev. Lett.* **66** (1991) 1406.
18. L. Carson *Nucl.Phys.* **A535** (1991) 479.
19. B. E. Stern, *Yad.Fiz.* **49** (1989) 530.
20. T. S. Walhout and J. Wambach, *Phys. Rev. Lett.* **67** (1991) 314; Preprint *KFA-IKP(TH)-1992-15*
21. V. Thorson and I. Zahed, *Phys. Rev.* **D45** (1992) 965.
22. J. Kunz and P. J. Mulders, *Phys. Lett.* **215B** (1988) 449.
23. V. B. Kopeliovich and B. E. Stern, *NORDITA preprint 89/34 N* (1989).
24. V. B. Kopeliovich, *Yad.Fiz.* **51** (1990) 241.
25. V. B. Kopeliovich, B. Schwesinger and B. E. Stern, *Phys.Lett.* **242B** (1990) 145.
26. V. B. Kopeliovich, *Phys. Lett.* **259B** (1991) 234.
27. I. Klebanov, *Nucl. Phys.* **B262** (1985) 133.
28. H. Weigel, J. Schechter, N. W. Park and U.-G. Meissner, *Phys. Rev.* **D42** (1990) 3177.
29. N. W. Park, J. Schechter and H. Weigel, *Phys. Rev.* **D43** (1991) 869.
30. J. Schechter and H. Weigel, *Phys. Rev.* **D44** (1991) 2916.
31. B. Schwesinger and H. Weigel, *Phys. Lett.* **B267** (1991) 438.
32. G. S. Adkins and C. R. Nappi, *Nucl. Phys.* **B233** (1984) 109.
33. J. Balog, *Phys. Lett.* **149B** (1984) 197.
34. D. I. Dyakonov and M. I. Eides, *Pis'ma v ZhETF* **38** (1983) 358.
35. I. R. Kozhevnikov, Yu. P. Rybakov and M. B. Fomin *Theor.Math.Phys.* **75** (1989) 353.
36. V. B. Kopeliovich, *Yad.Fiz.* **47** (1988) 1495.
37. V. A. Nikolaev and O. G. Tkachev, *Yad.Fiz.* **55** (1992) 179.
38. T. S. Walhout, *Phys. Lett.* **227B** (1989) 10.
39. G. S. Adkins, C. R. Nappi and E. Witten, *Nucl.Phys.* **B228** (1983) 552.
40. E. Guadagnini, *Nucl. Phys.* **B236** (1984) 35.
41. V. B. Kopeliovich, B. Schwesinger and B. E. Stern, *Nucl.Phys.* **A** (1992) in print.
42. H. Yabu and K. Ando, *Nucl. Phys.* **B301** (1988) 601.
43. B. Moussalam and D. Kalafatis, *Phys. Lett.* **272B** (1991) 196.
44. A. I. Issinsky, V. B. Kopeliovich and B. E. Stern, *Yad.Fiz.* **48** (1988) 209.

SOLITON DYNAMICS OF THE B=2 SECTOR OF THE SKYRME MODEL

N. J. Snyderman
University of California
Lawrence Livermore National Laboratory
Livermore, CA 94550

ABSTRACT

A class of periodic classical bound state solutions to the B=2 sector of the Skyrme model have been found by numerical simulation in 3 spatial dimensions. These solutions are quantized by the Bohr-Sommerfeld quantization condition. The Bohr-Sommerfeld quantization prescription is the lowest approximation to a systematic quantization framework for soliton bound states in quantum field theory due to Dashen, Hasslacher, and Neveu. This framework is extended to include the treatment of internal dynamical collective coordinates by Manton. The general quantum dynamics of solitons is naturally split between, on one hand, the finite number of quantum degrees of freedom (collective coordinates) associated with the general classical dynamical solution, as advocated by Manton, and illustrated by the dynamics of magnetic monopoles, and, on the other hand, the infinite number of quantum fluctuation degrees of freedom, as exemplified by the DHN and Fadeev and Korepin analysis of the quantum corrections to the kink dynamics in the 1-D quantum sine-Gordon model. The quantum fluctuation problem has new conceptual issues in the context of the nonrenormalizable Skyrme model. Guidance for the resolution of these issues comes from the analyses by Weinberg of effective low energy gauge theories, and of the quantum corrections to π-π scattering in the nonlinear σ-model. The quantum scattering problem is also discussed. The DHN framework is generalized to deal with the new features of 3-D, and with internal dynamical collective coordinates. Arguments are also given for why the Skyrme model may be a good low energy approximation to QCD.

1. Introduction

A dual description of QCD has been suggested by Witten[1] based on the 1/N expansion[2], in which the baryons and nuclei arise as topological solitons of a string-like theory of meson dynamics. Such a description of QCD, exact in principle, could lead to a simple fundamental description of low energy nuclear physics if 1/N is effectively a small expansion parameter, even for N=3. This is because 1/N plays the role of $\hbar$ for the soliton dynamics, which is therefore semiclassical if 1/N is effectively small. Until the string theory of meson dynamics is determined, one can begin to explore this framework with the Skyrme model[3]. This model is an effective chiral Lagrangian for low energy pion dynamics[4], but remarkably also has nucleons[5], baryon resonances[6], and nuclei[7] as soliton excitations. Is the Skyrme model just an inspired vision, or does it offer a description of the low energy dynamics of baryons and nuclei?

In this talk I will discuss two general questions: For understanding nuclear structure and low energy scattering and reactions, can the Skyrme model be justified as an adequate low energy approximation to the meson string theory of QCD?, and, How does one do semiclassical quantum bound state and soliton scattering calculations within the context of the Skyrme model? The first question can not yet be answered in any serious way, but I will offer some opinions on why the Skyrme model may be a good low energy approximation to QCD. As for the second, more technical question, I believe there are now sufficiently powerful computers and numerical techniques, as well as a sufficiently well established conceptual framework, that one can seriously embark on this endeavor. There are significant technical and conceptual problems remaining, but there are also three very important examples that serve as a guide to the problems of the systematic semiclassical quantization of soliton dynamics in the Skyrme model, considered as a quantum field theory.

First, there is the analog of the sine-Gordon breather quantization. The sine-Gordon field theory in 1-space and 1-time dimension has a set of classical bound state solutions of the kink and antikink solitons. These classical solutions were quantized by Dashen, Hasslacher, and Neveu[8]. They developed the general framework for systematically calculating the semiclassical bound state spectrum in quantum field theory. Their calculations were the field theory equivalent to the WKB approximation in quantum mechanics, the field theory generalization of Bohr-Sommerfeld quantization plus 1-loop quantum corrections. Remarkably, this level of quantization gave the exact solution to the sine-Gordon quantum field theory. Apart from simplifications due to only 1-spatial dimension, they relied on completely analytic techniques that unfortunately do not naturally generalize to problems that must be studied by numerical methods. A different technical method for the evaluation of the quantum corrections to the time dependent sine-Gordon breather solution, due to Fadeev and Korepin[9], may be implementable numerically.

The second important example that serves as a conceptual guide to the analysis of the Skyrme model is the classical and quantum dynamics of the BPS magnetic monopoles[10], due to Manton[11], Atiyah and Hitchin[12], Gibbons and Manton[13], and Schroers[14]. This is a problem in 3-spatial dimensions, but one for which there are remarkable analytic properties. This problem is especially useful in understanding how to deal with the quantum dynamics of internal collective coordinates. In the Skyrme model, the Bohr-Sommerfeld quantization of the classical soliton dynamics gives energies of order N in the 1/N expansion. There are internal rotor degrees of freedom that on quantization give a fine structure to the Bohr-Sommerfeld levels, splitting the states of different spin and isospin by order 1/N. The quantum theory of the classical dynamics and rotor dynamics is that of a finite number of quantum degrees of freedom. The BPS monopole dynamics shows how the internal charge degree of freedom can be excited in scattering processes. For momopole scattering, this corresponds to charge exchange in which the monopoles acquire electric charge and become dyons of opposite charge. The total electric charge is conserved, but the relative charge can change in scattering. Also, since there is additional field angular momentum associated with the dyon excitations, the motion, for which the total angular momentum is conserved, can be quite complicated. Analogously, the excitation of the relative isospin collective coordinates of two Skyrmions leads to Δ excitation. Because the Δ is a spin as well as isospin excitation, the transfer of angular momentum from orbital to spin can lead to complex dynamical motion analogous to that of the dyons. As a consequence of this relative isospin collective coordinate, it is plausible that there are additional classes of bound orbits of two Skyrmions that are different from the ones we have already studied. In these orbits the the relative isospin rotor could have an angular velocity of order 1, rather than order 1/N.

There are additional quantum corrections of order 1 and also of order 1/N and higher, that are associated with an infinite number of degrees of freedom characteristic of quantum field theory. Consequently there are additional conceptual issues associated with renormalization in a nonrenormalizable model. Guidance for the analysis of this problem comes from the third important example, that of quantum loop corrections to π-π scattering in the nonlinear σ-model, due to Weinberg[4] and Gasser and Leutwyler[15]. This work shows how one deals with the divergences that arise in a field theory that is non-renormalizable. At the 1-loop level, one must add counter terms that go beyond the nonlinear σ-model Lagrangian in order to absorb the infinities. The Skyrme term is one of these counterterms. While the counterterms that renormalize the pion dynamics should also renormalize the divergences in the soliton sector, as they do in renormalizable models with topological solitons, there are nevertheless serious new conceptual and physics questions in the soliton sector of the Skyrme model. I suggest that for 1-loop calculations in the Skyrme model, one must begin with an effective Lagrangian that contains all the 1-loop counterterms of the Skyrme Lagrangian. One must then be able to show that these extra terms, when evaluated for the classical solutions to the Skyrme model, are perturbatively small compared to the Skyrme model terms. This will require the renormalized couplings of these terms to be appropriately small. In addition to the absorption of divergences, renormalization introduces a scale parameter, μ. The physical results must be independent of μ, so under a change of μ, while the values of all couplings change, this change must be compensated. This is the renormalization group of Gell' Mann and Low[16]. In scattering amplitudes, as Weinberg[4] shows for π-π scattering, there are Log(E/μ) terms that compensate the change of couplings with μ. For the Skyrmion mass, there must also be explicit μ dependence to compensate the change in couplings with μ.

I will take the opportunity of a forum of this kind to raise a number of issues that are not understood, but nevertheless, topics I feel must be properly understood. The understanding of the collective coordinate dynamics in the Skyrme model is currently very incomplete. I feel this most important problem is, however, relatively tractable. The renormalization and renormalization group aspects of the $B \neq 0$ sector, though, are possibly more difficult problems. These are problems for which models simpler than Skyrme model woud be extremely useful. The SU(2) -> U(1) gauge theory is my favorite model for all of these issues. For the renormalization issues, one can ask about finite renormalizations in the soliton sector of the full model where the renormalizability is manifest. One could also ask the analogous questions in the effective U(1) gauge theory below M_W. This effective gauge theory is the prototype for thinking about almost all questions about the Skyrme model. For this effective gauge theory, though, even the classical soliton dynamics of the magnetic monopoles is far from clear, (although the topological aspects will be discussed below), let alone the renormalization issues in the magnetic monopole sector.

An outline of the contents of this paper are as follows: In the first part we discuss the B=2 soliton dynamics. We will first discuss the Bohr-Sommerfeld quantization of a class of periodic orbits of B=2 in the Skyrme model[17]. We will then discuss the collective coordinates of the general B=2 classical dynamics. After bringing out the analogy with magnetic monopole dynamics, we will then discuss the quantization of the collective coordinates of the classical Skyrmion dynamics. This discussion naturally breaks up into separate discussions of the limit of large separation, where the product ansatz is valid, and for which the spin and isospin of the separate nucleons is conserved, and then the shorter distance regime where the Skyrmions strongly polarize one another, and for which the virtual Δ excitation is possible. While the large separation region is becoming understood, the small separation region requires both additional numerical simulations of the relative isospin collective coordinate dynamics, as well as additional group theoretic insight. The quantization of the collective coordinates dynamics is greatly simplifyed if the classical motion is adiabatic, in which case Hamiltonian quantum mechanics methods apply. Our numerical simulations on finite grids with massless pions indicate that even for the weakly bound deuteron, the classical dynamics is not completely adiabatic. To the extent that the classical dynamics is approximately adiabatic, though, we can extract the necessary inertia tensors for the collective coordinate Lagrangian by our numerical simulations. We also briefly discuss the path integral formalism for the collective coordinate quantization, which, while technically more difficult than the Hamiltonian approach, is more natural for quantizing numerical dynamical solutions. After this long discussion of the collective coordinate quantization, we will then give a brief qualitative discussion of the properties of the deuteron that emerge from quantizing this class of periodic orbits. We then comment on possible additional classes of B=2 periodic orbits in the Skyrme model. We also briefly discuss the scattering problem. We complete the discussion of soliton dynamics with brief discussions of quantum fluctuation corrections for the B=1 Skyrmion. Part of the quantum correction, the determinant of the pion propagator in the background Skyrmion field, is calculable from the famous π - N scattering phase shifts[6]. There should also be a contribution of the same order from the Δ resonance. Finally, we give some thoughts on renormalization. In the second part, we attempt to justify the Skyrme model. We discuss the important effective gauge theory analogy, and then attempt to understand from meson theory why the Skyrme model may be a good low energy approximation to the QCD string theory. We end with some conclusions, and challanges for the future.

2. Soliton Field Theory

2.1 Bohr-Sommerfeld Quantization

Quantum bound states can be obtained from classical periodic orbits by Bohr-Sommerfeld quantization. In quantum mechanics,

$$\int_{x_1}^{x_2} p\, dx = \left(n+\frac{1}{2}\right) \hbar, \tag{1}$$

where x_1 and x_2 are the turning points of the orbit, and the 1/2 is a correction for small principle quantum number, n. This expression can be re-expressed in a way that generalizes immediately to field theory,

$$2\int_{x_1}^{x_2} p\, dx = S(T(E)) + E\, T(E), \tag{2}$$

where the classical relation between action and energy, $E = - dS(T)/dT$, gives the functional relation between period and energy, T(E). The systematic framework in field theory that justifies Bohr-Sommerfeld quantization, and shows how quantum corrections to it arise, is due to Dashen, Hasslacher, and Neveu[8]. The quantum bound states are described by the poles of

$$\mathrm{tr}\frac{1}{E-H} = -i\int_0^\infty dT\, e^{iET}\, \mathrm{tr}\, e^{-iHT}, \tag{3}$$

where, in quantum field theory, tr exp($- i/\hbar$ H T) can be expressed as a functional integral over fields that are periodic in T, $U(\mathbf{x},t+T) = U(\mathbf{x},t)$. From the functional integral representation, the semiclassical approximation is seen to correspond to the stationary phase approximation. DHN show that this expression is proportional to

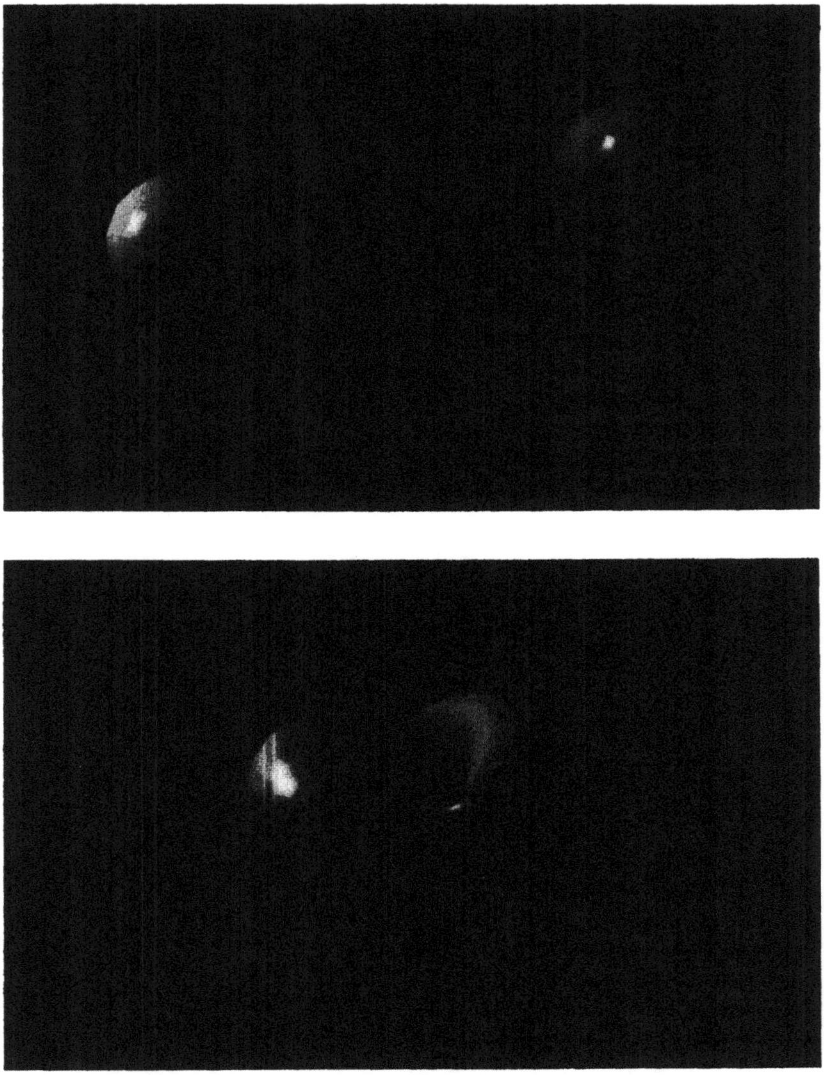

Figure 1. Time snapshots of a dynamical orbit. Shown are contours of the energy density at 1/2 maximum. The original separated Skyrmions polarize one another into a toroidal configuration, and then scatter at 90^0.

$$\frac{1}{1+e^{\frac{i}{\hbar}W(E)}}\,, \tag{4}$$

where

$$W(E) = S(T(E)) + E\,T(E) + \text{quantum corrections}\,, \tag{5}$$

The plus sign in the denominator, which gives the 1/2 in Bohr-Sommerfeld quantization, arises from the phase of a determinant associated with the quantum corrections. The bound state poles arise from the condition,

$$W(E) = (2\,n+1)\,\pi\,\hbar\,, \tag{6}$$

which is the field theory equivalent of the Bohr-Sommerfeld quantization condition. The systematic calculation of the quantum corrections will be discussed later.

2.2 Class of Periodic Orbits

A class of periodic classical orbits, continuously relating the static toroidal soliton[18] of lowest energy for B=2 to seperated B=1 Skyrme solitons, were found by numerical simulation[17]. See Fig.1.Such a class of solutions was conjectured by Manton[19].

A set of periodic field configurations is obtained from different initial separations of the Skyrmions, with relative isospin orientation in the most attractive channel. Each such configuration has a different total energy and period. From this set of classical solutions, the quantization function, W=S+ET, is obtained.

When the Skyrme model is expressed in terms of dimensionless scaled variables, x and t are measured in units of $1/ef_\pi$, and E measured in units of f_π/e, the dimensionless action is rescaled to S/e^2. Because $e^2 \sim 1/N$, we see that 1/N plays the role of $\hbar$, and therefore if 1/N is effectively small, then the semiclassical approximation is valid. From these rescalings, the expression $S+ET \rightarrow 1/e^2\,(S+ET)$. The Bohr-Sommerfeld condition becomes $S+ET=(2n+1)\pi\,\hbar\,e^2$. The magnitude of e^2 therefore determines the number of quantum states and their energies. Our results[17,20] for W vs E are presented in Fig.2. The different curves are obtained from sets of simulations on

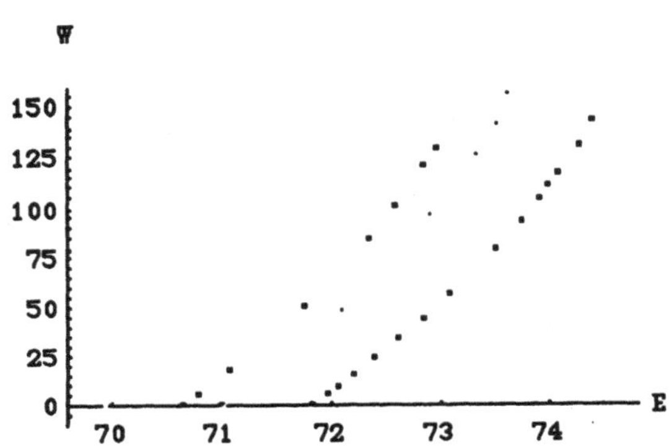

Figure 2. For a given coupling e^2, the intersection of horizontal lines at $(2n+1)\pi e^2$ with the curve W(E) give the energies of the bound states. From the right, the grids, (which cover 1/8 of space, reflection symmetries completing the field configuration over all space), are: 60x60x30, $\delta x=.143$, so in natural units, 8x8x4; 80x80x40, $\delta x=.12$, or 9.6x9.6x4.8 natural units; 68x68x34, $\delta x=.143$, or 9.8x9.8x4.9 natural units; and 80x80x40, $\delta x=.143$, or 11.4x11.4x5.7 natural units.

lattice grids of different size. Each curve extends from energies corresponding to the minimum of the potential well, the torus, to close to the continuum of two non interacting Skyrmions. Higher energy corresponds to starting higher up in the potential with larger initial separations of the Skyrmions. For each grid, our highest energies correspond to the most widely separated initial configuration of Skyrmions that fits on the lattice. For higher energy the Skyrmions spend proportionally more time near their end points of maximum separation than near the minimum of the potential energy. For e^2 appropriately large, there is only one solution. In Nature the deuteron is the only bound state, and is very weakly bound. The torus is only a good basis for quantization for small e^2, in which case there would be many B=2 bound states.

Large values for the coupling e^2 may make one uncomfortable about the semiclassical approximation. While the quantization condition gives a value for $W \lesssim 100$, (for the ANW value, $e=5.45$, $\pi e^2 = 93.3$), the action is much larger, $S \gtrsim 3000$. We show the action verses period in Fig.3. Therefore, $S/e^2 \gtrsim 100 >> 1$, suggesting the semiclassical approximation should be valid. The validity of the semiclassical approximation also depends on the corrections being small.

2.3 Collective Coordinates Characterizing Classical Dynamics

The class of periodic orbits we have picked out evolved from the product ansatz at large separation. In order to properly understand the collective coordinates, we return to the product ansatz[21]. The field configuration has the form, for large separation,

$$U(\mathbf{x}) = A\, U_1(\mathbf{x}-\mathbf{x}_1)A^{-1}\, B\, U_1(\mathbf{x}-\mathbf{x}_2)B^{-1}\,, \tag{7}$$

where U_1 is a B=1 Skyrmion hedgehog, A and B are global isospin rotation matrices, and $\mathbf{x}_1$ and $\mathbf{x}_2$ are the positions of the separate Skyrmions. There are 12 collective coordinates characterizing this field configuration: 3 position coordinates for each Skyrmion, and 3 isospin Euler angles for each Skyrmion. These coordinates should be redefined. Instead of the position coordinates there should be the CM position, $\mathbf{R} = (\mathbf{x}_1 + \mathbf{x}_2)/2$, and the relative separation, $\mathbf{r} = (\mathbf{x}_1 - \mathbf{x}_2)$. Also, the isorotations should be re expressed: with $B= A\, C^{-1}$, A becomes an overall global isorotation, and C is the relative isospin. Since the CM position and motion are unimportant, there are only 9 collective coordinates,

$$U(\mathbf{x}) = A\left(U_1\!\left(\mathbf{x}+\frac{\mathbf{r}}{2}\right) C^{-1}U_1\!\left(\mathbf{x}-\frac{\mathbf{r}}{2}\right)C\right) A^{-1}\,. \tag{8}$$

The energy of this field configuration depends on $\mathbf{r}$ and C, and equals 2 M, where M is the mass of a hedgehog Skyrmion, plus a potential energy of interaction that varies with relative isospin orientation. The interaction energy in the product ansatz[21], for large separation, is

$$V(\mathbf{x},C) \approx \frac{3(\mathbf{c}\cdot\mathbf{x})^2 - \mathbf{c}^2 r^2}{r^5}$$
$$= -\left(\cos(\alpha+\gamma)\cos^2\left(\frac{\beta}{2}\right) - \cos\beta\right)\frac{2}{r^3}, \tag{9}$$

where $C = c_0 + i\,\mathbf{c}\cdot\boldsymbol{\tau}$, $c_0^2 + \mathbf{c}^2 = 1$, or in terms of Euler angles,

$$C = e^{-i\alpha\frac{\tau_3}{2}}\, e^{-i\beta\frac{\tau_2}{2}}\, e^{-i\gamma\frac{\tau_3}{2}}\,. \tag{10}$$

The configuration with one Skyrmion rotated in isospin space by π about an axis perpendicular to their separation, $\beta = \pi$, minimizes the interaction energy. There are 4 zero modes of this field configuration, the 3 Euler angles characterizing A, and 1 Euler angle of C, $\alpha - \gamma$, corresponding to rotations of the axis of $\mathbf{c}$ about the axis of $\mathbf{r}$. This rotation is a simultaneous spatial rotation and isorotation about the body fixed axis, $\mathbf{r}$, by $\psi=(\alpha - \gamma)/2$.[22] The remaining 5 collective coordinates characterize the dynamics: r, the relative separation in the body fixed frame, θ and ϕ, characterize the orientation of the body fixed axis relative to the space fixed axes, and 2 Euler angles of the relative isospin orientation, $\alpha + \gamma$, and β. For large separation, the energy cost for varying β away from π, and varying $\alpha + \gamma$, is small.

However, for small separation, where the product ansatz is not valid, there is a very large energy cost in varying the relative isospin away from the minimum energy values. For the field configuration of minimum energy derived from two hedgehogs on top of one another, and having $\alpha=0$, $\beta=0$, $\gamma=0$, the energy is ~ 3 M,[18] while the minimum energy toroidal configuration, that derived from relaxing a configuration with $\beta=\pi$, has energy ~10% less than 2 M. Therefore, there is a potential well of over M deep with minimum for $\beta=\pi$. Excitations in this potential well will correspond physically to virtual Δ excitations in the deuteron. This implies the second derivative of the relative isospin potential evaluated at the minimum should be $\sim (M_\Delta - M_N)^2$. Since the potential well for binding the deuteron has a depth, $2M - M_{torus} << M_\Delta - M_N << M$, the virtual excitation of Δ should be a small perturbation for the deuteron. This suggests it will be a good first approximation to neglect the relative isospin excitation away from its minimum value when the Skyrmions are close together.

2.4 Magnetic Monopole Analogy

Additional important insight can be gained into the dynamics of the Skyrmions from the dynamics of the BPS monopoles. Magnetic monopoles[10] are topological solitons of Yang-Mills Higgs field theories. The magnetic monopoles have a tower of electrically charged dyon[24] excitations, arising from quantizing a phase collective coordinate that is the analog of the global isorotation of the Skyrmion. In the PBS limit, there is a massless Higgs scalar field as well as a massless photon field. Because the scalar interaction is attractive and has the same strength as the repulsive magnetic Coulomb interaction of the monopoles, the BPS monopoles do not have static interactions. There are analytic classical solutions in this limit for all values of magnetic charge.[25] For the 2 monopole problem, the solution depends on the collective coordinates $\mathbf{r}$, a relative phase angle ψ, that is the analog of the relative isospin coordinates, C, and a global phase χ, that is the analog of the overall global isospin coordinates, A. When the collective coordinates become time dependent, the time derivative of the global phase χ is conjugate to the total electric charge of the system, which is conserved, and so χ plays no role in the dynamics. As a function of r, the classical solution looks like two separate monopoles at large separation, and for smaller separation the monopoles polarize one another, and eventually form a toroidal configuration! For all values of r, the potential energy is the same as that for widely separated monopoles! The dynamics of this system is then due only to the time dependence of the collective coordinates: r, θ, φ, and ψ. The classical motion is geodesic motion on a 4-D spatial surface[11], a Kaluza-Klein geometry (the extra dimension parameterized by ψ is a circular dimension), corresponding to a particular gravitational instanton metric, the Atyiah-Hitchin metric[12]. The Lagrangian is of the form,

$$L = f(r)\,\dot{r}^2 + a(r)^2\omega_1^2 + b(r)^2\omega_2^2 + c(r)^2\omega_3^2\,, \tag{11}$$

where ω_1=- dθ/dt sinψ+dφ/dt sinθ cosφ, ω_2=dθ/dt cosψ+dφ/dt sinθ sinψ, and ω_3=dψ/dt +dφ/dt cosθ . The angular dependence is that of a triaxial rotor, but with an inertia tensor that depends on r. For the scattering of magnetic monopoles initially on the z axis, as r decreases a toroidal configuration forms, and the monopoles scatter through 90^0, with θ changing from 0 to π/2.[12] This dynamic solution, exactly analogous to the Skyrmion scattering, suggests the Skyrmion solution can be considered to be characterized by the same spatial collective coordinates, r, θ, and φ.

The quantum mechanics of the finite number of collective coordinates is most easily obtained from the canonical quantization of the Hamiltonian[13] derived from this Lagrangian,

$$H = -\frac{1}{abcf}\frac{\partial}{\partial r}\left(\frac{abc}{f}\frac{\partial}{\partial r}\right) + \left(\frac{K_1^2}{a^2} + \frac{K_2^2}{b^2} + \frac{K_3^2}{c^2}\right), \tag{12}$$

where the K_i are the body fixed angular momentum operators conjugate to the ω_i.

The qualitative similarity between monopole scattering and Skyrmion scattering at zero impact parameter suggests there is a corresponding collective coordinate approach to Skyrmion dynamics. This approach has been pursued by Manton[13], and Walhout[22], Wambach[23]. The definition of the radial coordinate is somewhat arbitrary (as it is in General Relativity), but given a definition for the constrained field configuration, one could determine the analog of $a^2(r)$ and $b^2(r)$. The determination of f(r) is somewhat more difficult, but if the constrained static configurations could be matched to the dynamic solutions, then from the momentum in the radial motion one could determine f(r). The rotor for the non zero mode Euler angles of the relative isospin rotation is now 2-D, as opposed to the 1-D phase in the monopole case. Because the 2 rotor angles parameterize a sphere, the Kaluza-Klein model describing the Skyrmion dynamics may contain hidden SU(2) Yang-Mills interactions. Just as the monopoles can have their relative phase excited, corresponding to electrically charged dyons, due to the excitation of the charged W fields at short distance, Skyrmions corresponding to nucleons at large distance can exchange charge through the π fields, and excite the relative isospin rotor, corresponding to N-Δ and Δ-Δ excitation. It has recently been argued that such a contribution is actually important for the N-N interaction[26]. Despite the wonderful similarities between the BPS monopole problem and the Skyrmion problem, the most obvious difference between them is the importance of the potential energy in the Skyrmion problem. Because of the potential energy, the two Skyrmion problem has bound state solutions. The monopole problem also has bound states[13], even without the potential, but only for total electric charge =1, corresponding to a monopole and a dyon at long distance. (Because of charge exchange at short distance, the relative electric charge is not conserved.)

2.5 Quantization of Skyrmion Collective Coordinates

2.5.1 Collective Coordinate Lagrangian for r -> ∞

For r -> ∞, the field should reduce to the product ansatz,

$$U(\mathbf{x},t) = A(t)\left[U_1\left(\mathbf{x} + \frac{\mathbf{r}(t)}{2}\right) C^{-1}(t)\, U_1\left(\mathbf{x} - \frac{\mathbf{r}(t)}{2}\right) C(t)\right] A^{-1}(t)\,. \tag{13}$$

The lagrangian of this U is,

$$L = -2\,M + 2\,I\,\mathrm{tr}\,\dot{A}\dot{A}^{-1} + I\,\mathrm{tr}\,\dot{C}\dot{C}^{-1} + 2\,I\,\mathrm{tr}\,A^{-1}\dot{A}C^{-1}\dot{C} + \frac{1}{2}M\,\dot{r}^2 - V(r,C)\,, \tag{14}$$

where I is the B=1 Skyrmion moment of inertia. The A rotor will be seen to be an integral spin rotor; the moment of inertia of 2 I reflects the fact that the A rotor corresponds to the isospin of the combined system. The relative isospin rotor, C, will be seen to be 1/2 integral, and the moment of inertia of I reflects the fact that only a single Skyrmion is being rotated. There is also a term that couples the two rotors; because the rotors are nonabelian, the eigenvalues of the operators conjugate to dC/dt and dA/dt are coupled. The potential, V, although negligibly small for very large r, is shown to emphasize the existence of the C dependence, which will become important at small r. Also shown is the kinetic energy of relative motion. Neglected in this asymptotic expression is a term that couples the relative isospin rotor, dC/dt , with dr/dt.[27] This coupling will allow for dynamic excitation of the relative isospin rotor. With this asymptotic expression we can more easily see the quantization of the rotor degrees of freedom.

2.5.2 Wave Functions for r -> ∞

The wave function for the rotor degrees of freedom of two Skyrmions, in the original form of the product ansatz, is,

$$\psi(A,B) = (-1)^{j+s_z} D^{j}_{i_z,-s_z}(A)\,(-1)^{j'+s'_z} D^{j'}_{i'_z,-s'_z}(B)\; C^{\,j\ j'\ I}_{\,i_z\ i'_z\ I_z}\; C^{\,j\ j'\ J}_{\,s_z\ s'_z\ J_z}\,, \tag{15}$$

which can be re expressed, after the change of variables with $B=A\,C^{-1}$,

$$\psi(A,C) = (-1)^{j+j'+s_z+s'_z}\left(D^{j}_{i_z,-s_z}(A)\, D^{j'}_{i'_z,s}(A)\; C^{\,j\ j'\ I}_{\,i_z\ i'_z\ I_z}\right) D^{j'}_{s,-s'_z}(C^{-1})\; C^{\,j\ j'\ J}_{\,s_z\ s'_z\ J_z}\,, \tag{16}$$

where the A rotor wavefunction can now be recoupled,

$$D^{j}_{i_z,-s_z}(A)\, D^{j'}_{i'_z,s}(A)\; C^{\,j\ j'\ I}_{\,i_z\ i'_z\ I_z} = D^{I}_{I_z,-s'_z+s}(A)\; C^{\,j\ j'\ I}_{\,s'_z\ s\ I_z}\,. \tag{17}$$

For j and j' 1/2 integers, I is integer. Although the A rotor was originally associated with 1/2 integer angular momentum when it referred to isorotation of a single Skyrmion, as a global rotation of the combined system it is integer valued. The relative isospin rotor is 1/2 integer valued, since C -> - C corresponds to rotating one Skyrmion by 2π. The wave functions are further constrained by the symmetry under interchange of the two Skyrmions. In the case of magnetic monopoles, they are classically identical; once the solitons form the torus, there is no way to identify the scattered solitons with the initial ones. For this reason the Skyrmions are presumably also classically indistinguishable, and this constrains the quantum wave functions. Under this interchange[23], $C \to C^{-1}$, and the wavefunctions become,

$$\psi^{jj'}_{II_z,JJ_z}(A,C) = (-1)^{j+j'+s_z+s_z'}\left(D^{j'}_{s,-s_z'}(C^{-1}) - D^{j'}_{s,-s_z'}(C)\right) C\begin{pmatrix} j & j' & J \\ s_z & s_z' & J_z \end{pmatrix} D^{I}_{I_z,-s_z+s}(A)\; C\begin{pmatrix} j & j' & I \\ -s_z & s & I_z \end{pmatrix} . \quad (18)$$

The minus sign is if j and j' are 1/2 integer, and are therefore fermions. This wavefunction vanishes for I=J =0. Also, if $C = i\,\tau_2$, it implies that if S=1, then I=0, and similarly if S=0, then I=1.

With this wave function, one can project onto physical states of given quantum numbers from the general functional integral,

$$\int dA'dA''dC'dC''\,\psi^{jj'\,*}_{I,I_z,J,J_z}(A'',C'')\int_{A'}^{A''} \mathcal{D}A(t)\int_{C'}^{C''}\mathcal{D}C(t)\,e^{i\int_0^T dt\,L(A,C)}\,\psi^{jj'}_{I,I_z,J,J_z}(A',C') , \quad (19)$$

where the path integrals over the rotor collective coordinates correspond to propagators for motion on the group manifolds. We will return to this DHN framework for the dynamics below, but first consider a potentially simpler approach.

If the Skyrmion dynamics can be approximated as adiabatic, then a collective coordinate quantization exactly analogous to that of the monopole dynamics is appropriate. The full wave function then involves the spatial collective coordinates. These wave functions are of the form,

$$\psi^{jj'}_{II_zJJ_z}(r,\theta,\phi,A,C) = u_{LS}(r)\,\psi^{jj'}_{II_zSS_z}(A,C)\,Y_{LM}(\theta,\phi)\,C\begin{pmatrix} L & S & J \\ M & S_z & J_z \end{pmatrix} . \quad (20)$$

This wave function, however, does not properly take into account the constraint relating the zero mode of C with a combined rotation and overall isorotation about the body fixed axis. This implies the spherical harmonics should be generalized, $Y_{LM}(\theta,\phi)$ ->

$D^L_{M0}(\theta,\phi,\psi) \to D^L_{MN}(\theta,\phi,\psi)$, where the N index is coupled to an index of the $D^I(A)$ matrix to implement the constraint. There should be a Clebsch-Gordan coefficient coupling L, I and j', because a relative isospin rotation is equivalent to a combined spatial and overall isospin rotation.

2.5.3 Collective Coordinate Lagrangian for finite r

For the general product ansatz at finite r, the collective coordinate Lagrangian is more complicated. But this Lagrangian is an important analog to Manton's large separation Lagrangian[11] for BPS dyon dynamics. We give the structure only, for the nonlinear σ–model term of the Lagrangian,

$$L = I(U_C)\,\mathrm{tr}\,\dot{A}\dot{A}^{-1} + \mathrm{tr}\,U_C^{-1}\dot{U}_C U_C^{-1}\dot{U}_C + 2\,\mathrm{tr}\,\dot{A}\dot{A}^{-1}\left[\dot{U}_C, U_C^{-1}\right], \tag{21}$$

where $U_C = U_+C^{-1}U_-C$, where U_+ is an abbreviation for $U_1(\mathbf{x}+\mathbf{r}/2)$, and U_- is $U_1(\mathbf{x}-\mathbf{r}/2)$. Here U_1 should be the polarized single Skyrmion field, determined by Walhout and Wambach[23]. For the polarized Skyrmion field, though, the isospin moment of inertia becomes a tensor, (see below). The time derivatives of U_+ gives $1/2\ d\mathbf{r}/dt\cdot\nabla U_1$, and similarly for U_-, with a minus sign. The moment of inertia in the first term is that for the full field U_C, and as $r \to \infty$, approaches 2 I. The second term gives,

$$\begin{aligned} \mathrm{tr}\,U_C^{-1}\dot{U}_C U_C^{-1}\dot{U}_C &= 2\left(\mathrm{tr}\,C\dot{C}^{-1}C\dot{C}^{-1} + \mathrm{tr}\,\dot{C}C^{-1}U_-^{-1}\dot{C}C^{-1}U_-\right) \\ &+ 2\left(\mathrm{tr}\,\dot{C}^{-1}CU_+^{-1}\dot{U}_+ + \mathrm{tr}\,C\dot{C}^{-1}\dot{U}_-U_-^{-1} + \mathrm{tr}\,\dot{C}C^{-1}U_-^{-1}CU_+^{-1}\dot{U}_+C^{-1}U_- + \mathrm{tr}\,\dot{C}C^{-1}U_-^{-1}\dot{U}_-\right) \\ &+ \left(\mathrm{tr}\,U_+^{-1}\dot{U}_+U_+^{-1}\dot{U}_+ + \mathrm{tr}\,U_-^{-1}\dot{U}_-U_-^{-1}\dot{U}_- + \mathrm{tr}\,CU_+^{-1}\dot{U}_+C^{-1}U_-^{-1}\dot{U}_-\right). \end{aligned} \tag{22}$$

The first term corresponds to the rotor for a single Skyrmion, U_-, rotating with the isospin matrix, C. The second term couples the relative isospin rotor and the Skyrmion motion. This is a dynamic feature that has not yet been included in the numerical dynamic simulations. It will allow for Δ virtual excitation. The last term is the kinetic energy of the relative motion of the Skyrmions. Due to the mutual polarization of the Skyrmions, there will be additional radial dependence to the coefficient of $(dr/dt)^2$, as well as radial dependence to the angular velocity terms. The moment of inertia becomes a tensor. The coefficients of ω_1^2 and ω_2^2 will in general be different, and functions of r.

This structure is analogous to that in the monopole problem.[13] For large r, where these inertia components are equal, the terms in $\omega_1^2 + \omega_2^2$ correspond to the usual orbital angular momentum potential. At short distance, where the product ansatz is not valid, the simulations show how the inertia tensor components match on to those of the torus. For large r, the dynamics of dθ/dt and dϕ/dt corresponds to the motion of the body fixed axis joining the Skyrmions, relative to the space fixed axes. For the torus, the dynamics of dθ/dt and dϕ/dt corresponds to 2 of the rotational zero mode degrees of freedom of Bratten and Carson[18].

Finally, there is the term that couples the total isospin rotor to the relative motion,

$$\begin{aligned} &2\,\mathrm{tr}\,\dot{A}\dot{A}^{-1}\left[\dot{U}_C, U_C^{-1}\right] \\ &= 2\,\mathrm{tr}\,\dot{A}\dot{A}^{-1}\left(U_+\dot{C}^{-1}CU_+^{-1} + U_+C^{-1}U_-\dot{C}C^{-1}U_-^{-1}CU_+^{-1} - C^{-1}U_-^{-1}C\dot{C}^{-1}U_-C - C^{-1}\dot{C}\right) \\ &+ 2\,\mathrm{tr}\,\dot{A}\dot{A}^{-1}\left(\dot{U}_+U_+^{-1} + U_+C^{-1}\dot{U}_-U_-^{-1}CU_+^{-1} - C^{-1}U_-^{-1}CU_+^{-1}\dot{U}_+C^{-1}U_-C - C^{-1}U_-^{-1}\dot{U}_-C\right). \end{aligned} \tag{23}$$

The first term reduces, for $r \to \infty$, to the term with moment of inertia 2 I that couples the two rotors. For finite r, if one includes the mutual polarization of U_+ and U_-, the moment of inertia will become a tensor. The second term, that couples the overall isorotation to the motion of the pair of Skyrmions, will vanish due to total isospin conservation.

Including the dynamics of the collective coordinates, the general structure of the product ansatz, especially when Skyrmion polarization is included to correct the hedgehog fields, as was done for the static potential by Walhout and Wambach[23], is very much analogous to the long distance structure of the monopole problem studied by Manton[11]. The full dynamics of this generalized product ansatz is very much worth further study. One could set up the Schrodinger equation for the generalized product ansatz that is the analog of the Gibbons and Manton equation[13]. A start in this direction was made by Walhout[22].

2.5.4 Inertia Tensors

The collective coordinate Lagrangian could also be obtained from the numerical simulations. The inertia tensors are, in the notation of Bratten and Carson[18],

$$U_{ij} = \frac{1}{8}\int d^3x \, \mathrm{tr}\left\{U^{-1}\left[\frac{\tau_i}{2},U\right]U^{-1}\left[\frac{\tau_j}{2},U\right] + \left[U^{-1}\nabla U, U^{-1}\left[\frac{\tau_i}{2},U\right]\right]\cdot\left[U^{-1}\nabla U, U^{-1}\left[\frac{\tau_j}{2},U\right]\right]\right\}, \tag{24}$$

$$\begin{aligned} V_{ij} = \frac{1}{8}\int d^3x \, \mathrm{tr}\Big\{ & U^{-1}\left(x\times\frac{1}{i}\nabla\right)_i U U^{-1}\left(x\times\frac{1}{i}\nabla\right)_j U + \\ & \left[U^{-1}\nabla U, U^{-1}\left(x\times\frac{1}{i}\nabla\right)_i U\right]\bullet\left[U^{-1}\nabla U, U^{-1}\left(x\times\frac{1}{i}\nabla\right)_i U\right]\Big\}, \end{aligned} \tag{25}$$

and

$$W_{ia} = \int d^3x \, \frac{1}{2}\varepsilon_{ijk} \, x_j \, J_k^a , \tag{26}$$

where $J_k{}^a$ is the isospin current,

$$J_k^a = \frac{i}{2}\,\mathrm{tr}\left\{U^{-1}\left[\frac{\tau^a}{2},U\right]U^{-1}\nabla_k U + \left[U^{-1}\left[\frac{\tau^a}{2},U\right], U^{-1}\nabla_i U\right]\left[U^{-1}\nabla_k U, U^{-1}\nabla_i U\right]\right\}. \tag{27}$$

All inertia tensors have units of $1/e^3 f_\pi$. For fixed relative isospin corresponding to the most attractive channel, these expressions can be simplified because of the discrete symmetries of the classical solution[18]. All the tensors are diagonal, and for the symmetry axis of torus in the x direction, $W_{22}=W_{33}=0$, for the entire orbit.

The inertia tensors for the dynamic soliton field are time dependent. In Figs. 3. and 4. we show the U and V tensor components as a function of time. For large initial separation the Skyrmions interact weakly, and for initial separation on the z-axis,

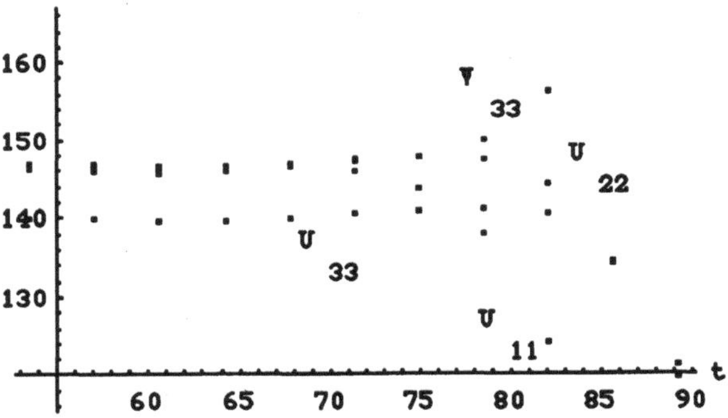

Figure 3. Moments of inertia verses time, on a finite grid, for overall isospin rotations of the entire field configuration, and for spatial rotations about the z-axis. The Skyrmions approach along the z-axis and scatter onto the y-axis; the torus is in the y-z plane. The maximum Skyrmion separation occurs at t=59.9. The minimum of the potential energy occurs at t=86.6, at which point the torus relations between tensor components are satisfied, $U_{22}=U_{33}$, and $U_{11}=V_{11}/4$, where V_{11} is shown in Fig.4. For large separation these moments should approach 2 I, where I is the Skyrmion moment of inertia. The continuum value for $2\,I \approx 212\ (1/e^3 f_\pi)$.

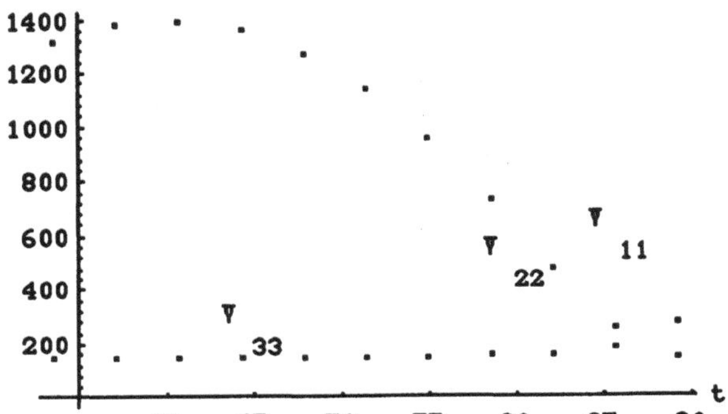

Figure 4. Rotational inertia tensors, for the same field solution as Fig. 3. V22 = V33 at t=86.6, corresponding to the torus.

$U_{11} \sim U_{22} \sim U_{33} \sim V_{33} \sim 2$ I, where I is the B=1 moment of inertia. (There are large finite grid size corrections to the simulated inertia tensors[20]. The numerical inequality of U moments for large separation is presumably due to the non symmetric box, with half the number of grid points perpendicular to the scattering plane.) Initially, with the separation on the z-axis, the rotational tensor components $V_{11} \sim V_{22}$ are very large due to the large moment arm. As the Skyrmions move together, the isospin tensor shows little sign of interaction until the solitons rapidly deform toward the torus, and the isospin inertia tensor components correspondingly rapidly decrease in magnitude, while the rotational inertia tensor components increase in magnitude. At the time corresponding to the minimum of the potential well, the tensor components $U_{22}= U_{33}$, $V_{22}=V_{33}$, and $V_{11} = 4 U_{11} = 2 W_{11}$, associated with the cylindrical symmetry of the torus[18] in the y-z plane. However, the isospin tensor components continue to decrease, not reaching their minimum values until slightly later in time. The more the total energy of the simulation, the larger the kinetic energy at the bottom of the well, and correspondingly the longer the time past that of the potential energy minimum it takes for the isospin tensors to reach their minimum values. For V_{11}, which is more symmetric about the minimum value, the minimum of V_{11} is seen to be shifted to times slightly after the potential energy minimum. The field configuration continues to collapse non symmetrically due to the kinetic energy, as the potential energy increases from the minimum. The field configuration then rebounds from this compression, with the isospin inertia tensors overshooting their large separation values until again approaching these values. In the calculation of other moments of the field configuration, this same asymmetry of the dynamic field configurations is observed[20].

If this nonadiabaticity can be neglected, the inertia tensor components needed for the Hamiltonian collective coordinate quantization can be obtained from a definition of the radial collective coordinate, that can be simulated as a function of time. An example is the mean square charge radius. See Fig.5.(A better choice might be the distance between topological centers of the Skyrmions, the points where the field U = -1.) Given r(t) from the simulations, the time dependence of the inertia tensors can be replaced by r dependence. The inertia coefficient of dr/dt can also be determined from the dynamic simulations. To complete the determination of the collective coordinate Lagrangian, we must also be able to simulate the relative isospin dynamics. (Current simulations have the relative isospin fixed in the most attractive channel.) This will be practical when the Adaptive Mesh Refinement technique is implemented in the numerical simulation code. This AMR technique will enable numerical simulations to be implemented for many more complex problems in Skyrmion dynamics.

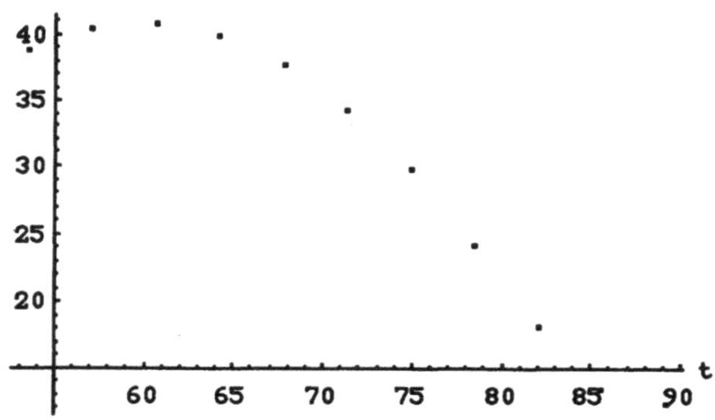

Figure 5. r^2 moment of the baryon density. For an isosinglet state, like the deuteron, the mean square charge radius is determined from 1/2 of this baryon moment. See Eq.38.

2.5.5 Wave Functions for r -> 0

The wave function, $\psi^{ii'}_{II_zJJ_z}(r,\theta,\phi,A,C)$, would be appropriate for $r \to \infty$, where the product ansatz is a good approximation. For large r, the term in the Lagrangian coupling dC/dt and dr/dt is very small, and so both the spin and orbital angular momentum are separately conserved. This will not be true at small r, where there can be a transfer of angular momentum between orbital and spin[27], conserving only the total angular momentum.

For r -> 0, both the separate monopoles and Skyrmions polarize one another into toroidal configurations. For the monopoles, the inertia tensor components b^2 and c^2 become equal for the torus in the y-z plane; the inertia component, a -> 0 as the torus is approached. For the Skyrmions, there is a possible analog. As Bratten and Carson have shown[18], for rotations about the symmetry axis of the torus, the field configuration is invariant under a combined rotation and isorotation. If the x-axis is the symmetry axis of the torus, the terms in the rotor Lagrangian corresponding to this symmetry rotation are,

$$\frac{1}{2}\left(\omega_1^{rot}\right)^2 V_{11} + \frac{1}{2}\left(\omega_1^{iso}\right)^2 U_{11} - \omega_1^{iso}\omega_1^{rot} W_{11}, \tag{28}$$

where V_{11} is the rotational inertia for rotations about the x-axis, U_{11} is the isorotational inertia, and W_{11} is the mixed rotation and isorotation inertia component. For the torus, $V_{11} = 4\,U_{11} = 2\,W_{11}$. Also, the invariance of the torus requires $\omega_1{}^{iso} = 2\,\omega_1{}^{rot}$; that is, for a rotation about the symmetry axis, the isospin is rotated through twice the angle[18]. For this constraint on the angular velocities, and with the relations between inertia components for the torus, these terms in the Lagrangian vanish. This suggests that there may be a set of angular variables, analogous to ω_1 for the monopole problem, for which the inertia coefficient vanishes for the torus in the Skyrme model.

The overall symmetry of the Skyrmion dynamics is $SU(2)_I \times SU(2)_{J=L+S}$, but with a constraint between I and S, and a further constraint because the two Skyrmions are identical. This must generalize the monopole symmetry, $SO(3) = SU(2)/Z_2$. The total angular momentum of the monopole system is conserved, but this combines the orbital angular momentum with the angular momentum in the field of the dyons. With the constraint due to the identical nature of the classical monopoles, the symmetry is SO(3). For the monopole problem, this symmetry implies the appropriate angular variables, and correspondingly, the appropriate harmonics for the wave functions, eigenfunctions of the angular momentum operators conjugate to these angular variables. The rotation matrices, functions of θ, ϕ, and ψ, that are the appropriate wave functions could be expressed in terms of the spherical harmonics[28], appropriate for large separation where the orbital angular momentum is approximately conserved,

$$D^{j\,*}_{mn}\left(\theta,\phi,\psi\right) = (-1)^{j-m} \sum_{l} [l] \begin{pmatrix} j & j & l \\ m & -n & n-m \end{pmatrix} H_{j,l}(\psi)\, Y_{l,m-n}\left(\theta,\phi\right), \tag{29}$$

where,

$$H_{j,l}(\psi) = l! \left[\frac{(2j-l)!}{(2j+l+1)!}\right]^{\frac{1}{2}} (2i \sin\psi)^l\, C^{l+1}_{2j-l}(\cos\psi), \tag{30}$$

where the C's are Gegenbauer polynomials. We need to determine the corresponding appropriate set of angular variables and associated harmonics for the Skyrmion dynamics.

2.5.6 Path Integral Collective Coordinate Quantization

If the adiabatic approximation to the Skyrmion dynamics is not appropriate, then the collective coordinate quantization can still be carried out within the DHN framework. Above, we considered the path integral projected with the collective coordinate wave functions. For the simpler example of a single rotor, the path integral[29] over A is easy to write down from the eigenfunction representation of the Hamiltonian problem,

$$\int_{A(0)=A'}^{A(T)=A''} DA(t)\; e^{i\int_0^T I\,\mathrm{tr}\,\dot{A}\dot{A}^{-1}\,dt} = \sum_{j,m,n} \frac{2j+1}{2\pi^2} D^j_{mn}(A'') D^j_{mn}(A')^* \, e^{-i\frac{j(j+1)}{2I}T} . \tag{31}$$

The rotor quantization for constant I can be generalized to the time dependent case $I(t) = I[U_{cl}(\mathbf{x},t)]$. The derivation is outlined in the Appendix. The result is that

$$\frac{1}{I} \to \frac{1}{T}\int_0^T \frac{dt}{I(t)} . \tag{32}$$

The effect of a rotor with time dependent moment of inertia is to shift the action for the classical dynamic solution by,

$$S(T) \to S(T) - \frac{1}{2}\hbar \nu(T) , \tag{33}$$

where,

$$\nu(t) = \int_0^T \frac{j(j+1)}{I(t)} dt . \tag{34}$$

The reciprocal moment of inertia is replaced by its time average. This will correct the quantization condition in the following way[30].

The proper-time integral over T will now have a phase, $S(T) - 1/2\,\hbar\,\nu(T) + E\,T$. Doing the integral by stationary phase gives the condition,

$$-\frac{\partial S}{\partial T} = E_{cl} = E - \frac{1}{2}\hbar \frac{\partial \nu}{\partial T} , \tag{35}$$

from which the quantization condition becomes,

$$\begin{aligned} &S(T(E_{cl})) - \frac{1}{2}\hbar\nu(T(E_{cl})) + \left(E_{cl} + \frac{\hbar}{2}\frac{\partial \nu}{\partial T}\right) T(E_{cl}) \\ &= W(T(E_{cl})) - \frac{\hbar}{2}\left(\nu(T(E_{cl})) - \frac{\partial \nu}{\partial T} T(E_{cl})\right) = (2n+1)\pi\hbar\, e^2 . \end{aligned} \tag{36}$$

The quantization condition determines the bound state quantum values for E_{cl}; the actual quantum energies are,

$$E = E_{cl} + \frac{1}{2}\hbar\frac{\partial\nu}{\partial T}, \tag{37}$$

where the T derivative is evaluated at the period, $T(E_{cl})$.

2.6 Deuteron

One can now qualitatively understand the small splitting between the s=1, i=0 deuteron and the s=0, i=1 state. For the classical orbit that satisfies the quantization condition, the skyrmions spend most of their time at large separation. The correction term in the quantization condition, 1/2 (ν(T(E)) - dν(T)/dT T(E)), is very small for the large period simulations. The energy shift is then essentially, 1/2 dν(T)/dT, and dν(T)/dT is approximately 1/T ν(T), proportional to the time average of the reciprocal of the inertia moment. The product ansatz is a good approximation in the regime of large separation. The rotor quantization in this region is essentially the same as that for two separated Skyrmions. For this region of large separation there is essentially no splitting of the states of different spin and isospin. A very small splitting comes from the potential term. At short distance, the relative isospin gets locked into the most attractive channel. The torus analysis of Bratten an Carson[18] is possibly a good approximation. The isospin moments decrease for the torus relative to their values for large Skyrmion separation, by about 10%. The rotational moments, however increase from their large separation values by a similar amount. The increased rotational moments decreases the energy in the rotational rotor, decreasing the energy of the s=1; the decreased isospin moments increases the energy in the isospin rotor, increasing the energy of the i=1 state. Because the splitting of these states is essentially a time average over the period of the motion, and because the time spent near the torus is small, the splitting of these states will be very small relative to the splitting calculated from the torus.

For the deuteron magnetic moment, a similar description applies. For all the static electromagnetic properties of the deuteron, the mean square charge radius, the quadrupole moment, and the magnetic moment, the matrix elements of operators with respect to deuteron wave functions is required. A general semiclassical expression for matrix elements is,

$$\int DU(x)\,|\psi_n[U(x)]|^2\, \mathcal{O}[U(x)] \cong \frac{1}{T_n}\int_0^{T_n} dt\, \mathcal{O}[U_{cl}^{(n)}(x,t)]\,, \tag{38}$$

where the matrix elements of an operator functional of the fields with respect to wave functionals for the quantum state n is, semiclassically, equal to the time average of the operator as a functional of the dynamic classical field that satisfies the quantization condition; the time T_n is the period for that dynamic classical field that satisfies the quantization condition for the state n. There are many examples in quantum mechanics for this relation showing its accuracy in solvable cases. As a consequence of the time average, the magnetic moment is also dominated by the large separation region where the Skyrmions spend most of their time. In this region the magnetic moment is essentially just the sum of proton and neutron moments in the model. At short distance, since the rotational inertia increases, the contribution to the magnetic moment decreases. Since the time spent in this short distance region is small, the magnetic moment will be close to, but slightly less than, the sum of Skyrmion proton and neutron moments.

2.7 General Classical Solution

For large r, where the product ansatz is valid, the rotors A and C are slow, in the sense that dA/dt and dC/dt ~ 1/N. Since I ~ N, the rotor terms in the Lagrangian are of order 1/N, and therefore a perturbation on the dynamics, which is of order N. However, as the Skyrmions get close, and polarize to the torus, dC/dt can become ~ 1, rather than 1/N. In this regime, the relative isospin rotor is a fundamental part of the order N terms in the Lagrangian, along with the Skyrmion mass terms. In the Bohr-Sommerfeld quantization of the classical orbits, this distinction will affect our previous analysis depending on how much of the time the orbit spent in region of rapid variation of C. For the long period orbits, this correction is presumably small, since most of the time in the orbit is spent at large separation. However, this raises the important issue of weather there are other dynamic periodic orbits that are as important, or even more important than the set we have studied so far. There orbits would presumably look more like what we normally think of as orbits, as opposed to these breather -like orbits that interpolate between the maximally attractive separated Skyrmion pair, and the potential energy minimum torus. A new class of orbits would be dynamically dependent on exciting the relative isospin rotor. They would also be dynamically dependent on the orbital angular momentum rotors associated with dθ/dt and dϕ/dt. My guess is that exciting these rotors as order 1 degrees of freedom will lead to orbits with much larger action than the ones we have considered. This problem will, of course, be studied when we acquire the necessary numerical simulation computing capability.

2.8 Scattering Theory

In the collective coordinate approach, one deals with a Hamiltonian with a finite number of quantum coordinates. The Schrodinger Hamiltonian could be solved for scattering states, just as well as for bound states. In the more systematic approach of DHN quantization, the scattering problem is very much more complex than in the 1-D quantum field theories they studied. First, one begins by considering Tr $(1/(E - H + i\varepsilon)\, P_Q(\theta))$, where we must have a projection operator for fixed angle scattering as well as for the quantum numbers of the asymptotic states. The reason for fixed angle scattering will become clear when we discuss the required dynamic simulations. From fixed angle scattering, we can project with Legendre polynomials to get the usual partial wave scattering phase shifts. To get phase shifts, take the imaginary part to get the density of states,

$$\rho_Q(E;\theta) = -\frac{1}{\pi}\,\mathrm{Im}\,\mathrm{Tr}\left(\frac{1}{E - H + i\varepsilon}\,P_Q(\theta)\right), \tag{39}$$

then project onto partial waves,

$$\rho_{Q,L}(E) = \int_{-1}^{1} d\cos\theta\;\rho_Q(E;\theta)\,P_L(\cos\theta)\,. \tag{40}$$

From the partial wave density of states, the phase shifts are,

$$\frac{1}{\pi}\frac{\partial}{\partial E}\delta_{Q,L}(E) = \rho_{Q,L}(E) \ . \tag{41}$$

Knowing the density of states as a function of energy the phase shifts can be determined by integration[31].

The calculation of scattering phase shifts therefore begins with the path integral representation for Tr (1/(E-H)). We must find periodic scattering orbits. These orbits will go through the walls of the box, which has periodic boundary conditions. The simplest periodic scattering is for 90^0 scattering. We know that for the Skyrmions in the most attractive channel, separated along the z axis with relative isospin, $C=i\,\tau_x$, if given some initial kinetic energy will scatter through 90^0, go through the walls of the box, and return to their original configuration. Since the classical dynamics is deterministic, changing the relative isospin will give scattering at an angle different from 90^0. Correspondingly, if we change the impact parameter to a non zero value, then for initial relative isospin, $C=i\,\tau_x$, the Skyrmions will not scatter through 90^0. For this relative isospin, the scattering angle will decrease with increasing impact parameter. For each impact parameter there will be an initial relative isospin orientation that will give 90^0 scattering. To determine the density of states for 90^0 scattering, for a given energy E, we add up the amplitudes, $\exp\{i(S+ET)\}$, weighted by the wave functions for two Skyrmions of good spin and isospin quantum numbers , with isospin orientations, A and B, where the relative isospin orientation is, $C = A\,B^{-1}$. In this way we can carry out the projection onto states of fixed angle and quantum numbers, Q. This procedure for calculating the quantum scattering phase shifts is quite time consuming, requiring for each scattering angle and energy a numerical simulation of the scattering. This is like a real scattering experiment! Given the classical result for the dynamic field configuration, from which the action S[U] is calculated, the quantum corrections can be calculated systematically, using the DHN framework.

2.8 Quantum Corrections to Skyrmion Mass

The O(1) correction to the Skyrmion mass can be determined[8,32] from the π -Skyrmion scattering phase shifts[6],

$$\Delta M = \int d\omega\, \frac{1}{2}\hbar\omega\, (\rho(\omega) - \rho_0(\omega)) - (B{=}0 \text{ counterterms})\ , \tag{42}$$

where the free particle density of states is $\rho_0(\omega) = \omega^2/2\pi^2c^3$, and where the change in the density of states of the pion due to the Skyrmion is related to the scattering partial wave phase shifts through

$$\rho(\omega) - \rho_0(\omega) = \frac{1}{\pi}\sum_{JLI} (2J+1)(2L+1)(2I+1)\frac{\partial}{\partial\omega}\delta_{JLI}(\omega)\ . \tag{43}$$

The O(1) renormalization of the Skyrmion mass follows from the O(1) pion-nucleon phase shifts, which should include the Δ- resonance. A sharp resonance contributes $\partial\delta/\partial\omega \sim \delta(\omega-\omega_{resonance})$. While the sum over resonant phase shifts is positive, not all phase shift contributions are positive, but mainly the renormalization counterterm subtraction should produce a net negative result. This is just what happens for the kink in the ϕ^4 and sineGordon models in 1+1 dimensions.[41] Consequently, if there is a negative correction to the Skyrmion mass, this would allow f_π to be larger, closer to its experimental value for π decay and π-π scattering. The order N classical B=1 mass, proportional to f_π/e, can now be larger, because the nucleon mass would be reduced by the non-zero mode corrections of order 1, proportional to e f_π . The value for e^2 would only be reduced slightly, because the N - Δ mass difference is proportional to $e^3 f_\pi$. This slight reduction in the value of e^2 would still allow e to be large, as is needed to understand the weak binding of the deuteron (as well as justify why the effective Lagrangian at 4th order in derivatives can be valid up to energies ~ m_ρ, and not just to f_π).

2.8.1 *π-N Scattering Born Terms*

We commented that the Δ resonance should also contribute to π-N scattering at O(1).[33] This is because the zero mode corrections are a problem in degenerate perturbation theory. In fact, perturbation theory is invalid, since the size of the perturbation and the energy denominator are of the same magnitude. The degeneracy is that of the N and Δ, brought about by the neglect of the zero modes. The translation zero modes allow for the propagation of the soliton in intermediate states. The isospin zero modes allow for the excitation of the Δ at the π-N vertex. The Born diagrams have the form,

$$\frac{g_{\pi N\Delta}^2}{M^2}\,\omega^3\left(\frac{1}{\omega - \Delta M + i\frac{\Gamma_\Delta}{2}} + \frac{1}{9}\,\frac{1}{\omega + \Delta M} - \frac{8}{9}\,\frac{1}{\omega}\right), \tag{44}$$

where $\Delta M = M_\Delta - M_N$, and where $\Gamma_\Delta \sim g_{\pi N\Delta}{}^2/M^2\,(\Delta M)^3 \sim 1/N^2$. The first two terms are the s and u channel Δ exchange, and the last term is the u channel N exchange. If ΔM were neglected altogether, then the amplitude vanishes. The N counting follows, first, from $g_{\pi N\Delta}{}^2/M^2 \sim N$; then for $\omega \sim 1$, the amplitude $\sim g_{\pi N\Delta}{}^2/M^2\,\omega\,\Delta M \sim 1$; if $\omega \sim 1/N$, then we need to include the Δ width, and the amplitude $\sim g_{\pi N\Delta}{}^2/M^2\,\omega^3/\Gamma_\Delta \sim 1$. For higher ω, there is a form factor that suppresses the ω^3 , preventing a ridiculous behavior.

2.8.2 *Renormalization*

The semiclassical approximation includes one-loop quantum fluctuation corrections to the background classical field. The divergences associated with this quantum correction are absorbed into renormalizations of the parameters of the model. The divergent renormalizations are of course the same as those of the B=0 sector for meson scattering; the finite parts of the renormalizations, though, differ in general for different B. The first term of the effective Lagrangian, of 2nd order in derivatives, the nonlinear σ-model, has one-loop divergences absorbed by two couplings of 4th order terms[4]. The Skyrme term is one of these; the other term has the form of the square of the nonlinear σ-model,

$$L^{(4)} = \gamma\left(\mathrm{tr}\,\partial_\mu U \partial^\mu U^{-1}\right)^2 . \tag{45}$$

The classical solitons are solutions of $L_{\text{Skyrme model}}$, and the quantum corrections come from π-fluctuations about these solutions using $L_{\text{Skyrme model}}$. The divergences of $L_{\text{Skyrme model}}$ come from both the nonlinear σ-model term and the Skyrme term. While the divergences of the 2nd order term are absorbed by 4th order terms, the divergences of the 4th order Skyrme term must be absorbed by terms of 6th and 8th order. The consistency of this analysis requires these terms be included in the effective Lagrangian, but that the parameters associated with these terms be sufficiently small that $L^{(4)}$, $L^{(6)}$, and $L^{(8)}$ can be dynamically neglected. That is, the full effective Lagrangian is,

$$L_{\text{eff}} = L_{\text{Skyrme model}} + L^{(4)} + L^{(6)} + L^{(8)} + \cdots, \tag{46}$$

where

$$L_{\text{Skyrme model}} = \frac{f_\pi^2}{16}\, \mathrm{tr}\,\partial_\mu U \partial^\mu U^{-1} + \frac{1}{32e^2}\, \mathrm{tr}\left[\, \partial_\mu U U^{-1}, \partial_\nu U U^{-1}\right]^2 . \tag{47}$$

If the physics of interest is well approximated by $L_{\text{Skyrme model}}$, then the one-loop divergences can be absorbed into the coefficients of $L^{(4)}$, $L^{(6)}$, and $L^{(8)}$, etc., with these terms contributing only as perturbations.

Not only are divergences absorbed as a consequence of renormalization, but also the renormalized couplings acquire a renormalization scale, μ, dependence. While physical results must be independent of μ, there is required an explicit dependence of physical quantities on μ to compensate the change of couplings resulting from changing μ.

3. Justification for the Skyrme Model

3.1 Effective Field Theories

Strongly linked to these renormalization issues is the meaning of effective field theories. How one interprets these renormalization issues determines how one interprets the Skyrme model as an effective field theory. Suppose one could only do experiments with pions with center of mass energies less than the mass of the ρ meson, but one has a theory of more massive degrees of freedom that affect the pion dynamics. This is an example of a completely general situation. I will give a different, more extreme example to illustrate this general idea. The example is one of effective gauge theories, and is a specific version of the Wilson renormalization group idea of effective field theories, due to Weinberg[34]. Consider the standard model below the mass of the W boson, the world as we knew it experimentally until recently. It is a theory of leptons, 5 quark flavors, photons and gluons. At low energy, this theory is dominated by the renormalizable interactions, like QED and QCD, but there are also the nonrenormalizable interactions due to the low energy effects of the W, Z, t-quark, Higgs bosons, etc. Weinberg[34] shows explicitly how to integrate out the heavy particles in the functional integral, maintaining gauge invariance for light fields in the low energy effective theory. The W and Z give the familiar Fermi interactions of quarks and leptons, while the t-quark gives an infinite set of weak couplings of photons and qluons. The parameters of this effective low energy field theory are not the parameters measured in low energy experiments. The parameters are appropriate for experiments near the W mass scale. To relate these parameters to those appropriate for low energy experiments, one has to calculate quantum loop corrections involving the light particles. These loop corrections renormalize the parameters of the effective theory, at a scale, μ, enabling the parameters to be rescaled from M_W to a low energy scale, μ. As shown by Hall et. al.[35] in their analysis relating the W and Z masses to experimental measurables like the fine structure constant and Fermi constant, the effective Lagrangian, with which one calculates amplitudes in a tree approximation, is like an operator product expansion,

$$L = \sum_i C_i(g,\mu)\, O_i \,, \tag{48}$$

where the O_i are a set of nonrenormalizable operators describing the effective interactions of the low energy that resulted from integrating out the heavy fields, (in their case 4-Fermi interactions), and the coefficients C_i depend on a set of coupling, written in vector form as g, and on the renormalization scale μ. The coefficient functions satisfy the Callan-Symanzik equations[36],

$$\left[\left(\mu\frac{\partial}{\partial\mu} + \beta(g)\cdot\frac{\partial}{\partial g}\right)\delta_{ij} + \gamma_{ij}(g)\right] C_j(g,\mu) = 0 \,. \tag{49}$$

The vector of β - functions describe the way the couplings change with μ, and the γ_{ij} are a matrix of anomalous dimensions of the operators O_i, where operators of the same symmetry and dimension can mix under renormalization. These equations express the physical equivalence of different renormalization scales, μ. Nevertheless, an appropriate choice of μ for a given physical problem can make the corrections small.

In the Skyrme model, the parameters in the Lagrangian should be interpreted as parameters for which $\mu \sim m_\rho$, and quantum corrections will enable the scale to be changed, in order to determine the parameters from real experiments on π-π scattering.

There are two major new issues to be understood in the Skyrme model that the known examples of effective field theories have not addressed. One is that the usual effective gauge theories have both renormalizable as well as nonrenormalizable interactions. The couplings that enter into the renormalization of the coefficients, C_i, are the renormalizable couplings of the low energy theory. This is presumably just a technical problem. The second problem is the renormalization group aspects of the soliton sector of the effective low energy theory. For this problem, insight can be gained by studying the effective gauge theory of SU(2) -> U(1).

Finally, I would like to comment on the topology of effective field theories, to gain insight into the topology of the Skyrme model. The nonlinear σ-model is the limit, $m_\sigma \to \infty$, of the linear σ-model. This seems to imply that an effective field theory with cutoff above m_σ would not have the required topology for stable baryons. The topology of the effective field theory must ultimately follow from the underlying QCD meson string theory, which we are far from understanding. But the issue of any effective low energy field theory having the correct topology of the full underlying theory can be understood from a simple example. For the SU(2) Yang-Mills theory spontaneously broken to U(1), the theory has a massless photon, heavy $W^\pm$, and a Higgs boson. The magnetic monopole is a topological soliton of the model. For the effective low energy theory below the Higgs and W masses, one has a nonlinear theory of photon dynamics. Such a theory,

$$L = -\frac{1}{4}F_{\mu\nu}^2 + a\frac{\alpha^2}{M_W^4}\left(F_{\mu\nu}F^{\mu\nu}\right)^2 + b\frac{\alpha^2}{M_W^4}\left(F_{\mu\nu}\tilde{F}^{\mu\nu}\right)^2 + \cdots \,, \tag{50}$$

superficially has only a U(1) symmetry, and consequently no topology in 3-D. But the electromagnetic field is actually built out of the SU(2) gauge and Higgs fields. The electromagnetic field tensor is actually the 't Hooft tensor[10],

$$\begin{aligned} F_{\mu\nu} &= \hat{\phi}^a G^a_{\mu\nu} + \frac{1}{e}\varepsilon_{abc}\hat{\phi}^a\left(D_\mu\hat{\phi}\right)^b\left(D_\nu\hat{\phi}\right)^c \\ &= \partial_\mu a_\nu - \partial_\nu a_\mu + \frac{1}{e}\varepsilon_{abc}\hat{\phi}^a\partial_\mu\hat{\phi}^b\partial_\nu\hat{\phi}^c \end{aligned}, \tag{51}$$

where $G_{\mu\nu}$ is the SU(2) Yang-Mills field, D_μ is the covariant derivative, ϕ is the unit isovector Higgs field, and the electromagnetic vector potential is, $a_\mu = \mathbf{A}_\mu . \phi$. The magnetic flux from the last term is the topological magnetic charge, corresponding to mapping $S^2_{space} \to S^2_{minV}$, where "minV" refers to the minimum of the Higgs potential. The effective nonlinear theory of photon dynamics should still have magnetic monopole solitons, still corresponding to configurations of $A_\mu{}^a$ and ϕ^a.

It is remarkable that the effective Lagrangian for the chiral SU(3) Skyrme model with Wess-Zumino term[1] has explicitly all the same topology as QCD[1]. Despite the larger strange quark mass, and the zero strangeness quantum numbers of nuclei, because of this remarkable property of the Skyrme-Witten model, the K meson degrees of freedom may have some importance for understanding the consistency of this picture of nuclear structure.

3.2 Skyrme Model Approximation to QCD 1/N Meson Theory

The application of the Skyrme model to nuclear structure and low energy scattering and reactions requires first a justification for a perturbative meson string theory for QCD[2,1], and then a justification of the Skyrme model as an adequate approximation to this meson string theory. While these problems are far from understood, the following arguments offer hope that this justification may be possible.

In the quantum theory of QCD the gauge field coupling, g, is not an expansion parameter for masses or cross sections. If the theory is defined from the continuum limit of the lattice theory, all masses have the form,

$$M = \#\frac{1}{a}e^{-1/g^2}\left(1+O\left(g^2\right)\right) , \tag{52}$$

where the pure number #, is independent of g; all masses have the same dependence on the lattice spacing, a, and g, and differ only in #. If the SU(3) gauge symmetry is generalized to SU(N), then N enters # as a series in 1/N, and also enters with the coupling like g^2/N. This suggests that 1/N is a hidden perturbative expansion parameter. Quark confinement and asymptotic freedom, spontaneous chiral symmetry breaking, and an infinite tower of mesons are all believed to be features of the $N \to \infty$ limit. The 1/N expansion gives an ordering of importance of physical effects that is consistent with strong interaction phenomenology. For many features of hadronic phenomena the 1/N expansion gives the only explanation from QCD. For example, the suppression of quark loops in the 1/N expansion gives a reason why the quark model should give such a good classification of hadronic states, given that the quarks are very strongly interacting for small momentum transfer.

The low energy limit of the meson string theory of 1/N QCD must reduce to a field theory for the low mass mesons. With the electroweak interactions turned off, QCD has a large mass gap as a consequence of exact chiral symmetry realized as a spontaneously broken symmetry, with massless pion excitations. The renormalization group implies that below m_ρ, the effective Lagrangian of QCD (with only u and d quarks) is based only on the chiral field, $U(\mathbf{x},t)=\exp[2i\pi^a(\mathbf{x},t)\tau^a/f_\pi]$. The Skyrme model corresponds to the first two terms in the derivative expansion of the exact chiral effective Lagrangian below m_ρ. Since the 1/N QCD counting rules give $f_\pi \sim \sqrt{N}$, if $e \sim 1/\sqrt{N}$,

$L_{\text{Skyrme model}}$ satisfies all the QCD 1/N perturbation theory counting rules for pion dynamics. Expanding U in powers of π/f_π shows that $L_{\text{Skyrme model}}$ has terms of every order in 1/N. But to each order in 1/N there are additional terms neglected in the Skyrme model. The derivative expansion series may converge rapidly if effects of the higher derivative terms are actually suppressed for fields varying slowly in space or time on the scale $\hbar/m_\rho c \sim 1/4$ fm, rather than just $1/f_\pi$.

For π-π scattering this is the case. The first term of $L_{\text{Skyrme model}}$, the nonlinear σ-model, gives

$$\frac{1}{f_\pi^2}\ \pi\times\partial_\mu\pi\,.\,\pi\times\partial^\mu\pi\,, \tag{53}$$

which is ~1/N. The interaction is of J=1, I=1 currents, and the QCD string theory implies,

$$\frac{1}{f_\pi^2}=\frac{g^2_{\rho\pi\pi}}{m_\rho^2}+\frac{g^2_{\rho'\pi\pi}}{m_{\rho'}^2}+\frac{g^2_{\rho''\pi\pi}}{m_{\rho''}^2}+\cdots\,, \tag{54}$$

since there are an infinite number of mesons of each IJ^P. This is the low energy limit of the pions interacting through not only the ρ, but also its higher mass counterparts of the same quantum numbers. QCD thus implies this correction to the KFSR relation; the ρ' (1450) and ρ'' (1700) couple more weakly to (they have smaller branching ratios to) $\pi\,\pi$ than the ρ does, so these corrections are small, but they are in the right direction to numerically improve the relation. This effective Lagrangian gives the leading low energy behavior, $\sim s/f_\pi^2$. The corrections of order s^2 arise from both the 1-loop radiative corrections of the nonlinear σ-model, of order $1/N^2$, and from the Skyrme term, of order 1/N, and also from another term that must be added to the effective Lagrangian corresponding to the square of the nonlinear σ-model, also of order 1/N. These give a contribution of order $\sim s^2/f_\pi^4\ \text{Log}\,s/\mu^2$, where the terms of order $s^2/(e^2 f_\pi^4)$ have been absorbed into the scale of the Log. This behavior describes π-π scattering all the way up to the ρ resonance[36].

The terms of higher order in derivatives in the effective Lagrangian correspond to the low energy limit of the exchange of higher spin resonances, as well as higher partial waves of the exchange of lower spin mesons. The dominance of the exchange of trajectories of mesons is the picture given by the 1/N string theory[2]. Meson loop diagrams are suppressed by powers of 1/N in the amplitude. The exchange of higher spin terms are suppressed both by powers of masses of heavier mesons and by their weaker coupling to $\pi\,\pi$. For example, the spin-2 $f_2(1270)$ coupling to $\pi\,\pi$ is much smaller than the $\rho\,\pi\,\pi$ coupling. The f_2 exchange is one of the contributions associated with the Skyrme term. Also contributing to the Skyrme term, and to $L^{(4)}$, is the spin zero isosinglet $f_0(975)$, that couples to $\partial_\mu\pi\,.\partial^\mu\pi$. The narrow width of this resonance, ~34 MeV, implies a weak coupling to $\pi\,\pi$. The most important contribution to the Skyrme term is the first finite momentum transfer correction to ρ exchange. The ρ propagator has the low energy expansion,

$$\int\frac{d^4q}{(2\,\pi)^4}\,\frac{e^{-iq.(x-x')}}{q^2-m_\rho^2}=\frac{1}{-m_\rho^2}\left(1-\frac{\partial^\mu\partial_\mu}{m_\rho^2}+\cdots\right)\delta^4(x-x')\,. \tag{55}$$

This propagator between the pion currents generates higher derivative terms. Furthermore, multiple pion couplings to higher spin mesons are suppressed by powers of 1/N. For example, the ω coupling to 3 π is 1/N suppressed relative to ω-> ρ π. The ω meson is almost degenerate with the ρ, a fact understood by the 1/N expansion[37], and couples to the baryon number current, $g_{\omega\pi\pi\pi}\,\omega^{\mu}B_{\mu}$. The effective interaction resulting from ω-exchange is,

$$L^{(6)} = \frac{g^2_{\omega\pi^3}}{m^2_\omega} B^2_\mu \,, \tag{56}$$

where

$$B^\mu = \frac{1}{24\pi^2}\,\varepsilon^{\mu\nu\alpha\beta}\,\mathrm{tr}\, U^{-1}\partial_\nu U\, U^{-1}\partial_\alpha U\, U^{-1}\partial_\beta U \,. \tag{57}$$

For the consistency of the Skyrme model, at the scale $\hbar/m_\omega c \sim \hbar/m_\rho c$,

$$\frac{g^2_{\omega\pi^3}}{m^2_\omega}\frac{1}{\left(\frac{\hbar}{m_\omega c}\right)^6} << \frac{1}{e^2}\frac{1}{\left(\frac{\hbar}{m_\rho c}\right)^4}\,, \tag{58}$$

or

$$g^2_{\omega\pi^3} << \frac{1}{e^2}\,. \tag{59}$$

The 1/N expansion suggests ω->$\rho\pi$ should dominate the ω decay, having a width of order 1/N, while the direct decay ω->$\pi\pi\pi$ should be suppressed, giving a width of order $1/N^2$. Therefore, the inequality may be satisfied empirically.

Even if the Skyrme model is a good approximation to the pion dynamics up to the ρ meson scale, the solitons are much more massive than the ρ meson. However, since nuclear structure is insensitive to quantum fluctuations on the scale of 1/4 fm, this effective Lagrangian may be adequate for quantitative nuclear structure calculations. Only the nucleon structure may be quantitatively sensitive to quantum fluctuations on the scale of 1/4 fm, although asymptotic freedom sets in at ~1/5 fm.

4 Conclusions

All qualitative features of the deuteron are reproduced by the semiclassical quantization of the B=2 soliton dynamics of the Skyrme model. There is a single bound state for large Skyrme coupling, e^2. For values suggested by either the quantization of the B=1 sector, or better, the more complete chiral perturbation theory analyses in the B=0 sector, this bound state is close to the continuum. There is a small fine structure splitting of the Bohr-Sommerfeld levels; for the orbit that satisfies the quantization condition for large e^2, the spin-isospin splitting separating the s=1, i=0 deuteron from the s=0, i=1 state is very small, and in the correct direction. The deuteron magnetic moment is slightly less than the Skyrmion proton plus neutron moments. This class of periodic orbits for the Skyrme model deuteron has a large quadrupole moment, and gives an excellent value for $Q/\langle r^2\rangle$.[19]

A more complete semiclassical quantization is clearly required, one that completes the collective coordinate quantization of the classical dynamics, and one that includes the properly renormalized quantum fluctuations. The 't Hooft-Polyakov magnetic monopole dynamics should be studied numerically. Because of the BPS limit, the collective coordinate dynamics is understood, and because of the renormalizability of the model the quantum fluctuation problem is free of conceptual issues. Nevertheless, the numerical implementation of the quantization of that system would be an enormously useful guide to the Skyrmion problem. Also, the generalized product ansatz dynamics of the Skyrme model, including the Skyrmion polarization corrections of Walhout and Wambach[28], deserves a more complete study.

With additional theoretical understanding of the semiclassical dynamics of the Skyrme model, we could be justified in encouraging new experimental studies of low energy π-π scattering, and in studying more completely the meson resonances of higher mass and spin that couple to $\pi\,\pi$..

Finally, the qualitative success of the semiclassical approximation to the quantum dynamics of the B=2 sector of the Skyrme model gives further support to the hope that 1/N, even for N=3, is a good perturbative expansion parameter for QCD.

5 Acknowledgments

I would particularly like to acknowledge and thank W. Crutchfield, who developed the numerical computer simulation capability that is the foundation of this work. I would also like to thank V. Brown, L. Carson, J. Hughes, S. Libby, M. Prasad, N. Walet, T. Walhout, J. Wambach, and L. Wilets for valuable discussions. I would finally like to express my gratitude to G. Holzwarth for organizing this most enjoyable and useful conference. This work was performed under the auspices of the U.S. Department of Energy by Lawrence Livermore National Laboratory under Contract No. W-7405-Eng-48.

6 Appendix

In this Appendix we indicate the modifications of Schulman's semiclassical path integral quantization of the SU(2) rotor, when the moment of inertia is time dependent. If the SU(2) matrix, A, is parameterized by Euler angles,

$$A = e^{-i\frac{\phi}{2}\tau_3}\, e^{-i\frac{\theta}{2}\tau_2}\, e^{-i\frac{\psi}{2}\tau_3}\,, \tag{A1}$$

the action has the form,

$$\int_0^T \frac{1}{2}\left(\frac{ds}{dt}\right)^2 dt\,, \tag{A2}$$

where the arc length is,

$$(ds)^2 = I(t)\left(d\theta^2 + d\phi^2 + 2\cos\theta\; d\phi\; d\psi\right), \tag{A3}$$

where the moment of inertia acts like a conformal factor in front of the metric on S^3. The action can be expressed in terms of an invariant arc length. From the eigenfunction representation expression for the propagator, the sum on the magnetic quantum numbers gives the character for the jth representation,

$$\sum_{m,n} D^j_{mn}(A'') D^j_{mn}(A')^* = \frac{\sin(2j+1)\frac{\Gamma}{2}}{\sin\frac{\Gamma}{2}}, \tag{A4}$$

where,

$$\cos\frac{\Gamma}{2} = \frac{1}{2}\,\mathrm{tr}\, A'' A'^{-1}$$
$$= \cos\frac{(\theta'' - \theta')}{2}\cos\frac{(\phi'' - \phi')}{2}\cos\frac{(\psi'' - \psi')}{2} - \cos\frac{(\theta'' + \theta')}{2}\sin\frac{(\phi'' - \phi')}{2}\sin\frac{(\psi'' - \psi')}{2}. \tag{A5}$$

Γ is the coordinate independent arc length on the sphere. Because of the symmetry of the sphere, one can choose a coordinate system to simplify the analysis, and at the end re express the results in invariant form. Choose the coordinate system, $\theta = \pi/2$, ψ = const., $\cos \Gamma/2 = \cos(\phi'' - \phi')/2$, or $\Gamma = \phi'' - \phi' + 4\pi n$. Then in this coordinate system the action becomes,

$$S = \int_0^T \frac{1}{2} I(t)\,\dot{\phi}^2\, dt = \frac{1}{2}\int_{\phi'}^{\phi''} \left(I\,\dot{\phi}\right) d\phi. \tag{A6}$$

Now the Euler-Lagrange equations give,

$$\frac{d}{dt}\left(I\,\dot{\phi}\right) = 0. \tag{A7}$$

Call the conserved angular momentum, J; then integrating the equation of motion gives,

$$\phi'' - \phi' = J\int_0^T \frac{dt}{I(t)}. \tag{A8}$$

The action now becomes,

$$S = \frac{1}{2} J\left(\phi'' - \phi'\right) = \frac{(\Gamma + 4\pi n)^2}{2\int_0^T \frac{dt}{I(t)}}. \tag{A9}$$

Since in the semiclassical approximation the prefactor of exp i S , the Van Vleck determinant, is computed from derivatives with respect to the end point angles of S, the only difference between the constant I case and the current time dependent case is the replacement of 1/I by its average over a period.

7 References

1. E.Witten, Nucl.Phys. **B160** (1979) 57; Nucl. Phys. **B223** (1983) 422 ; Nucl. Phys. **B223** (1983) 433.
2. G. 't Hooft, Nucl.Phys. **B72** (1974) 461 ; **B75** (1974) 461.
3. T. H. R. Skyrme, Proc. Roy. Soc. **A260** (1961) 127.
4. S. Weinberg, Phys. Rev. Lett. **18** (1967) 188 ; Physica **96A** (1979) 327 .
5. G. S. Adkins, C. R. Nappi and E. Witten, Nucl. Phys. **B228** (1983) 552 .
6. M. Mattis and M. Karliner, Phys. Rev. D **31** (1985) 2833; G. Eckart, A. Hayashi and G. Holzwarth, Nucl. Phys. **A448** (1986) 732 .
7. E. Braaten and L. Carson Phys. Rev. Lett. **56** (1986) 1897; E. Braaten, S. Townsend and L. Carson Phys. Lett. **B235** (1990) 147.
8. R. Dashen, B. Hasslacher, and A. Neveu, Phys. Rev. D **10** (1974) 4114; Phys. Rev. D **11** (1975) 3424.
9. L. D. Fadeev and V. E. Korepin, Phys. Rep. **42** (1978) 1.
10. G. 't Hooft, Nucl. Phys. **B79** (1974) 276; A. M. Polyakov, JETP Lett. **20** (1974) 194; JETP **41** (1974) 988; M. K. Prasad and C. M. Sommerfield, Phys. Rev. Lett. **35** (1975) 760; E. B. Bogomolny, Sov. J. Nucl. Phys. **24** (1976) 449.
11. N. S. Manton, Phys. Lett. **B110** (1982) 54; Phys. Lett. **B154** (1985) 397.
12. M. F. Atyiah and N. J. Hitchin, Phys. Lett. **107A** (1985) 21; Phil. Trans. Roy. Soc. Lond. **A 315** (1985) 125.
13. G. W. Gibbons and N. S. Manton, Nucl. Phys. **B274** (1986) 183; N. S. Manton, Phys. Lett. **B198** (1987) 226.
14. B. J. Schroers, Nucl. Phys. **B367** (1991) 172.
15. J. Gasser and H. Leutwyler, Ann. Phys. (N.Y.) **158** (1984) 142.
16. M. Gell' Mann and F. Low, Phys. Rev. **95** (1954) 1300.
17. W. Y. Crutchfield, N. J. Snyderman, and V. R. Brown, Phys. Rev. Lett. **68** (1992) 1660.
18. V. B. Kopeliovitch and B. E. Shtern, JETP Lett. **45** (1987) 203 ; J. Verbaarschot, T. Walhout, J. Wambach, and H. Wyld, Nucl. Phys. **A468** (1987) 520; A. J. Schramm, Y. Dothan, and L. C. Biedenharn, Phys. Lett. **B205** (1988) 151; E. Braaten and L. Carson, Phys.Rev. **D38** (1988) 3525.
19. N. S. Manton, Phys. Lett. **192B** (1987) 1.
20. W. Y. Crutchfield, N. J. Snyderman, V. R. Brown, and J. Hughes, "Properties of the Deuteron in the Skyrme Model", in preperation.
21. A. Jackson, A. D. Jackson, and V. Pasquier, Nucl. Phys. **A432**, 567 (1985); R. Vinh Mau, M. Lacombe, B. Loiseau, W. N. Cottingham, and P. Lisboa, Phys. Lett. **B150**, 259 (1985).
22. T. S. Walhout, Nucl. Phys. A
23. T. S. Walhout and J. Wambach, Phys. Rev. Lett. **67** (1991) 314.
24. B. Julia and A. Zee, Phys. Rev. **D11** (1975) 2727.
25. N. J. Hitchin, Comm. Math. Phys. **83** (1982) 579.
26. N. R. Walet, R. D. Amado, and A. Hosaka, Phys. Rev. Lett. **67** (1992) 3849.
27. D. O. Riska and E. M. Nyman, Phys. Lett. **183B** (1987) 7.
28. J. D. Talman, "Special Functions", W. A. Benjamin, Inc., New York (1968).
29. Schulman, Phys. Rev. **176**, 1558 (1968)[22].
30. For a review, see, for example, R. Rajaraman, Phys. Rep. **21C**, 277 (1975).
31. R. Dashen, S-k Ma, and H. J. Bernstein, Phys. Rev. **187** (1969) 345.
32. B. Moussallam and D. Kalafatis, Phys. Lett. **B272** (1991) 196; B. Moussallam, proceedings of this conference.
33. G. Holzwarth, proceedings of this conference.

34. S. Weinberg, Phys. Lett. **91B** (1980) 51.
35. S. Dawson, J. S. Hagelin, and L. Hall, Phys. Rev. **D23** (1981) 2666.
36. C. G. Callan, Phys. Rev. **D2** (1970) 1541; K. Symanzik, Comm. Math. Phys. **18** (1970) 227
36. H. Lehmann, Phys. Lett. **41B** (1972) 529.
37. S. Okubo, Supl. Prog. Theor. Phys. **63** (1978) 1.

NUCLEAR MATTER ASPECTS OF SKYRMIONS

ANDREAS WIRZBA
Institut für Kernphysik, Technische Hochschule Darmstadt, Schloßgartenstr. 9
D-W-6100 Darmstadt, Germany

ABSTRACT

As an alternative approach to the infinite-array description of dense matter in the Skyrme model, we report about the properties of a single skyrmion on a compact 3-sphere of finite radius. The density of this matter can be increased by decreasing the hypersphere radius. As in the array calculations one encounters a transition to a distinct high density phase characterized by a delocalization in energy and baryon charge and by increased symmetries. We will argue that the high density phase has to be interpreted as chirally restored one. The arguments are based on the formation of complete chiral multiplets and the vanishing of the pionic massless Goldstone modes in the high-density fluctuation spectrum. We show that the restoration of chiral symmetry is common to any chirally invariant extension of the usual Skyrme model - whether via higher-order contact terms or via the introduction of stabilizing vector mesons which act over a finite range.

1. Introduction

In this lecture we will present some studies about the high-density behavior of baryonic configurations in the Skyrme model[1] and its variants[2,3]. These models belong to a class of effective models which treat the baryon stabilization (and hence the baryon structure) and the interaction between baryons on the same footing. There is no principal difference between the stabilization and the interaction mechanism, the models just act in different topological sectors: in the baryon (=winding) number $B = 1$ sector for the structure physics[4,3] and in the $B > 1$ sector for the interaction physics[5]. For $B \to \infty$ one has a new parameter (the density) at ones disposal in order to tune the interplay between the stabilization and the interaction mechanisms. Because of these features the Skyrme model and its variants have the inherent possibility to allow for radical changes when baryonic matter is compressed to high densities: the baryons which at low densities are well-separated and clearly defined objects might completely loose their identity and the baryon matter can become uniform. There are at least two distinct ways for investigating the high-density behavior of Skyrme-type models.

The first approach which was pioneered by Igor Klebanov[6] uses infinite periodic arrays of skyrmions with baryon number (winding number) of one per unit cell in real space-time. The details of the lattice structure and the periodic boundary conditions to which the mesonic fields are subject can be chosen as to avoid nearest neighbor frustration at low densities. The aim is to find the classical static field configuration which minimizes the energy per cell volume for a fixed baryon charge per array cell. The equations of motion (which are non-linear partial differential equations) are solved numerically on a lattice grid subject just to two constraints: the choice of the crystal and the form of the *twisted* periodic boundary conditions which guarantee the periodic structure and minimize the frustrations in the field gradients (and thus the energy) between neighboring cells in the asymptotic low density region. The boundary conditions are then extrapolated without alteration to high densities as well. At low densities the skyrmions in the periodic arrays are well-separated, ensembled in a phase of weakly interacting baryons. With increased density, however, they grow in size as measured by their r.m.s. radius until at a critical density - as first observed by Wüst, Brown and Jackson[7] - they "melt" to a distinct high-density phase where the skyrmions completely lose their identity. The new phase is characterized by the approximate uniformity in the baryon as well as the energy density, by the fact that the averages of the σ- and π-fields vanish over the cell volume and that there appears an additional symmetry at the critical density: the "half-skyrmion symmetry" [a] of Goldhaber and Manton.[8]

The short-comings of these calculations are: they are by fiat of numerical nature, the rotational and translational symmetries are explicitly broken down to discrete ones by the artificial crystalline grid structure. Finally the calculations depend on the form of the crystal and the twisted periodic boundary conditions which are extracted in the asymptotic low density regime, but nevertheless applied at high densities. For instance, the order of the phase transition can depend on these choices: For the same simple cubic lattice it is first order for Klebanov's choice of boundary terms, while it is second order for the modified rectangular boundary terms of Jackson and Verbaarschot[9] where a preferred direction in space is singled out. But the fact that there are phase transitions are common for all these calculations - and even the critical densities are approximately the same. Klebanov's crystal is not of minimal energy, a face-centered cubic arrangement of skyrmions (at low densities) is preferable. Again, as the density increases there is a second-order phase transition and the half-skyrmion symmetry emerges.[10] The crystal of minimal energy is in this higher density phase and has energy per baryon only of 3.8% above the topological lower bound. Further generalizations are the inclusion of temperature and the derivation of the equation of state for the Skyrme matter in the case of the simple cubic lattice[11] as well as the fcc structure[12] by T.S. Walhout. He has also studied periodic arrays using an ω-stabilized variant of the Skyrme model.[13]

Nevertheless, the question might arise whether these phase transitions are lattice artifacts or not. Fortunately, there is an alternative approach to study the same phenomena by replacing the periodic arrays in flat space $\mathbf{R}^3$ by few-skyrmion-systems (or even a single skyrmion) on the compact manifold $\mathbf{S}^3(L)$ pioneered by Nick Manton.[14,15] The finite baryon number on $\mathbf{S}^3(L)$ corresponds to a finite baryon density in $\mathbf{R}^3$ (with infinite baryon number), so that one obtains a model for skyrmion matter which is appreciably simpler to study than any lattice model. The density of this matter can be increased by decreasing the hypersphere radius L, the radius of

[a] This symmetry is characterized by families of planes on which the scalar field, σ is zero. The pionic fields have a reflection symmetry about these planes while the scalar field is reflection antisymmetric.

the 3-sphere in four dimensions. In the limit $L \to \infty$ a skyrmion localized on the hypersphere has basically the same properties as an isolated skyrmion in flat space or in a periodic array for infinite separation. For a large hypersphere it is still well localized and dominated by its stabilization mechanism, but because of the curvature and finite size effects of the hypersphere it acts as if the tail of another skyrmion (here of course nothing else than itself seen via the opposite pole of the 3-sphere) is present. By shrinking the hypersphere radius this interaction mechanism will become stronger and stronger. Finally it can become so strong that the stabilizing mechanism cannot localize the skyrmion any longer: the skyrmion will be smeared (=delocalized) over the whole 3-sphere. The connection between both approaches can be made by identifying the averaged baryon densities: in the periodic array calculation the baryon density is averaged over the cell volume, on the hypersphere it is given by the ratio $B/(2\pi^2 L_{\text{dim}}^3)$ of the baryon charge B and the (surface-) volume of the 3-sphere in terms of the dimensioned radius L_{dim}. The $\mathbf{S}^3(L)$ approach has the advantage that it is technically far simpler than the period array one[b] and that it allows for mathematically rigorous results for the properties of the "high-density" (or "small L") phase of a single skyrmion on $\mathbf{S}^3(L)$. Furthermore neither the (continuous) rotational nor the translational symmetry are broken. The disadvantage is obvious: The physical space is not a hypersphere - at least not a small one. Both approaches supplement each other in a complementary way: the first is realized in normal space-time and for infinite systems (a precondition for a phase transition), but is technically complicated and plagued by its crystalline nature, by the ambiguities in choosing the array itself and the boundary terms; the other is simple and exact, but is set up in an unphysical world and limited to finite systems which therefore should be not identified with an infinite periodic array, but just with *one* array cell.

The talk is organized as follows. In sect. 2 we will give Manton's general description of the Skyrme model on the 3-sphere and review some of the $B = 1$ properties. In sect. 3 we present arguments why the phase transition has to be interpreted as chiral restoration. In sect. 4 we report on the generalization to variants of the Skyrme-model which involve higher-order contact terms or which are stabilized by vector mesons. Sect. 5 contains a discussion about the order of the phase transition. We end the talk with a short discussion section.

2. The Skyrme Model on the Hypersphere

As mentioned there is considerable interest in studying the Skyrme model

$$\mathcal{L}_{2,4} = \frac{f_\pi{}^2}{4}\mathrm{Tr}\left(\partial_\mu U^\dagger \partial^\mu U\right) + \frac{\epsilon_4{}^2}{4}\mathrm{Tr}\left[U^\dagger\partial_\mu U, U^\dagger\partial_\nu U\right]^2, \tag{1}$$

on a 3-dimensional hypersphere of radius L, $\mathbf{S}^3(L)$.[14,15] We will report here especially on the findings of refs.[15,16,17]. In order to get parameterization-independent results it is useful to relate the quaternion representation U of the Skyrme model to a cartesian

[b]In fact most of the calculations can be done analytically or involve at most the numerical task of solving ordinary non-linear differential equations.

representation $\{\Phi^\alpha\} = (\Phi^0, \Phi^z, \Phi^x, \Phi^y)$ with the imposed constraint $\Phi^\alpha\Phi^\alpha = 1$ as $U = \Phi^0 + i\vec{\tau}\cdot\vec{\Phi}$ and to introduce the strain tensor[15]

$$\begin{aligned} K_{ij} &\equiv \partial_i\Phi^a\partial_j\Phi^a \qquad \text{(with} \quad \{\Phi^\alpha\} = (\sigma, \pi^z, \pi^x, \pi^y)) \\ &= -\tfrac{1}{4}\mathrm{Tr}\{U^\dagger\partial_i U, U^\dagger\partial_j U\}_+, \end{aligned} \tag{2}$$

where i and j label the space coordinates. This is a symmetric 3×3 matrix with three positive eigenvalues which we will denote as λ_a^2, λ_b^2, λ_c^2. Manton[15] has shown that the λ_i have a simple geometrical interpretation. They correspond to the length changes of the images of any orthonormal system in a given space manifold (here $\mathbf{R}^3$ or $\mathbf{S}^3(L)$) under the conformal map, U, onto the group manifold, here $SU(2) \cong S_3$. Therefore the name strain tensor which refers to such a general "rubber-sheet" geometry. Among the invariants of K_{ij}, three are fundamental[c] and have a simple geometric meaning:

$$\begin{aligned} \mathrm{Tr}(K) &= \lambda_a^2 + \lambda_b^2 + \lambda_c^2 &&= \textstyle\sum \text{length}^2 \\ \tfrac{1}{2}\left\{(\mathrm{Tr}(K))^2 - \mathrm{Tr}(K^2)\right\} &= \lambda_a^2\lambda_b^2 + \lambda_b^2\lambda_c^2 + \lambda_c^2\lambda_a^2 &&= \textstyle\sum \text{area}^2 \\ \det(K) &= \lambda_a^2\lambda_b^2\lambda_c^2 &&= \text{volume}^2\ . \end{aligned} \tag{3}$$

The first one measures the sum of the squared length changes of the mapped orthonormal frame, the second one the sum of the squared area changes and the third the squared volume change. With the help of Eq.(2) it is easy to see that these invariants are (modulo normalization factors) the static energy densities of the second-order non-linear σ model, $(f_\pi^2/2)\mathrm{Tr}(\partial_i U\partial_i U^\dagger)$, the fourth-order Skyrme term, $-(\epsilon_4^2/4)\mathrm{Tr}[U^\dagger\partial_i U, U^\dagger\partial_j U]$, and the sixth-order term proportional to the square of the baryon density (see Eq.(24) in section 4). The static energy of a general skyrmion configuration on the hypersphere has in this language the form [d]

$$\begin{aligned} E &= \int_{S^3(L)} dV\,(\lambda_a^2 + \lambda_b^2 + \lambda_c^2 + \lambda_b^2\lambda_c^2 + \lambda_c^2\lambda_a^2 + \lambda_a^2\lambda_b^2) \\ &= \int_{S^3(L)} dV\,(\lambda_a - \lambda_b\lambda_c)^2 + (\text{cycl. perm.'s}) + 6\int_{S^3(L)} dV\,\lambda_a\lambda_b\lambda_c. \end{aligned} \tag{4}$$

The last term is just $12\pi^2 B$ with the baryon (winding) number B, since $\lambda_a\lambda_b\lambda_c/(12\pi^2)$ is the Jacobian of the map $\mathbf{S}^3(L)\to\mathbf{S}^3(1)\cong SU(2)$ - in other words the baryon (winding) number density. Because of the positive definiteness of the terms in the last expression of (4) it is obvious that any skyrmion configuration has to respect the topological bound

$$E \geq 12\pi^2 B. \tag{5}$$

Furthermore, one can immediately see that there is exactly one possibility to satisfy the topological bound: The insertion of the identity map, $\lambda_a = \lambda_b = \lambda_c$, with $\lambda_i = 1$.

[c] All other invariants can be constructed from them.

[d] We have adopted a dimensionless form of the Skyrme lagrangian. To obtain the dimensioned quantities, one divides the λ_i's and multiplies L by $2\sqrt{2}\epsilon_4/f_\pi$ and multiplies E by $\sqrt{2}\epsilon_4 f_\pi$.

which corresponds to an isometric mapping from the spacial manifold into the target manifold and which has baryon (winding) number $B = 1$. Specifying to the case at hand where the target manifold is $SU(2) \cong \mathbf{S}^3(1)$, the unit 3-sphere, we see that the only way for saturating the topological bound (the absolute minimum of any skyrmion configuration with a non-zero winding number) is the isometric mapping of the spatial $\mathbf{S}^3(L)$ with $L = 1$ onto the target iso-spin sphere $\mathbf{S}^3(1)$. We can therefore conclude that the topological bound can nether be saturated by any $B > 1$ system or by any skyrmion configuration in flat space, $\mathbf{R}^3$.[e] After minimizing $E[\lambda_a, \lambda_b, \lambda_c]$ under the constraint

$$\int_{S^3(L)} dV\, \lambda_a \lambda_b \lambda_c = 2\pi^2 B \tag{6}$$

Manton[15] and Loss[16] found that for $L < 1$ there is even a stronger bound on any $B = 1$ skyrmion configuration,

$$E_{\text{identity}} = 6\pi^2 \left(L + \frac{1}{L} \right), \tag{7}$$

which is still satisfied by the identity map $\lambda_a = \lambda_b = \lambda_c = 1/L$ independently of the parameterization of the $B = 1$ configuration. In other words for $L \leq 1$ the identity map is the absolute minimum of any $B = 1$ configuration.

Finally, it is now simple to show in full generality[15,16] that for $L < \sqrt{2}$ the identity map is stable for all allowed ($\delta B = 0$) small-amplitude perturbations $\lambda_i \to (1/L) + \delta_i(x)$, whereas it becomes a saddle for $L > \sqrt{2}$.[f] Under the small perturbation given above the energy is given to second-order as

$$E[\lambda_a, \lambda_b, \lambda_c] = E_{\text{identity}} + \left(\frac{2}{L} + \frac{4}{L^3} \right) \int I_1 + \left(1 + \frac{2}{L^2} \right) \int I_2 + \frac{4}{L^2} \int I_3 + O(\delta^3) \tag{8}$$

with $I_1 = \delta_a + \delta_b + \delta_c$, $I_2 = \delta_a^2 + \delta_b^2 + \delta_c^2$ and $I_3 = \delta_a \delta_b + \delta_b \delta_c + \delta_c \delta_a$. Using the B-constraint (6) we have up to third order corrections the relation

$$\frac{1}{L^2} \int_{S^3(L)} dV\, I_1 = -\frac{1}{L} \int_{S^3(L)} dV\, I_3 + O(\delta^3).$$

After inserting this relation in Eq.(8) and using the fact that the δ_i's are local variations, one can finally find that the identity map is stable against small-amplitude fluctuations, *i.e.* $E[\lambda_a, \lambda_b, \lambda_c] > E_{\text{identity}}$, as long as the following local inequality holds

$$\left(1 + \frac{2}{L^2} \right) (\delta_a^2 + \delta_b^2 + \delta_c^2) - 2(\delta_a \delta_b + \delta_b \delta_c + \delta_c \delta_a) > 0, \tag{9}$$

---

[e] Of course, we cannot exclude the possibility that such a configuration might come infinitesimally close to the topological bound.

[f] That means that the identity map is at least a local minimum for $1 < L < \sqrt{2}$.

in other words as long as $L < \sqrt{2}$. This concludes Manton's general proof - independent of any parameterization - that the identity map is at least a local minimum up to $L_c = \sqrt{2}$ and becomes a saddle for $L > \sqrt{2}$.

So far we have not specified any parameterization. In the following we will specialize to the familiar hedgehog parameterization. It has the following properties: (a) according to a general theorem by Palais[18] about reduced variational equations for symmetric fields a variational solution of hedgehog form is a solution of the full set of Euler-Lagrange equations of the Skyrme model, (b) the hedgehog solution is the energetically lowest solution known so far in the $B = 1$ sector in $\mathbf{R}^3$ and (c) it is numerically proven that the $B = 1$ hedgehog solution is at least a local minimum. Hedgehog configurations on the hypersphere are conveniently described in terms of the conventional 'polar' coordinates μ, θ, ϕ on $\mathbf{S}^3(L)$ with $0 \le \mu, \theta \le \pi$ and $0 \le \phi \le 2\pi$, so that a typical point on $\mathbf{S}^3(L)$ has Cartesian coordinates

$$(L\cos\mu,\ L\sin\mu\cos\theta,\ L\sin\mu\sin\theta\cos\phi,\ L\sin\mu\sin\theta\sin\phi). \tag{10}$$

In these coordinates the metric is $ds^2 = L^2(d\mu^2 + \sin^2\mu\, d\theta^2 + \sin^2\mu\, \sin^2\theta\, d\phi^2)$ and the volume element is

$$dV = L^3 \sin^2\mu\, d\mu\, \sin\theta\, d\theta\, d\phi\ . \tag{11}$$

The field components of a hedgehog of baryon number B on $\mathbf{S}^3(L)$ have then the form

$$\begin{aligned} \Phi^0 &= \cos f(\mu) \\ \Phi^z &= \sin f(\mu)\cos\theta \\ \Phi^x &= \sin f(\mu)\sin\theta\cos\phi \\ \Phi^y &= \sin f(\mu)\sin\theta\sin\phi \quad , \end{aligned} \tag{12}$$

where the "radial" profile function $f(\mu)$ is subject to the boundary condition $f(0) = 0$ and $f(\pi) = B\pi$, B integer. The topological winding-number B is of course the baryon charge of the corresponding field configuration. The field components, $f_\pi\Phi^\alpha$ (f_π is the pion-decay constant) should be identified with the familiar σ, π^z, π^x, π^y fields. In the normal quaternion representation the hedgehog ansatz reads

$$U \equiv \Phi^0 + i\vec{\tau}\cdot\vec{\Phi} = \exp(i\vec{\tau}\cdot\hat{r}(\theta,\phi)\, f(\mu)) \tag{13}$$

where $\hat{r}(\theta,\phi)$ is the usual radial $\mathbf{S}^2$ unit vector. Note that for the hedgehog configuration the eigenvalues of the strain tensor are simply $\lambda_a^2 = f'^2/L^2$, $\lambda_b^2 = \lambda_c^2 = \sin^2 f/(L^2\sin^2\mu)$ where f' stands for $\frac{df}{d\mu}$.

The energy of the hedgehog field configuration (13) on $\mathbf{S}^3(L)$ is [g]

$$\begin{aligned} E = 4\pi L \int_0^\pi \sin^2\mu d\mu \left(f'^2 + 2\frac{\sin^2 f}{\sin^2\mu} \right) \\ + \ 4\pi\frac{1}{L}\int_0^\pi \sin^2\mu d\mu \left(\frac{\sin^2 f}{\sin^2\mu}\left[2f'^2 + \frac{\sin^2 f}{\sin^2\mu}\right]\right) . \end{aligned} \tag{14}$$

[g] Note in order to obtain the dimensioned quantities, one multiplies L in (14) by $2\sqrt{2}\epsilon_4/f_\pi$ and E by $\sqrt{2}\epsilon_4 f_\pi$.

Variation of (14) with respect to the radial profile function $f(\mu)$ leads us to the Euler equation

$$\frac{d^2 f}{d\mu^2}+\frac{\sin 2\mu}{\sin^2\mu}\frac{df}{d\mu}-\frac{\sin 2f}{\sin^2\mu}$$
$$+\frac{2}{L^2}\frac{\sin^2 f}{\sin^2\mu}\frac{d^2 f}{d\mu^2}+\frac{\sin 2f}{L^2\sin^2\mu}\left\{\left(\frac{df}{d\mu}\right)^2-\frac{\sin^2 f}{\sin^2\mu}\right\} = 0\,. \tag{15}$$

Again because of Palais' general theorem[18], solutions of (15) yield solutions of the full set of Euler-Lagrange equations of the Skyrme model on $\mathbf{S}^3(L)$ when substituted into (12) or (13). It is very simple to check that one solution of Eq.(15) (in the case $B = 1$) is the map $f(\mu) = \mu$ which corresponds to the uniform mapping of $\mathbf{S}^3(L)$ onto the isospin manifold $\mathbf{S}^3(1) \cong SU(2)$. One can easily see that for the uniform hedgehog map, $f(\mu) = \mu$, the eigenvalues of the strain tensor (2) are simply $\lambda_a^2 = \lambda_b^2 = \lambda_c^2 = 1/L^2$. In other words it is a special parameterization of the identity map. The uniform map, $f(\mu) = \mu$, is a solution for all L, but has no finite $L \to \infty$ limit. The energy associated with the uniform map is of course given by (7). As mentioned for $1 < L < \sqrt{2}$, the uniform map is still at least a local minimum. For $L > \sqrt{2}$ there exist a lower energy configuration which represents a skyrmion localized about one point, *i.e.* $f(\mu) \neq \mu$ in the hedgehog parameterization (12).[15,17] In fact in this parameterization the identity map bifurcates into two solutions of the same energy concentrated around the north or south pole of the hypersphere which are related by[17]

$$f_N(\mu) = \pi - f_S(\pi - \mu). \tag{16}$$

In the limit $L \to \infty$, one recovers the usual flat space skyrmion.[17] The identity map has O(4) symmetry (*see* Eq.(12)) and is therefore completely uniform in energy and baryon density. For $L > \sqrt{2}$ the symmetry of the energetically most favorable solution is broken to O(3) indicating that the solution is localized in energy and baryon distribution. The bifurcation at $L = \sqrt{2}$ is of the standard "pitchfork" type corresponding to a second order phase transition. In Landau's scheme the bifurcation is characterized by a symmetric quartic polynomial whose quadratic term changes sign.

In order to discriminate the delocalized (high-density) phase from the localized (low-density) phase in a way which can be generalized to few-skyrmion systems on the hypersphere or to flat space array-calculations, several candidates for an order parameter were studied in ref.[17]: One candidate is the integrated squared deviation of the local energy/baryon density from the averaged density normalized by the squared averaged density times the hypersphere volume. It converges to the value 1 for the localized solution in the limit $L \to \infty$, for smaller L it decreases and shows a square-root bifurcation at $L = \sqrt{2}$ to the value zero belonging to the identity map. The interpretation would be that this order parameter signalled deconfinement in the energy and baryon distribution. This is, however, a special property of the $B = 1$ system on $\mathbf{S}^3(L)$. For few-skyrmion systems[17,19] and flat space arrays[9,10] this parameter never becomes zero in the high-density phase, although there is still a sizable decrease in its value from the low-density to the high-density phase. The second suggestion of ref.[17] on the other hand works also in these cases. It is the squared chiral expectation value

$$\langle \sigma/f_\pi \rangle^2 + \langle \vec{\pi}/f_\pi \rangle^2, \tag{17}$$

where the σ and pion fields are averaged over the hypersphere volume (or the cell volume for periodic arrays). Because of the residual $O(3)$ symmetry for the hedgehog (or discrete symmetries in the array calculations) which cause $\langle \vec{\pi} \rangle$ to be zero it has the same content as the parameter $\langle \sigma \rangle$ alone. For the identity map it is obvious (*see* Eq.(12)) that the parameter (17) vanishes. In the case of the localized solution it approaches the value 1 with increasing L, since in most of the space the U field of the localized skyrmion is approximately close to the vacuum value $U_0 = 1$ and only markedly deviates from this in the small region where the skyrmion is located. The interpretation is that a zero in this order parameter signals chiral symmetry restoration or at least "chiral democracy", since there is no escape from the "chiral-circle" constraint $\sigma^2 + \vec{\pi}^2 = f_\pi^2$. But nevertheless as shown for few-skyrmion systems on the hypersphere[17,19] and for flat space arrays[9,10] the σ and pion profiles of the high-density phase are so equally distributed about the chiral circle that their averages vanish.

Let me summarize the $B = 1$ results on $\mathbf{S}^3(L)$: There is a second-order transition at a baryon density $1/(2\pi^2 L_c{}^3)$ for $L_c = \sqrt{2}$ from a localized and chirally broken low-density solution to a "chirally restored" high-density solution. Restoring dimensionful parameters into the Skyrme lagrangian (by using the empirical value of the pion decay constant $f_\pi = 93$ MeV and an $\epsilon_4 = 0.0743$ guaranteeing a reasonable value for g_A) one finds

$$\rho_c = \left(\frac{f_\pi}{2\sqrt{2}\epsilon_4}\right)^3 \frac{1}{\sqrt{2}^3}\frac{1}{2\pi^2} \approx 0.20\text{fm}^{-3}, \tag{18}$$

which is far too low.[h](remember that the nuclear matter density is $\rho_{\rm nm} = 0.16\text{fm}^{-3}$). The important result, however, is that the $\mathbf{S}^3(L)$ value for ρ_c is more or less the same than the one found in the periodic array calculations[7,9,10] under the same input parameters, approximately $\rho_c = 0.17\text{fm}^{-3}$. Furthermore, the corresponding transition densities ρ_c in the few-skyrmion calculation[17,19] interpolate between these two numbers. Thus the identification of the hypersphere formalism with the periodic array calculation seems to be justified even quantitatively.

[h]This might improve if the stabilization term is replaced by more realistic ones. Furthermore all contributions from kinetic terms are neglected. Finally, there is the problem of the missing central attraction[5] which is probably linked to loop corrections[20] or missing $1/N_c$ corrections[21]. Any additional attractive term tends to localize the skyrmion and will therefore move the transition point up to higher densities.

3. Indications for a Chiral Symmetry Restoration

In this section I will present some arguments (*see* refs.[22,23] for more details) why the transition reviewed in the preceeding section should be interpreted as chiral symmetry restoration. As mentioned the complete uniformity of the high-density $B = 1$ solution on $\mathbf{S}^3(L)$ is linked to the strong condition $f(\mu) = \mu$ and does not generalize to few-skyrmion systems[17,19] on $\mathbf{S}^3(L)$ or periodic arrays[7,9,10] in flat space. However, the "chiral democracy" as discussed at the end of the preceeding section can even be achieved via a weaker condition

$$f(\mu) = \pi - f(\pi - \mu) \tag{19}$$

which indicates that in the high-density phase there is a symmetry about the hypersphere equator between the "northern" and the "southern" hemisphere. In fact this is just the half-skyrmion symmetry[8] which is a common signal of the high-density phase for few-skyrmion systems on the hypersphere[17,19] as well as for periodic arrays in flat space[8,24]. So the chiral symmetry restoration seems to be prior to the delocalization. In fact this can be tested by adding a (pion mass) term which explicitly breaks chiral symmetry:[22]

$$\mathcal{L}_{\mathrm{SB}} = \frac{m_\pi{}^2 f_\pi^2}{4} \mathrm{Tr}(U + U^\dagger - 2). \tag{20}$$

When this term is added, the phase transition is not a sharp one any longer, but it is smeared out. The chiral order parameter (17) still approaches zero for higher and higher densities, but never actually becomes exactly zero. Of course, this is just a consequence of the explicit breaking of the chiral symmetry. A second effect of adding the term (20) is a small shift of the now approximate phase transition to higher densities, since the term induces an attractive force.

In ref.[22] it is furthermore argued that the vanishing of the order parameter $\langle\sigma\rangle$ (for the Skyrme model without explicit symmetry breaking of course) implies the vanishing of the matrix element for pion decay $\langle 0|A^a_\mu(x)|\pi^b\rangle$ in case the axial current A^a_μ is calculated in the mean-field approximation in the frame-work of a σ-model, and therefore the vanishing of the "effective" pion decay constant and the quark condensate.

The most convincing arguments why the phase transitions in the Skyrme model ought to be identified with a chiral symmetry restoration are the following two: In the $B = 1$ case on $\mathbf{S}^3(L)$ it can be shown that the transition from the low-density phase to the high-density phase is accompanied

(i) by the formation of complete chiral multiplets in the fluctuation spectrum
(ii) and by the vanishing of the three pionic Goldstone modes from the spectrum.

These two points are treated at length in ref.[23] where the local stability of the uniform *and* the localized skyrmion solution on $\mathbf{S}^3(L)$ and thus their small-amplitude normal modes are investigated. If all normal modes have positive energy, the solution is locally stable. If some normal modes have negative energy, the solution is a saddle point. The study of the small-amplitude fluctuations about the $B = 1$ solutions on the hypersphere revealed the following scenario:

(i) In general, the modes about a hedgehog solution have just the usual $O(3)$ degeneracies, *i.e.* a degeneracy $2N + 1$ in terms of the principal quantum number $N \geq 1$ of the modes.

(ii) The modes about the identity map on the other hand can be classified according to the following representations of the group $O(4)$: the symmetric tensor representation $(N, 0)$ which has $(N + 1)^2$ degenerate states and as static eigenvalues [i]

$$\lambda_{N,0} = L\{N(N+2) - 4\} + 2\{N(N+2) - 2\}/L, \tag{21}$$

and the $(N, 1)$ representation with a $2((N + 1)^2 - 1)$ degeneracy and the static eigenvalues

$$\lambda_{N,1} = \{N(N+2) - 3\}(L + 1/L). \tag{22}$$

Thus their degeneracy exceeds by far the one of the modes about localized hedgehog solutions. Furthermore for each mode about the identity map with a given parity, there exists at least one degenerate mode of opposite parity. Using the covering group $SU(2)_L \times SU(2)_R$ of $SO(4)$ it can be shown (see ref.[23]) that the $(N, 0)$ modes belong to a trajectory based on the four-fold degenerate $(1/2, 1/2)$ representations of $SU(2)_L \times SU(2)_R$, while the $(N, 1)$ modes belong to a trajectory based on the six-fold degenerate $(1, 0) + (0, 1)$ representation. Thus the modes form complete multiplets of chiral symmetry either by the additional $(\sigma, \vec{\pi})$ degeneracy of the $(N, 0)$ modes (21) or by the parity doubling of the $(N, 1)$ modes (22). Note that in ref.[23] the $(N, 0)$ modes were identified as purely "electric" grand-spin modes (but with a degeneracy between *e.g.* grand spin $\mathbf{K} = 0^+$ and $\mathbf{K} = 1^-$ modes), while the $(N, 1)$ modes were identified as degenerate "magnetic" and "electric" modes of the same grand spin $\mathbf{K}$, but opposite parity. The $(N = 1, 1)$ modes are the sixth zero modes about the identity map.

(iii) At high densities ($L < \sqrt{2}$) all normal modes about the identity map have positive energies, the identity map is stable. There are four degenerate low-lying modes in the spectrum, the four $(N = 1, 0)$ modes, one with σ quantum numbers and three with pion ones indicating that the σ and all three π-fields are degenerate and treated on the same footing. They have the static eigenvalue

$$\lambda_{N=1,0} = -L + 2/L. \tag{23}$$

(iv) The four $(N = 1, 0)$ modes cross zero at $L = \sqrt{2}$ and become negative for $L > \sqrt{2}$ signalling there the instability of the identity map.

(v) At $L = \sqrt{2}$ there is the bifurcation into the normal localized hedgehog solution with the reduced $O(3)$ symmetry. All modes about the localized solution are

---

[i] We use the phrase "static eigenvalue" as a short-hand notation for the change of the static energy under the small perturbation of the fluctuation-mode.

stable for $L \geq \sqrt{2}$. The modes with $N \geq 1$ have (with the exception of some accidental degeneracies for special values of L) only the usual $O(3)$ degeneracy $2N+1$. However, there exist 9 zero modes.[j] The three additional zero modes (compared with the number of zero modes for the identity map) correspond to the three unstable pionic ($N = 1, 0$) modes of the identity map. Their fourth partner, the σ-like mode, corresponds to a positive stable excitation about the localized hedgehog (to an infinitesimally small conformal variation in the profile function $f(\mu) \to f(\mu) + \delta(\mu)$). The breakdown of the continuous symmetry $O(4)$ of the identity map at $L = \sqrt{2}$ to the $O(3)$ symmetry of the localized for $L > \sqrt{2}$ energetically favorable hedgehog solution is therefore closely linked to the appearance of three new zero modes.

(vi) These three zero modes should be in fact interpreted as the three pionic Goldstone modes linked to the spontaneous breakdown of chiral symmetry. The arguments for this identification are the following: Whereas six of the nine zero modes stay normalizable even in the limit $L \to \infty$, three have no finite $L \to \infty$ limit. This is consistent with the fact that the flat space hedgehog has only six zero modes. Furthermore when the chiral symmetry breaking term (20) is added to the Skyrme lagrangian, the above mentioned normalizable zero modes stay zero modes, whereas the three additional zero modes are shifted in energy by the pion mass. These are the three pionic Goldstone modes which become massive when an explicitly chiral symmetry breaking term is added and which show up as non-normalizable plain wave excitations of the vacuum in the flat space limit.

Since the three Goldstone modes disappear for the high-density delocalized phase and the high-density modes form complete chiral multiplets, the chiral symmetry restoration is established.

4. Generalizations

4.1. *Contact Terms*

In the following we will show that the form of the Skyrme model (1) and especially the stabilization by the fourth-order Skyrme term is not a necessary precondition for the very existence of the chiral phase transition. Let us for instance replace the fourth-order stabilizing term in (1) by a stabilizing term of sixth-order (*see e.g.* ref.[25])

$$\mathcal{L}_{2,6} = \frac{f_\pi^{\,2}}{4} \mathrm{Tr}\left(\partial_\mu U^\dagger \partial^\mu U\right) - c_6 B_\mu B^\mu \tag{24}$$

(note $c_6 > 0$) where B_μ is the topological baryon current

$$B^\mu = \frac{\varepsilon^{\mu\nu\alpha\beta}}{24\pi^2} \mathrm{Tr}\left((U^\dagger \partial_\nu U)(U^\dagger \partial_\alpha U)(U^\dagger \partial_\beta U)\right). \tag{25}$$

[j] Note that a hedgehog in flat space has only 6 zero modes: 3 rotational (which cannot be separated from iso-rotational ones because of the hedgehog symmetry) and 3 translation ones.

As reported in section 2 the fourth-order Skyrme term as well as the sixth-order term and the second-order non-linear sigma model term have a geometrical meaning when expressed in the strain tensor language. One can apply therefore Manton's general machinery[15] to show that the sixth-order term allows for the identity map as a solution and to construct the critical value L_c where the identity map of the model (24) becomes unstable and bifurcates into a localized solution. In fact all qualitative phenomena show up as before. Naturally the value of L_{min} for the minimum of the static energy is in general not the same for both models and more importantly the critical hypersphere radius L_c where the chiral restoration occurs is

$$L_c = (3)^{\frac{1}{4}} L_{\text{min}} \tag{26}$$

for the model (24) instead of $L_c = \sqrt{2} L_{\text{min}}$ as it was the case for the usual Skyrme model (1).[17]

As shown in ref.[26] any lagrangian which (a) is a polynomial of the fundamental invariants (3) of the strain tensor, (b) which has the usual vacuum properties and (c) which allows for (locally) stable $B = 1$ Skyrmion solutions, leads to a chirally restored phase at high densities. In addition the position of the phase transition can be uniquely derived just from the lagrangian without solving any equation of motion. Let us discuss for this purpose a lagrangian of the analytic form

$$\mathcal{L}_{\mathcal{G}} \equiv -\mathcal{G}(-\mathcal{L}_2, \mathcal{L}_4, \mathcal{L}_6) = -\sum_{j_1, j_2, j_3 = 0}^{\infty} A_{j_1 j_2 j_3} (-\mathcal{L}_2)^{j_1} (-\mathcal{L}_4)^{j_2} (-\mathcal{L}_6)^{j_3} \tag{27}$$

where j_1, j_2, j_3 are integer indices and $A_{j_1 j_2 j_3}$ Taylor coefficients. The arguments of the analytic function $\mathcal{G}$ and the signs are chosen in such a way, that the corresponding static energy density has the simple form

$$\mathcal{E} \equiv \mathcal{G}(\mathcal{E}_2, \mathcal{E}_4, \mathcal{E}_6) = \sum_{j_1, j_2, j_3 = 0}^{\infty} A_{j_1 j_2 j_3} \, \mathcal{E}_2{}^{j_1} \mathcal{E}_4{}^{j_2} \mathcal{E}_6{}^{j_3}. \tag{28}$$

The function $\mathcal{G}$ should be of course positive definite to ensure a positive energy density and $\mathcal{G}(0,0,0) \equiv 0$ (*e.g.*, $A_{000} = 0$) to ensure triviality in the $B = 0$ sector. By inserting $f = 0$ or $f' = 0$ one furthermore learns that $\mathcal{G}(x, 0, 0) \geq 0$ and $\mathcal{G}(x, x^2/4, 0) \geq 0$ for any real $x \geq 0$. The first inequality signals that a possible negative term expressed solely by the second-order term can nether be compensated by adding fourth- or sixth-order terms in *any* combination to the static energy density. Therefore Skyrme-type models which at fourth-order are not positive definite cannot be saved by the addition of a sixth-order term proportional to $B_\mu B^\mu$ or any power of this term.

Note that the energy density (28) is the most general symmetric function of the eigenvalues, λ_i^2, of the strain tensor K_{ij} (2) and includes all forms which can come form lagrangians which involve arbitrary combinations of powers of first derivatives of the fields to even order.

In the chirally restored (uniform) regime on the hypersphere the energy densities of the second-, fourth- and sixth-order term, $\mathcal{E}_2$, $\mathcal{E}_4$ and $\mathcal{E}_6$, have the simple form (modulo a prefactor which can be incorporated into the Taylor coefficients $A_{(j_1 j_2 j_3)}$)

$$\mathcal{E}_{2i} = \frac{3}{L^{2i}}, \qquad i = 1,2,3. \tag{29}$$

In ref.[26] the eigenvalues of all possible small amplitude (static) perturbations around the *identity* map were obtained as the following analytical result:

$$\lambda^{\mathcal{G}}_{N,a} = \left\{ \sum_{i=1,2,3} \lambda_{i,N,a} \frac{\partial}{\partial \mathcal{E}_{2i}} + \lambda_{3,N,a} \frac{2L^2}{3} \left(\frac{\partial}{\partial \mathcal{E}_2} + \frac{2}{L^2} \frac{\partial}{\partial \mathcal{E}_4} + \frac{3}{L^4} \frac{\partial}{\partial \mathcal{E}_6} \right)^2 \right\} \mathcal{G}(\mathcal{E}_2, \mathcal{E}_4, \mathcal{E}_6) \tag{30}$$

with

$$\lambda_{i,N,a} = i \left\{ N(N+2) + a^2 - 4 - \frac{i-1}{2} \left(\{a(N+1)\}^2 - 4 \right) \right\} L^{3-2i}. \tag{31}$$

where the indices $N \geq 1$ and $a = 0, 1$ characterize the allowed $O(4)$ representations:[k] $(N, a = 0)$, the symmetric tensor representations which have $(N+1)^2$ degenerate states and the $(N, a = 1)$ representations which have a $2\{(N+1)^2 - 1\}$ degeneracy (*see* refs.[23,26] for further details). Note that in the case $a = 1$ the expression for $\lambda^{\mathcal{G}}_{N,a=1}$ simplifies to

$$\lambda^{\mathcal{G}}_{N,a=1} = L\{N(N+2) - 3\} \left(\frac{\partial}{\partial \mathcal{E}_2} + \frac{1}{L^2} \frac{\partial}{\partial \mathcal{E}_4} \right) \mathcal{G}(\mathcal{E}_2, \mathcal{E}_4, \mathcal{E}_6). \tag{32}$$

A negative value for any eigenvalue $\lambda_{N,1}$ in (32) would mean that infinitely many of the $(N, 1)$ modes would pass through zero energy *simultaneously*. This is probably indicative of a non-perturbative configuration of lower dimension, *e.g.*, a string-like configuration. Fortunately, one can exclude this pathological instability, since it violates the constraint that the model (27) should have the same vacuum properties as the nonlinear σ model (see ref.[26] for a discussion about this point). So the $(N, 1)$ fluctuations cannot lead to an instability. In case the identity map is (locally) stable in the high-density region (*i.e.* all $\lambda^{\mathcal{G}}_{N,a} \geq 0$), the monotonic increase of the eigenvalues $\lambda^{\mathcal{G}}_{N,a}$ of the small-amplitude normal modes with increasing N is guaranteed. Then, the critical L_c where the identity map becomes unstable against small-amplitude perturbations is given by the value of L where the first $\lambda^{\mathcal{G}}_{N,0}$ in Eq.(30) becomes negative.

Finally, let

$$\mathcal{G}(\mathcal{E}_2, \mathcal{E}_4, \mathcal{E}_6) \to \mathcal{E}_2 \qquad \text{for } \mathcal{E}_{2i} \to 0 \tag{33}$$

and

$$\mathcal{G}(\mathcal{E}_2, \mathcal{E}_4, \mathcal{E}_6) \to C_1 \mathcal{E}_2{}^2 + C_2 \mathcal{E}_4 \qquad \text{for } \mathcal{E}_{2i} \to \infty \tag{34}$$

where i runs from 1 to 3 and $C_1 \geq 0$ and $C_2 > 0$ are fixed positive coefficients. The condition (33) enforces that asymptotically for low densities (large L) the lagrangian (27) is dominated by the non-linear sigma model lagrangian, whereas for high densities

[k]*See* section 3 for the corresponding results for the usual Skyrme model (1).

(small L) it scales as a free Fermi gas (*i.e.* $\mathcal{E}_{\mathcal{G}} \propto 1/L^4$). The low density behavior is of course indisputable, the high density constraint on the other hand requires the additional input that such effective models can be applied even for densities where the underlying theory, QCD, becomes to leading order a free Fermi gas.

Back to the stability analysis: Note that the signs of the static eigenvalues $\lambda^{\mathcal{G}}_{N,0}$ and therefore the stability of the uniform regime follow from the signs of the derivatives acting on $\mathcal{G}$ and from the signs of the coefficients $\lambda_{i,N,0}$ (31). From all possible $\lambda_{i,N,a}$ with $i = 1,2,3$, $N \geq 1$ and $a = 0,1$ only the term $\lambda_{i=1,N=1,a=0}$ is negative, all the other coefficients are guaranteed to be larger than zero or at most equal zero. (To the latter category belong the six $O(4)$ zero modes with $N = 1$ and $a = 1$. Furthermore all the terms $\lambda_{i=3,N,1}$ are zero indicating that the sixth-order term itself has infinitely many zero modes.) Now taking the high-density (34) and the low-density behavior (33) into account, we can conclude that at high densities the uniform $B = 1$ skyrmion solution on the hypersphere is bound to be stable whereas at sufficiently low densities, where the non-linear σ model term becomes the dominant one, the uniform solution is unstable since $\lambda^{\mathcal{G}}_{N=1,0}$ becomes negative eventually. Furthermore we have to take into account that in the stable regime - even infinitesimally close to the instability - all $\lambda^{\mathcal{G}}_{N,0}$ have to increase monotonically with N, that the $\lambda^{\mathcal{G}}_{N,1}$ modes have to be stable by fiat and that the eigenvalues of the modes have to be smooth functions of L. Then the existence of a critical value of L is guaranteed where $\lambda^{\mathcal{G}}_{1,0}$ becomes negative such that the uniform high density phase becomes unstable. There is a bifurcation from the uniform solution which is chirally restored to the usual, localized and chirally broken $B = 1$ hedgehog solution. So even for the most general *geometric* lagrangian (27) built form first order field derivatives the existence of the chiral phase transition (as discussed at length in the preceeding chapters for the normal Skyrme model) is guaranteed under the assumption of a few reasonable constraints (the vacuum structure, Eqs. (33) and (34)) on the form of the lagrangian.

4.2. *Vector Meson Stabilization*

We have discussed so far lagrangians which ensured the stabilization of skyrmions by higher-order (contact) terms in the field derivatives. Naturally the question arises whether the existence of the chiral phase transition is linked to the stabilization by *contact* terms or whether vector mesons can be added which act over a finite range, as *e.g.* ω (and ρ) mesons. Let us take as an example a Skyrme-type variant with ω meson stabilization. (See ref.[13] for a periodic array study of this model.) The corresponding lagrangian has the structure[27]

$$\mathcal{L}_{2,\omega} = \frac{f_\pi^2}{4}\mathrm{Tr}(\partial_\mu U^\dagger \partial^\mu U) - \frac{1}{4}\omega_{\mu\nu}\omega^{\mu\nu} + \frac{m_\omega^2}{2}\omega_\mu\omega^\mu + g_\omega \omega_\mu B^\mu \qquad (35)$$

where the first part is the usual non-linear σ model term, the second one the ω kinetic term expressed through the ω field-strength tensor $\omega_{\mu\nu} = \partial_\mu\omega_\nu - \partial_\nu\omega_\mu$, the third part is the ω mass term and the last one is the coupling of the ω field to the topological baryon current (25). As shown by Adami[28] the lagrangian (35) has stable $B = 1$ solitons.

By putting an ω-stabilized $B = 1$ hedgehog on the hypersphere one can easily find that one solution of the equations of motion is always the uniform solution $f(\mu) = \mu$ for the hedgehog profile and

$$\omega_0 = -\frac{g_\omega}{m_\omega^2} B_0 = -\frac{g_\omega}{m_\omega^2} \frac{1}{2\pi^2 L^3} \tag{36}$$

for the ω_0-component. The ω_0-component is in this case spatially constant and depends only on the hypersphere radius L. (As in flat space the spatial components ω_i are identically zero because of the static hedgehog form of the soliton profile.) When the uniform solutions $f(\mu) = \mu$ and (36) are reinserted into the corresponding energy density of the lagrangian (35), one recovers the same structure as for a model stabilized by a sixth-order term with $\mathcal{L}_6 = -(g_\omega^2/2m_\omega^2)B_\mu B^\mu$. This should not come as a surprise, since in the limit $m_\omega, g_\omega \to \infty$, under a constant ratio g_ω/m_ω, the lagrangian (35) converges to the lagrangian (24) with $c_6 = (g_\omega^2/2m_\omega^2)$. Both lagrangians have the same high-density phase in the static sector. The physical reason for the identical behavior of the static solutions of both lagrangians is that at high densities it is energetically favorable that the ω_0-component becomes spatially constant, otherwise the "costs" in energy from a finite spatial gradient term in this component becomes higher and higher with increasing density. This together with the equation of motion for the ω_0-component necessarily lead to Eq.(36) which in turn guarantees the identical high-density behavior of both lagrangians. However, the ω-stabilized model bifurcates at a smaller L_c and the low-density behavior of both lagrangians is different. Under increasing parameters g_ω and m_ω with the ratio g_ω/m_ω kept constant the critical density and the low density phase of the ω stabilized model will approach the corresponding quantities of the model (24). So the inclusion of finite range vector mesons leads to an increase in the transition density in comparison to the corresponding contact-term model, see also ref.[13]. In summary, also the ω-stabilized Skyrme model has a critical hypersphere radius L_c where the uniform (chirally restored) phase on the hypersphere becomes unstable and bifurcates into a localized (chirally broken) phase which has the usual $B = 1$ hedgehog as flat space limit.

A similar behavior follows when the $\mathcal{L}_{2,\omega}$ model is extended by the introduction of ρ mesons. This can be done in different ways. For our purposes the only essential precondition is that the total lagrangian is still explicitly chiral symmetric. For simplicity let us consider in the following the hidden ρ meson coupling à la Bando et al.[29] in a minimal way[30], *i.e.*

$$\begin{aligned} \mathcal{L}_{2,\omega,\rho} &= \frac{f_\pi^2}{4}\mathrm{Tr}(\partial_\mu U^\dagger \partial^\mu U) - a\frac{f_\pi^2}{4}\mathrm{Tr}(\xi^\dagger\partial_\mu\xi - \xi\partial_\mu\xi^\dagger - 2ig\rho_\mu)^2 - \frac{1}{4}\rho_{\mu\nu}\rho^{\mu\nu} \\ &\quad - \frac{1}{4}\omega_{\mu\nu}\omega^{\mu\nu} + \frac{m_\omega^2}{2}\omega_\mu\omega^\mu + g_\omega\omega_\mu B^\mu + (\omega\rho\pi \text{ coupling terms}) \end{aligned} \tag{37}$$

with $\xi \equiv \sqrt{U}$, and $\rho_{\mu\nu} = \partial_\mu\rho_\nu - \partial_\nu\rho_\mu - ig[\rho_\mu, \rho_\nu]$, the non-abelian field-strength tensor. The second term in (37) is responsible for the generation of the ρ mass and the $\rho\pi\pi$ coupling with a coupling constant g. (The standard choice for the parameter

a is $a = 2$, *see* ref.[29].) Note that under the static hedgehog ansatz only the ***spatial*** ρ_i components are excited. Again it is easy to show that the equations of motion of this model are satisfied by the uniform profiles

$$f(\mu) = \mu \tag{38}$$

$$\omega_0 = -\frac{g_\omega}{m_\omega^2} B_0 \tag{39}$$

$$\rho_i = \frac{1}{2ig}(\xi^\dagger \partial_i \xi - \xi \partial_i \xi^\dagger) \qquad \text{with } \xi = \exp(i\tau \cdot \hat{r}\mu/2). \tag{40}$$

The last equation guarantees that the second term in (37) vanishes and that the ρ kinetic term behaves as fourth-order Skyrme term with a coefficient $\epsilon_4^2 = 1/(8g^2)$. The lagrangian (37) (without extra $\omega\rho\pi$ couplings) has therefore the same high density behavior as the simple contact lagrangian

$$\mathcal{L}_{2,4,6} = \frac{f_\pi^2}{4}\mathrm{Tr}(\partial_\mu U^\dagger \partial^\mu U) + \frac{1}{32g^2}\mathrm{Tr}[U^\dagger \partial_\mu U, U^\dagger \partial_\nu U]^2 - \frac{1}{2}\left(\frac{g_\omega}{m_\omega}\right)^2 B_\mu B^\mu. \tag{41}$$

This behavior presented here for the simplest $\omega\rho\pi$-model is generic – just the coefficient of the sixth-order term in (41) may change when $\omega\rho\pi$-coupling terms are introduced. For all models with vector-meson stabilization which have been studied so far[2,31,32] the following is true: It can be numerically shown that

(i) the only existing $B = 1$ solution at high densities is the uniform (delocalized) hedgehog one, the static solutions are identical to those of contact-term lagrangians with suitable coefficients and the static vector meson fields are equal to their driving pionic currents (*see e.g.* (36) or (40)),

(ii) there exists a critical model-dependent hypersphere-radius L_c where the uniform solution becomes unstable and bifurcates into the usual localized configuration and

(iii) at low densities the solutions on the hypersphere $\mathbf{S}^3(L)$ converge asymptotically to the corresponding solutions in the flat space.

The essential conditions on these models are two-fold: they should be explicitly chiral symmetric and they should allow for locally stable $B = 1$ hedgehog solitons. In case these two conditions are met, there will be necessarily a chiral phase transition at high densities with the same features as already discussed. Still, under realistic values for the mesonic input parameters (*e.g.* f_π, ϵ_4, c_6, *see* ref.[25], or g_ω and g, *see* refs.[30,32,2])) the transition densities ρ_c come out systematically too low, $\rho_c \approx \rho_{\text{nm}}$ (0.16 fm^{-3}). Apparently these models lack terms which generate additional attraction. Note that the hedgehog mass under such realistic parameters is systematically too high ($\sim$ 1.6 GeV) whereas there is not sufficient central attraction in the $B = 2$ system.[5] If one would tune the stabilizing parameters – while keeping $f_\pi = 93$MeV fixed – such that the classical hedgehog mass would be 0.87 GeV (consistent with a nucleon mass of 0.94 GeV), then the transition density ρ_c would be approximately three times the nuclear matter density.

5. Second- Versus First-Order

Whereas in the periodic array calculations second as well as first order phase transitions were found depending on the choice of the lattice and the twisted boundary conditions, we have so far encountered only second order transitions between the delocalized and the localized phase on the 3-sphere. This does not need to be the case in general: All the Skyrme-type lagrangians presented are explicitly not scale invariant and do not "know" (yet) about the trace anomaly in QCD. Let us therefore try to incorporate the same scaling behavior as in QCD into these lagrangians (*see* refs.[33,34,35]), *e.g.* consider the lagrangian

$$\begin{aligned}\mathcal{L}_{2,4,\chi} &= \frac{\chi^2}{\chi_0^2}\frac{f_\pi^2}{4}\mathrm{Tr}(\partial_\mu U^\dagger \partial^\mu U) + \frac{\epsilon_4^2}{4}\mathrm{Tr}[U^\dagger\partial_\mu U, U^\dagger\partial_\nu U]^2 + \frac{1}{2}(\partial_\mu\chi)^2 \\ &\quad - B_B\left(1 + \left(\frac{\chi}{\chi_0}\right)^4 \log\left(\frac{\chi^4}{e{\chi_0}^4}\right)\right)\end{aligned} \tag{42}$$

as simplest extension of the usual Skyrme lagrangian. The scalar field χ with the vacuum expectation value $\chi_0 = \langle 0|\chi|0\rangle$ is introduced with the purpose of making the first term in (42) scale invariant (the Skyrme term and the χ-kinetic term are already scale invariant), whereas the last term in (42), the χ-potential term, is adjusted to fit the trace anomaly of QCD. B_B is the "bag constant" which can be expressed in terms of the gluon condensate as $B_B = (9/32)\langle 0|(\alpha_s/\pi)G^2|0\rangle$. For values of the parameters χ_0 and B_B *see* ref.[34]. (Note that the fluctuations of the χ-field correspond to glueball-excitations.) In ref.[36] it was numerically shown that the lagrangian (42) again possesses a phase transition to a uniform delocalized solution at high densities with $f(\mu) = \mu$ and a constant χ profile which – depending on the choice of parameters – can be even $\chi = 0$. Because of the insufficient accuracy in the numerics, the authors of ref.[36] missed the fact that the phase transition is of first order. The latter is more or less obvious from the form of the χ-potential term which is adjusted to be minimal for $\chi = \chi_0$. Thus it is impossible to find a smooth transition between the constant $\chi < \chi_0$ (high-density) profile and the localized (low density) χ profile (which at the south pole on the 3-sphere is exactly χ_0) without violating the χ equation of motion:

$$\chi'' + 2\frac{\cos\mu}{\sin\mu}\chi' - \frac{\chi}{\chi_0^2}\frac{f_\pi^2}{L^2}\left(f'^2 + 2\frac{\sin^2 f}{\sin^2\mu}\right) - 4B_B\frac{\chi^3}{\chi_0^4}\log\left(\frac{\chi^4}{\chi_0^4}\right) = 0. \tag{43}$$

For a non-vanishing nonlinear sigma term (as in the present case since $f(\mu) \neq 0$) the χ field cannot both be constant and equal to χ_0. So the χ field can only do the transition from the low- to the high-density phase and vice versa by a jump. Thus by incorporating a scalar field à la Schechter[33] in the Skyrme model (or its extensions) the second order phase transition can be easily changed to a first order one. *See* also ref.[22] for a different mechanism to achieve the same thing.

It is rather easy to extend the lagrangian (27) to the most general with the QCD trace anomaly consistent geometrical form involving only first derivatives to even order:

$$\mathcal{L}_{\mathcal{G},\chi} = -\left(\frac{\chi}{\chi_0}\right)^4 \mathcal{G}\left(-(\frac{\chi_0}{\chi})^2\mathcal{L}_2, -(\frac{\chi_0}{\chi})^4\mathcal{L}_4, -(\frac{\chi_0}{\chi})^6\mathcal{L}_6\right)$$
$$+\frac{1}{2}(\partial_\mu\chi)^2 - B_B\left(1+\left(\frac{\chi}{\chi_0}\right)^4 \log\left(\frac{\chi^4}{e\chi_0{}^4}\right)\right). \tag{44}$$

Note that the restoration of scale invariance, $\chi = 0$, is only consistent with the lagrangian (44) if $\mathcal{G}$ has the asymptotics (34), *i.e.*

$$\left(\frac{\chi}{\chi_0}\right)^4 \mathcal{G}\left(-(\frac{\chi_0}{\chi})^2\mathcal{L}_2, -(\frac{\chi_0}{\chi})^4\mathcal{L}_4, -(\frac{\chi_0}{\chi})^6\mathcal{L}_6\right) \to C_1(-\mathcal{L}_2)^2 + C_2(-\mathcal{L}_4) \quad \text{if } \chi \to 0. \tag{45}$$

We know from what was said before that one can expect in this case a first-order phase transition to a chirally restored phase. However, the χ-profile in the high-density phase does not need to become zero, it can equally well be just a constant $\chi < \chi_0$. Scale invariance implies chiral restoration, but not vice versa. Nevertheless, just the possibility to have a scale invariant limit at high densities leads to strong constraints on the form of the effective model. If a lagrangian is considered which includes *any* term higher than fourth-order in the derivatives of the pion fields, terms of *all orders* have to be included as well both to satisfy the high-density asymptotic behavior (45) and to ensure the stability of the chiral symmetric high-density phase.[26] The necessity for infinitely many terms is consistent with the large N_c philosophy.[37,38] Because of asymptotic freedom, the effective model which follows from QCD to leading order in the $1/N_c$ expansion must involve infinitely many mesons. The hope is – as shown for the ω and ρ meson case – that the high density phase of this infinite tower of meson resonances can be approximated by infinitely many suitable contact terms. This might lead to "dreams" of Regge trajectories.

6. Discussion

Even effective models of the Skyrme class (which do not possess any quark degrees of freedom) can have a phase transitions to a high-density phase with restored chiral symmetry. There are essentially just two conditions on such a model: It must have the relevant symmetries (especially the lagrangian must be explicitly chirally invariant) *and* it must treat the baryon structure and interaction on the same footing. Normally, the phase transitions are of second order, but one can easily change them to first-order ones by incorporating the QCD trace anomaly for instance. This allows also - depending on the input parameters - for a scale invariant high density limit. Scale invariance implies chiral symmetry restoration, but not vice versa.

Under the same parameter input (from the meson sector) the periodic array calculations as well as the 3-sphere calculations give approximately the same values for the transition densities. The predictions, however, are for both systematically too low, $\rho_c \approx \rho_{\rm nm}$. This problem may be connected with two other deficiencies which plague the Skyrme-like models: the missing central attraction in the $B = 2$ channel and the predicted high value for the hedgehog mass. Quantum (higher loop or higher $1/N_c$) corrections may have a chance to cure all three problems at the same time by providing extra non-local terms which generate attraction.

Acknowledgements

The author would like to thank A.D. Jackson, L. Castillejo and C. Weiss for useful discussions.

References

1. T.H.R. Skyrme, *Nucl. Phys.* **31**, 556 (1962).
2. U.-G. Meißner, *Phys. Rep.* **161**, 213 (1988), and references therein.
3. *for a review and further references see:*
   I. Zahed and G.E. Brown, *Phys. Rep.* **142**, 1 (1986);
   U.-G. Meißner and I. Zahed, *Adv. Nucl. Phys.* **17**, 143 (1986);
   G. Holzwarth and B. Schwesinger, *Rep. Prog. Phys.* **49**, 825 (1986);
   G. S. Adkins *in: Chiral Solitons*, ed. K.F. Liu (World Scientific, Singapore, 1987);
   various contributions in these proceedings.
4. G.S. Adkins, C.R. Nappi and E. Witten, *Nucl. Phys.* **B228**, 552 (1983).
5. A. Jackson, A.D. Jackson and V. Pasquier, *Nucl. Phys.* **A432**, 567 (1985).
6. I. Klebanov, *Nucl. Phys.* **B262**, 133 (1985).
7. E. Wüst, G.E. Brown and A.D. Jackson, *Nucl. Phys.* **A468**, 450 (1987).
8. A.S. Goldhaber and N.S. Manton, *Phys. Lett.* **198 B**, 231 (1987).
9. A.D. Jackson and J.J.M. Verbaarschot, *Nucl. Phys.* **A484**, 419 (1988).
10. M. Kugler and S. Shrikman, *Phys. Lett.* **208 B**, 491 (1988) and *Phys. Rev.* **D40**, 3421 (1989); L. Castillejo, P.S. Jones, A.D. Jackson, J.J.M. Verbaarschot and A. Jackson, *Nucl. Phys.* **A501**, 801 (1989).
11. T.S. Walhout, *Nucl. Phys.* **A484**, 397 (1988).
12. T.S. Walhout, *Nucl. Phys.* **A519**, 816 (1990).
13. T.S. Walhout, *Phys. Lett.* **227 B**, 10 (1989).
14. N.S. Manton and P.J. Ruback, *Phys. Lett.* **181 B**, 137 (1986).
15. N.S. Manton, *Commun. Math. Phys.* **111**, 469 (1987).
16. M. Loss, *Lett. Math. Phys.* **14**, 149 (1987).
17. A.D. Jackson, A. Wirzba and L. Castillejo, *Nucl. Phys.* **A486**, 634 (1988) and *Phys. Lett.* **198 B**, 315 (1987).
18. R.S. Palais, *Commun. Math. Phys.* **69**, 19 (1979).
19. A.D. Jackson, N.S. Manton and A. Wirzba, *Nucl. Phys.* **A495**, 499 (1989).
20. I. Zahed, A. Wirzba and U.-G. Meißner, *Phys. Rev.* **D33**, 830 (1986).
21. B. Moussallam and D. Kalafatis, *Phys. Lett.* **272 B**, 196 (1991); B. Moussallam, these proceedings; G. Holzwarth, these proceedings.
22. H. Forkel, A.D. Jackson, M. Rho, C. Weiss, A. Wirzba and H. Bang, *Nucl. Phys.* **A504**, 818 (1989).
23. A. Wirzba and H. Bang, *Nucl. Phys.* **A515**, 571 (1990).

24. A.D. Jackson, C. Weiss, A. Wirzba and A. Lande, *Nucl. Phys.* **A495**, 499 (1989).
25. G.E. Brown, M. Rho and W. Weise, *Nucl. Phys.* **A454**, 669 (1986); A. Wirzba and W. Weise, *Phys. Lett.* **188 B**, 6 (1987).
26. A.D. Jackson, C. Weiss and A. Wirzba, *Nucl. Phys.* **A529**, 741 (1991).
27. G.S. Adkins and C.R. Nappi, *Phys. Lett.* **137 B**, 251 (1984).
28. C. Adami, Diplomarbeit, Bonn University (1988), unpublished.
29. M. Bando, T. Kugo, S. Uehara, K. Yamawaki and T. Yanagida, *Phys. Rev. Lett.* **54**, 1215 (1985); M. Bando, T. Kugo and K. Yamawaki, *Nucl. Phys.* **B259**, 493 (1985); Y. Igarashi, M. Johmura, A. Kobayashi, H. Otsu, T. Sato and S. Saweda, *Nucl. Phys.* **B259**, 721 (1985); T. Fujiwara, T. Kugo, H. Terao, S. Uehara and K. Yamawaki, *Prog. Theor. Phys.* **73**, 926 (1985).
30. U.-G. Meißner, N. Kaiser, A. Wirzba and W. Weise, *Phys. Rev. Lett.* **57**, 1676 (1986).
31. N.N. Scoccola, M. Rho and D.P. Min, *Nucl. Phys.* **A489**, 612 (1988).
32. P. Jain, R. Johnson, U.-G. Meißner, N.W. Park and J. Schechter, *Phys. Rev.* **D37**, 3252 (1988); U.-G. Meißner, N. Kaiser, H. Weigel and J. Schechter, *Phys. Rev.* **D39**, 1956 (1989).
33. H. Gomm, P. Jain, R. Johnson and J. Schechter, *Phys. Rev.* **D33**, 3476 (1986); P. Jain, R. Johnson and J. Schechter, *Phys. Rev.* **D35**, 2230 (1987).
34. B.A. Campbell, J. Ellis and K.A. Olive, *Nucl. Phys.* **B345**, 57 (1990) and *Phys. Lett.* **235 B**, 325 (1990).
35. G.E. Brown and M. Rho, *Phys. Rev. Lett.* **66**, 2720 (1991).
36. H. Reinhardt and B.V. Dang, *Phys. Rev.* **D38**, 2881 (1988).
37. G. 't Hooft, *Nucl. Phys.* **B72**, 461 (1974), *Nucl. Phys.* **B75**, 461 (1974).
38. E. Witten, *Nucl. Phys.* **B160**, 57 (1974).

NUCLEON-ANTINUCLEON ANNIHILATION IN THE SKYRME-MODEL

B. STOETZEL and G.HOLZWARTH
Physics Department, Siegen University
59 Siegen, Germany

ABSTRACT
An attempt is made to calculate the branching ratios for the annihilation at rest of $N\bar{N} \rightarrow 2\ meson$ final states. The interaction vertices are obtained by expanding a $\pi\rho\omega$-effective lagrangian to second order in the mesonic fluctuations around the soliton-antisoliton configuration. Ambiguities in the necessary input are discussed. The results are qualitatively unsatisfactory and indicate that other processes may have to be included.

1. Introduction

The antiproton-proton interaction is a field of intensive research, although the first definite observation of an antiproton dates back nearly forty years[1]. On the experimental side the advent of the LEAR-facility at CERN led to a wealth of new data concerning especially the antiproton-proton annihilation[2] and the branching ratios to the final meson-channels in this process[3]. One of the most striking features of the annihilation-process is the occurrence of dynamical selection rules, by which some final channels, in principle allowed by general conservation laws, are suppressed. These selection rules provide a stringent test for all theoretical attempts to understand the annihilation branching ratios.

In the Skyrme-model description of baryons as topological solitons, the concept of a soliton-antisoliton system coupled to mesonic fluctuations seems to provide a rather natural basis for investigating the process $N\bar{N} \rightarrow$ *mesons*. To our knowledge, there have been three attempts to make use of this concept: two of them rather describe methods than explicit results[4], whereas the third one, by numerically following the time-dependent classical equations of motion, shows how the initial soliton-antisoliton configuration dissolves into pionic fluctuations[5]. Although intuitively pleasing, such an approach misses the quantum nature of the process, and cannot provide answers to questions like the branching ratios for individual final channels.

In the present work, which is related to that of Zahed and Brown[4], the quantized collective coordinates of the soliton-antisoliton configuration are used to define the initial $N\bar{N}$ state. The final multimeson state is constructed as usual in the Fock space of the quantized mesonic fluctuations. Expansion of the lagrangian in increasing powers of time-dependent fluctuations around the soliton-antisoliton configuration defines the coupling terms which connect in tree approximation the initial $N\bar{N}$ state with the final multimeson state.

Each power of the fluctuations introduces a factor $1/\sqrt{N_c}$ into the coupling terms of the lagrangian. Therefore we would roughly expect dominance of two-meson final states although, admittedly, with $N_c = 3$ the suppression of n-meson states with increasing n is not very strong. Still, the experimental results indicate that two-meson states act by subsequent decay as "doorway" states for a large portion of the strength found in multimeson final states[6]. The experimental analysis of these resonant decays for $N\bar{N}$-annihilation at rest has provided branching ratios for these two-meson doorway states[7]. Even for the three lightest mesons π, ρ, and ω the nine possible channels show variations in the branching ratios of about two orders of magnitude (cf.tab.1). So this appears to be an ideal testing ground for theoretical concepts. The expansion of the effective lagrangian up to second power in mesonic fluctuations is sufficiently

Table 1. Experimental branching ratios for two-meson states[3] (in units of 10^{-3})

$\pi^+\pi^-$	$\pi^0\pi^0$	$\pi^\pm\rho^\mp$	$\pi^0\rho^0$	$\pi^0\omega^0$	$\rho^+\rho^-$	$\rho^0\rho^0$	$\rho^0\omega$	$\omega\omega$
3.2	0.25	30.1	16.0	5.2	< 95	4	39.4	14
$\pm$ 0.2	$\pm$ 0.03	$\pm$ 3.0	$\pm$ 1.0	$\pm$ 0.5		$\pm$ 3	$\pm$ 5.9	$\pm$ 6

simple and well known from meson-baryon scattering. A crucial difference is that the kinetic energy of the initial meson in the latter process can be close to zero which may justify the use of low-energy effective lagrangians. In the annihilation process the final mesons share about 2 GeV. Still, with regard to the successful application of effective lagrangians to meson-baryon dynamics up to several hundred MeV we may hope that the commonly used effective lagrangians are not completely inappropriate for the present purpose. In this sense our attempt to describe $N\bar{N}$-annihilation in terms of the soliton model can only be of a rather explorative nature.

Our choice of the effective π, ρ, ω-lagrangian is given in section 2. Another source of ambiguities is located in the choice of the classical soliton-antisoliton configuration: For large separations a product ansatz of individual $B = \pm 1$ hedgehogs certainly is appropriate. To which extent this holds for the S and P protonium bound states at or below 1 fm relative distance is not known. But even the product ansatz requires specification of a wavefunction for the relative orientation of the hedgehogs in isospace. This wavefunction may be coupled in a complicated way with the relative distance, and the chiral angles $F(r)$ of the individual hedgehogs may depend on both, relative orientation and distance. Without any reliable information on these matters we choose in section 3 the most simple configuration which meets the geometrical requirements. Then, in sections 4 and 5, the quantum states for the initial $N\bar{N}$ and final two-meson configurations can be constructed.

Unfortunately there is still another source of uncertainty in the theoretical procedure: there are many different ways for introducing the pionic fluctuations. In an exact method the final result should of course not depend on this choice. However, in a first order plane wave approximation which we will use here, the results do depend on this choice. In section 5 we choose to introduce the pions as local chiral rotations of the whole soliton-antisoliton configuration at point $\boldsymbol{x}$. (Other possibilities would be to introduce them as local chiral rotations of the individual soliton or antisoliton field at their respective positions, or as fluctuations of the chiral angles in the hedgehog or antihedgehog.) This choice affects the form of the coupling terms when the lagrangian is expanded in powers of the fluctuations. Again our choice is motivated by its simplicity. Still, the actual calculation of transition matrix elements demonstrated in section 6 is very tedious in detail.

2. The effective lagrangian.

The failure of the pure statistical model[8] in describing the annihilation at rest shows, that both, structure of the nucleons and the dynamics of the interaction are important in this process. The Skyrme-model, successful in describing many phenomena in the low-energy region, includes both aspects: the nucleon structure is reflected in the baryon density, while the interaction dynamics is determined by the effective lagrangian used. For the final channels of tab.1 it is sufficient to stay within the SU(2) framework and include only ρ and ω fields in addition to the pions. The usual method to incorporate vector mesons into the Skyrme-model treats them as massive gauge-bosons of the $SU(2)_L \otimes SU(2)_R \otimes U(1)_V$ symmetry and leads, after appropriate elimination of the a_1-meson, to a chirally symmetric effective lagrangian, containing pions, ρ- and ω-mesons[9]. For our explorative purpose, we here use a simplified version, which does not preserve the full chiral symmetry, but is constructed by gauging the vector-symmetry $U(2)_V$ alone[10]. This simplification ought to be sufficient to demonstrate the essential steps of the method.

Our choice for the effective pion-plus-vector-meson lagrangian is:

$$\begin{aligned}
\mathcal{L} \; = \; & \frac{f_\pi^2}{4}\mathrm{tr}(\partial_\mu U \partial^\mu U^\dagger) + \frac{f_\pi^2 m_\pi^2}{4}\mathrm{tr}(U + U^\dagger - 2) - \frac{1}{8}\mathrm{tr}(\rho_{\mu\nu}\rho^{\mu\nu}) \\
& + \frac{m_\rho^2}{4}\mathrm{tr}(\rho_\mu \rho^\mu) - \frac{1}{4}\omega_{\mu\nu}\omega^{\mu\nu} + \frac{m_\omega^2}{2}\omega_\mu\omega^\mu + N_c g_\omega \omega_\mu B^\mu \\
& - \frac{i g_\rho f_\pi^2}{2}\mathrm{tr}[\rho_\mu(L^\mu - R^\mu)] - \frac{g_\rho^2 f_\pi^2}{4}\mathrm{tr}([\rho_\mu, U][\rho^\mu, U^\dagger]) \\
& + \frac{i N_c g_\omega g_\rho}{8\pi^2}\varepsilon^{\mu\nu\alpha\beta}\partial_\mu\omega_\nu \mathrm{tr}[\rho_\alpha(R_\beta + L_\beta) - i g_\rho \rho_\alpha U^\dagger \rho_\beta U] \qquad (1)
\end{aligned}$$

with

$$\rho_\mu = \rho_\mu^a \tau^a \qquad \rho_{\mu\nu} = \partial_\mu\rho_\nu - \partial_\nu\rho_\mu - i g_\rho[\rho_\mu, \rho_\nu] \qquad (2)$$

$$\omega_{\mu\nu} = \partial_\mu\omega_\nu - \partial_\nu\omega_\mu \qquad (3)$$

$$L_\mu = U^\dagger \partial_\mu U \qquad R_\mu = \partial_\mu U U^\dagger \qquad (4)$$

and the baryon-density

$$B^\mu = \frac{1}{24\pi^2}\varepsilon^{\mu\nu\alpha\beta}\mathrm{tr}(L_\nu L_\alpha L_\beta). \qquad (5)$$

This lagrangian has apart from the terms, which arise from gauging the non-linear σ-model, kinetic and mass-terms for the gauge-fields, added by hand, and terms from gauging the Wess-Zumino-action[11]. These latter parts are the only ones which provide coupling for the ω-mesons. It is, however, already at this point evident, that these coupling terms (being only linear in the ω-field) will not allow for two-ω final states in tree approximation.

3. Soliton-Antisoliton Configuration and Collective Coordinates.

As we have discussed in the introduction we will use the product ansatz for the basic soliton-antisoliton configuration and parametrize it by collective coordinates analogously to the description of $N\bar{N}$-systems in the $B = 2$ sector[12]. Starting with the static hegdehog solution for $B = 1$: $U_H = \exp(i\boldsymbol{\tau} \cdot \hat{\boldsymbol{r}} F(r))$ and the corresponding antihedgehog $U_{\bar{H}} = U_H^\dagger$ for $B = -1$, we then have

$$U = U_H(\boldsymbol{x} - \boldsymbol{R} - \frac{\boldsymbol{r}}{2})U_H^\dagger(\boldsymbol{x} - \boldsymbol{R} + \frac{\boldsymbol{r}}{2}) \tag{6}$$

where $\boldsymbol{R}$ is the centre of mass of the system and $\boldsymbol{r}$ is the relative distance of soliton and antisoliton. Each skyrmion for itself has as further collective degrees of freedom the rotations in isospace. This is accounted for in the ansatz (??) by rotating the hedgehog relatively to the antihedgehog, and then rotating globally the product of both. This leads to

$$U_{N\bar{N}} = A(\boldsymbol{t})S(\boldsymbol{s})U_H(\boldsymbol{x} - \boldsymbol{R} - \frac{\boldsymbol{r}}{2})S^{-2}(\boldsymbol{s})U_H^\dagger(\boldsymbol{x} - \boldsymbol{R} + \frac{\boldsymbol{r}}{2})S(\boldsymbol{s})A^\dagger(\boldsymbol{t}). \tag{7}$$

The global rotation A creates, as in the $B = 1$ case, the isospin of the whole system, if quantized appropriately. Correspondingly, quantization of the collective coordinate $\boldsymbol{s}$, which parametrizes the relative rotation, leads to additional quantum numbers, which can be identified with an intrinsic relative spin. With the definiton of the angular velocity $\boldsymbol{\Omega}$

$$A\dot{A}^\dagger = \frac{i}{2}\boldsymbol{\tau} \cdot \boldsymbol{\Omega} \tag{8}$$

we get the isospin from

$$\boldsymbol{t} = \frac{\partial L}{\partial \boldsymbol{\Omega}}. \tag{9}$$

Defining the intrinsic spin through

$$\boldsymbol{\sigma} = \hat{\boldsymbol{s}} \times \frac{\partial L}{\partial \dot{\hat{\boldsymbol{s}}}} \tag{10}$$

and analogously the relative angular momentum by

$$\boldsymbol{l} = \hat{\boldsymbol{r}} \times \frac{\partial L}{\partial \dot{\hat{\boldsymbol{r}}}} \tag{11}$$

leads to the total angular momentum

$$\begin{aligned} \boldsymbol{j} &= -\int d^3x \frac{\partial \mathcal{L}}{\partial \dot{U}}(\boldsymbol{x} \times \nabla)U \\ &= \boldsymbol{l} + \boldsymbol{\sigma} - \boldsymbol{t}D \end{aligned} \tag{12}$$

with $D_{ia} = \frac{1}{2}\mathrm{tr}(\tau_i A \tau_a A^\dagger)$.

Ansaetze for the ρ- and ω-parts of the soliton-antisoliton system are suggested by the requirement, that the lagrangian be invariant under the G-parity transformation. This transformation exchanges soliton and antisoliton, which means for the collective coordinates:

$$\begin{aligned} \boldsymbol{r} &\rightarrow -\boldsymbol{r} \\ \boldsymbol{s} &\rightarrow -\boldsymbol{s} \end{aligned} \tag{13}$$

and therefore

$$U^G_{N\bar{N}} = AS^{-1}U_H(\boldsymbol{x} + \frac{\boldsymbol{r}}{2})S^2 U^\dagger_H(\boldsymbol{x} - \frac{\boldsymbol{r}}{2})S^{-1}A^\dagger = U^\dagger_{N\bar{N}}. \tag{14}$$

(The centre of mass coordinate is unimportant here and will be omitted.) For the baryon-density (5) we get from this

$$B^G_\mu = -B_\mu. \tag{15}$$

The term $\omega_\mu B^\mu$ is then invariant under G-parity, if ω_μ transforms as

$$\omega^G_\mu = -\omega_\mu \tag{16}$$

which corresponds also to the intrinsic G-parity of the ω-meson. Such a behaviour can be enforced by the ansatz

$$\omega^\mu_{N\bar{N}} = \omega^\mu_H(\boldsymbol{x} - \frac{\boldsymbol{r}}{2}) - \omega^\mu_H(\boldsymbol{x} + \frac{\boldsymbol{r}}{2}). \tag{17}$$

Similar considerations for the ρ-meson yields

$$\rho^\mu_{N\bar{N}} = AS\rho^\mu_H(\boldsymbol{x} - \frac{\boldsymbol{r}}{2})S^{-1}A^\dagger + AS^{-1}\rho^\mu_H(\boldsymbol{x} + \frac{\boldsymbol{r}}{2})SA^\dagger. \tag{18}$$

(The index H denotes the $B = 1$ hedgehog configuration.) In the intrinsic system we therefore have:

$$\begin{aligned} U_{H\bar{H}} &= S(\boldsymbol{s})U_H(\boldsymbol{x} - \boldsymbol{R} - \frac{\boldsymbol{r}}{2})S^{-2}(\boldsymbol{s})U^\dagger_H(\boldsymbol{x} - \boldsymbol{R} + \frac{\boldsymbol{r}}{2})S(\boldsymbol{s}) && (19) \\ \rho^\mu_{H\bar{H}} &= S(\boldsymbol{s})\rho^\mu_H(\boldsymbol{x} - \boldsymbol{R} - \frac{\boldsymbol{r}}{2})S^{-1}(\boldsymbol{s}) + S^{-1}(\boldsymbol{s})\rho^\mu_H(\boldsymbol{x} - \boldsymbol{R} + \frac{\boldsymbol{r}}{2})S(\boldsymbol{s}) && (20) \\ \omega^\mu_{H\bar{H}} &= \omega^\mu_H(\boldsymbol{x} - \boldsymbol{R} - \frac{\boldsymbol{r}}{2}) - \omega^\mu_H(\boldsymbol{x} - \boldsymbol{R} + \frac{\boldsymbol{r}}{2}). && (21) \end{aligned}$$

Inserting Eqs.19 - 21 into the lagrangian (1) defines the baryon-antibaryon dynamics in terms of the collective degrees of freedom $\boldsymbol{R}$, $\boldsymbol{r}$, $\boldsymbol{s}$, and $\boldsymbol{\Omega}$. If we would seek for static solutions of this collective lagrangian, we would find that the profile-functions F, G, and ω not only depend on $\boldsymbol{x}$, as usual, but also on the collective coordinates $\boldsymbol{s}$ and $\boldsymbol{r}$. Walliser, Holzwarth and Hayashi[13] showed, that in the pure Skyrme-model (non-linear σ-model plus Skyrme-term) and for vanishing relative distance between soliton and antisoliton the usual profile-function $F(x)$ is obtained for the choice $s = \pi/4$. Furthermore it was shown, that the wave function for the collective motion of the relative angle s is concentrated within a region close to this value (see also next section). We therefore use the usual static profile-functions for the individual solitons in our calculation.

4. $N\bar{N}$-States.

In general an $N\bar{N}$-state is characterized by its isospin T, spin S, orbital angular momentum L and total angular momentum $J = L + S$. Because the isospin and spin of proton and antiproton is $\frac{1}{2}$ respectively, T and S both can take the values 0 or 1. The orbital angular momentum is not restricted in general but in the annihilation process only states with $L = 0, 1$ are important[14]. Therefore, we have to deal with four S-states ($L = 0$) and eight P-states ($L = 1$) which we denote as

$$|N\bar{N}\rangle = |^{(2T+1)(2S+1)}L_J\rangle \tag{22}$$

These states have the following symmetries:

$$\begin{array}{lrcl} \text{(i) parity} & P|N\bar{N}\rangle & = & (-1)^{L+1}|N\bar{N}\rangle \quad (23) \\ \text{(ii) charge conjugation} & C|N\bar{N}\rangle & = & (-1)^{L+S}|N\bar{N}\rangle \quad (24) \\ \text{(iii) G-parity} & G|N\bar{N}\rangle & = & (-1)^{L+S+T}|N\bar{N}\rangle \quad (25) \end{array}$$

To construct these states within the ansatz (7) we use the collective degrees of freedom introduced, namely: motion of the centre of mass, relative motion of soliton and antisoliton, global isospin-rotation and relative rotation in isospace. The motion of the centre of mass is unimportant for the annihilation problem and is described simply by a plane wave, which yields energy- and momentum conservation in the calculation of matrixelements. The relative motion leads to a wavefunction $\Psi(\boldsymbol{r}, \boldsymbol{s})$, which is not known, but its angular dependence can of course be expanded in terms of functions $\Psi_{\sigma L}(s, r) Y_{\sigma\sigma_3}(\hat{\boldsymbol{s}}) Y_{LL_3}(\hat{\boldsymbol{r}})$. The global isospin-rotation yields a D-function, if parametrized by Euler-angles.

The state with the quantum numbers (22) in the representation of Euler-angles $[\alpha]$, relative distance $\boldsymbol{r}$, relative rotation $\boldsymbol{s}$ and centre of mass $\boldsymbol{R}$, then is

$$\begin{aligned} &\langle [\alpha]; \boldsymbol{s}; \boldsymbol{r}; \boldsymbol{R} |^{(2T+1)(2S+1)}L_J; \boldsymbol{P}\rangle \\ &= \frac{\sqrt{(2T+1)E_{N\bar{N}}}}{2\pi\sqrt{2J+1}} \sum_{\sigma,\sigma_3} \Psi_{\sigma L}(s,r) Y_{\sigma\sigma_3}(\hat{\boldsymbol{s}}) \sum_k (-1)^k D^T_{-k,0}([\alpha]) \sum_{L_3} Y_{LL_3}(\hat{\boldsymbol{r}}) \\ &\quad \cdot \sum_{M,S_3} (Tk\sigma\sigma_3|SS_3)(SS_3LL_3|JM) \exp(-i[E_{N\bar{N}}t - \boldsymbol{P}\cdot\boldsymbol{R}]). \end{aligned} \tag{26}$$

The state (26) satisfies (23) and (24), but (25) requires a restriction of the summation over σ such that $(-1)^\sigma = (-1)^{S+T}$.. With this additional requirement the $N\bar{N}$-state is fixed.

The wavefunctions $\Psi_{\sigma L}(s,r)$ are not yet determined. To do this, in principle one had to insert the ansaetze (19 - 21) into the lagrangian (1) to get the dynamics of the collective coordinates s and r. Then from the resulting Schroedinger-equation the wavefunctions could be obtained. Here, we proceed however in a very simplified way: Proton and antiproton form a Coulomb-boundstate before annihilating[15]. The radius of this state (mainly $L=0$) is about $58fm$. In the annihilation region of about $1fm$ such a wavefunction is essentially a constant. This constant of course determines the absolute values of the annihilation amplitudes. Because of the strong annihilation interaction, the real wavefunction will differ appreciably from the Coulomb-function at small distances and using the latter would yield a much too high absolute value for the transition amplitudes. But this holds for all final states in the same way and as we are interested in the branching ratios only, it is hoped, that the stated approximation is not too bad.

Now we are left with the function $\Psi_{\sigma L}(s)$. Walliser, Holzwarth and Hayashi[13] have calculated Ψ in a special case, namely for the pure Skyrme-model and vanishing distance of soliton and antisoliton. One of their results is, that the relative rotation-angle s is restricted to a small region about $\pi/4$, where also the profile-functions are the usual ones (see foregoing section). As a first approximation it therefore may be sufficient to take just this value $\pi/4$ for s, which means that the wavefunction Ψ is simply a delta-function for s equal $\pi/4$.

5. Two-Meson States.

The final two-meson states are classified like the initial $N\bar{N}$-states by their isospin I, spin S, angular momentum l and total angular momentum $J = l + S$. As in Eqs. 23-25 we can determine the symmetry properties of these states.

Especially simple is the G-parity, as we take only hyperneutral mesons into account, which are eigenstates of G. If we denote the intrinsic G-parity of the meson M_a by $\eta_G(M_a)$, the G-parity of a meson-pair is the product

$$G(M_a, M_b) = \eta_G(M_a) \cdot \eta_G(M_a). \tag{27}$$

For the parity we get analogously to the $N\bar{N}$-case

$$P(M_a, M_b) = \eta_P(M_a) \cdot \eta_P(M_b) \cdot (-1)^l \tag{28}$$

with the relative angular momentum l and the intrinsic parities η_P. The charge conjugation can be defined for states with $I_3 = 0$ through the G-parity:

$$C(M_a, M_b) = (-1)^I \cdot G(M_a, M_b). \tag{29}$$

A special case are those states, where one meson is the antiparticle to the other one, e.g. $\pi^+\pi^-$. Then we get as in (24)

$$C(M_a, M_b) = (-1)^{l+S}. \tag{30}$$

Another special case is, that both mesons are neutral and therefore eigenstates to C. Then

$$C(M_a, M_b) = \eta_C(M_a) \cdot \eta_C(M_b). \tag{31}$$

Because of the conservation of parity, G-parity, charge conjugation, isospin and total angular momentum, there are strong selection rules for protonium annihilation, which can be determined with the help of Eqs.23-25 and 27-31. For the process $N\bar{N} \to \pi^+\pi^-$, for example, only one state is allowed, namely the $^{33}S_1$-state.

The meson-states are now written as plane wave product states in Fock-space. Let $a^{\dagger I_3}_{S_3}(\boldsymbol{q})$ and $b^{\dagger I_3}_{S_3}(\boldsymbol{q})$ be the creation operators of the mesons M_a and M_b with isospin-projection I_3, spin-projection S_3 and momentum $\boldsymbol{q}$. Then we have for the two-meson state with total angular momentum J, isospin I, spin S, relative angular momentum l, momentum $\boldsymbol{p}$ and relative momentum q:

$$\begin{aligned} |M_a, M_b; \boldsymbol{p}, q; J, I, S, l\rangle = & \frac{1}{\sqrt{2J+1}} \sum_{M,S_3} \sum_{m} (SS_3 lm|JM) \int d\Omega_q Y_{lm}(\hat{\boldsymbol{q}}) \\ & \cdot \sum_{m_a,m_b} \sum_{z_a,z_b} (S_a m_a S_b m_b|SS_3)(I_a z_a I_b z_b|I0) a^{\dagger z_a}_{m_a}(\boldsymbol{q}_1) b^{\dagger z_b}_{m_b}(\boldsymbol{q}_2)|0\rangle \end{aligned} \tag{32}$$

with $\boldsymbol{p} = \boldsymbol{q}_1 + \boldsymbol{q}_2$ and $\boldsymbol{q} = \boldsymbol{q}_2 - \boldsymbol{q}_1$. S_i is the spin of meson M_i, I_i the isospin. The integration over the space-angle Ω_q projects on the desired relative angular momentum. For the operators a and b the usual commutation relations for bosons hold:

$$[a^{I_3}_{S_3}(\boldsymbol{q}), a^{\dagger I_3'}_{S_3'}(\boldsymbol{q}')] = 2E_q(2\pi)^3 \delta_{S_3 S_3'} \delta_{I_3 I_3'} \delta_3(\boldsymbol{q} - \boldsymbol{q}') \tag{33}$$

with $E_q = \sqrt{\boldsymbol{q}^2 + m_a^2}$.

6. Transition Matrixelements.

As we have discussed in the introduction we introduce pions as local chiral rotations (Schnitzer-ansatz[16]) of the whole soliton-antisoliton field. Most conveniently this is done in the intrinsic frame, i.e.

$$U = Ae^{i\boldsymbol{\eta}\cdot\boldsymbol{\tau}} U_{H\bar{H}} e^{i\boldsymbol{\eta}\cdot\boldsymbol{\tau}} A^\dagger. \tag{34}$$

For the ρ- and ω-fluctuations we write

$$\begin{aligned} \rho^\mu &= A(\rho^\mu_{H\bar{H}} + u^\mu)A^\dagger, & (35) \\ \omega^\mu &= \omega^\mu_{N\bar{N}} + w^\mu. & (36) \end{aligned}$$

These ansaetze are now inserted into the lagrangian (1), which then is expanded to second order in the fluctuations to get the coupling V_{int} of the $N\bar{N}$-system to two-meson final channels. Terms of first order correspond to either kinematically impossible transitions or such with one virtual off-mass-shell meson, which will not be considered in the following. As we mentioned before, for ω-mesons the expansion ends already at the first order, while for ρ-mesons there are nontrivial second-order couplings, but no higher order terms. The expansion to higher order in the pion fields is rather tedious, but possible in principle and leads to final states with more than two pions.

In PWBA we have to calculate matrixelements of the form $\langle M_a M_b | V_{int} | N\bar{N} \rangle$ with the quantum states constructed in the last two sections. To evaluate the matrixelements, we express the fluctuations by the corresponding creation operators. For the pion-fluctuations we write

$$\eta(\boldsymbol{x}, t) = \frac{1}{2f_\pi} \int d_L^3 k \boldsymbol{c}_\pi^\dagger(\boldsymbol{k}) \exp(i[E_k t - \boldsymbol{k} \cdot \boldsymbol{x}]) + c.c. \tag{37}$$

The factor $1/2f_\pi$ has its origin in the factor $f_\pi^2/4$ in front of the lagrangian (1) which must cancel out for the kinetic energy of the free pions $\boldsymbol{\eta}$. The operators $\boldsymbol{c}_\pi$ differ from the $\boldsymbol{a}_\pi$ in (32) by an isorotation, transforming from the laboratory system, where the states (32) are defined, to the intrinsic system, where the ansaetze (19 - 21) are used. Further $d_L^3 k$ is an abbreviation[17] for

$$d_L^3 k = \frac{d^3 k}{2E_\pi (2\pi)^3} \tag{38}$$

corresponding to the normalization (33). The fluctuations of the ρ- and ω-mesons are a little bit more complicated because of their spin $S = 1$. This is described with the help of a polarization-vector. With that, we have

$$w_\mu(\boldsymbol{x}, t) = \int d_L^3 k \sum_m \varepsilon_\mu^*(\boldsymbol{k}, m) c_\omega^{\dagger m}(\boldsymbol{k}) \exp(i[E_k t - \boldsymbol{k} \cdot \boldsymbol{x}]) + c.c. \tag{39}$$

$$u_\mu(\boldsymbol{x}, t) = \int d_L^3 k \sum_{m,a} \varepsilon_\mu^*(\boldsymbol{k}, m)(-1)^a c_{\rho,a}^{\dagger m}(\boldsymbol{k}) \exp(i[E_k t - \boldsymbol{k} \cdot \boldsymbol{x}]) + c.c. \tag{40}$$

(m and a take the values $-1, 0, 1$). The polarization-vector is

$$\varepsilon_\mu(\boldsymbol{k}, m) = \left(\frac{\boldsymbol{k} \cdot \boldsymbol{e}(m)}{M}, \boldsymbol{e}(m) + \frac{\boldsymbol{k}(\boldsymbol{k} \cdot \boldsymbol{e}(m))}{M(E_k + M)} \right) \tag{41}$$

with the mesonmass M and the unit-vectors

$$\boldsymbol{e}(\pm 1) = \frac{1}{\sqrt{2}}(\mp 1, -i, 0) \quad , \quad \boldsymbol{e}(0) = (0, 0, 1). \tag{42}$$

As a typical example, we now demonstrate for the process $N\bar{N}(^{33}S_1) \to \pi^+\pi^-$ (actually, $|^{33}S_1\rangle$ is the only possible initial state for annihilation into two pions), how to calculate the corresponding matrixelements. The two-pion state has relative angular momentum $l = 1$ and spin $S = 0$ and from (32) we get

$$\begin{aligned} &|\pi^+\pi^-; \boldsymbol{p}, q; J = 1, S = 0, l = 1\rangle \\ &= -\frac{1}{\sqrt{6}} \sum_{M,a} (-1)^a a \int d\Omega_q Y_{1M}(\hat{\boldsymbol{q}}) a_\pi^{\dagger a}(\boldsymbol{q}_1) a_\pi^{\dagger -a}(\boldsymbol{q}_2) |0\rangle. \end{aligned} \tag{43}$$

Correspondingly, the $N\bar{N}$-state (26) with quantum numbers $^{33}S_1$ is:

$$\begin{aligned}&\langle[\alpha]; \boldsymbol{s},\boldsymbol{r},\boldsymbol{R}|^{33}S_1; \boldsymbol{P}\rangle \\ &= N\sum_{\substack{\sigma\sigma_3 \\ (\sigma=0,2)}} Y_{\sigma\sigma_3}(\hat{\boldsymbol{s}}) \sum_{M,k}(-1)^k D^1_{-k,0}([\alpha])(1k\sigma\sigma_3|1M)\exp(-i[E_{N\bar{N}}t - \boldsymbol{P}\cdot\boldsymbol{R}]). \quad (44)\end{aligned}$$

The constant N is fixed later, by normalizing the 2π-branching ratio. The interaction terms, quadratic in the fluctuations, are of the form

$$\mathcal{L}_{int} = T_{1ab}\eta^a\eta^b + T^i_{2ab}\eta^a\partial^i\eta^b + T_{3ab}\dot{\eta}^a\dot{\eta}^b + T^{ij}_{4ab}\partial^i\eta^a\partial^j\eta^b \quad (45)$$

The T's depend on the soliton-antisoliton configuration (19 - 21). All time-derivatives on collective coordinates are neglegted, that means, the collective velocities are assumed to be small. This approach corresponds to the adiabatic approximation in pion-nucleon-scattering[9], where the dynamical effects of the collective motion (isorotation in this case) are also omitted.

The two-pion state (43) contains a projection on angular momentum $l = 1$, so only terms with an odd power of $\boldsymbol{q}$ are different from zero in matrixelements, calculated with this state. After the Fouriertransformation of the fluctuations (37), every spatial derivative gives one power of $\boldsymbol{q}$ in the matrixelement. Therefore only the term $T^i_{2ab}\eta^a\partial^i\eta^b$ from (45) is left and we have to calculate the matrixelement

$$\mathcal{M} = \langle\pi^+\pi^-|\int dt d^3x T^i_{2ab}\eta^a\partial^i\eta^b|N\bar{N}\rangle. \quad (46)$$

With (43) and (37) we get

$$\begin{aligned}\mathcal{M} &= \frac{i}{4\sqrt{6}f_\pi^2}\sum_{M,a}(-1)^a a\int d\Omega_q Y^\star_{1M}(\hat{\boldsymbol{q}})\int dt d^3x \sum_{i,b,c}\langle 0|a_\pi^{-a}(\boldsymbol{q}_2)a_\pi^a(\boldsymbol{q}_1) \\ &\cdot T^i_{2bc}\int d^3_L k_1 d^3_L k_2 k_{2i}\sum_{j,k} D^\star_{bj}D^\star_{ck}a_\pi^{\dagger j}(\boldsymbol{k}_1)a_\pi^{\dagger k}(\boldsymbol{k}_2) \\ &\cdot\exp\Big(i[(E_{k_1}+E_{k_2})t - (\boldsymbol{k}_1+\boldsymbol{k}_2)\cdot\boldsymbol{x}]\Big)|N\bar{N}\rangle \quad (47)\end{aligned}$$

The integrations over $\boldsymbol{k}_1$ and $\boldsymbol{k}_2$ can be done easily by use of the commutation-relations (33) to get

$$\begin{aligned}\mathcal{M} &= \frac{i}{4\sqrt{6}f_\pi^2}\sum_{M,a}(-1)^a a\int d\Omega_q Y^\star_{1M}(\hat{\boldsymbol{q}})q_i\int d^3x dt \\ &\cdot\langle 0|T^i_{2bc}D^\star_{b,a}D^\star_{c,-a}\exp\Big(i[(E_{q_1}+E_{q_2}) - \boldsymbol{p}\cdot\boldsymbol{x}]\Big)|N\bar{N}\rangle. \quad (48)\end{aligned}$$

Integration over Ω_q puts the index i on M and with (44)

$$\begin{aligned}\mathcal{M} &= \frac{iNq}{6f_\pi^2}\sqrt{\frac{\pi}{2}}\sum_{M,M'}\sum_{a,k}\sum_{b,c}\sum_{\substack{\sigma\sigma_3 \\ (\sigma=0,2)}}(-1)^{a+k}a\int d^3x dt(1k\sigma\sigma_3|1M') \\ &\cdot\int d[\alpha]D_{-k,0}D^\star_{b,a}D^\star_{c,-a}\int d\Omega_s Y_{\sigma\sigma_3}(\hat{\boldsymbol{s}})\int d^3r\int d^3R\, T^M_{2bc}(\boldsymbol{x} - \boldsymbol{R},\boldsymbol{r},\boldsymbol{s}) \\ &\cdot\exp\Big(-i[(E_{N\bar{N}} - E_{q_1} - E_{q_2})t + \boldsymbol{p}\cdot\boldsymbol{x} - \boldsymbol{P}\cdot\boldsymbol{R}]\Big)\delta_3(\boldsymbol{x} - \boldsymbol{R}). \quad (49)\end{aligned}$$

The δ-function originates from the matrixelement of the $N\bar{N}$-configuration-function T_2 between the $N\bar{N}$-state and the vacuum-state and can be understood by remembering that (with $\boldsymbol{p} = \boldsymbol{q}_1 + \boldsymbol{q}_2$) $\boldsymbol{x}$ in (48) plays the role of the centre of mass coordinate of the meson-pair which must coincide with the c.m. coordinate $\boldsymbol{R}$ of the $N\bar{N}$ pair. The integration over $\boldsymbol{R}$ can be performed immediately and the subsequent integration over $\boldsymbol{x}$ yields the conservation of momentum $\delta_3(\boldsymbol{P} - \boldsymbol{p})$, whereas the integration over the time t leads to energy-conservation $\delta(E_{N\bar{N}} - E_{q_1} - E_{q_2})$. After integration over the Euler-angles $[\alpha]$, we get for the matrixelement in the centre of mass system $\boldsymbol{P} = 0$:

$$\mathcal{M} = -\frac{iNq}{f_\pi^2}\left(\frac{2\pi}{3}\right)^2\sqrt{\frac{\pi}{2}}\sum_{M,M'}\sum_{b,c,k}\sum_{\substack{\sigma\sigma_3\\(\sigma=0,2)}}(1k\sigma\sigma_3|1M')\varepsilon^{kbc}\int d\Omega_s\int d^3r\, Y_{\sigma\sigma_3}(\hat{\boldsymbol{s}})T^M_{2bc}(\boldsymbol{r},\boldsymbol{s}). \quad (50)$$

The last task is now, to perform the integrals over $\boldsymbol{r}$ and $\boldsymbol{s}$ for all the vertexfunctions T. That is not difficult in principle, but more or less cumbersome due to the complicated structure of the $N\bar{N}$-configuration (19 - 21). As an example, we investigate here the contribution from the non-linear σ-term

$$\mathcal{L}_\sigma = \frac{f_\pi^2}{4}\mathrm{tr}(\partial_\mu U \partial^\mu U^\dagger). \quad (51)$$

First we need the expansion, quadratic in the fluctuations. With $U = AU_0A^\dagger$ and

$$U_0 = e^{i\boldsymbol{\eta}\cdot\boldsymbol{\tau}}U_{H\bar{H}}e^{i\boldsymbol{\eta}\cdot\boldsymbol{\tau}} \quad (52)$$

we have to second order

$$U_0 = U_{H\bar{H}} + \delta U_1 + \delta U_2 \quad (53)$$

with

$$\delta U_1 = i\{\boldsymbol{\eta}\cdot\boldsymbol{\tau}, U_{H\bar{H}}\} \quad (54)$$

$$\delta U_2 = -(\boldsymbol{\eta}^2 U_{H\bar{H}} + \boldsymbol{\eta}\cdot\boldsymbol{\tau}U_{H\bar{H}}\boldsymbol{\eta}\cdot\boldsymbol{\tau}). \quad (55)$$

With (52) and (53) the interaction-terms, which are quadratic in the fluctuations, are

$$\mathcal{L}_\sigma^{int} = \frac{f_\pi^2}{4}\mathrm{tr}(\partial_\mu U_{H\bar{H}}\partial^\mu \delta U_2^\dagger + \partial_\mu U_{H\bar{H}}^\dagger \partial^\mu \delta U_2 + \partial_\mu \delta U_1 \partial^\mu \delta U_1^\dagger). \quad (56)$$

Because we must have one spatial derivative on fluctuations and antisymmetric indices (note the ε-tensor in (50)) the only nonvanishing term is

$$\mathcal{L}_\sigma^{int} = T_{2ab}^i \eta^a \partial^i \eta^b \quad (57)$$

with the vertex-function

$$T_{2ab}^i = \frac{f_\pi^2}{2}(-1)^i \varepsilon_{-a,-b,c}\mathrm{tr}(\tau_{-c}[U_{H\bar{H}}, \partial_{-i}U_{H\bar{H}}^\dagger]). \quad (58)$$

Insertion of this T_2 in (50) yields after carrying out all the angular integrations and summations

$$\mathcal{M} = -\frac{3Nq}{10}\left(\frac{8\pi}{3}\right)^4 (2+3\sqrt{2}) \int drr(rF' - 2\sin 2F)\sin^2 F. \tag{59}$$

(The δ-functions, describing four-momentum conservation, are omitted here.)

Besides the non-linear σ-term $\mathcal{L}_\sigma$ there are more contributions from other terms of the lagrangian, namely

$$-ig_\rho f_\pi^2 \operatorname{tr}(\rho_\mu[U^\dagger, \partial^\mu U]) \rightarrow T^i_{2ab}\eta^a\partial^i\eta^b \tag{60}$$

with the vertex-function

$$T^i_{2ab} = 2ig_\rho f_\pi^2 C^2_{H\bar{H}}\, \varepsilon_{-a,-b,c}(-1)^i \operatorname{tr}(\rho^{-i}_{H\bar{H}}\, \tau_{-c}) \tag{61}$$

and the abbreviation

$$C_{H\bar{H}} = \frac{1}{2}\operatorname{tr}(U_{H\bar{H}}). \tag{62}$$

This term originates from the gauged non-linear σ-term. The explicit evaluation of (62) shows, that $C_{H\bar{H}}$ is even both in $\hat{\boldsymbol{r}}$ and in $\hat{\boldsymbol{s}}$. Because $\rho_{H\bar{H}}$ is odd in $\hat{\boldsymbol{r}}$, the $\boldsymbol{r}$-integral over (61) vanishes and $\mathcal{M} = 0$.

A last contribution to $N\bar{N}(^{33}S_1) \rightarrow \pi^+\pi^-$ comes from the gauged Wess-Zumino-term

$$\frac{iN_c g_\rho g_\omega}{32\pi^2}\varepsilon^{\mu\nu\alpha\beta}\partial_\mu\omega_\nu \operatorname{tr}(\rho_\alpha\{U^\dagger, \partial_\beta U\}) \rightarrow T^i_{2ab}\eta^a\partial^i\eta^b \tag{63}$$

with

$$T^i_{2ab} = \frac{N_c g_\rho g_\omega}{32\pi^2}(-1)^{a+b}\varepsilon^{-i,-j,-k}\partial_j\omega_0^{H\bar{H}} \operatorname{tr}(\rho_k^{H\bar{H}}\{\{\tau_{-a}, U^\dagger_{H\bar{H}}\}, \{\tau_{-b}, U_{H\bar{H}}\}\}). \tag{64}$$

From this term we get the matrix element

$$\begin{aligned}\mathcal{M} &= -\frac{6NN_c g_\omega q}{35}\left(\frac{32\pi}{9f_\pi}\right)^2 \int drr\omega' G \sin^2 F \\ &\quad \cdot [7(4+\sqrt{2}) - 4(9+2\sqrt{2})\sin^2 F]\end{aligned} \tag{65}$$

7. Results

After calculating all matrixelements in the same way, as shown in the last section for one example, we get the following results: There are only three nonvanishing matrixelements, namely for the final states $\pi^+\pi^-, \pi^\pm\rho^\mp$ and $\pi^0\omega$. All the other transition-amplitudes are exactly zero. This has different origins for different final channels. For the $\omega\omega$ there is no vertex at all in the lagrangian (1), independent of any assumptions or approximations used in the calculation. For $\omega\rho$, the existing couplings have the wrong parity, which means, that they describe transitions to meson-pairs with even relative angular momentum only, which do not occur in the annihilation-process (at least, if we look at initial S-states). For $\rho\rho$, there is a similar problem: the lagrangian has only vertices describing transitions from initial states with odd angular momentum, but not from S-states. The same is true also for the $\pi^0\rho^0$-case, which means, that the iso-singlet state is suppressed. Actually, experimentally there is a strong isospin-dependent suppression in the $\pi\rho$-channel (the so called "$\pi\rho$-puzzle"), but unfortunately it is the other way round: the iso-triplet is the suppressed channel[18].

Finally we are left with the three nonvanishing amplitudes stated above. For these we find that the ratio of 2π to $\pi\rho$ has the correct order of magnitude (~ 0.1), but that the $\pi\omega$-amplitude is much too small (about a factor of 10^{-4}). We also have calculated the matrixelements for the annihilation from initial P-states. There appear similar problems as for initial S-states. If one actually uses Coulomb-wavefunction in the $N\bar{N}$-state (26), the ratio of S- to P-wave annihilation has the correct order of magnitude ($\sim 10^5$). (But, of course, as stated already in section 4, the absolute values of the transition amplitudes are much too high (about a factor of 10^5)).

Obviously the results we have obtained are quite unsatisfactory. As previously discussed, the theoretical procedure which we have used involves many ambiguities and therefore, there are many possibilities to change something. One could try to construct other $N\bar{N}$-configurations or to introduce the mesonic fluctuations in other ways. One might also determine the wavefunctions for the collective coordinates $\boldsymbol{r}$ and $\boldsymbol{s}$ in a more appropriate way, than was done here. Other possibilities are to allow for vertices with time-derivatives on the collective coordinates, which we have ommitted here due to the adiabatic assumption. All these changes may improve the results, but it seems unlikely that they will lead to the qualitatively different results needed. Actually we tried another way to introduce the fluctuations (Callan-Klebanov-ansatz[19]), but this gave no qualitative improvement. Especially for the 2ω-final state, the above stated changes will not have any effect.

One conclusion therefore could be that the effective lagrangian must contain other terms which are quite different from what is obtained by the usual gauging procedures. This is also not a pleasant conclusion. All of this suggests that it may be necessary to consider with the second order tree graphs, (fig.1), which we have dis-

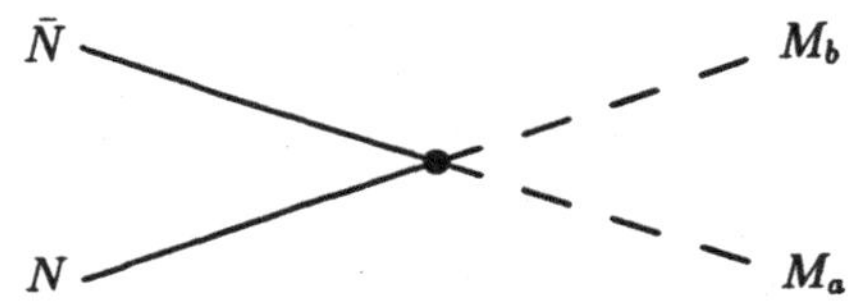

Fig. 1. Second order tree graph for $N\bar{N} \to 2$ *mesons*.

cussed here, also the iterated first order graphs indicated in fig.2. In addition to the linear $N\bar{N} \to 1$ *meson* matrixelements they involve also linear vertices for the

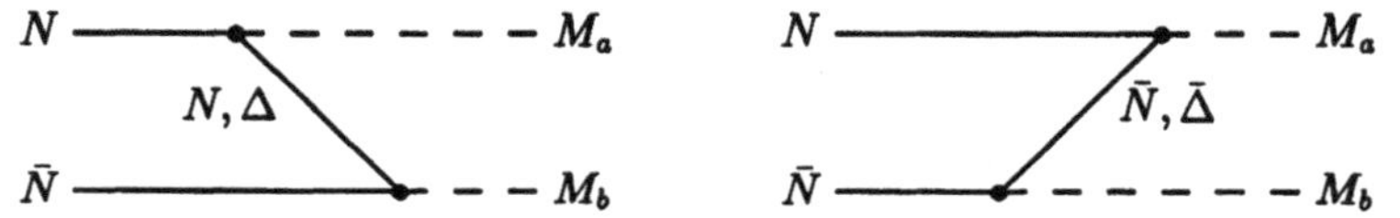

Fig. 2. Iterated first order graphs for $N\bar{N} \rightarrow 2$ *mesons.*

individual soliton or antisoliton. The structure and strength of the latter depends very sensitively on the form of the individual (anti-)soliton, especially on the extent by which it differs from the static solution for the free nucleon. Such an approach would correspond to the work of Mull, Haidenbauer, Hippchen and Holinde[20] in the standard boson fermion picture. So, after all, we have to conclude, that presently we do not yet have a reliable method to describe meson production in the B=0 sector of the soliton model. Similar arguments will in principle also apply to the B=2 sector, although in that case the situation is more favourable, because there the final mesons can be produced with small energies and the distortion of the soliton-soliton configuration is not additionally enhanced by the annihilation process.

References

1. O. Chamberlain, E. Segrè, C. Wiegand, T. Ypsilantis, *Phys.Rev.* **100**,947 (1955).
2. C. Amsler, G. Backenstoss, R. Klapisch, C. Leluc, D. Simon, L. Tauscher (ed.), *Physics at LEAR with Low Energy Antiprotons* (Harwood Academic Publisher, Chur, 1988).
3. P. Bluem, G. Bueche, H. Koch, *KfK-Primaerbericht*, 14.02.01.P29A (1988)
   C. Amsler, F. Myhrer, *Ann.Rev.Nucl.Sci.* **41**, 219 (1991).
4. I. Zahed, G.E. Brown, *Z.Phys.* **A335**, 349 (1990)
   M.M. Musakhanov, I.V. Musatov, *Phys.Lett.* **B273**, 309 (1991).
5. H.M. Sommermann, R. Seki, S. Larson, S.E. Koonin, *Phys.Rev.* **D45**, 4303 (1992).
6. ASTERIX-Collaboration, *Nucl.Phys.* **B8**, 218 (1989).
   J. Vandermeulen, *Z.Phys.* **C37**, 563 (1988).
7. M. Bloch, G. Fontaine, E. Lillistøl, *Nucl.Phys.* **B23**, 221 (1970).
   Fukui-INS-KEK-Kyoto Sangya-Osaka-Tokyo Metropolitan Collaboration, *Phys.Rev.* **B202**, 447 (1988).
   R. Bizzarri, M. Foster, Ph. Gavillet, G. Labrosse, L. Montanet, R. Salmeron, P. Villemoes, C. Chesquière, E. Lillestøl, *Nucl.Phys.* **B14**, 169 (1969).
   C. Baltay, P. Franzini, G. Luetjens, J.C. Severiens, J. Steinberger, D. Tycka, D. Zanello, *Phys.Rev.* **B140**, 1042 (1965).
8. M. Kretzschmar, *Ann.Rev.Nucl.Sci.* **11**, 1 (1961).
9. B. Schwesinger, H. Weigel, G. Holzwarth, A. Hayashi, *Phys.Rep.* **173**, 173 (1989).
10. U.G. Meissner, N. Kaiser, *Z.Phys.* **A325**, 267 (1986).

11. O. Kaymakcalan, S. Rajeev, J. Schechter, *Phys.Rev.* **D30**, 594 (1984).
12. A. Jackson, A.D. Jackson, V. Pasquier, *Nucl.Phys.* **A432**, 567 (1985).
13. H.Walliser, G. Holzwarth, A. Hayashi, *Nucl.Phys.* **A456**, 717 (1986).
14. G. Reifenroether, E. Klempt, *Nucl.Phys.* **A503**, 885 (1989).
15. C.J. Batty, *Rep.Prog.Phys.* **52**, 1165 (1989).
16. H.J. Schnitzer, *Phys.Lett.* **B139**, 217 (1984).
17. H.M. Pilkuhn, *Relativistic Particle Physics* (Springer-Verlag, New York Inc., 1979).
18. V. Girija, *in ref.2*, p.443.
19. C.G. Callan, I. Klebanov, *Nucl.Phys.* **B262**, 365 (1985).
20. V. Mull, J. Haidenbauer, T. Hippchen, K. Holinde, *Phys.Rev.* **C47**, 1337 (1991)

www.ingramcontent.com/pod-product-compliance
Lightning Source LLC
LaVergne TN
LVHW060542180926
848329LV00007B/37
* 9 7 8 9 8 1 0 2 1 4 0 8 1 *